The Geochemical Origin of Microbes

This textbook covers the transition from energy-releasing reactions on the early Earth to energy-releasing reactions that fueled growth in the first microbial cells. It is for teachers and college students with interests in microbiology, geosciences, biochemistry, and evolution. The scope of the book is a quantum departure from existing "origin-of-life" books in that it starts with basic chemistry and links energy-releasing geochemical processes to the reactions of microbial metabolism. The text reaches across disciplines, providing students of the geosciences an origin/biology interface and bringing a geochemistry/origin interface to students of microbiology and evolution. Beginning with physical chemistry and transitioning across metabolic networks into microbiology, the timeline documents chemical events and organizational states in hydrothermal vents, the only known environments that bridge the gap between spontaneous chemical reactions that we can still observe in nature today and the physiology of microbes that live from H_2, CO_2, ammonia, phosphorus, inorganic salts, and water. Life is a chemical reaction. What it is and how it arose are two sides of the same coin.

Key Features

- Provides clear connections between geochemical reactions and microbial metabolism
- Focuses on chemical mechanisms and transition metals
- Is richly illustrated with color figures explaining reactions and processes
- Covers the origin of the Earth, the origin of metabolism, the origin of protein synthesis, and genetic information, as well as the escape into the wild of the first free-living cells: bacteria and archaea

The Geochemical Origin of Microbes

William F. Martin and Karl Kleinermanns

CRC Press
Taylor & Francis Group
Boca Raton London New York

CRC Press is an imprint of the
Taylor & Francis Group, an **informa** business

Designed cover image: An image of a Lost City vent showing a calcium-magnesium carbonate flange and ca. 70°C warm water emerging from it. Printed with permission courtesy of Susan Q. Lang, U. of South Carolina/NSF/ROV Jason 2018 © Woods Hole Oceanographic Institution

First edition published 2024
by CRC Press
2385 NW Executive Center Drive, Suite 320, Boca Raton FL 33431

and by CRC Press
4 Park Square, Milton Park, Abingdon, Oxon, OX14 4RN

CRC Press is an imprint of Taylor & Francis Group, LLC

© 2024 William F. Martin and Karl Kleinermanns

Library of Congress Cataloging-in-Publication Data
Names: Martin, William F., Professor Dr., author. | Kleinermanns, Karl, author.
Title: The geochemical origin of microbes / William F. Martin, Karl Kleinermanns.
Description: First edition. | Boca Raton, FL : CRC Press, 2024. | Includes bibliographical references and index.
Identifiers: LCCN 2023044107 (print) | LCCN 2023044108 (ebook) | ISBN 9781032457727 (hardback) | ISBN 9781032457673 (paperback) | ISBN 9781003378617 (ebook)
Subjects: MESH: Bacteria—metabolism | Bacteria—cytology | Origin of Life | Chemical Phenomena | Energy Metabolism | Environmental Microbiology
Classification: LCC QR74.8 (print) | LCC QR74.8 (ebook) | NLM QW 52 | DDC 579.3—dc23/eng/20240228
LC record available at https://lccn.loc.gov/2023044107
LC ebook record available at https://lccn.loc.gov/2023044108

ISBN: 9781032457727 (hbk)
ISBN: 9781032457673 (pbk)
ISBN: 9781003378617 (ebk)

DOI: 10.1201/9781003378617

Typeset in Times
by Apex CoVantage, LLC

Dedication

*The authors have separate scientific paths
and met only late in their careers.*

Hence we have separate dedications.

*William Martin dedicates his efforts towards the first edition
of this book to his mentors during three career phases,
Rüdiger Cerff, Heinz Saedler, and John F. Allen.*

*Karl Kleinermanns dedicates his efforts towards the first edition of
this book to his longstanding scientific cooperation partners and
friends Peter Andresen, Allan Luntz and Mattanjah de Vries.*

Contents

Preface

Welcome to the first college textbook on the geochemical origin of microbial cells. Just about every college student of the natural sciences has wondered at one time or another, "Life is so different from non-living matter, how was it possible that life arose on our planet?" Students of biology ask that question in the context of the structure and function of cells. Students of chemistry ask about origins in the context of the chemical constituents of life. Students of the geosciences ask about the role that the environment played in the transition from non-life to life. But philosophers, physicists, and students of all disciplines tend to ask the question about origins and for good reason: it is an excellent question, and humans generally want to know where everything began, including our most ancient ancestors. At first glance, the gap between non-life and life seems insurmountable. To bridge that gap we will need information from all of the natural sciences—biological, geological, chemical, and physical—and we will need a pinch of mathematics to make sure that everything adds up.

There is good news and bad news. The bad news is, if we look at the thousands of chemical reactions that comprise a living cell, they seem in sum to be so numerous and complex that it appears impossible that they could have arisen *all by themselves*. The good news is that the organic–chemical reactions of life did not arise *all by themselves*. They arose with the help of catalysts. Today, those catalysts are enzymes that accelerate reaction rates by many orders of magnitude and introduce exquisite specificity into the nature of the intermediates and products. Enzyme catalysis is founded in amino acid side chain geometry in the enzyme active site, typically excluding water molecules from the reaction and stabilizing the transition state in the reaction mechanism. In the beginning, there were no enzymes but there were catalysts in abundance: transition metals abound in minerals. Transition metals can also accelerate reactions by orders of magnitude, and they can introduce a great deal of specificity concerning the nature of products even in the absence of enzymes. They do that with the help of unfilled *d* electron shells in Fe, Ni, Mo, and Co, and that is the reason that such transition metals are often found in the active site of modern enzymes. Transition metals in enzymes are especially common when the reaction is catalytically demanding and the average amino acid side chain will not do the job, for example, in the enzyme nitrogenase, which fixes N_2 with the help of seven iron atoms coordinated by nine sulfur atoms at its active site. Modern chemical industry would be unthinkable without the use of transition metals as catalysts. A large field of study behind catalysts is organometallic chemistry. The transition metals used by industry (often entailing metals that life does not use, such as Ru, Rh, Pd, and Pt) can be bound by ligands in such a way as to modulate the catalytic properties of their *d* or *f* orbitals so as to elicit exquisite specificity as catalysts. However, catalysis is a property of the transition metals themselves. Their catalytic activity existed from day one on planet Earth, and there was really no way to turn it off.

Another piece of good news is that the chemical reactions of life did not arise all at once. They arose in sequence.

From what starting point? The chemistry of life starts from CO_2. The same was almost certainly true at origins. That gives us a vector of chemical complexity that orders information and thoughts. All ecosystems on Earth (no exceptions) are based on what are called primary producers, organisms that obtain their carbon from CO_2. They provide the food of the food chain. At origins, starting from CO_2 is the decisive step on the way to understanding how the origin of metabolism—the reactions that give rise to the stuff that cells are made of—was possible. The sequence of reactions that CO_2 will undergo on a transition metal catalyst is anything but random. They are highly specific and highly reproducible owing to the properties of the *d* electrons of transition metals. A typical reaction of CO_2 on transition metals in water using H_2 as a reductant (source of electrons) generates formate (C1), acetate (C2), and pyruvate (C3), the methyl compound methanol, and some methane, and that's it. Those are not random chemicals; they are the starting point of microbial metabolism. There is something very natural about metabolism, and the reactions of carbon on transition metal catalysts are the reason. Catalysts do not alter the thermodynamic equilibrium of a reaction (the nature of the most stable products), but catalysts exert a huge effect on the kinetics of a reaction and can determine which among possible alternative products are formed first. With the right catalysts, plenty of CO_2 as pure starting material, and a source of electrons like H_2 to make organic molecules, the origin of metabolism is maybe easier to understand than we thought. From metabolism, cells unfold all by themselves, given 1000 enzymes as catalysts. That leaves us with the question of how we get from CO_2 to those 1000 enzymes and what reactions they catalyze. For that we need a book, this book.

Our seven chapters do not explain exactly how life arose, but they do span the full spectrum of organizational states from CO_2 to cells that harness energy like some cells still do today. A reductant (H_2) is required to get organic chemistry going. That puts fairly strict constraints on where the first organic molecules could have been synthesized and hence where life could have started. The source of H_2 both on the early Earth and today is provided by a fascinating natural geochemical reaction that takes place between water and the rocks underneath hydrothermal vents. The process is called serpentinization; it generates H_2 day-in, day-out, 24/7 and has done so since there was first water on Earth. That connects life's first reactions to serpentinization in hydrothermal systems, which today still synthesize formate and methane in chemical reactions that look all the world like a carbon copy of primitive metabolism. The difference is that metabolism is a carbon copy of those geochemical reactions, which take place because they release energy, just like in metabolism.

Serpentinization is the starting point for geochemical origins. The endpoint is the emergence of free-living cells that obtain their carbon and energy from the reaction of H_2 and CO_2 but without the help of light. Such cells are called chemolithoautotrophs (a mouthful) because in their mode of nourishment (trophy) they obtain their energy from chemical (chemo) reactions, not light; their electrons from an inorganic (litho) donor, H_2, not organic molecules; and their carbon from CO_2 (auto) rather than from organic molecules. The organisms that live from the reaction of H_2 with CO_2 are called acetogens and methanogens; they are strict anaerobes that live in the dark and that are present in the two most ancient domains of life, bacteria and archaea, collectively called prokaryotes. They are by all measures of physiology the most ancient forms of life on the Earth. Their emergence as free-living cells culminates in the origin process.

Life started with small, simple, and imperfect units that had the capacity to evolve. This is not only true for organisms on their long way from microbes to humans but also for the molecular machinery of life. The evolution of translation and the genetic code provides an example. We know now that the ribonucleic acids (RNAs) used for the transfer of amino acids (tRNAs), the AARS enzymes used to energize and specifically bind amino acids to tRNAs, and the ribosome that forms peptides from the amino acid loaded tRNAs all started small and from imperfect precursors. The initial function of these RNA-peptide aggregates was probably not to provide heritable information but simply to enhance random peptide formation and stabilize the polymers by aggregation against chemical attack, especially hydrolysis, for longer survival. Sequences were duplicated and mutated by chemical reactions, the transfer RNA grew in length permitting to specific binding to a messenger RNA (mRNA) template. The AARS enzymes had to enlarge accordingly to bind the larger tRNAs, and the short RNA sequence of the protoribosome formed peptides step by step. Translation from mRNA sequence to amino acid sequence in peptides became increasingly specific and the information deposited in nucleic acids could be stably transmitted. Once amino acid sequences were programmed, increasingly specific biochemical functions became possible.

Our approach to the topic takes into account standard origins topics such as Earth history, abiotic synthesis, biochemical syntheses, physical effects, and chemical principles of organization, but it also includes energy harnessing, redox chemistry, catalysis, autocatalysis and autocatalytic networks, the origin of reproducible catalysis, the origin of heritable information, and an outline of early physiological evolution.

Our approach is not that of a comprehensive review. Our goal is to take the reader from simple, energy-releasing (exergonic) reactions on the early Earth to the origin of cells that still use those same reactions today to obtain their carbon and energy for growth. We cannot possibly cover all ideas on origins. We focus on energy-releasing (exergonic) reactions. Self-replicating RNA molecules play little or no role in this textbook—for two reasons: there are no self-replicating RNA molecules in cells, and there are dozens of fine books out there on self-replicating RNA in the laboratory. We have tried to keep the connections between prebiotic chemistry and real biochemistry in real cells as tight as possible. In that sense, the main RNA molecule in cells is by far and away the ribosomal RNA, the molecule at the heart of the ribosome, and that makes up about 20% of an average cell's dry weight. The ribosome is where enzymes, the catalysts of cells, are made. The role of the RNA molecules involved in translation is to synthesize proteins. The origin of translation marks perhaps the most dramatic transition in early evolution. How that came to pass is one of the big questions in evolution, and some solutions are in sight, even if all the details of that process are not fully resolved. We provide students with a path from chemical reactions on mineral surfaces to cells with DNA as memory, RNA to make protein, and proteins as catalysts to provide life-supporting physiology. That is admittedly a tall order, but it is also where our textbook differs most distinctly from other books on the topic.

The list of people we *wish* to thank is too long to print. The list of people we *have* to thank includes Martina Preiner, John F. Allen, Dan Graur, John Baross, Deborah Kelley, Susan Q. Lang, Rolf Thauer, Bernhard Schink, Georg Fuchs, Peter Schönheit, Wolfgang Buckel, Volker Müller, Julia Vorholt, Ines Cardoso Pereira, Christiane Dahl, Takashi Gojobori, Yoichi Kamagata, Shino Suzuki, Harun Tüysüz, Joseph Moran, Oliver Trapp, Thomas Carrell, Mario Trieloff, Filipa Sousa, Mike Steel, Dave Bryant, Pete Lockhart, Tal Dagan, Heinz Saedler, Rüdiger Cerff, Michael Schmitt, Markus Gerhards, Mattanjah de Vries, Eyal Nir, Rainer Weinkauf, Sergey Kovalenko, Nikolaus Ernsting, Verena Zimorski, Max Brabender, Natalia Mrnjavac, Manon Schlikker, Oliver Kraft, and the entire team at the Institute for Molecular Evolution, who all helped us along the way. Our thanks to Anna Sophia Janssen, who helped us by obtaining permissions for images. The most help, of course, came from our families. K.K. thanks Jutta and Arne; W.F.M. thanks Annette, Lilli, and Hannah for constant support and understanding.

We hope that students and teachers find our book useful. We will be grateful for any and all feedback.

**William F. (Bill) Martin and
Karl (Carlo) Kleinermanns**
Düsseldorf

About the Authors

William F. Martin received his undergraduate (diplom) degree in biology in 1985 from the Technische Universität Hannover, Germany, and his Ph.D. in genetics in 1988 from the Max-Planck Institute for Breeding Research in Cologne, Germany, on molecular evolution. He completed postdoctoral work at the Technische Universität Braunschweig, Germany, with a habilitation on the evolution of primary metabolism. Since 1999 he has held a chair in biology at the University of Düsseldorf, Germany, as head of the Institute of Molecular Evolution. Author of more than 350 scientific publications that have been cited over 45,000 times, he has served as a referee for 135 different journals and 50 different funding agencies, as an editor for 16 different journals, and as a panelist on various advisory boards. He is an elected fellow of the Nordrhein-Westfälische Akademie der Wissenschaften, the American Academy of Microbiology, and the European Molecular Biology Organisation, and a corresponding foreign member of the Accademia delle Scienza di Bologna. His distinctions for work on molecular evolution include the Heinz Maier-Leibnitz Prize of the Deutsche Forschungsgemeinschaft in 1990, a Julius von Haast Fellowship from the New Zealand Ministry for Science in 2006, the Klüh Foundation Award for Science and Research in 2018, and the Motoo Kimura Award of the Motoo Kimura Trust Foundation in 2023. Since 2009, his work has been funded by three advanced grants from the European Research Council. His main scientific interests are chemical, physiological, and biochemical evolution, endosymbiosis, microbial evolution, and early evolution including the origin of microbial cells.

Karl Kleinermanns received his diplom degree in chemistry at the RWTH Aachen and his Ph.D. in physical chemistry at the Max-Planck in Göttingen in laser-induced chemical reactions. He then completed a postdoctoral fellowship at the IBM research laboratories in San Jose, California, USA, and a habilitation in kinetics and dynamics of elementary oxidation reactions at the University of Heidelberg, Germany. He received the Heinz Maier-Leibnitz prize of the Deutsche Forschungsgemeinschaft (DFG) for his work in chemical reaction dynamics in 1984. He followed a call to the chair of Physical Chemistry at the University Düsseldorf as head of the Institute of Molecular Spectroscopy and Nanosystems in 1989. He had several honorary offices as a lead consultant of the Deutsche Forschungsgemeinschaft, chair of the Molecular Physics Department of the Deutsche Physikalische Gesellschaft, and member of its board, member of the prize committee of the Bunsengesellschaft für Physikalische Chemie, member of the panel of the European Research Society, chair of the Scientific Advisory board of the Max-Born institute in Berlin, and member of the Editorial Advisory Board of ChemPhysChem—A European Journal of Chemical Physics and Physical Chemistry. He is the author of more than 200 peer-reviewed scientific publications and editor and coauthor of the textbook on experimental physics *Gase, Nanosysteme, Flüssigkeiten*. His main scientific interests in Düsseldorf were the spectroscopic investigation of intermolecular interactions between nucleobases of RNA and DNA and short peptides and their aggregates using lasers of high spectral resolution and femtosecond lasers. Further research interests were the development of solar cells based on inorganic nanoparticles and environmental projects with industrial partners fostered by a dedicated laboratory for organic trace analysis in complex mixtures.

1 The Early Earth Setting and Chemical Fundamentals

1.1 STARS, PLANETS, AND CHEMICAL ELEMENTS SET THE STAGE FOR LIFE

1.1.1 THE BIG BANG THEORY DESCRIBES THE ORIGIN OF THE UNIVERSE

Very early in the Earth's history, physical and chemical processes brought forth the first forms of life. To understand the early Earth as a setting for life's origin, understanding the process of how the Earth formed is key. That means we have to think about the solar system, the origin of the elements in stars and supernovae, and how those elements came to be distributed on the Earth. This chapter provides a bit of context for those very early events.

According to the *big bang* theory, at the beginning of the universe, gravity had attracted all matter to a small volume until repulsion forces prevailed and the extremely hot and dense conglomerate exploded. After this explosion (the big bang), space and time developed and expanded together with the fragments of the explosion. The big bang is dated to roughly 13.8 billion years ago (13.8 Ga, Giga-annum). What happened before the big bang is unknown.

The estimated age of the universe is a fairly well-known number. Less well known is how one obtains such an estimate. The traditional method is based on the size distribution of stars. Large stars are short lived; small stars are long lived. Stars the size of the sun burn for about 9 billion years. A star twice as large as the sun will burn for about 800 Ma (million years); a star that is 10 times larger than the sun will burn for only about 20 million years. A star half the size of the sun would burn for 20 billion years. The oldest star clusters known have no stars smaller than about 0.7 solar masses, which by their mass are not older than roughly 14 Ga. The age of the oldest stars can thus be estimated from their size, and this constrains the age of the universe. More modern approaches to date the age of the universe use the Hubble constant, which describes the present rate of expansion of the universe and allows extrapolation back to the start of the acceleration, but this requires additional information regarding the estimated amount of mass in the universe among other variables.

Information about the nature of matter and energy during the early phases (the first second or so) of the big bang comes from high-energy collision experiments in modern particle accelerators. Matter as we know it now did not exist at the onset of the big bang. Elementary particles like *protons*, *electrons*, and *neutrons* are not stable at the ultrahigh temperatures of the big bang; they immediately decay and transform. After the initial expansion, the universe gradually cooled by radiation and other mechanisms, and the first elementary particles were formed. Subatomic, atomic, and molecular particles emerged: protons (+ charged), electrons (– charged), neutrons (uncharged, neutral n), hydrogen atoms containing one electron orbiting a proton (chemical symbol H), hydrogen molecules with two covalently joined hydrogen atoms (H–H) and light (photons) from collisions between these particles.

Astronomers observe galaxies, stars, planets, nebulas, and interstellar gas with sophisticated instruments; however, they seem to see only a small part of the matter in the universe: invisible objects deflect the light trajectories of galaxies. This *dark matter* probably attracted and collected the early giant clouds of hydrogen. Dark matter was co-discovered by Vera Rubin, who noticed that spiral galaxies spin at the same angular velocity near their center as they do near their edge. This contrasts with our solar system, where

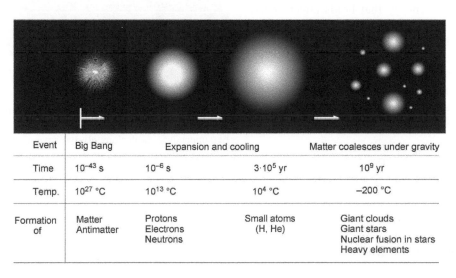

Event	Big Bang	Expansion and cooling		Matter coalesces under gravity
Time	10^{-43} s	10^{-6} s	$3 \cdot 10^5$ yr	10^9 yr
Temp.	10^{27} °C	10^{13} °C	10^4 °C	−200 °C
Formation of	Matter Antimatter	Protons Electrons Neutrons	Small atoms (H, He)	Giant clouds Giant stars Nuclear fusion in stars Heavy elements

FIGURE 1.1 **The origin of the universe.** Currently accepted principles of physics can explain the evolution of the universe shortly after the *big bang*. What happened before the big bang is unknown.

DOI: 10.1201/9781003378617-1

1

the more distant objects like Neptune orbit the sun at a much slower angular velocity (165 years per orbit) than Mercury (88 days). The only known force that could hold spiral galaxies together like a coin is gravity, but the mass (or matter) required to exert that gravity is not visible by current methods; hence the existence of dark matter is inferred to account for that additional mass.

In the wake of the big bang, gravity contracted the clouds further and the temperature inside the clouds increased until protons and neutrons fused to helium (chemical symbol He) cores. This *nuclear fusion* released energy as heat and radiation and gave birth to early stars. Early stars were very large and very short lived. Inside giant stars, the extreme force of gravity compressed the helium cores for further fusion reactions and heavier elements were formed. Two protons and two neutrons fuse to helium cores, and three helium cores fuse to carbon (C) if temperatures and pressures are high enough. Carbon cores coalesce to oxygen (O), oxygen to silicon (Si), and silicon to iron (Fe). The nuclear reactions that generate elements heavier than iron (Fe) consume more energy than they release; therefore, such reactions do not take place normally in burning stars. The synthesis of atoms heavier than iron (element number 26; see Table 1.3) therefore generally does not take place even in the largest stars. The energy needed to synthesize elements heavier than Fe via nuclear fusion comes from the explosion of supernovae. The fact that heavy elements like copper, silver, gold, lead, and uranium exist on the Earth tells us that our planetary disc was formed from material that went through at least one supernova before our solar system was formed.

A *supernova* forms when a high-mass star comes to the end of its life. After consumption of nuclear fuel, the star collapses under the force of its own gravity, heats up during collapse, and finally explodes in a supernova. Temperature and pressure in the shock wave of a supernova are high enough to breed elements heavier than iron. The curious reader might ask why elements lighter than iron form in stars while elements heavier than iron require supernova energy input for formation. In simple terms, this has to do with the energy that holds atomic nuclei together (the strong nuclear force). In brief, when protons and neutrons join to form new nuclei, the mass of the new nucleus is less than the mass of the component protons and neutrons. The missing mass corresponds to an amount of energy according to $E = mc^2$. When light elements like helium form, the missing mass is released as energy, the energy of nuclear fusion.

1.1.2 THE EARTH FORMED UNDER THE FORCE OF GRAVITY BY ACCRETION

Interstellar clouds of hydrogen, helium, and other ionized, neutral, or molecular gases and cosmic dust (*solar nebulae*) become denser because of gravitation. In the wake of supernovae, heavier material ejected by the explosion forms denser regions. If the nebulae become dense enough, they can form stars that can host orbiting planets formed from the remaining material (Figure 1.2). The Earth was formed

that way at about 4.54 Ga. See Section 1.1.8 about radiometric dating and how we know the age of the Earth. The dust, grains, small bodies, and planetesimals that accreted to form the Earth were mainly composed of O, Si, Al, Fe, and Ni along with all other natural elements that we find on the Earth today.

The elements formed in supernovae are ejected as elements. No bonds as they exist between atoms in molecules could survive the high energy of a supernova explosion. In nebulae, and on the solid surface of grains, the elements or atomic ions can meet and form bonds in the cold depths of space. Hydrogen is the most common element in the universe. When two hydrogen atoms meet, they can form H_2, the simplest molecule. Helium is a noble gas; it does not form bonds with other atoms. When hydrogen atoms meet oxygen atoms in space, they can form bonds and generate water, H_2O. The water that was brought to Earth during accretion and from comets was made in space. One reason scientists are so interested in exploring the chemical composition of comets is that they were formed when the solar system was formed and are a kind of frozen time capsule that contains compounds that already existed in space as our sun and planets were formed.

FIGURE 1.2 An artist's impression of a protoplanetary disk. Stars develop from the gravitational collapse of interstellar clouds. The loss of potential energy during collapse heats the cloud such that kinetic energy is gained. Clouds generated by supernovae explosions always have a small amount of rotation, an angular momentum around some random axis, because the explosion is never perfectly omnidirectional. Gravity flattens the initially spherical rotating cloud along the rotation axis (the "pizza dough" effect). As mass is collected toward the sun, angular momentum is conserved, and the cloud rotates faster (the "ice skater" effect) so that a rotating *accretion disk* forms. Compression from gravity heats the proximal nebula until nuclear fusion (hydrogen into helium, then heavier elements) begins. From the remaining distal material, planets form by gravity. Image credit: STEP-ANI-MOTION Studio für Computertrick GmbH, Cologne, Germany.

1.1.3 Earth's Water Stems from Accretion and Comets from the Outer Solar System

Earth's water originally came from ice frozen in the solid objects of the protoplanetary disk from which the Earth was made or from more distant celestial bodies from the outer part of the solar system, mainly from the *asteroid belt* between Mars and Jupiter. Most of this water was probably not bulk ice but bound in hydrated silicate minerals which are abundant in many meteorites. If the water came with the accretional building blocks of the Earth, then it originated from a time before the moon-forming impact. Volcanic eruptions and large impacts later released the bound water into the Earth's atmosphere.

The early Earth grew by heat-releasing accretion of heavier interplanetary objects until its interior was hot enough to melt heavier elements like iron (melting point Mp = 1538°C) and nickel (Mp = 1455°C). Radioactive decay also contributed heat. Once the Earth was molten, metallic Fe and Ni sank to the center of the molten Earth due to their higher density (Fe 7.87 g/cm³ and Ni 8.91 g/cm³) and formed a metallic *core*. Al, Mg, and the silicates that we commonly know as rocks, or magma in their molten state, were displaced toward the surface. This process of separation into core and mantle in the molten state is called *differentiation*. Magma typically consists of molten iron magnesium silicate with a density of about 2.5 g/cm³; hence, it was readily displaced by Fe and Ni during differentiation. The core still consists of heavy elements, mainly iron and nickel, and is surrounded by a mantle of light elements like silicon, aluminum, magnesium, oxygen, nitrogen, and sulfur. Differentiation did not separate heavy metals from light ones completely, because elements like silicon, sulfur and carbon also exist in the core, estimated at roughly 6%, 2%, and 0.2%, respectively. Iron and nickel are still present in the mantle and crust as well, mainly in the form of oxides and silicates. The layered structure of the Earth was formed only 10 million years after its formation. A second round of differentiation occurred following the moon-forming impact.

1.1.4 Heat Radiation Cooled the Upper Mantle

The upper mantle cooled by heat radiation to space and generated a thin crust of floating tectonic plates (Figure 1.3). Mantle convection is a consequence of heat flow from the Earth's interior to the surface and drives plate tectonics. The mantle of the early Earth was much hotter than today, and convection in the mantle was faster. Subduction of colliding plates into the mantle was probably more common than today, and therefore tectonic plates were smaller than modern continents. There is still considerable debate about when the first continents were formed. Continents consist of rocks with lower iron content than oceanic crust. As a consequence of their lower density, they have higher buoyancy and rise higher out of the magma than the oceanic crust. But, as we will see, the existence of continents is not a prerequisite for the geochemical origin of the first forms of life.

FIGURE 1.3 **The early Earth before the Moon-forming impact.** Artist's impression. Credit: STEP-ANI-MOTION Studio für Computertrick GmbH, Cologne, Germany.

1.1.5 The Moon-Forming Impact Transformed Carbon and Nitrogen into CO_2 and N_2

Analysis of isotopes from the Moon's rocks traditionally suggests that the moon is 4.48 billion years old and thus was formed 70–100 million years after the origin of the solar system, although newer findings based on the distribution of volatile elements suggest that the Moon may have formed somewhat later, about 4.35 billion years ago. The timing of the Moon formation, its low density, its small metallic core, and the same relative abundance of the oxygen isotopes as on the Earth all indicate that some of Earth's crust and mantle were incorporated by the Moon, suggesting that a collision must have taken place. The *giant impact hypothesis* for the origin of the Moon is widely accepted: a body the size of Mars (called *Theia*) hit the proto-Earth with grazing incidence. The hit was strong enough to melt both bodies and vaporize some of the Earth's crust and mantle. The material was ejected in an orbit around the Earth and condensed into a spherical body within a short time: the Moon. Some of its material was from the Earth's outer layers explaining the low density and small proportion of metals on the moon. The impact gave the Earth's axis its 23° tilt relative to the solar orbital plane.

The moon-forming impact was very important for the chemical setting that followed, the setting in which life eventually arose. The impact caused the Earth to be surrounded by a hot silicate vapor atmosphere for about 1000 years with a surface temperature exceeding 2500°C. The rock vapor quickly cooled to the liquid state and returned to the magma ocean. The magma ocean persisted for 1–10 million years. In this phase, carbon from space that had been sequestered during accretion was converted to CO_2. The same conversion occurs today with organic material in sediment that is subducted into the magma of the mantle; it returns to the surface through volcanoes as CO_2. The moon-forming impact transferred volatiles such as water, carbon, nitrogen, and sulfur compounds that came to

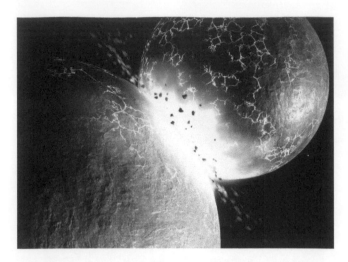

FIGURE 1.4 **The Moon-forming impact.** An artist's impression of the collision of Theia with Earth. Credit: STEP-ANI-MOTION Studio für Computertrick GmbH, Cologne, Germany. A corresponding short film on the origin of life is freely available at www.molevol.hhu.de/en/movies.

Earth during accretion into the atmosphere as gases, mainly water vapor, carbon dioxide, nitrogen, and sulfur gases. After cooling of the magma ocean to a solid crust at about 4.4 Ga, the water vapor condensed to oceans and left behind a ~100 bar, ~230°C atmosphere consisting primarily of CO_2, as calculations show. At 100 bar pressure, water is still liquid at 230°C. Atmospheric carbon dioxide was gradually dissolved by the liquid water on the terrestrial surface and ended up as rock carbonates that were eventually subducted into the mantle. Until that process was complete, the CO_2 content of Earth's atmosphere remained high, but levels oscillated vastly due to new surfaces created by crust–mantle cycles.

1.1.6 THE LATE HEAVY BOMBARDMENT MIGHT OR MIGHT NOT HAVE TAKEN PLACE

Isotope dating of rocks from lunar craters sampled during the Apollo mission implied that most of the impact melts date back to about 4.1 to 3.8 billion years. This points to a relatively large number of asteroid hits on the moon in a rather short time period. This impact should have affected the Earth also, giving rise to the idea of a *Late Heavy Bombardment* (LHB) on the Earth about 4 billion years ago. The bombardment was proposed to have been caused by objects dislodged from the asteroid belt between Jupiter and Mars or from the Kuiper belt in the outer solar system into the orbits of the terrestrial planets. The cause and consequences of the LHB have been debated, but the whole theory that the LHB took place at all is also now disputed. Some scientists argue that the narrow range of impact melt ages is an artifact of sampling from a single lunar crater or, perhaps more likely, the result of a single massive impact event at the origin of the Mare Imbrium crater and lava plain that spread material of the same age all over the surface of the moon, creating the impression that many independent impacts had occurred at the same time when in fact only one event had occurred.

1.1.7 VOLCANIC ACTIVITY AND HYDROTHERMAL SOURCES WERE ABUNDANT AT THE END OF THE HADEAN ERA

Following the moon-forming impact, the Earth radiated heat to space and cooled. The water in the atmosphere condensed, giving rise to a primordial ocean that was about twice as deep as today's. That inference comes from the observation that the modern crust and mantle together bind about one ocean volume of water. That surplus ocean volume of water was in the primordial ocean before rock–water interactions in the primordial crust commenced.

Mantle convection is a consequence of heat flow from the Earth's interior to the surface and drives plate tectonics. If two plates collide, one is pushed under the other (subduction), forcing magma upwards at the subduction zone (see Figure 1.5). The magma can push through the crust and erupt as volcanoes. The subduction of the oceanic plate compresses the material of the continental crust and folds rocks into mountains. Magma emerges at

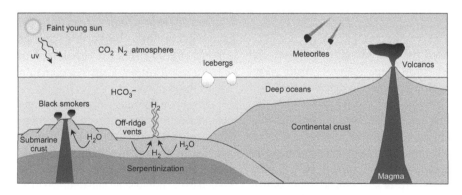

FIGURE 1.5 **Late Hadean ocean floor spreading, hydrothermal mounds, and vents.** The young sun was faint (~¾ of its current luminosity) and occasional icing probable. Ice cover, if any, was thin, because Hadean heat flow was large and greenhouse gases in the atmosphere, especially carbon dioxide and water vapor, were potent. Ultraviolet radiation, X-rays, and solar wind were much more intense than today, especially in the absence of a UV-absorbing ozone layer. The mantle in the late Hadean period was hotter than today, and convection in the mantle was faster. Volcanic activity and formation of hydrothermal vents were more common. Modified from Arndt N. T., Nisbet E. G. (2012). Processes on the young Earth and the habitats of early life, *Annu. Rev. Earth Planet. Sci.* 40: 521–549; Nisbet E. G., Fowler C. M. R. (2004). The early history of life. In *Biogeochemistry*, Schlesinger W. H. ed., Vol. 8, Academic, Oxford, pp. 1–41.

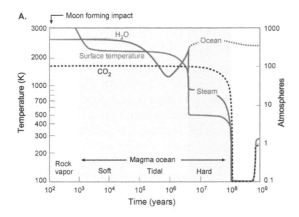

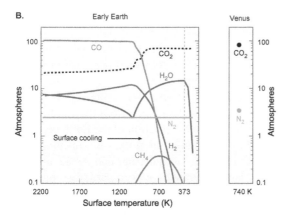

FIGURE 1.6 **The Earth was molten after the Moon-forming impact.** **(A)** A timeline of cooling at the surface. The time scale starts with the moon-forming impact. For the first 1000 years, the Earth had an atmosphere from rock vapor, which contained CO_2, water vapor, and other gases. The rock vapor condensed to a magma ocean, which stayed hot for about 2 million years due to the greenhouse effect and tidal heating. "Soft" and "Hard" indicate the progressive solidification of the magma. After the mantle solidified to a "crust", the steam atmosphere condensed to a ~500 K water ocean under ~100 bars of CO_2; hot water is liquid at such pressures. Water is somewhat soluble in silicate melt but the magma-bound water was expelled and contributed to the water ocean. CO_2 pressure in the atmosphere is assumed to decrease at 10^8 years because of subduction into the mantle but that may have taken place earlier (see also panel B for a different CO_2 estimate). After the loss of the greenhouse atmosphere from CO_2 and H_2O, surface temperature decreased, and the Earth's surface was probably partially ice covered. Finally, CO_2 pressure in the atmosphere increased again. The left scale indicates surface and radiating temperatures (red curves). The right scale indicates the reservoir size of water and CO_2 in atmospheres. The dark blue curve "Ocean" indicates the reservoir size of water in an atmospheric volume.

(B) The chemical evolution of the Earth's early atmosphere as a function of surface temperature. The graph summarizes the amounts of gases estimated to have existed in the primordial atmosphere based on a relative ratio of Fe^{3+} to Fe^{2+} of $Fe^{3+}/(Fe^{3+} + Fe^{2+}) = 0.037$ in primordial magma oceans. At a surface temperature of roughly 2500°C, the atmosphere likely contained appreciable amounts of CO, CO_2, H_2, H_2O, and N_2. As the surface cooled to the point that water precipitated into the liquid phase (ca. 373 K), the main gases remaining in the primordial atmosphere were CO_2 and N_2. Their relative proportions were, according to the model, very similar to those in modern Venus (right panel in B). (A) Redrawn from Zahnle K. *et al.* (2007) Emergence of a habitable planet, *Space Sci. Rev.* 129: 35–78. (B) Redrawn from Sossi P. A. *et al.* (2020) Redox state of Earth's magma ocean and its Venus-like early atmosphere, *Science Adv.* 6: 48.

spreading zones on the ocean floor, mid-ocean ridges, and deep-sea volcanoes (Figure 1.5). Ocean floor spreading currently accounts for 75% of the magma discharge of the Earth. Due to magma upwelling and faulting, mantle rocks become exposed on the seafloor, forming new crust. If the mantle rocks contain peridotite, which consists mainly of pyroxene and the iron–magnesium silicate mineral group olivine, a process called *serpentinization* can take place. The iron in the olivine minerals reacts with hot seawater to produce hydrogen and the minerals serpentine, brucite, and magnetite. Serpentinization is important for geochemical origins; we will discuss it in detail in Section 2.2.

At the end of the Hadean period (the time from Earth's formation to 4 Ga), volcanic activity and ruptures of the crust were abundant. Fractures on the seafloor, from which geothermally heated water escapes, are called *hydrothermal vents*. Fissures emitting a hydrothermal fluid arise near volcanically active sites, mid-ocean spreading centers, and magma hotspots. The region between an oceanic trench and the accompanying volcanic arc, called *forearc region* (Figure 1.5), contains the uplifted oceanic crust/upper mantle material.

A summary of the possible series of events following the moon-forming impact, which marks the beginning of the Hadean, is given in Figure 1.6. The most recent estimates point to a Venus-like atmosphere of the Earth in the late Hadean with several 10 bars CO_2 and a few bars N_2 and a high Fe^{2+}/Fe^{3+} ratio in the mantle-derived rocks. Hence when prebiotic organic synthesis started 4.2–4 billion years ago, by the time that water had cooled to the liquid state, the Earth's atmosphere consisted mainly of CO_2 and N_2 and was not particularly reducing. Mantle-derived rocks, however, contained much more Fe^{2+} (ferrous iron) than Fe^{3+} (ferric iron), and Fe^{2+} can reduce water to H_2 in a process called s*erpentinization*. As explained later, H_2 produced by serpentinization within the Earth's crust served as the reductant for the fixation of CO_2 to organic molecules. That is, there was no need for the atmosphere to contain H_2 for CO_2 reduction to take place because, as we will see in later chapters, H_2 was produced by exergonic geochemical processes within the crust at sites where CO_2 existed and where catalysts for H_2 activation and CO_2 reduction were being formed.

1.1.8 GEOLOGICAL AGES ARE OBTAINED FROM RADIOMETRIC DATING

The ages of events are important for understanding early Earth history. How are the geologic times on the scale of millions and billions of years obtained? The main experimental method is radioisotope dating. Elements that have different numbers of neutrons in the nucleus are called *isotopes* of this element; the number of nucleons (protons plus neutrons) in the isotope's nucleus is written in superscript before the element's symbol but is pronounced after the element. For example, the most common isotope of carbon ^{12}C is spoken as "C-twelve". Some isotopes are stable, such as ^{13}C, which has one more neutron than ^{12}C and does not decay. Others are inherently unstable and decay to form other isotopes of the same element or, more commonly, to form isotopes of other elements. The *decay rate*, the decrease of the parent isotope concentration P with time, is proportional to P

$$\frac{dP(t)}{dt} = -\lambda \cdot P(t)$$

where λ is the isotope-specific decay constant and t is the decay time. Integration gives

$$P(t) = P(0) \cdot exp(-\lambda \cdot t)$$

The daughter isotope concentration D obtained upon decay of the parent isotope is equal to the difference between the parent isotope concentration before decay at time zero $P(0)$ and what is left from the parent isotope P after decay time t, namely, $P(t)$

$$D(t) = P(0) - P(t) = P(0) \cdot (1 - exp(-\lambda \cdot t))$$

$$\frac{D(t)}{P(t)} = \frac{P(0) \cdot (1 - exp(-\lambda \cdot t))}{P(0) \cdot exp(-\lambda \cdot t)} \cdot \frac{exp(\lambda \cdot t)}{exp(\lambda \cdot t)} = \frac{exp(\lambda \cdot t) - 1}{1}$$

$$\frac{ln\left(1 + \frac{D(t)}{P(t)}\right)}{\lambda} = t$$

Hence by measuring the number of atoms of the parent and the daughter isotope in the sample, the age t of the sample is obtained. How does that tell us the age of the rock, that is, the time at which the rock was formed? A crucial aspect of radiometric dating minerals this way is additional information that is provided by inorganic chemists and mineralogists. They can tell us how much, at most, daughter isotope the mineral contained at its formation (crystallization) during cooling from the magma melt. This information about the exclusion of the initial daughter isotope content is essential to obtain a reliable date. In an ideal isotope pair, the daughter isotope will not form the mineral at all (complete exclusion) such that we can be sure that the sample contained, at the time of its crystal formation, *only* the parent isotope. In that case *all* of the daughter isotope in the sample arose from the parent isotope in the sample after its formation. In other words, t is the time at which the rock or mineral cooled to a temperature such that diffusion of the parent or daughter isotopes out of the system was no longer possible. At a certain *closure temperature*, the crystal structure has formed sufficiently to stop any diffusion of isotopes into the sample.

Uranium-lead pairs are widely used for very ancient rocks. The two most common uranium isotopes ^{235}U and ^{238}U decay to different isotopes of Pb at different rates. For example, the decay of ^{235}U (number of protons and neutrons) to ^{207}Pb has a half-life of about 704 million years. The half-life $t_{1/2}$ is the time at which the parent isotope has decayed to half of its initial concentration: $\frac{1}{2} = 1 - exp(-\lambda \cdot t_{1/2})$. Hence $\lambda = ln2/t_{1/2} = 9.846 \times 10^{-10}$/years. For a sample with a measured ratio of $D/P = 55/1$, we obtain $t = ln\ 56/(9.846 \times 10^{-10}$/years$) = 4.09 \times 10^9$ years. The sample including ^{235}U

crystallized at 4.09 Ga. The other U isotope in the sample, ^{238}U, decays with a half-life of 4.47 billion years to ^{206}Pb. Here, the ratio of daughter to parent isotope should be $D/P = (exp(\lambda \cdot t)) - 1 = (exp(ln2/t_{1/2}) \cdot t)) - 1 = (exp(ln2/4.47 \times 10^9$ years$) \cdot 4.09 \times 10^9$ years$) - 1 = 0.885$. The measurement of both the ^{235}U and ^{238}U decays is a good crosscheck of the accuracy of age determination. The dating of the ^{238}U decay is more accurate because the D/P ratio is smaller and can be measured more accurately. Because modern analytic techniques allow very precise measurements of U and Pb isotopes in a sample and because the exclusion of Pb from U minerals at their formation is extremely high, U/Pb can give very precise dates on very ancient rocks, sometimes to four significant digits. This is how we know that the Earth is 4.54 ± 0.05 billion years old (rather than roughly 4 to 5).

This age represents the age of the material from which the Earth was formed and was obtained from dating of meteorite material considered to represent samples from the accreting solar disk; dating from lunar samples gave consistent results. Most geological samples from the Earth, however, cannot directly date the formation of the Earth because they have mixed by differentiation in the core, mantle, and crust and by plate tectonics, weathering, and subduction. Parent or daughter isotopes may have been partially removed from the sample by these processes.

The oldest minerals on Earth are zircons. Until recently, the oldest-known rocks on Earth were dated at 3.8 Ga by uranium-lead isotope measurements in solid zircon inclusions in rocks. Uranium atoms can replace zirconium in the mineral zircon ($ZrSiO_4$) but lead as a substitute is excluded. Therefore, radiometric dating is not complicated by any daughter isotope initially present: all lead isotopes are from radioactive decay, thus the zircon inclusions solidified 3.8 Ga ago. A 4.4-Ga-old zircon was found as an inclusion in a 3.8-Ga-old rock from Jack Hills, Australia. At normal atmospheric pressure, zirconium is stable up to a temperature of 1855°C; hence, zircons could remain intact in ~980–1260°C molten basalt or ~1215–1260°C molten granite. Isotope dating of zircons does not exclude very hot Earth surface temperatures with abundant molten basalt and granite at 4.4 Ga, but it does indicate that solid minerals had formed by 4.4 Ga.

Generally, the decay half-life of the isotopes selected for age determination should be in the same order as the expected age. We can take very recent archaeological dating as an example. For very recent archaeological dating, ^{14}C (6p, 8n) decays to ^{14}N (7p, 7n) with a half-life of 5730 years by emitting a β particle. A β particle is an electron that is emitted from the decay of a neutron. The decay converts a neutron into a proton (8n→7n + p + β), changing the identity of the element in this example from ^{14}C to ^{14}N. Biological material cannot incorporate carbon after death; hence, this *closed system* is suitable for isotope dating. Whereas the U isotopes used to date ancient rocks were formed in the supernova that produced our heavy elements, ^{14}C is constantly formed in the upper atmosphere from ^{14}N through the capture of a neutron from cosmic radiation with the expulsion of a proton from the same nucleus. Thus, ^{14}C is constantly incorporated by CO_2-fixing organisms and can be used to date organic material

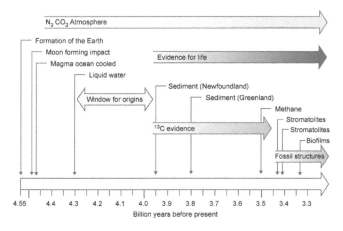

FIGURE 1.7 Major events in the early history of Earth and life. The Earth is about 4.55 billion years old. The oldest evidence for liquid water comes from oxygen isotopes in zircons dated at 4.3 Ga, but there may have been liquid water on the Earth before that (see Figure 1.6). The first evidence for life is preserved as depletion of light carbon isotopes ^{13}C in sediments 3.8 billion years old from Isua in Greenland and sediments 3.95 billion years old from eastern Canada, both lacking associated fossils (see also discussion about carbon isotopes in Section 1.3.5). Methane of probable methanogen origin goes back 3.5 billion years. Stromatolites (fossilized microbial communities from tidal pools) start appearing about 3.4 billion years ago. The sparse record for early life is mainly due to the fact that rocks older than 3.5 billion years are extremely rare. The oldest sedimentary rocks known harbor evidence for life. That leaves a fairly narrow window of time, roughly 200–300 million years, in which life arose. Data from Arndt N. T., Nisbet E. G. (2012) Processes on the young earth and the habitats of early life, *Annu. Rev. Earth Planet Sci.* 40: 521–549.; Tashiro T. *et al.* (2017) Early trace of life from 3.95 Ga sedimentary rocks in Labrador, Canada, *Nature* 549: 516–518.

back to about 50,000 years (roughly 10 half-lives, leaving about 10^{-3} of the original ^{14}C in the sample). If an organism died 114,000 years ago, corresponding to 20 ^{14}C half-lives, there is only 10^{-6} of the original amount of ^{14}C left to measure, too little to accurately determine the age of the sample.

The time in Earth's history from the Moon-forming impact to 4 Ga is called the Hadean (from Greek *Hades*, the underworld). Figure 1.7 shows a scheme of the geological time scale up to 3.3 billion years.

1.1.9 The Internal Structure of the Earth Is Revealed by Seismic Waves

The Earth's crust has changed continuously since its formation due to subduction, melting in the mantle, and resurfacing of magma from the upper mantle at spreading zones. By contrast, the Earth's core has remained nearly unaltered since differentiation because the higher density of iron and nickel versus magma keeps them in place.

The structure of the Earth is mainly investigated by measuring the echo of intense soundwaves emitted into the Earth's interior. The reflected intensity and arrival time depend on rock material and density. The echo of soundwaves of different wavelengths gives information about

chemical depth profiles. Earthquakes release intense shock waves that penetrate far into Earth's interior such that very deep-reaching profiles can be obtained from their analysis. More direct information was obtained by the analysis of samples from holes drilled up to 12 km into the crust. Further information about Earth's density distribution and material composition derives from measurements of Earth's gravitational and magnetic fields. In laboratory experiments, rock material and crystalline solids were exposed to high temperatures, pressures, and chemically active fluids to study their physicochemical behavior. From these measurements a coherent picture of Earth's shell structure has emerged (Figure 1.8).

Earth's *inner core* consists of roughly 80% iron and 20% nickel and is solid despite temperatures of around 6000°C. The pressure of 3.6 million bar from the surrounding mass is high enough to keep the inner core crystalline. The outer core is mainly iron and molten at 3000–5000°C. The rotation of the Earth causes the conductive metallic fluid in the outer core to flow. The Earth's magnetic field arises from electric currents in the conductive iron- and nickel-containing liquid outer core. The magnetic field protects the Earth by deflecting the charged particles from solar wind

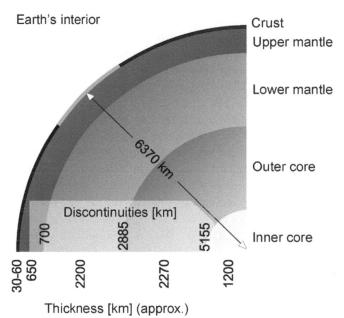

FIGURE 1.8 Schematic structure of the modern Earth. The Earth has a shell-like structure. Inner core: Solid, 80% iron and 20% nickel, pressure 3.6 million bars, temperature 6000°C. Outer core: Iron and nickel, traces of sulfur and oxygen, temperature 3000–5000°C, pressure smaller than in the inner core—therefore liquid. Lower mantle: Highly viscous with heavy silicates containing magnesium and iron. Transition zone to upper mantle. Upper mantle: Rigid with partially molten rocks (magma) consisting mainly of a magnesium and iron-containing silicate mineral called peridotite. Peridotite consists primarily of the mineral olivine and some pyroxene. The portion of olivine in melted upper mantle material can increase up to 98%. Crust: Solid tectonic plates that swim on the partially fluid upper mantle. The oceanic crust is 10 kilometers thick at the maximum and covered by seawater. The continental crust is 35–70 km thick and lighter than the oceanic crust because the proportion of lighter aluminum in its quartz is higher. The continents float on the mantle.

that would otherwise strip off the atmosphere, including the ozone layer that shields us from UV light.

The lower and upper *mantle* surrounds Earth's core and consists mainly of silicates. The temperatures are lower here, and the silicate material is predominantly solid, but behaves like highly viscous liquid rock on geological time scales. Molten rocks have a lower density than solid material and ascend in the mantle. Upon cooling in the upper mantle, the *magma* fluid descends again. Powerful convection currents pull through the mantle.

Solid tectonic plates swim on the fluid upper mantle and form the *crust*. The basaltic *oceanic crust* is heavier than the granitic *continental crust* and therefore plunges deeper in the upper mantle. *Basalt* is a magnesium- and iron-rich (*mafic* from *ma*gnesium and *f*erric) silicate mineral and therefore quite heavy. More than 80% of the Earth's volcanic rock (*lava*) is basalt. *Granite* is a silicate mineral enriched in the lighter elements of silicon, oxygen, aluminum, sodium, and potassium (*felsic* from feldspar and silica). Felsic magma is more viscous than mafic magma. The oceanic crust is about 10 km thick and covered with seawater. The continental crust is 35–70 km thick and extends further into the mantle than the oceanic crust. Oceanic crust is constantly created at spreading zones, where magma from the mantle emerges, creating new ocean floor crust. This pushes the older crust under the continents at subduction zones (Figure 1.5). In this way, submarine crust is constantly recycled through the mantle, a process called Wilson cycles. This goes faster than one might think. The oldest crust in the open ocean dates to only about 250 Ma. The oldest oceanic crust is in the Mediterranean and is dated at roughly 340 Ma. That is, the primordial oceanic crust that initially covered the Earth at its formation is gone and has been replaced 10–20 times over through Wilson cycles. Continental crust is lighter, swims on top, and rides high at subduction zones; it can reach ages of nearly 4 billion years. Geology provided the setting for early life. Life is made of cells, cells are made of molecules, and molecules are made of atoms. The next section starts there.

1.2 CHEMISTRY DESCRIBES THE PROPERTIES OF ATOMS AND MOLECULES

1.2.1 COVALENT AND NONCOVALENT BONDS DETERMINE THE STRUCTURE AND STABILITY OF MOLECULES

All matter consists of atoms as the smallest unit. Without carbon atoms, life cannot exist. Carbon (chemical symbol C) has unique capabilities to form chains with other carbon atoms, leading to millions of organic molecules. The C atom consists of an atomic core of six positively charged *protons* (Figure 1.9) and six uncharged *neutrons*. Six negatively charged electrons orbit the atomic core. The positive and the negative charges compensate—the C atom is uncharged outward. Two inner electrons orbit near the atomic core and are tightly bound due to the strong attraction by the positively charged core. Their energy is low—they do not participate in chemical reactions. The four outer electrons, however, can bind to other atoms. They are called *valence electrons*.

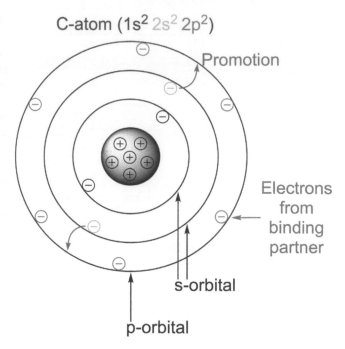

C-atom ($1s^2\,2s^2\,2p^2$)

Promotion

Electrons from binding partner

s-orbital

p-orbital

FIGURE 1.9 Electron shells of the carbon atom. The atomic core of carbon (C) consists of six protons (+) and six neutrons (not shown here). Six electrons orbit around the atomic core. The two inner electrons (drawn in black) move in a spherical **s**-orbital in shell **1** (electron configuration **1s²**). The next two electrons (drawn in blue) orbit in shell 2 (**2s²**). The remaining two electrons (drawn in red) circulate in the dumbbell-shaped p-orbitals in shell 2 (**2p²**). If four electrons from binding partners (drawn in green) populate the outer orbitals of carbon, then the most stable carbon configuration of eight electrons is reached in shell 2, which is then completely filled (closed shell). Even greater stability of carbon's orbitals is achieved by promotion of one of the 2s-electrons to the 2p-orbital upon carbon bonding.

1.2.1.1 Covalent Bonds

In a chemical bond of carbon, its four outer electrons spend some time in the outer orbitals of the binding partners, while four outer electrons of the binding partners spend some time in the outer orbitals of carbon. Thus, when bonded, eight electrons populate the valence orbitals of the C atom—at least temporarily. With eight electrons the outer orbitals of C are completely filled in a symmetric spherical shell *noble gas* configuration. This is an especially stable low-energy state. If the bonding partners are stabilized similarly by this sharing of electrons, *covalent* bonds arise and stable *molecules* form.

The other atoms in organic molecules also attempt to populate eight electrons in their outer shells (*octet rule*). Two inner and five outer electrons circulate the atomic core of *nitrogen* (chemical symbol N) in a $1s^2$[Helium: He]$2s^22p^3$ electron configuration. Hence for an N atom to achieve the low-energy configuration of eight outer electrons, it must share three electrons with bonding partners in three covalent bonds. Oxygen (O) can form two chemical bonds and hydrogen (H) one chemical bond (Figure 1.10). Depending upon their oxidation state, nitrogen (N) and phosphorus (P) can form three or more chemical bonds, and sulfur (S) can form two or more chemical bonds. The reaction of C, N, O, and H with hydrogen atoms generates CH_4 (methane), NH_3 (ammonia), H_2O (water), and H_2 (molecular hydrogen).

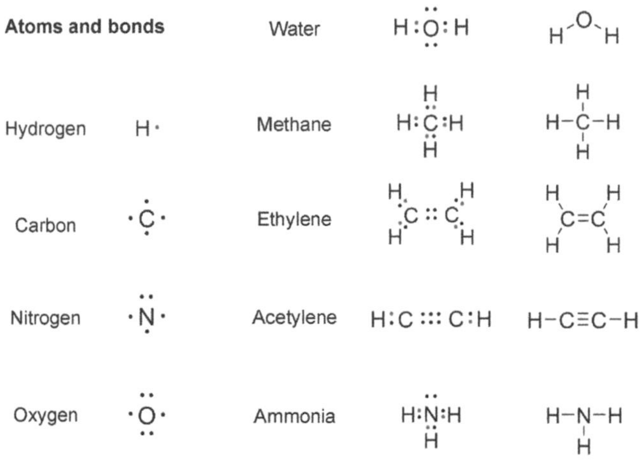

FIGURE 1.10 **Dot representation of electrons in atoms and molecules and chemical bonds.** Electrons are illustrated as dots; electrons contributed by hydrogen are shown as red dots. Carbon can form double bonds (as in ethylene) and triple bonds (as in acetylene). In the chemical formulas on the right, every dash corresponds to a single bond formed by two electrons (dots). Covalent bond energies of the bonds shown in the figure are on the order of 160–1070 kJ/mol.

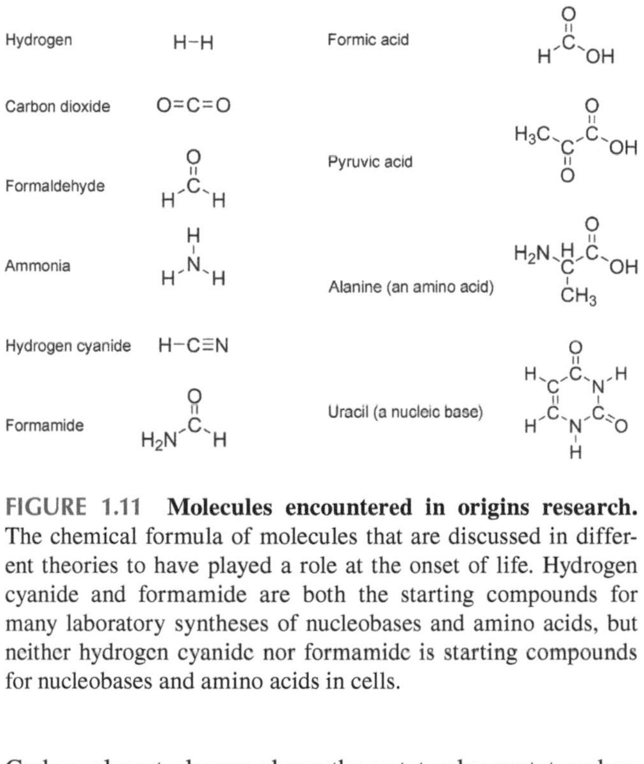

FIGURE 1.11 **Molecules encountered in origins research.** The chemical formula of molecules that are discussed in different theories to have played a role at the onset of life. Hydrogen cyanide and formamide are both the starting compounds for many laboratory syntheses of nucleobases and amino acids, but neither hydrogen cyanide nor formamide is starting compounds for nucleobases and amino acids in cells.

Carbon almost always obeys the octet rule; sextet carbon (carbenes) is common in industrial chemistry but almost non-existent in biology.

More complicated molecules like amino acids also have exactly the number of bonds that stabilize their atoms most: carbon 4, nitrogen 3, oxygen 2, and so on. Consider the molecules in Figure 1.11 and count the number of bonds.

Some molecules have C, N, or O with other numbers of bonds. For example, carbon binds oxygen in carbon monoxide with three bonds instead of four (triple bond; Figure 1.12). Sulfur and phosphorus do not obey the octet rule at all because they have a third shell with d-orbitals that are unfilled with electrons but that can hybridize with lower

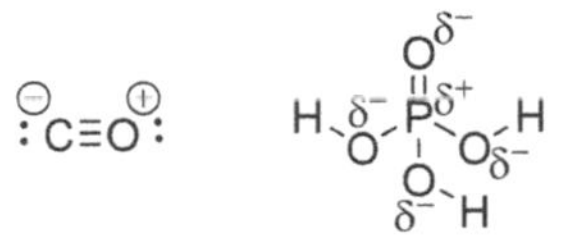

FIGURE 1.12 **Carbon monoxide and phosphoric acid.**

orbitals. P and S when bonded to oxygen in their highest oxidation states yield phosphate (PO_4^{3-}) and sulfate (SO_4^{2-}), which are typically written as forming five and six chemical bonds, respectively. The electrons are not uniformly distributed in these molecules. Oxygen attracts electrons generally stronger than carbon, phosphorus, and sulfur and is therefore partially negatively charged (δ^-) in molecules with a C=O or C–O bond, in H_3PO_4 and in H_2SO_4, whereas carbon, phosphorous, and sulfur are partially positively charged (δ^+) there. Carbon monoxide is an exception; here carbon carries a formal charge of −1.

These *partial charges* play an important role in chemical reactions. For example, if an atom in a molecule has a partial negative charge, then it can be attracted to a positively charged atom in another molecule by electrostatic forces. This can cause electrons from the species with the partial negative charge to interact with the nucleus of the atom having the partial positive charge. The molecules might react to form new products upon this kind of *nucleophilic attack*.

Reactions with hydrogen or involving the addition of electrons from hydrogen or another atom or molecule are called *reduction* reactions; reactions with oxygen or with the removal of electron density by oxygen or by another species are called *oxidation* reactions. The terms *reduction* and *oxidation* come from metallurgy. Metal ores are typically oxides; the oxygen has to be removed to obtain the pure metal. This can be done with charcoal or coke and high heat. Removing the oxygen from the melt *reduces* its weight and delivers the pure metal, in the reduced state. Carbon reacts with oxygen to carbon dioxide CO_2 because carbon can form four covalent bonds and oxygen two. Carbon forms two bonds with each oxygen atom: O=C=O. These *double bonds* are very stable and CO_2 therefore sluggish in reaction. The bonds that chemists draw with lines "–" represent electron pairs in covalent bonds. Covalent chemical bonds have very different energies depending upon the atoms bonded and their environment. For example, the average energy to break 1 mol (6.02×10^{23} molecules) of C–H bonds is 413 kJ, 347 kJ for C–C bonds, 614 kJ for C=C double bonds, and 839 kJ for C≡C triple bonds.

Covalent bonds are spatially directed. For example, methane (CH_4) could, in principle, form three bonds at right angles after the *promotion* of a 2s electron to the 2p orbital using its p_x-, p_y-, and p_z orbitals (Figure 1.13) and a fourth bond in some arbitrary direction using its spherical s-orbital, but this is not what is observed.

Experiments show that methane holds four bonds of equal strength pointing in tetrahedral directions. Linus Pauling, who coined the terms *covalent* and *ionic* to describe bonds, explained this by assuming that in the presence of four hydrogen atoms the s- and p-orbitals combine into four equivalent *hybrid orbitals*, each termed sp³ in Figure 1.13. These hybrid orbitals are directed along the four tetrahedral

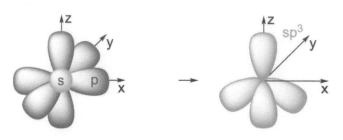

Promotion: $1s^2 2s^2 2p^2 \longrightarrow 1s^2 \underline{2s^1 2p^3}$
Hybrid orbitals

FIGURE 1.13 Hybridization of carbon orbitals to form directed chemical bonds. The p-orbitals of carbon are directed in the *x*-, *y*-, and *z*-direction (drawn pink) and the s-orbital is spherical (drawn green). The s-orbital and the three p-orbitals hybridize into four equivalent sp³-hybrid orbitals which point in the direction of the corners of a tetrahedron (hybridization).

CH-bonds of methane. The energy necessary for promoting one electron from the 2s-orbital to the 2p-orbital (*promotion* in Figure 1.9) is compensated by the energy released by forming the tetrahedrally oriented CH bonds. The symmetry of the four C–H bonds in methane leads to a particularly stable configuration and a very high bond energy of 439 kJ/mol. The energy of an average C–F bond is higher still at 485 kJ/mol, which is why Teflon (polytetrafluorethylene, PTFE) is so unreactive and heat stable.

The *hybridization model* can explain the structure of organic molecules with double and triple bonds of carbon as well. For example, the structure of ethylene can be explained by three tridiagonal sp²-hybrid orbitals forming σ-bonds with hydrogen and carbon and one unhybridized p_z-orbital shaping the π-bond between the carbon atoms (Figure 1.14). Similarly, the bonds in acetylene can be described by two linear sp-hybrid orbitals for the C–H σ-bonds and the C–C σ-bond and unhybridized p_y- and p_z-orbitals shaping the two π-bonds between the carbon atoms to form a triple bond (1 C–C σ-bond and 2 C–C π-bonds).

In ammonia (NH₃), three of the four sp³ hybrid orbitals form σ-bonds with hydrogen, whereas the fourth sp³ orbital is occupied by two *nonbonding* electrons (a *lone pair of electrons* or a *free electron pair*). In water (H₂O), two of the four sp³ hybrid orbitals form σ-bonds with hydrogen, whereas the two remaining sp³ orbitals are occupied by nonbonding electrons. The electron pairs are drawn as red dots in Figure 1.15.

FIGURE 1.15 Ammonia and water. Nitrogen and oxygen are reduced by hydrogen in these molecules. The red dots indicate free electron pairs; see the text for details.

Electrons are not always paired. Unpaired electrons in outer orbitals are called *radicals*. Most atoms or molecules that exist as *radicals* are highly reactive and have short lifetimes because the atom is one electron short of the stable octet (noble gas) configuration or has one electron in surplus. Radicals can be synthesized in biological systems by radical-generating enzymes. The radical *superoxide* O_2^-, for example, is generated by the immune system to destroy invading microorganisms; the gaseous radical *nitric oxide* NO regulates many biochemical processes such as control of vascular tone and hence blood pressure. The unpaired electron needed to generate a radical is typically donated by a metal in the enzyme that donates a single electron and thereby undergoes a change in oxidation state by +1.

1.2.1.2 Noncovalent Bonds

Noncovalent interactions are weaker than covalent bonds but are often vital for processes impacting the origin of life. *Electrostatic interactions, hydrogen bonds, dipole,* and *induced dipole interactions* have different bond energies, specificity, and behavior toward water solvents.

1.2.1.3 Electrostatic Interactions

When two atoms or molecules of very different electron affinity approach, an electron may hop to the atom of higher electron affinity. For example, sodium (chemical symbol Na) must detach the electron in the 3s¹-orbital (drawn red in Figure 1.16) to achieve the most stable configuration of its outer shell, whereas chlorine (Cl) has to capture one electron to attain that. Hence an electron will hop from Na to Cl. Na⁺ and Cl⁻ *ions* arise and attract each other electrostatically such that a *sodium chloride salt* will be formed.

The attractive interaction energy E between two ions of opposite charge is given by

$$E = -\frac{1}{4\pi\varepsilon 0} \frac{e^2}{Dr} \tag{1.3}$$

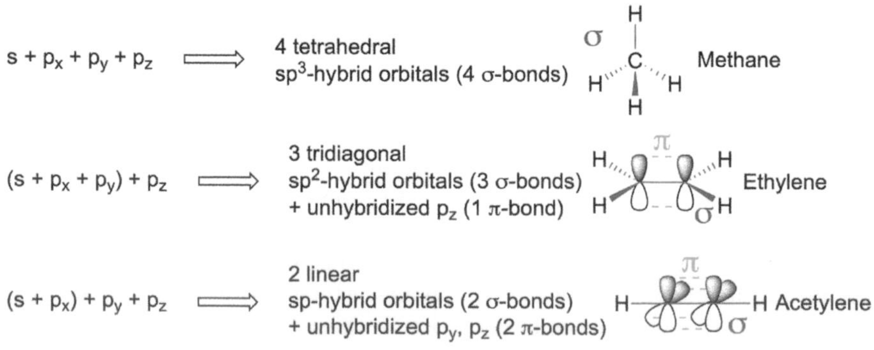

s + p_x + p_y + p_z ⟹	4 tetrahedral sp³-hybrid orbitals (4 σ-bonds)	Methane
(s + p_x + p_y) + p_z ⟹	3 tridiagonal sp²-hybrid orbitals (3 σ-bonds) + unhybridized p_z (1 π-bond)	Ethylene
(s + p_x) + p_y + p_z ⟹	2 linear sp-hybrid orbitals (2 σ-bonds) + unhybridized p_y, p_z (2 π-bonds)	Acetylene

FIGURE 1.14 Hybridization model for σ- and π-bonds. Hybridization describes the different bond situations in molecules.

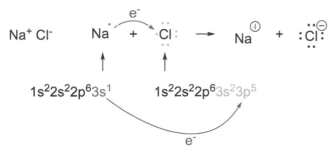

FIGURE 1.16 **Ionic bond between atoms of very different electron affinity.** The outer electron of the electron donor is drawn red and the outer electrons of the acceptor are drawn blue. Na$^+$ and Cl$^-$ ions attract each other.

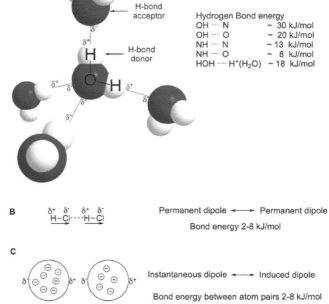

FIGURE 1.17 **Noncovalent interactions.** (A) Schematic of the network of hydrogen bonds in water. The hydrogen bonds between the oxygen and hydrogen atoms are indicated by dotted lines and the partial charges at the O- and H-atoms by δ^- and δ^+, respectively. (B) Interaction between permanent dipoles. (C) Interaction between induced dipoles (interatomic or intermolecular).

where $\varepsilon 0$ is the dielectric field constant 8.85×10^{-12} C^2/J/m, $+e$ is the charge of the *cation* and $-e$ of the *anion* in units of the electronic charge e of 1.602×10^{-19} C, D is the relative dielectric constant accounting for the charge screening capability of the solvent ($D = 80$ for water), and r is the distance between the two ions.

Electrostatic attraction between ions of different charges plays an important role in biochemical processes. For example, doubly charged magnesium ions (Mg^{2+}) bind to negatively charged phosphate groups and help to catalyze their condensation to energy-storing polyphosphates like adenosine triphosphate (ATP). Consider that polyatomic anions like polyphosphates cannot be approximated as spheres, their specific structures must be considered to calculate the interaction energies. In a nonpolar or low water activity environment, the attraction energies between ions of opposite charge can be substantial and can approach the energies of weak covalent bonds.

Salts influence *water activity* (effective concentration of water) by electrostatic binding of the salt ions to water. By neutralizing charged groups on the outside of proteins, salts can determine protein folding (Section 4.2).

1.2.1.4 Hydrogen Bonds

Oxygen atoms attract electrons stronger than hydrogen atoms. They have higher *electronegativity*, another term introduced by Pauling. Hence the electron distribution in water is polarized with positive partial charges at the hydrogen atoms and negative partial charges at the oxygen atom (Figure 1.17). Opposite charges attract each other such that the oxygen atoms interact with hydrogen atoms and are thus connected by hydrogen bonds –O···H–O–H···O– in liquid water and ice. Hydrogen bonds also have a weak covalent contribution from electron orbital overlap. Hydrogen bonds are an important component of the forces that determine the structure of RNA, DNA, and proteins.

1.2.1.5 Dipole and Induced Dipole Interactions

Molecules such as water have permanent dipoles. Molecules with permanent dipoles can attract each other electrostatically (Figure 1.17 B). Even molecules with unpolar bonds can attract each other by inducing instantaneous dipoles through spatially fluctuating electron densities (Figure 1.17C). These interactions are generally weak with

energies between 2 and 8 kJ/mol. In a large molecule or polymer, however, there are often very many fluctuating dipole interactions (also called *induced dipole*, *dispersion*, or *van der Waals* interaction), which can strongly influence molecular structure. Consider, for example, the *helix* ("spiral") structure of DNA strands. The nucleobases in the DNA are flat, circular molecules. The delocalized electrons in the stacked rings of the nucleobases in a DNA helix avoid each other because they are equally charged. Hence dipoles arise for a short time, which attract each other: $+\to - +\to -$ (Figure 1.17C). The maximal attractive interaction between the nucleobases is achieved in a DNA *double helix* where the nucleobases stack on top of each other with a maximum possible overlap like the staggered steps in a spiral chaircase. These stacking forces are induced dipole interactions. Similarly, the induced dipole interactions between nonpolar amino acid side chains play an important role in protein folding.

1.2.1.6 Hydrophobic Interactions

Nonpolar molecules attract each other due to induced dipole interaction. The attraction is greatly enhanced in water where *hydrophobic* (water-hostile or water-repellant) molecules aggregate and form droplets. Oil is an example of a hydrophobic substance. Oil droplets repel water because the released water molecules are less ordered in *bulk* water than surrounding the oil molecules (Figure 1.18). The overall decrease of order (an increase of entropy) favors the aggregation of nonpolar molecules in water by a decrease of the Gibbs free energy of the process (see Section 1.2.5). Hydrophobic interactions are very important for the folding

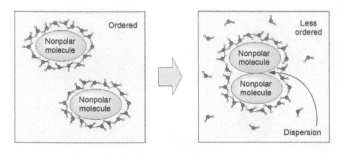

FIGURE 1.18 Aggregation of hydrophobic molecules in water. Nonpolar molecules aggregate by induced dipole interaction (dispersion) and expel water from their ordered envelope to their less ordered alignment in bulk water. This process is favorable due to the increase of entropy (measure of disorder) leading to a decrease of the Gibbs free energy.

of globular proteins, which generally have nonpolar amino acid side chains interacting with one another in the interior of the molecule and charged, hydrophilic side chains at the surface, which interact with the universal solvent of cells, water.

1.2.2 A Few Functional Groups Account for Most of the Properties of Biomolecules

Only a few combinations of C-, H-, N-, and O-atoms form the functional groups of bioorganic molecules (Table 1.1).

The *hydroxy* (–OH) group belongs to the most important ones. The OH-group is polar ($O^{\delta-}$ $H^{\delta+}$) and develops hydrogen bonds to water. Hydroxyl groups make alcohols miscible with water, and in sugars like ribose they can form ester bonds with organic acids or phosphate. The polymer backbone of RNA and DNA is built up of the sugar ribose and phosphate by ester condensation.

The *carbonyl* (–C=O) group is the functional group of aldehydes and ketones. Carbonyl groups are generally reactive. The carbonyl bond is polar ($C^{\delta+}=O^{\delta-}$) and sensitive to a *nucleophilic* attack on the carbon atom. For example, the negatively polarized oxygen of OH ($O^{\delta-}$) can attack the positively polarized carbon in C=O ($C^{\delta+}$) and add to the C=O double bond. Organic molecules with a hydroxy group and a remote carbonyl group tend to cyclize to sugars. Examples are the sugar *glucose* with a six-membered ring and *ribose* with a five-membered ring. Other examples of nucleophilic additions or substitutions are given in the next section.

The *carboxyl* (–COOH) group is the functional moiety of organic acids. It can relinquish its proton H^+ because the carboxylate anion –COO$^-$ is stabilized by electron delocalization.

Furthermore, the detached proton is stabilized by solvation with water, forming H_3O^+ or $(H_2O)_nH^+$. The carboxylate anion is stabilized similarly. Carboxyl groups are both hydrogen-bond acceptors (–C=O) and hydrogen-bond donors (–OH) and therefore readily form hydrogen bonds. Carboxylic acids are generally weak acids. In a

TABLE 1.1
Functional Groups

Chemical Group	Structure	Name of Compound	Example	Functions and Properties
Hydroxy		Alcohol Sugar	Ethanol Ribose	Polar H-bonding to water Reaction to form esters
Carbonyl		Aldehyde Ketone	Acetone Acetaldehyde	Cyclization to aldehyde sugars (aldose) or keto sugars (ketose)
Carboxyl		Carboxylic acid	Acetic acid Formic acid	Acidic

(Continued)

TABLE 1.1 Functional Groups (*Continued*)

Chemical Group	Structure	Name of Compound	Example	Functions and Properties
Amino	$\overset{\displaystyle H}{\underset{\displaystyle H}{/N\backslash}}$	Amine	Amino acid glycine	Zwitterion (—NH₂ ⇌ H₂O ⇌ ⁺N–H + OH⁻)
Sulfhydryl	$/S{\scriptstyle\diagdown}H$	Thiols (thioalcohols)	Amino acid cysteine	—SH + HS— → —S–S— + H₂ —S–S– disulfide bridges stabilize proteins. Coordination of Fe and Ni in FeS-clusters
Phosphate	$\overset{\displaystyle O}{\underset{\displaystyle \ominus}{\overset{\|}{-O-P-O}}}{\scriptstyle\ominus}$	Organic phosphates	Acetyl-phosphate	Negative charge in organic molecules. Exergonic hydrolysis. Phosphate transfer to activate organics
Methyl	$\overset{\displaystyle H}{\underset{\displaystyle H}{/C{\scriptstyle\diagdown}H}}$	Methylated compounds	Nucleobase thymine	Critical CO₂ fixation. Intermediate in the acetyl-CoA pathway. Base modifications in tRNA and rRNA. Methylation of uracil to thymine lowers reactivity of C=C double bond

Note: **A** list of functional groups that are responsible for characteristic bioorganic reactions. The functional groups of a molecule allow prediction of its reactive behavior.

dilute aqueous solution, less than 1% of the molecules of acetic acid, CH_3COOH, will exist in the dissociated state as CH_3COO^-.

The *amine* ($-NH_2$) group binds a proton from water ($H_2O \leftrightarrow H^+ + OH^-$; $-NH_2 + H^+ \leftrightarrow -NH_3^+$) such that OH^- is abundant and the aqueous solution becomes *alkaline*. Amino acids both have a carboxyl group and an amine group and therefore can transfer a proton internally to generate a *zwitterion*: $NH_3^+–CHR–COO^-$.

The *sulfhydryl* or *thiol* ($-SH$) group resembles a hydroxy group with the oxygen atom replaced by a sulfur atom. The amino acid cysteine has a thiol in its side chain.

Crosslinking of the sulfur atoms of cysteines by oxidation stabilizes the three-dimensional shape of proteins. Proteins containing two or more cysteines can bind iron to form iron–sulfur clusters, which function as electron carriers in reactions involving the transfer of single electrons (one electron transfers) or donors of single electrons in radical reactions.

The *phosphate* ($-O–PO_3^{2-}$) group can form esters (phosphoesters) with hydroxyl groups, anhydrides with carboxyl groups as in acyl phosphates like acetyl phosphate, and anhydrides with other phosphates as in ATP. Phosphate forms bonds with oxygen atoms in organic molecules by

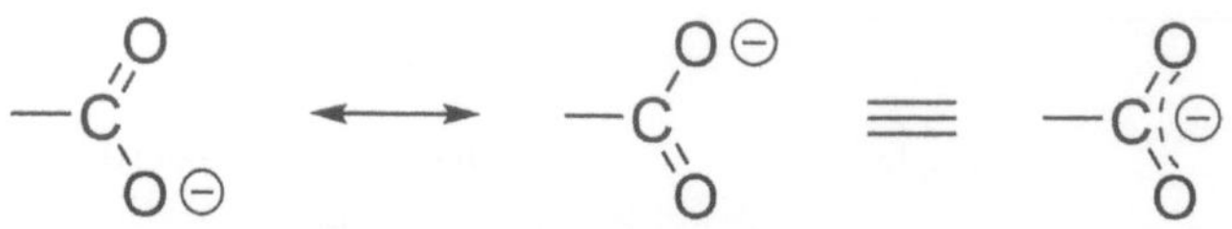

FIGURE 1.19 Resonance structures describe electron delocalization in the carboxylate anion COO⁻. The double arrow indicates the two energetically equivalent positions for the electron in the anion. The symbol "≡" indicates that drawing the two resonance structures is equivalent to drawing the electron in a delocalized manner (dotted line).

water elimination reactions. It is a good leaving group, meaning that organophosphate bonds are readily cleaved by nucleophiles and generally have a negative free energy of hydrolysis. The mixed anhydride bonds of phosphate with carboxylic acids –P–O–C– have a higher free energy of hydrolysis than phosphoanhydrides –P–O–P–. For example, the hydrolysis of acetyl phosphate releases enough energy to drive phosphate transfer in the reaction ADP (adenosine diphosphate) + acetyl phosphate → ATP (adenosine triphosphate) + acetic acid. ATP is the most common energy currency in biological systems. The phosphate group adds negative charge to an organic molecule and promotes binding to Mg^{2+}.

The *methyl* ($-CH_3$) group is an intermediate of the acetyl-CoA pathway—the only energy-releasing pathway of carbon dioxide fixation. Methyl groups can be enzymatically transferred to an organic compound to replace a hydrogen atom (*methylation*). Methylation is important in nucleic acid biochemistry. The genetic code requires *modified bases* to decode the information in genes. The site of the ribosome where proteins are synthesized is heavily methylated to facilitate ribosome–tRNA interactions. Methylation of bases in DNA has a wide variety of functions, including gene silencing, distinguishing the parent strand from the daughter strand at replication for repair of replication errors, and gene regulation via sequence-specific DNA-binding proteins (*transcription factors*). Methylation of the RNA-nucleobase *uracil* at the reactive C=C double bond converts uracil to the more stable DNA-nucleobase *thymine*.

1.2.3 ACID–BASE REACTIONS, REDOX REACTIONS, AND NUCLEOPHILIC ATTACKS ARE CENTRAL TO BIOCHEMISTRY

1.2.3.1 Acid–Base Reactions

Water can be acidic, neutral, or alkaline (basic) depending on the concentration of hydrogen ions H⁺ hydrated as $(H_2O)_n H^+$ in water (abbreviated H_3O^+ here). Concentrations are given in mol/l, abbreviated as the unit M. One mole is defined as the Avogadro number of 6.02×10^{23} particles. The unit mol/l is used because it is easier to handle than 6.02×10^{23} particles/l. The pH value indicates the acidity of an aqueous solution in terms of its proton concentration [H⁺] in mol/l:

$$pH = -\log \left[H^+ \right] \qquad (1.4)$$

For example, neutral water has a proton concentration of 10^{-7} M. The *logarithm*, abbreviated log, is the exponent of a number, here –7 for 10^{-7} mol/l H⁺. Hence, the pH of pure (neutral) water is 7 (pH 7). Concentrated acetic acid has a pH of 2.5 and wine pH of 4. *Sodium hydroxide* NaOH (*caustic soda*) is a very strong base which completely dissociates in water to Na⁺ and OH⁻. The anion OH⁻ reacts with protons H⁺ from dissociated acids to water:

$$OH^- + H^+ \rightleftarrows H_2O$$

Therefore, a high concentration of OH⁻ makes for a low concentration of H⁺—the pH of caustic soda is 13.5–14. High OH⁻ concentration implies a small proton concentration and hence high pH value. A high H⁺ concentration implies a small OH⁻ concentration and small pH. For example, in 0.01 M HCl, [H⁺] = 10^{-2} M, pH 2 and [OH⁻] = $(10^{-14}/10^{-2})$ M = 10^{-12} M because experimentally [H⁺] · [OH⁻] = 10^{-7} M × 10^{-7} M = 10^{-14} M².

Modern seawater has a pH of about 8. Hydrothermal systems can have very different pH values. The effluent of the submarine hydrothermal field *Lost City*, for example, has an alkaline pH of 9–11 because serpentinization of Lost City's rocks leads to the formation of hydroxides. In comparison, *black smokers* exhale sulfur-containing acids of pH 4 and higher.

Phosphoric acid (H_3PO_4) is an important acid in prebiotic chemistry and biological cells (Figure 1.20). H_3PO_4 can lose its protons in three steps:

The equilibrium constant for deprotonation of an acid HA according to HA $\rightleftarrows$ H⁺ + A⁻ is

$$K_a = [H^+][A^-]/[HA]$$

We can take logarithms of both sides of the equation without changing the equality of the two sides

$$\log K_a = \log([H^+][A^-]/[HA]) = \log[H^+] + \log([A^-]/[HA])$$

$$pH = pK_a + \log([A^-]/[HA])$$

where $pK_a = -\log K_a$. If the pH of the solution is equal to the pK_a, then [A⁻] = [HA] because $\log([A^-]/[HA]) = \log 1 = 0$: the acid is half-deprotonated. For example, at pH = pK_a = 7.2 inorganic phosphate in water is a nearly equal mixture of $H_2PO_4^-$ and HPO_4^{2-} ions (see pK_a values in Figure 1.20). Adding protons to the solution will not change the number of free protons (pH value) much, because the protons first react with HPO_4^{2-} to $H_2PO_4^-$ and

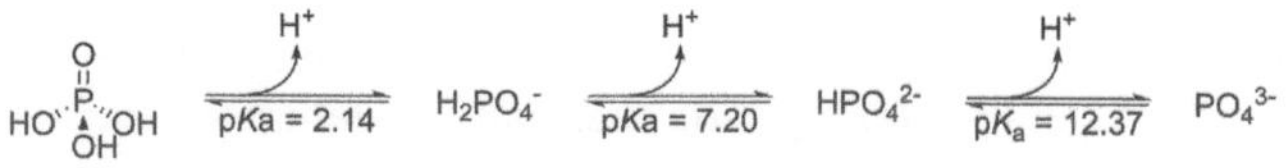

FIGURE 1.20 Schematic of H_3PO_4 deprotonation. With an increase of [OH⁻] in solution (increasing pH), phosphoric acid can relinquish up to three protons.

then to H_3PO_4. This means that dissolved phosphates are a pH buffer protecting acid-sensitive proteins from attack of protons, for example.

Inorganic phosphorus exists in different states of oxidation. Elemental phosphorus has two electrons in its outer 3s-orbital and three electrons in the 3p-orbitals ($[Ne]3s^23p^3$). Hence it can add 3 electrons to reach the favorable 8 electron configuration in the outer shells (*phosphine* PH_3) or lose 1–5 electrons until complete depletion of the third shell and 8 electrons in the second shell is reached (phosphate, PO_4^{3-}). The formal charge of phosphorus in the different compounds obtained by adhering to the convention that the oxidation state of oxygen is always –2, withdrawing electron density from phosphorus in every P–O bond, whereas hydrogen (oxidation state by convention always +1) adds electron density in every P–H bond. Of course, in O^- or OH, hydrogen has already added 1 electron to O such that only 1 electron has to come from phosphorus (phosphorus +1 for every OH or O^-).

For example, the phosphorus atom in *phosphite* $HP(OH)O_2^-$ has a charge of +4 from the three phosphorus–oxygen bonds and a charge of –1 from the P–H bond; its overall charge is +3. Phosphite is a reducing agent that reacts to form phosphate. The existence of phosphite-oxidizing bacteria shows that phosphite exists in the environment today.

Several minerals contain phosphorus. *Schreibersite* (Figure 1.21) consists of iron-nickel phosphide $(Fe,Ni)_3P$, with Fe, Ni, and P in the oxidation state 0 (P probably closer to –1). It is a very rare mineral, although it is present in most meteorites containing nickel and iron metal; that is, it forms from the elements in space but not on Earth. Phosphorylation of the nucleosides adenosine and uridine by synthetic schreibersite has been reported, probably via oxidation of schreibersite by water to magnetite (Fe_3O_4), H_2, and phosphorus oxyanions. *Apatite* $Ca_5(PO_4)_3(OH, F, Cl)_1$ contains phosphate and hydroxy, fluorine, or chlorine anions. For electrical neutrality, the charge of $5Ca^{2+}$ cations (charge +10) is compensated by the charge of three PO_4^{3-} anions and one OH^-, F^-, or Cl^- anion (charge –10). Tooth

enamel and bone mineral consist mainly of hydroxyapatite. Pure *calcium phosphate* $Ca_3(PO_4)_2$ does not exist in nature because phosphate minerals always contain further anions besides phosphate like in *apatite* or further cations besides calcium like in *Whitlockite*. Anhydrides of phosphates like acetyl phosphate and ATP are the main energy currencies of life. Anhydrides are typically formed when two acids bond to each other under water elimination (the "anhydrous" acid). Anhydride bonds have a very high free energy of hydrolysis, they are energy rich bonds. Acetyl phosphate is a mixed anhydride of an organic acid (acetate) and phosphate, ATP harbors phosphoanhydride bonds.

1.2.3.2 Redox Reactions

The transfer of electrons in redox reactions is *the central reaction* of living cells. All cells ultimately depend on chemical reactions in which atoms change their oxidation state. Redox reactions are the basis of *energy metabolism*, the energy-conserving reactions that generate ATP to fuel all other life processes. ATP synthesis by rhodopsin-based proton pumping via light absorption appears to constitute an exception but is merely a supplement to redox-based energy metabolism. Redox reactions are also the essential basis for all ecosystems. This is because all ecosystems depend on *primary production*, the synthesis of organic material from inorganic carbon, CO_2. To convert CO_2 into biological molecules, it has to be *reduced*, i.e. electrons have to be transferred to it. The electrons have to come from environmental donors, either with the help of light, which requires a very complicated photosynthetic machinery, or from environmentally available reductants, the most important of which on the early Earth, by far, was molecular hydrogen, H_2.

For example, the reduction of carbon dioxide with hydrogen to formic acid at pH 7

$$CO_2 + 2H^+ + 2e^- \rightarrow HCOOH$$

requires uptake of electrons from a suitable electron donor (see Section 1.2.5). Means of measuring and calculating electron energies are explained in Section 2.4.

1.2.3.3 Nucleophilic Reactions

In a nucleophilic reaction, an electron-rich reactant attacks a partially positive charged atom ("*nucleus*") and either substitutes a leaving group as in Figure 1.22A–C or adds to a double (or triple) bond as in Figure 1.22D.

The electron-rich nucleophile in Figure 1.22A is thioacetic acid CH_3COSH or its ester CH_3COSR with R designating an organic rest. The carbon of the CH_3-group has a partial negative charge (δ^-) as the resonance structures in Figure 1.22A show. This is because the carbon of the carbonyl group is positively charged (δ^+) by the electron-withdrawing oxygen. Hydrogen atoms on a carbon atom adjacent to a carbonyl group (called *α-hydrogen atoms*) are slightly acidic. The $C^{\delta-}$ attacks $C^{\delta+}$ of another thioester and substitutes the good leaving group RS^- (*thiol*) or SH^- to form a C–C bond (this reaction type is called *Claisen condensation*). A base like OH^- can catalyze the reaction by polarizing the $-CH_3$ group of the thioester for nucleophilic attack.

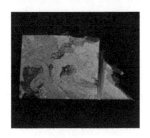

Schreibersite
$(Fe,Ni)_3P$

Rare, occurs in meteorites.
A phosphide.

Apatite
$Ca_3(PO_4)_2$

Common but poorly soluble.
Calcium phosphate,
often with F or Cl.

FIGURE 1.21 Minerals with phosphorus in different oxidation states. Left: Schreibersite embedded within an iron-nickel meteorite. The schreibersite has a slightly less smooth texture than the surrounding metal. Right: A crystal of fluoride-containing apatite. Images kindly provided by Matt Pasek, University of South Florida. Photo credits: Virginia Pasek.

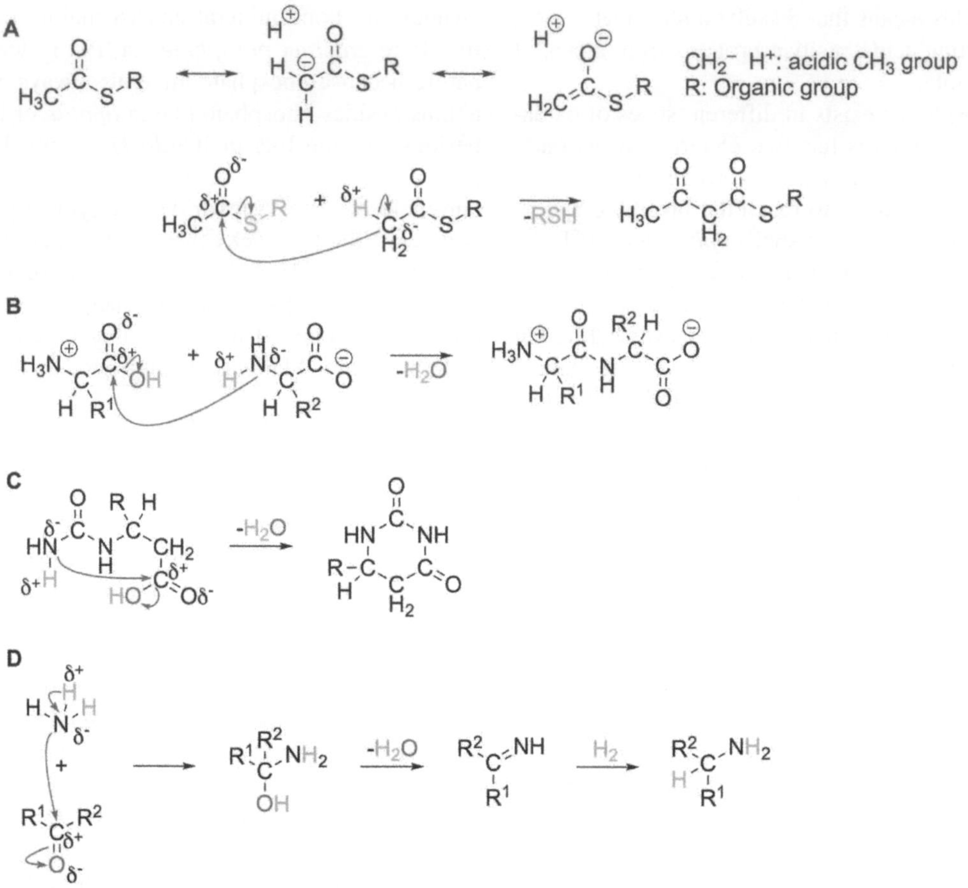

FIGURE 1.22 Nucleophilic substitution and addition in bioorganic reactions. (A) The Claisen condensation to form a C–C bond is catalyzed by a strong base. (B) Amino acid condensation to a peptide is catalyzed by a base or an acid which polarizes the C=O bond. (C) Formation of an intramolecular amide bond leads to cyclization. (D) The nucleophilic attack of activated ammonia on a keto group and reduction with H_2 results in an amine.

Figure 1.22B displays the nucleophilic attack of the negatively charged nitrogen atom of an amino acid on the positively charged carbon atom of the carboxyl group of another amino acid. Water is displaced in this reaction and a peptide bond is formed. A good leaving group such as a thiol (RS^-) instead of OH^- enhances the substitution. Again, the reaction is base-catalyzed. Figure 1.22C shows the formation of an intramolecular amide bond, leading to cyclization of the molecule. The mechanism resembles the chemistry of peptide bond formation in Figure 1.22B. During nucleotide biosynthesis, amino acids cyclize to form the bases of nucleic acids by a similar mechanism (Section 4.5). Figure 1.22D depicts the *reductive amination* of a keto group to an amine. Nucleophilic addition of ammonia (NH_3) to a carbonyl group (C=O) is followed by water elimination to form an imino group (C=N) that is reduced to the amine by the addition of H_2. This reaction takes place during the biological incorporation of ammonia into organic compounds by the glutamate synthase reaction, with the difference that H_2 is replaced by an organic *hydride* donor called FADH, reduced *flavin adenine dinucleotide*. In laboratory analogues of the reaction, more reactive forms of ammonia (hydrazine, N_2H_4, or hydroxylamine, NH_2OH) must be used for this reaction to proceed rapidly without the catalytic help of the enzyme (see Section 4.1). In biology, nitrogen always enters metabolism as NH_3, either via the amino group in glutamate or via the amido group in glutamine.

1.2.4 HYDROLYSIS CAN FACILITATE OR HINDER THE SYNTHESIS OF BIOORGANIC MOLECULES

Water is essential for all known forms of life. It is the solvent of biomolecules and provides protons (H^+) and hydroxyl groups (OH^-) for myriad reactions. If we take a top-down view of the chemical reactions that take place in anaerobic cells, the most common reactant is water, followed by protons, organic hydride donors, reactions involving phosphate, thiols, CO_2, NH_4, and the methyl donor S-adenosyl methionine. This provides a picture of the most common kinds of reactions that take place in biological systems and might provide clues as to the chemical environment within which the chemical reaction network of metabolism arose. This topic is treated in more detail in Chapter 3.

Chemical reactions are generally reversible, if thermodynamics permits (see Section 1.2.5). This is also true for the biochemical reactions of cells, which tend to occur within a fairly narrow range of free energy changes that are not far from zero such that the reverse reaction can also occur, in principle. Biological systems do, however, sometimes employ a few tricks to make some reactions irreversible under the physiological conditions that exist in cells, for example, by removing reaction products, thereby lowering their concentration such that the back reaction becomes very slow. Because reactions are reversible, the water molecules that are eliminated during condensation or polymerization reactions can be added in the reverse reaction. This is also true for the polymerization reactions that form essential bioorganic molecules like proteins or nucleic acids. Cleavage of a peptide bond by the addition of water is an example (reverse of the reaction depicted in Figure 1.22B).

Water molecules dissociate chemical bonds that were formed by water elimination and thereby break larger molecules or polymers into their components. This is called *hydrolysis*, the breaking of bonds by water. Both protons and hydroxide ions can catalyze hydrolysis reactions, making it a highly pH-dependent process. Inside a cell, hydrolysis of peptides and nucleic acids can also take place, and does take place, from time to time. Living systems counteract this by coupling the synthesis of proteins and nucleic acids to the main ATP-producing reaction of the cell so that the rate of synthesis far exceeds the rate of hydrolysis. When ATP synthesis comes to a halt, so does the life process such that polymerization reactions cease while hydrolysis reactions set it.

A classical question in prebiotic chemistry concerns the issue of how hydrolysis reactions could have been slowed or counteracted under prebiotic settings. The classical answer is to assume the existence of conditions in which water activities are lowered. The *activity* of a substance describes the proportion of its molecules per unit volume or weight that are available to undergo a chemical reaction. The activity of a substance often deviates from its behavior in an ideal (highly diluted) solution. For example, in a pure water solution, water is more or less free to react. In an aqueous solution containing high concentrations of polymeric polyalcohol polyethylene glycol, which binds water very efficiently, many or even most of the water molecules bind to the hydroxyl groups of the polyol via hydrogen bonds. Bound to the surface of the polymer, they are effectively trapped and unavailable for other kinds of reactions. The presence of water-binding compounds reduces the overall water activity of a system, and the 'effective concentration" decreases. This "bound water" will be called "water of hydration" here to distinguish it from "bulk water" that is free for reactions such as hydrolysis.

Proteins bind enormous amounts of water. The term *water of hydration* is used in chemistry to designate water in the crystal structure of a metal complex or salt which is not directly bound to the cation. Often the crystal properties are lost when this water is removed by heating. In a biological context, the importance of water of hydration is obvious. The study of the crystal structure of proteins has revealed that crystallized proteins consist of roughly 50% water by weight. If that water is removed, the enzymatic function of the protein is usually impaired or the structure of the active protein is changed or both. Moreover, cells are about 50–60% protein by dry weight. Clearly, most of the water in biological cells is water of hydration. In this book the concept of water of hydration is used in a broader sense than just water of crystallization to take effects in biological systems into account.

Water of hydration is considered here as all water that is not bound to the fluctuating network of bulk water (pure liquid water) but to other species in the aqueous phase. Often water of hydration is more strongly and more orderly bound than bulk water and therefore exhibits a

substantially different free energy. Water of hydration can be bound in the first or second solvation shell of the cation or anion of a solved salt, bound to the charged side chain(s) of amino acids in a protein, bound to hydroxyl groups in nucleic acids or polysaccharides, ordered to hydrophobic molecules or chemical groups, bound to a chain of water molecules in a membrane protein channel conducting protons, or bound to a solid surface. All these interactions can be classified according to their free energy, which can be quite large. Physical chemists doing experiments under vacuum conditions know, for example, that a high vacuum in a reactor can be only obtained by heating its walls well above 100°C for many hours during pumping (a process known as bake-out) to release the monolayers of water molecules that are tightly bound to the reactor's inner surface.

A simple means to obtain the water activity a_w is to measure the water vapor pressure above an aqueous solution. Fundamental thermodynamic relations show that

$$\frac{p_w}{p_w^*} = a_w = f_w \cdot x_w \tag{1.7}$$

where p_w is the water vapor pressure of the aqueous solution, p_w^* is the vapor pressure of pure water at the same external pressure and temperature as the solution, x_w is the mole fraction of water in the solution, and f_w is the dimensionless activity coefficient. The activities of ions dissolved in water (salt solution) are particularly affected by strong electrostatic interactions with each other and with the partial charges of water. This leads to low activities. In case of a sufficiently diluted, "ideal" solution, $f_w \approx 1$, and

$$\frac{p_w}{p_w^*} = x_w = 1 - x_2$$

$$x_2 = 1 - \frac{p_w}{p_w^*}$$

$$x_2 = \frac{(p_w^* - p_w)}{p_w^*} \tag{1.8}$$

According to this relation, the relative lowering of the vapor pressure of the solvent $(p_w^* - p_w)$ is equal to the mole fraction x_2 of the solved compound, for example, the salt in an aqueous salt solution (Raoult's empirical law from 1890).

DNA becomes disordered at low water activity. For example, DNA can no longer be used as a template for the synthesis of RNA at values of water activity (a_w) below 0.55; DNA strands break at $a_w = 0.53$. If the water activity of the medium surrounding a biological cell is very low, the cell must protect itself against osmosis by lowering its own internal (cytosolic) water activity. This can occur by adjusting the internal osmotic pressure of the cell via the synthesis of *compatible solutes*, low-molecular-weight organic compounds that are non-toxic at high intracellular concentrations but that bind water or by the active import of monovalent ions such as K⁺ or both. Typical compatible solutes used by bacterial cells include polyols such as glycerol,

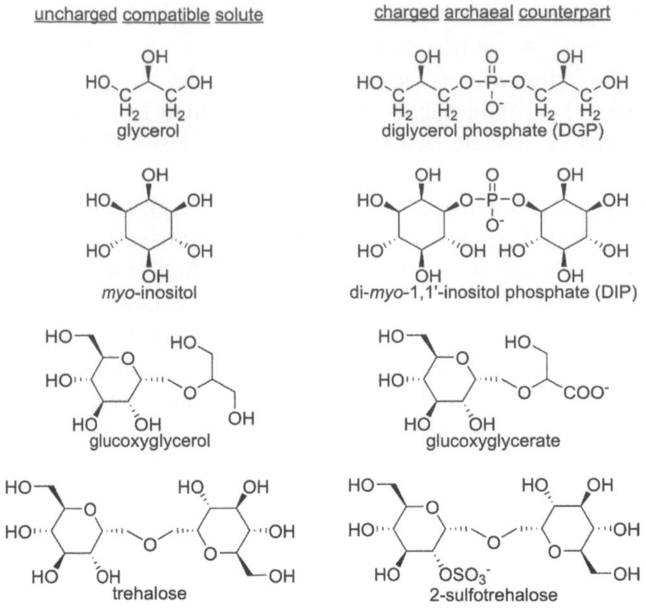

uncharged compatible solute

OH

HO. .OH
   C   C
   H₂ H₂
glycerol

OH

HO.   .OH

HO''   ''OH

OH
myo-inositol

HO   HO
      O
HO''
             OH
HO   OH
glucoxyglycerol

HO   HO   OH
      O
HO''   ''OH

HO   OH   OH
trehalose

charged archaeal counterpart

OH   O   OH

HO. . O-P-O. .OH
   C   C | C   C
   H₂ H₂ O⁻ H₂ H₂
diglycerol phosphate (DGP)

OH   O   OH

HO.   O-P-O.   .OH
         |
         O⁻
HO'' ''OH HO'' ''OH

OH   OH
di-myo-1,1'-inositol phosphate (DIP)

HO   HO
      O
HO''
             COO⁻
HO   OH
glucoxyglycerate

HO   HO   OH
      O
HO''   ''OH

HO   OSO₃⁻   OH
2-sulfotrehalose

FIGURE 1.23 **Compatible solutes are typically polyols.** Some common compatible solutes are used by bacterial and archaeal cells to regulate the water activity of the cytosol. The negative charges of the archaeal osmolytes are offset by the high K^+ concentrations of archaeal cells, which are often close to or exceeding 1M under normal growth conditions in low salt media. Redrawn from Roeßler M., Müller V. (2001) Osmoadaptation in bacteria and archaea: Common principles and differences, *Environ. Microbiol.* 3: 743–754.

mannitol, or inositol (Figure 1.23), while archaeal cells tend to use negatively charged variants thereof, the negative charges being offset by the high intracellular concentrations of K^+ that are typical of archaeal cells. For example, methanogens typically have cytosolic K^+ concentrations on the order of 1M.

Osmoregulation is the maintenance of ion homeostasis. Homeostasis is the maintenance of an optimal state of internal function in a cell. Osmoregulation requires energy expenditure. But the investment is worthwhile, for if a cell takes on too much water it will burst, and if it loses too much water, its proteins and nucleic acids will lose their biologically active structure (*denature*). An illustrative example of osmoprotection is found in the African resurrection plant *Craterostigma*, which can be completely desiccated and stored in the dry state for months but rehydrated and brought back to life. Photosynthetically active leaves of *Craterostigma* are about 80% water by fresh weight. Freeze-dried (lyophilized) leaf material of *Craterostigma* consists of 43% glycero-octulose by dry weight. In the desiccated state, the polyol supplies hydroxyl groups that protect proteins and nucleic acids from denaturation by effectively serving as a layer of organically bound water around the macromolecules.

1.2.5 THERMODYNAMICS DETERMINE WHETHER A CHEMICAL PROCESS CAN OR CANNOT TAKE PLACE

The molecules of living cells are in motion. Molecules move, attract each other, stick together, and react chemically, if and only if there is enough energy available for

these processes. The energy may be stored in the molecules themselves or come from the surroundings. The life process goes forward because energy is released. Life obeys the laws of thermodynamics.

The *First Law of Thermodynamics states that the total energy of a system* (a system is some subunit of the universe like a molecule or a biological cell) *and its surroundings* (the rest of the universe) *is constant*. The total energy is the sum of the *kinetic energy* and the *potential energy*. For example, molecules in motion represent a form of kinetic energy, regardless of whether this motion is random or directed. The kinetic energy of random molecular motion manifests as heat: faster motion implies greater heat. Alternatively, energy can be available as potential energy, be it the energy of a stone on top of a hill or the energy stored in a chemical bond. If a stone rolls down a mountain or an activated bond dissociates, then the potential energy is converted to kinetic energy—stone and molecular fragments move. The total energy however stays always constant.

Besides energy, another driving force for a process is its probability. Sugar will tend to distribute uniformly in a cup of coffee because this distribution is the most probable one—the one with the largest disorder or randomness. In thermodynamics, the *entropy S* is a measure of the degree of disorder. *The Second Law of Thermodynamics states that the total entropy of the universe* (system plus surroundings) *always increases:* $\Delta S_{total} > 0$. If, for example, the entropy in a biological cell decreases, because some highly ordered compartment has been synthesized in the cell, then this process is often accompanied by the release of heat from the cell, which enlarges the entropy of the surroundings, here the extracellular fluid. In quantitative terms,

$$\Delta S_{total} = \Delta S_{system} + \Delta S_{surroundings} \qquad (1.9)$$

In thermodynamics, the heat content is called *enthalpy* H. The release of heat from the system $-\Delta H_{system}$ increases the entropy of the surroundings more effectively in cold surroundings than at high temperatures where the surroundings are already highly disordered.

$$\Delta S_{surroundings} = -\Delta H_{system}/T \qquad (1.10)$$

where T is the temperature in Kelvin. Hence $\Delta S_{total} = \Delta S_{system} - \Delta H_{system}/T$. By multiplying both sides of the equation with $-T$, we obtain

$$-T\Delta S_{total} = -T\Delta S_{system} + \Delta H_{system}$$

In thermodynamics, $-T\Delta S_{total}$ is called *free enthalpy* or *Gibbs free energy* ΔG:

$$\Delta G = \Delta H_{system} - T\Delta S_{system} \qquad (1.11)$$

Note that the free enthalpy considers the total entropy—the entropy of the system and the entropy of the surroundings.

An increase of total entropy $\Delta S_{total} > 0$ implies a negative free enthalpy change $\Delta G = -T\Delta S_{total} < 0$. For the overall entropy of the universe to increase (Second Law of Thermodynamics), the free enthalpy change must be negative. A chemical reaction, for example, proceeds only, if $\Delta G < 0$. Otherwise, the entropy of the system plus the entropy of the surroundings does not increase. A chemical process with $\Delta G < 0$ is referred to as *exergonic*, a process with $\Delta G > 0$ is called *endergonic*.

Consider a single strand of RNA free to diffuse, rotate, and form different conformations by internal hydrogen bonds and stacking interactions. Its disorder, measured by the entropy, is high. For a double helix to form, two matched strands must aggregate and move together after aggregation. Furthermore, the double helix exists only in the helix conformation and does not internally fold to the numerous conformations of single strands. Hence, order increases upon the formation of the double helix, and entropy decreases. This is possible only if heat is released to the surroundings during aggregation. Double helix formation takes place if and only if $\Delta G < 0$, meaning the amount of heat released by the system $|\Delta H|$ must be larger than the amount of temperature-weighted entropy decrease in the system $|T\Delta S|$.

Similarly, a randomly folded protein exists in many different conformations, whereas an orderly folded functional protein exists only in one or a few conformations. Again, the transition from random folding to ordered structure implies a decrease in entropy, which is only possible, if enough heat is released. In contrast, the hydrophobic effect is connected to an increase in entropy. Hydrophobic molecules repel water and aggregate because the released water molecules are less ordered in bulk water than in the shell of the hydrophobic molecule. The overall increase of entropy favors the aggregation of nonpolar molecules in water. Even if the aggregation is endothermic, that is $\Delta H > 0$, the process will take place as long as $\Delta G < 0$, meaning $T\Delta S > \Delta H$.

Consider the reaction of CO_2 with hydrogen. The free reaction enthalpies depend on the state of CO_2 during the reaction (physisorbed in water or on the catalyst surface or dissociated as anion HCO_3^- or CO_3^{2-} depending on pH). In Section 2.4, we will see that the redox potential of hydrogen varies considerably with the physicochemical conditions (temperature, pH, H_2 partial pressure). This affects the free reaction enthalpies of CO_2 hydrogenation substantially. Anyway, experiments show that hydrogen is able to reduce CO_2 with good yields to formic acid at modest temperatures and pH values if the right catalyst is chosen (see Section 2.3).

The hydrolysis of acetyl phosphate is important for molecular activation by phosphate transfer (Figure 1.24) and its energetics will be discussed here in more detail.

The reaction is exergonic by -42.3 kJ/mol (Table 1.2) and will therefore proceed spontaneously in the presence of a suitable catalyst. The hydrolysis of ATP to ADP is exergonic by -30.5 kJ/mol. Accordingly, the energy released by the hydrolysis of acetyl phosphate is larger than the energy necessary for the reverse of the hydrolysis of ATP ($\Delta G^0 = +30.5$ kJ/mol). Hence, under standard physiological conditions (1 mol of reactants, 25°C, pH 7) ATP can be synthesized by the transfer of a phosphoryl group from acetyl phosphate to ADP (Figure 1.25).

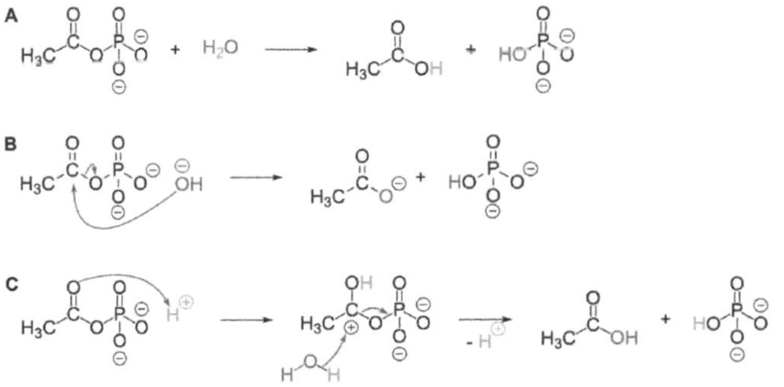

FIGURE 1.24 Scheme of hydrolysis of acetyl phosphate. Hydrolysis can occur in neutral water (A) or can be catalyzed by OH$^-$ in an alkaline aqueous medium (B) or can be catalyzed by H$^+$ in an acidic aqueous medium (C).

TABLE 1.2

Free Energy of Hydrolysis for Some Phosphorus and Sulfur-Containing Compounds

Compound	$\Delta G^{0\prime}$ for Hydrolysis (kJ/mol)
Phosphoenolpyruvate[a]	−61.9
Thioacetic acid[c]	−51.7
1,3-Biphosphoglycerate[a]	−49.4
ATP (to AMP + PP$_i$)[a]	−45.6
Creatine phosphate[a]	−43.1
Acetyl phosphate[a]	−42.3
Carbamoyl phosphate[b]	−39.3
Methylthioacetate[c]	−35.3
Acetyl-CoA[a]	−31.4
ATP (to ADP)[a]	−30.5
Glucose 1-phosphate[a]	−20.9
Pyrophosphate (PP$_i$)[a]	−19.3
Glucose 6-phosphate[a]	−13.8
Glycerol 3-phosphate[a]	−9.2

Note: Values are from (a) Berg *et al.* (2015), (b) Thauer *et al.* (1977), and (c) Chandru (2016). The superscript in $\Delta G^{0\prime}$ denotes the standard conditions commonly used in biology: 25°C, 1 atm pressure, 1 M of reactants, and pH 7.

ATP is the universal energy currency in biological cells. During the early molecular evolution, ATP was likely produced by stoichiometric phosphate transfer from high-energy phosphorylated organics (Table 1.2), a kind of ATP synthesis known as *substrate-level phosphorylation*.

We have already discussed that the free reaction enthalpies can be obtained from the difference of free formation enthalpies of products and reactants. Another means to obtain the energetics of a chemical reaction is to measure the concentrations of reactants and products at equilibrium. The concentration dependence of ΔG_R is

$$\Delta G_R = \Delta G_R^0 + RT \ln K' \qquad (1.12)$$

where ΔG_R^0 is the free reaction enthalpy, R is the gas constant, T is the absolute temperature, and $K' = [C][D]/[A][B]$

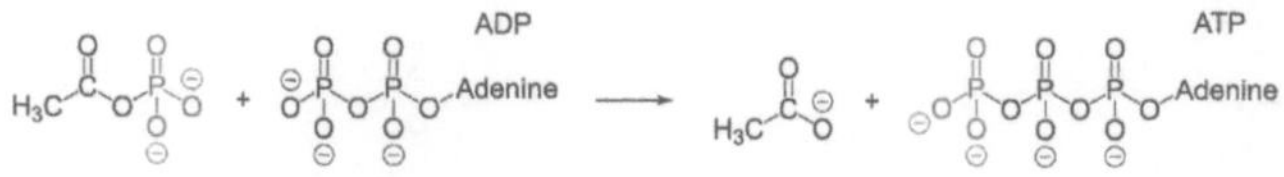

FIGURE 1.25 A phosphorylation reaction. Scheme of phosphate transfer from acetyl phosphate to ADP generating ATP (substrate-level phosphorylation).

is the concentration ratio for the reaction $A+B \rightleftarrows C+D$ with [] as the symbol for concentration in the units of mol/l. Hence, if product concentration is low and reactant concentration high, then $K' \ll 1$ and $\ln K'$ gets negative such that ΔG_R may become negative even if ΔG_R^0 is positive. Thus, a (slightly) endergonic reaction may become exergonic if pushed by high reactant and low product concentrations.

ΔG_R^0 is obtained as follows. At equilibrium, forward and backward reactions have the same rate, such that the concentrations of reactants and products do not change anymore with time. Hence at equilibrium, the free enthalpies for forward and backward reaction are equal (no net driving energy) and ΔG_R is zero

$$0 = \Delta G_R^0 + RT \ln K$$

$$\Delta G_R^0 = -RT \ln K \qquad (1.13)$$

where K is the concentration ratio at equilibrium, called *equilibrium constant*. If $\Delta G_R^0 < 0$, then $K > 1$ and products are favored over reactants. If $\Delta G_R^0 > 0$, then $K < 1$ and reactants are favored over products. For example, for the hydrolysis of acetyl phosphate at pH 7, we obtain with $\Delta G^{0'} = -42.3$ kJ/mol:

$$K = \exp(-\Delta G_R^0 / RT)$$
$$= \exp(+42300 \text{J/mol}/(8.314 \text{J/(K mol)} \cdot 298 \text{K}))$$
$$= 2.6 \cdot 10^7$$

Hence, $K \gg 1$ and the products acetate (Ac) and phosphate are favored by far over the reactant acetyl phosphate—the hydrolysis is nearly complete. The equilibrium lies on the side of acetate and phosphate. Conversely, we obtain $\Delta G_R^0 = -42.3$ kJ/mol from the measured acetyl phosphate and acetate (or phosphate) concentrations at equilibrium, if we can measure the small remaining quantity of acetyl phosphate sensitively enough.

The phosphorylation of ADP to ATP is endergonic by +30.5 kJ/mol (reverse of hydrolysis of ATP to ADP in Table 1.2). In thermodynamics, the *total free enthalpy change of chemically coupled reactions is equal to the sum of the free enthalpy changes of the individual reactions* (additivity rule). In our example, the energy-releasing hydrolysis of acetyl phosphate can drive the energy-consuming phosphorylation of ADP to ATP by (−42.3 +30.5) kJ/mol= −11.8 kJ/mol. Therefore, we get under equilibrium conditions:

$$K = \frac{[Ac][ATP]}{[AcP][ADP]} = e^{\frac{+11800\frac{J}{mol}}{8.314 \text{J}/(K\text{mol})\cdot 298K}} = 117$$

Thus, at a concentration of, say, 1mM AcP and 1mM ADP at equilibrium, we obtain approximately [Ac] = [ATP] = 10 mM. The phosphate transfer from acetyl phosphate to ADP is thermodynamically feasible. Of course, one can reach these equilibrium concentrations only if the reaction is *not kinetically hindered* (see next section).

Note that the free energies under standard conditions provide very useful guidelines, but under the conditions that exist in the environment or cells, the free energy changes of a chemical reaction can be substantially different than at standard conditions or even have the opposite sign.

1.2.6 KINETICS INDICATE HOW FAST A GIVEN PROCESS PROCEEDS

Thermodynamics tells us how much energy is released in a chemical reaction or must be expended to get the reaction going. A chemical reaction occurs spontaneously only if the process is exergonic, that is, $\Delta G_R^0 < 0$. Thermodynamics does not tell us how fast the reaction proceeds. The rate of a chemical reaction A + B → C + D is determined by

$$-d[A]/dt = -d[B]/dt = k\ [A][B] \qquad (1.14)$$

The rate −d[A]/dt is the decrease of the concentration of species A with time caused by the consumption of A in a chemical reaction. With increasing concentration of A and B, the probability of reactive collisions increases; hence, the rate is proportional to the concentrations of A and B. The proportionality constant k is called the *rate constant* and depends on temperature, nature, and concentration of ions in the solution, nature and surface area of an adsorbent or catalyst, pH, light intensity and wavelength in case of a photochemical reaction, and other factors. Rate equation 1.14 is called first order in species A and B because the exponents of their concentrations are 1, that is $[A]^1$ and $[B]^1$. The overall *reaction order* is the sum of these partial reaction orders, hence 2. More complex reactions can have non-integer experimental reaction orders.

If one of the reactants is in high excess like water in a hydrolysis reaction, then its concentration does not change appreciably during the reaction and is included in the rate constant, obtaining a *pseudo-first-order rate constant* $k' = k [H_2O]$. Then

$$-d[A]/dt = k'[A] \text{ for } [B] \gg [A]: k' = k[B]$$

Integration of $d[A]/[A] = -k'\ dt$ by using the relation $\int dx/x = \ln x$ results in

$$\ln([A]/[A_0]) = -k't \qquad [A]/[A_0] = \exp(-k't) \quad (1.15)$$

where $[A_0]$ is the initial concentration of molecule A and $[A] \equiv [A(t)]$ is the concentration of A at reaction time t. The half-life of the reaction $([A] = 1/2[A_0])$ is $\ln 2 = k'\ t_{1/2}$. Hence, as a rule of thumb, a (pseudo) first-order reaction with a rate constant of 10^{-3}/s will have a half-life $t_{1/2}$ of about 12 minutes. Hence, from the measured reaction half-life, we can obtain the rate constant k'. From the gradient of the plot $\ln([A]/[A_0])$ versus time t, we obtain a more exact value for k'. Most importantly, only the ratios of an analytical signal proportional to [A] must be measured as a function of time—the absolute concentrations of A have not to be known. Every second-order rate equation can be simplified to a pseudo-first-order equation by using an excess of one of the reactants in the experiment. This is the common way to obtain rate constants.

Consider the hydrolysis of acetyl phosphate (AcP) as an example. The pseudo-first-order rate constant is $k' = 4.8 \times 10^{-4}$/s at 25°C if the reaction is catalyzed by 0.1 M NaOH. Then at 5 mM initial acetyl phosphate concentration and 1 hour reaction time t (3600 s), we obtain

$$[AcP] = 5 \text{ mM} \cdot \exp\left(-4.8 \cdot 10^{-4} \text{s}^{-1} \cdot 3600 \text{ s}\right) = 0.89 \text{ mM}.$$

Therefore, $(5 - 0.89) = 4.11$ mM phosphate and the same concentration of acetate are generated after 1-hour hydrolysis.

1.2.6.1 Activation Energy

The rate constant k (or k') depends on temperature T according to the empirical *Arrhenius equation*

$$k(T) = A \exp\left(-E_a/RT\right) \qquad (1.16)$$

where A is *the Arrhenius factor*, E_a the *activation energy*, R the gas constant, and T the absolute temperature. The Arrhenius factor has been interpreted as the *collision frequency* of the reactants and the factor $\exp(-E_A/RT)$ as the proportion of collisions which leads to reaction. Hence $k(T)$ is the *frequency of reactive collisions* or probability of reactive collisions per second.

By taking the logarithm on both sites of the $k(T)$—equation and using the relations $\ln x \cdot y = \ln x + \ln y$ and $\ln e^x = x$ we obtain $\ln k = \ln A - E_a/RT$. Hence by plotting $\ln k$ versus $1/T$, the activation energy E_a can be obtained from the gradient of the plot and the Arrhenius factor A from the axis intercept. In so doing, de Meis and Suzano obtained an activation energy of 44.8 kJ/mol for the hydrolysis of acetyl phosphate.

The temperature dependence of the rate constant can also be obtained from theoretical considerations. During a chemical reaction, the atoms of the reactants rearrange their chemical bonds to form products. Let us assume that molecules A and B approach each other along a reaction coordinate. When A and B interact, the chemical bonds in A and B untighten. The free enthalpy of A and B increases up to a maximum value of $\Delta G_R^{0\ddagger}$ at the *transition state* (also called *activated complex*). Further motion along the reaction coordinate leads to the products C and D. The reaction displayed in Figure 1.26 is exergonic ($\Delta G_R^{0\ddagger} < 0$). Frequently, the more exergonic a reaction is, the lower the free reaction enthalpy to reach the transition state: $\Delta G_R^{0\ddagger}$.

Transition state theory connects the rate constant $k(T)$ with $\Delta G_R^{0\ddagger}$. It is essential that $\Delta G_R^{0\ddagger} = \Delta H_R^{0\ddagger} - T\Delta S_R^{0\ddagger}$; hence both enthalpic ($\Delta H_R^{0\ddagger}$) and entropic ($\Delta S_R^{0\ddagger}$) changes

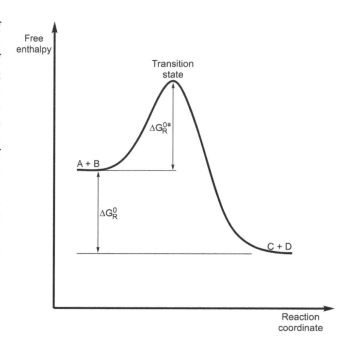

FIGURE 1.26 **Free enthalpy changes during a chemical reaction A + B → C + D**. $\Delta G_R^{0\ddagger}$ is the free activation enthalpy which is the difference between the free enthalpy of the transition state (activated complex) and the reactants. ΔG_R^0 is the difference between the free enthalpy of the products and the free enthalpy of the reactants (ΔG_R^0 is negative for an exergonic reaction). Forward and reverse reactions constantly occur, and the equilibrium concentrations are expressed by the equation $\Delta G_R^0 = -RT \cdot \ln K$.

are generally needed to reach the transition state. The fundamental equation of transition state theory is

$$\begin{aligned} k(T) &= (k_B T/h) \exp(-\Delta G_R^{0\ddagger}/RT) \\ &= (k_B T/h) \exp[-(\Delta H_R^{0\ddagger} - T\Delta S_R^{0\ddagger})/RT] \\ &= (k_B T/h) \exp(\Delta S_R^{0\ddagger}/R) \exp(-\Delta H_R^{0\ddagger}/RT) \end{aligned} \qquad (1.17)$$

where k_B, h, and R are the Boltzmann constant, Planck constant, and gas constant, respectively, and T is the temperature in Kelvin. The comparison with the Arrhenius equation shows that we can equate $\Delta H_R^{0\ddagger}$ with the experimental activation energy E_a and can calculate $\Delta S_R^{0\ddagger}$ from the experimental Arrhenius factor A.

1.2.6.2 Water Activity and Rate of Hydrolysis

Consider the influence of water activity on the OH$^-$-catalyzed hydrolysis rate of acetyl phosphate (Figure 1.24). The experiments of de Meis and coworkers have shown that, at very low water activity, the hydrolysis rate constant increased by three orders of magnitude and both E_a and $\Delta S^{0\ddagger}$ (determined from the experimental Arrhenius factor) decreased to nearly zero. The water activity was lowered by adding the organic solvents dimethyl sulfoxide (CH$_3$)$_2$S=O or ethylene glycol HO–CH$_2$–CH$_2$–OH, which binds water with strong hydrogen bonds. Water activity was determined from the measurement of water vapor pressures (see Section 1.2 about hydrolysis).

How can this apparent paradox of faster hydrolysis at lower effective water concentration be explained? Remember that the reaction is catalyzed by OH$^-$ ions. It is well known from spectroscopy that OH^- is stabilized by strong hydrogen bonds to surrounding water molecules in an aqueous medium (water solvation shell). For nucleophilic attack of OH^- on acetyl phosphate (Figure 1.24), this shell must be removed, and this costs energy. *The removal of the water shell around OH^- may well be the rate-limiting step during OH^--catalyzed hydrolysis of acetyl phosphate.* We propose that the decrease of water activity facilitates the removal of the water shell around OH^-, which dramatically lowers the activation energy of OH^--catalyzed hydrolysis and rises the hydrolysis rate. The decrease of activation entropy with decreasing water activity can also be explained by this model. A decrease of activation entropy implies an increase of order in the transition state region, which can be explained by the transfer of less ordered bulk water to highly ordered water surrounding the organic solvents, here dimethyl sulfoxide or ethylene glycol.

Similar to the above example, the activation energy and entropy of ATP hydrolysis and the free enthalpies of ATP and pyrophosphate hydrolysis decrease with decreasing water activity and the equilibrium constants shift to the product site. The rate of substrate-level phosphorylation may thus strongly depend on water activity. Probably, the OH^--catalyzed rate of phosphate transfer from acetyl phosphate to ADP rises at low water activity because the water shells around the anions become loose.

Interestingly, water activity might also be important in the F_1-ATPase of mitochondria, which squeezes ADP and inorganic phosphate to *tightly bound* ATP (see Chapter 7). ATP does not dissociate from the tightly bound complex. It has a low (very negative) free enthalpy of hydrolysis and a high hydrolysis rate, pointing to low water activity. In the subsequently formed *loosely bound* complex, dissociation of ATP from the complex is easy, the free enthalpy of hydrolysis is high, but the hydrolysis rate is slow, probably due to higher water activity.

Consider another example of the importance of water activity. Most medications that we use today are inhibitors of some sort. The mode of action in a medication is typically founded in simple molecular interaction. A chemical compound in the medication (the active ingredient) binds a protein and thereby changes its conformation, its ability to perform its normal reaction, or its ability to undergo normal interactions with other molecules. Binding inhibits its activity. The binding of the active ingredient is almost always driven by entropy. As the ingredient binds the protein, water molecules that are bound by the protein in an orderly manner are displaced by the active ingredient. This increases water entropy and is an important part of the driving "force" behind the binding. Consider that almost half of a typical protein's mass consists of water bound to the protein.

1.2.6.3 Kinetics on Surfaces

Many investigations into prebiotic chemistry have been performed using reactions in free solution. Very few reactions in cells occur in free solution. Almost all biological reactions occur at the interface between the aqueous phase of the cytosol and the solid phase of a protein (an enzyme). Residues of the protein interact with the reactants to hold them in such an orientation to one another that the reaction can take place. Rather than demanding random collisions with the

correct angle and velocity to permit a successful reaction, enzymes orient the reactants to one another in such a way as to promote their reaction, without themselves becoming altered during the reaction. Enzymes are catalysts; they are a special case of *heterogeneous catalysis*. In *heterogeneous catalysis*, the phase of reactants or products (gas or liquid) differs from the phase of the catalyst (liquid or solid). In *homogeneous* catalysis, both are in the same phase. Therefore, the adsorption of the reactants to the catalyst must be the first step in heterogeneous catalysis.

Before there were enzymes, chemical reactions were still possible. They had to have taken place, and because of this there can be no doubt: the building blocks of life that were ultimately needed to synthesize the first enzymes *cannot have arisen* without chemical reactions. Enzymes do not make impossible reactions possible; they just accelerate reactions that tend to occur anyway. The prebiotic (geochemical) reactions that took place before there were enzymes need not have occurred in free solution, though. It is more likely that they occurred on the surfaces of natural minerals.

Mineral surfaces, especially surfaces of transition metal compounds (see the next section), can catalyze many organic reactions and probably did so extensively before enzymes entered the picture of early evolution. Even at some distance from the surface of a catalyst, the reactant molecule can undergo weak interaction with surface atoms and become attracted, for example, through charge. The free energy decreases by weak van der Waals forces such as dipole–dipole interactions, induced dipole interactions, and dispersion (see Section 1.2 on noncovalent bonding). This kind of interaction is called *physisorption* and is drawn in blue in Figure 1.27. Physisorption describes molecular interactions between the liquid phase reactant and the solid phase surface that do not involve the sharing of electron pairs in covalent bonds.

Upon further approach, the electron clouds of surface atoms and adsorbates can begin to overlap. If the interacting atoms are so disposed, this stage of adsorption leads to *chemisorption* (drawn in red in Figure 1.27). Chemisorption describes binding between adsorbate and the surface atoms in the form of shared electrons that form chemical bonds. Chemisorption can leave the adsorbate intact (*molecular chemisorption*) or take place in such a way that one or more bonds break (*dissociative chemisorption*). For example, if H_2 binds to the surface of a ferrous catalyst, the H–H bond breaks and Fe–H bonds form (drawn red in Figure 1.27A). For this to occur, the decrease of free enthalpy due to the Fe–H formations must be larger than the energy necessary to break the H–H bond. However, an energy barrier can exist between the physisorbed and the chemisorbed state (Figure 1.27B), which affects the rate of adsorption. If the barrier to chemisorption is too large, the adsorbate can desorb from the surface because physisorption forces are generally weak on the order of 10–40 kJ/mol.

The chemisorbed atoms need not be held in place. If the surface structure permits, they can hop around on the surface by dissolving their bond to the surface and reforming a new, energetically equivalent bond to neighboring surface atoms. During this diffusion-like process in two dimensions, they can meet other adsorbed molecules or molecular

A

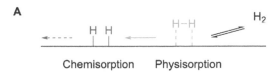

B

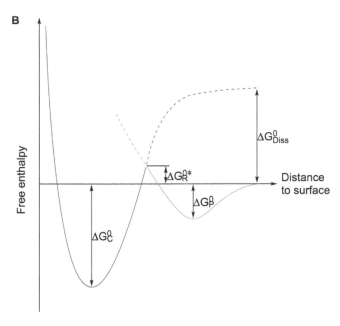

FIGURE 1.27 **Physisorption and dissociative chemisorption of molecules on surfaces.** (A). The physisorption of H_2 is drawn in blue, chemisorption in red, and diffusion of H atoms (H·) on the surface as a dashed red line. (B) The course of free enthalpy during physisorption of a diatomic molecule approaching a surface is drawn in blue, and dissociative chemisorption is drawn in red. The dashed red line indicates desorption of the chemisorbed atoms from the surface. ΔG_P^0, ΔG_C^0 and $\Delta G_R^{0\ddagger}$ are the free enthalpies of physisorption, chemisorption, and transition state, respectively. ΔG_{Diss}^0 is the free enthalpy of dissociation of the chemisorbed diatomic molecule to free atoms.

fragments on the surface and combine with them to form products, which then desorb. We will meet heterogeneous catalysis over and over again in this book.

With the help of scanning tunneling microscopy (STM), it is possible to observe reactions of molecules chemisorbed on catalytic surfaces, a technology pioneered by Gerhard Ertl. An instructive example of a surface reaction is the catalytic oxidation of carbon monoxide (CO) with oxygen atoms (O) to carbon dioxide (CO_2) on a single crystal platinum (111) surface, whereby the term (111) describes the plane of the crystal lattice that is imaged (Figure 1.28). The oxidation was monitored by STM. The images directly display the localization of the reacting atoms and molecules in separate domains and reaction only at the domain boundaries.

In living cells, almost all reactions are catalyzed by enzymes. Enzymes do not shift equilibria; they just lower the activation energy of a chemical reaction. They order reactants in three dimensions, not just two as in surface catalysis. At the origin of life, there were no enzymes to start, but catalysts were still required because reactions of randomly colliding molecules have too low rates for the reactions essential to life. We will talk about enzymes in Chapters 5 and 6. Here we look at simpler catalysts and catalysis on surfaces.

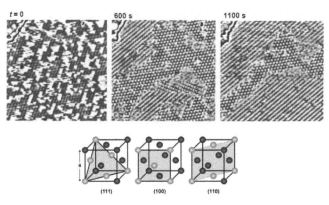

FIGURE 1.28 **Images of atoms on surfaces.** Scanning tunneling microscopy (STM) images of a submonolayer of adsorbed oxygen atoms and CO molecules at 247 K. The oxygen atoms are visible as dark dots forming small islands on a platinum (111) crystal at $t = 0$ seconds (s) and after a continuous supply of CO from the gas phase at $t = 600$ s and $t = 1100$ s. The images taken at 600 s and 1100 s show that CO and O are not randomly distributed but localized in distinct domains. The reaction between them is restricted to the boundaries between the domains. The area covered by oxygen atoms decreases during the reaction because of conversion to CO_2. The reaction product CO_2 desorbs directly upon formation and was not detected. The structure at the upper left edge of the images is an atomic step on the Pt surface. The lower panel shows the different faces of a cubic face-centered crystal lattice that are indicated by the terms (111), (100), and (110). Printed with permission from Wintterlin J. *et al.* (1997) Atomic and macroscopic reaction rates of a surface-catalyzed reaction, *Science* 278: 1931–1934.

1.2.7 TRANSITION METALS ACTIVATE INERT GASES LIKE H_2 AND CO_2 AND CATALYZE CHEMICAL REACTIONS

Transitions metals are elements of groups 3 to 12 of the periodic table (Table 1.3). These elements play an important role in the chemical industry as catalysts for reactions

TABLE 1.3

The Periodic Table of Elements

Note: Redox reactive transition metals important for biocatalysis are boxed in blue; they usually belong to the fourth period of the periodic table; Mo and W are exceptions. Lanthanides (marked by *) are now known in as cofactors of enzymes (see Daumann 2019).

involving carbon and nitrogen. They are also used by biological systems in enzymes. Transition metals were very probably (if not certainly) involved as catalysts in the chemical reactions from which life arose. As mineral catalysts in prebiotic reaction networks, they can activate molecules by molecular or dissociative chemisorption and catalyze chemical reactions. Typical transition metal catalysts in chemical research and industry are vanadium (V [Ar]$3d^34s^2$), chromium (Cr [Ar]$3d^54s^1$), manganese (Mn [Ar]$3d^54s^2$), iron (Fe [Ar]$3d^64s^2$), cobalt (Co [Ar]$3d^74s^2$), nickel (Ni [Ar]$3d^84s^2$), copper (Cu [Ar]$3d^{10}4s^1$), and ions of Zinc (Zn [Ar]$3d^{10}4s^2$). Transition metals have a partially filled d-shell (< 10 d-electrons) or generate cations with incomplete d-shells. The unfilled d-orbitals of transition metals have a complex geometry and can hybridize to form a variety of energetic states and electron orbital configurations. This natural property of transition metal atoms makes them powerful and versatile catalysts both in industry and in biological reactions.

Transient changes in oxidation state are typical for catalysis provided by transition metals, both in organic chemistry and in biological reactions. In biological systems, Zn always occurs as Zn^{2+} and is redox inert; that is, it never changes its oxidation state unlike most of the other metal catalysts. Iron is by far the metal most commonly used by enzymes. Copper is a widespread cofactor typical for O_2-reducing enzymes (including those in our own respiratory chain). The metals V, Mn, Co, and Ni are often used in enzymes of primitive anaerobes, and they tend to occur in reactions that were crucial in early biochemical evolution, for example, CO_2 reduction (Co, Ni), N_2 reduction (V), or O_2 synthesis (Mn). In almost all cases, transition metals from the fourth period (row) are used in biological catalysts; an exception is Cr (period 4, group 6). We know of no example in which Cr is used as a natural catalytic component of enzymes, although some enzymes can use chromium compounds as substrates. Below Cr in the periodic table, molybdenum (Mo [Kr]$4d^55s^1$) and tungsten (W [Xe] $4f^{14}5d^46s^2$) from the fifth and sixth periods are used by enzymes. We will encounter many examples in later chapters.

We have already seen that by the time there was water on the early Earth, the reservoir of carbon was mainly CO_2. We will see in the next chapter that H_2 from serpentinization was the main reductant available for converting CO_2 into organic molecules. H_2 was probably also the only reductant that was continuously available on a geological timescale to get organic chemistry started on a planet with oceans twice as deep as today's. We will see in later chapters that the most primitive forms of metabolism among strict anaerobes are built on the reduction of CO_2 with H_2 both for carbon and for energy metabolism. The reason is that in the reaction of H_2 with CO_2, the equilibrium lies on the side of reduced carbon compounds. The overall reaction of H_2 with CO_2 releases energy, but both gases are quite stable and not prone to react. If we put H_2 and CO_2 in an inert reactor in the laboratory and heat it to 100°C overnight, nothing will happen; no reaction will take place. But if we add a metal catalyst like elemental iron or an iron–nickel alloy or an iron oxide like Fe_3O_4 and some water, then we will get formate, acetate, and pyruvate in physiologically relevant amounts. The metals act as catalysts. This kind of reaction is important for understanding the interface between the geochemical and biochemical world, so

let us look at it in more detail from the standpoint of physical chemistry.

We consider the bonding between transition metals [M] and inert molecules like H_2 and CO_2 and start with the [M]–H_2 bonding. We will see that this bonding activates H_2 for a chemical reaction. The spatial distributions of electrons in d-orbitals of transition metals resemble cloverleafs pointing along and between the x, y, and z-axes of a three-dimensional coordinate system. As an example, Figure 1.29B shows the shapes of d-orbitals pointing in the z-direction (d_{z^2}) and between the x- and z-direction (d_{xz}). The orbitals of hydrogen are displayed in Figure 1.29A. If two H atoms are far apart, they have identical atomic orbitals 1s, each occupied by one electron. While approaching, the electron orbitals (wave functions) of the H atoms begin to overlap. However, two electrons are not allowed to have the same set of quantum numbers, that is, they are not allowed to occupy the same space and spin in the same direction. This is the

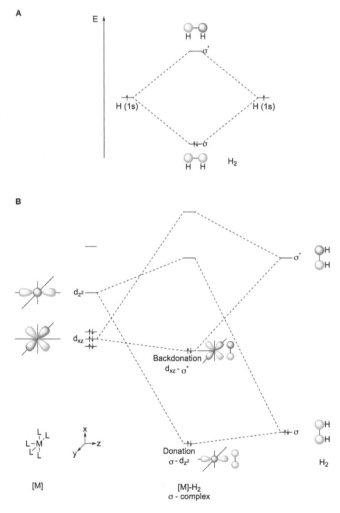

FIGURE 1.29 Transition metal–H_2 correlation diagram. (A) The correlation diagram of H_2. (B) The transition metal iron (Fe) with electron configuration [Ar]$3d^64s^2$ correlates with its lowest unoccupied dz^2 orbital to the doubly (electron) occupied σ-orbital of H_2 to form a bonding molecular orbital (electron donation σ→dz^2) in the side-on σ-complex M–H_2 (here M≡Fe). The complex is further stabilized by electron back donation from the highest-occupied d-orbital (d_{xz}) of Fe to the lowest-unoccupied orbital of H_2 (an antibonding σ*-orbital). The backdonation d_{xz}→σ* destabilizes the H_2 bond and may eventually lead to M–H_2 dissociation into reactive M–H complexes.

Pauli exclusion principle, which is a fundamental principle that organizes matter; it is the reason why electron shells fill up the way they do and why elements of the same group in the periodic table tend to have similar chemical behavior in terms of forming bonds.

Therefore, the spherical s-orbitals of hydrogen split into two *molecular orbitals*, one lower in energy (*bonding orbital* σ occupied by the two electrons spinning in opposite direction ↑↓) and one higher (antibonding orbital σ* with blue and red in Figure 1.29A, indicating a change of sign of the atomic orbitals). The stable σ orbital promotes the bonding between the two H atoms to H_2; the higher-energy σ* orbital opposes bonding if it is occupied (atomic orbitals with opposite signs do not overlap). In H_2, the two electrons occupy the lower-energy bonding orbital σ so that H_2 is more stable than the separate H atoms. Indeed, H_2 is a very stable molecule and chemically almost inert. Hence to make H_2 ready for reaction and decrease its bond strength, some electron density must be transferred into the antibonding σ* orbital and this is exactly what the transition metal does.

The energies and shapes of the transition metal [M] orbitals and of adsorbates like H_2, CO, CO_2, and N_2 can be plotted in the same diagram. Then, orbitals of the same symmetry are correlated to form the orbitals of the complex. An example is the formation of the σ—*complex* M–H_2 with H_2 side-on to M (Figure 1.29B). Electron donation σ→d_{z2} and backdonation d_{xz} → σ* leads to strong molecular chemisorption of H_2 on the iron surface, weakening of the H_2 bond, and eventually dissociation into two H atoms covalently bound to the metal as M–H complexes. The H-atoms (a proton with one electron) can hop on the transition metal surface (for example, iron: [M]≡Fe). In contrast to their parent H_2, the two H atoms (H·) are very reactive.

Transition metals M can catalyze chemical reactions on surfaces but are also embedded in clusters with ligand L like CO or H_2O (L_n in Figure 1.30A). Depending on M and L, the H_2 bond distances in the L_nM–H_2 complex increase to a different extent. A stable elongated H_2 complex can form or the H_2 bond in the complex can rupture to form a *dihydride*. The H–H bond cleavage can be *homolytic* ($H_2 \rightarrow 2H$) or *heterolytic* ($H_2 \rightarrow H^+ + H^-$) with a hydride ion H^- bond to M (Figure 1.30B). A hydride ion is a proton with an electron pair.

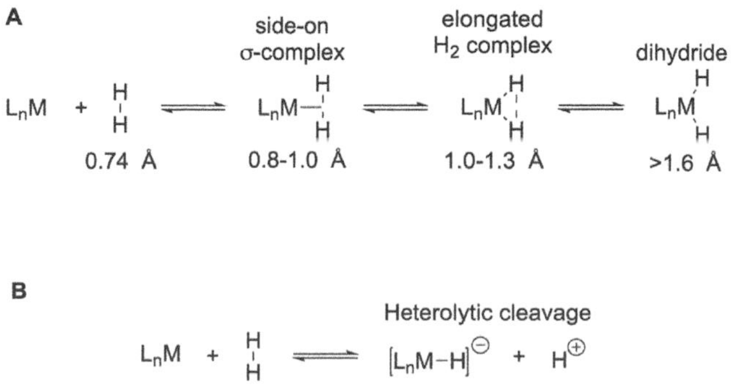

FIGURE 1.30 Homolytic (A) and heterolytic (B) cleavage of H_2 in a transition metal complex. Redrawn from Kubas G. J. (2007) Fundamentals of H_2 binding and reactivity on transition metals underlying hydrogenase function and H_2 production and storage, *Chem. Rev.* 107: 4152–4205.

Transition metal compounds ML_n can exist in many oxidation states due to the rather low energy gap between these states. They are known to form complexes with small molecules like H_2 and to catalyze homogenously in a common phase with the reactants or heterogeneously. Many active centers of enzymes are transition metal compounds ML_n, often with several metal atoms M (*metal clusters*). Organic chemists have mastered the art of constructing elaborate ligands that favor particular electronic configurations in the metal's outer shell, thereby generating specificity and efficiency in the way the metal complex reacts with given molecules. This is why transition metals are so widely employed as catalysts in industry.

Several factors contribute to the catalytic capabilities of transition metals. First, the concentration of reactants bound to a transition metal cluster or a small piece of solid is higher than in bulk solution or gas phase, because the reactants come into close contact with one another on the same surface. Adsorption to the catalyst not only activates the partners for reaction; it increases the rate of the reaction. Second, the internal bonds of the reactants are weakened upon adsorption because the atoms of the reactants share electrons with surface atoms. Because of that, the activation energy to form products is smaller.

So far, we have only considered the activation of H_2. What about the activation and reactivity of CO_2? The carbon atom in CO_2, due to its positive partial charge ($O^{\delta-}$–$C^{\delta 2+}$ –$O^{\delta-}$), is susceptible to nucleophilic attack by an electron-rich atom or molecule. Alternatively, an electron-poor species can perform an electrophilic attack on the negatively polarized oxygen atom. But, either way, CO_2 is quite inert and must be activated to react chemically. Enzymatic CO_2 fixation in nature involves a daunting molecular machinery, whether powered by light (photosynthesis) or chemical energy (chemosynthesis). Before there was life, how was it possible to activate CO_2 without sophisticated enzymes?

We now know that bonding to transition metals without protein backbone can perform this task also. Either the C–O bonds have to be weakened by bonding to the metal or the linear CO_2 molecule must be bent to increase its molecular energy. For the analysis of this activation process, we look closer at the molecular orbitals of carbon dioxide. Carbon has valence orbitals $2s^2 2p^2$ and oxygen $2s^2 2p^4$ (Figure 1.31).

While the atoms approach, their orbitals begin to overlap and form two bonding σ-orbitals $σ_s^b$ and $σ_x^b$ (from 2s of C and p_x of C with the p_x orbitals of O; b for "bonding"), two π-orbitals $π_z^b$ and $π_y^b$ (from the atomic orbitals p_z and p_y perpendicular to the C–O direction; shown in Figure 1.31 is $π_y^b$), two nonbonding π-orbitals $π_z$ and $π_y$, which are the highest occupied molecular orbitals (HOMOs), two antibonding π-orbitals $π_z^*$ and $π_y^*$, which are the lowest-unoccupied molecular orbitals (LUMOs), and two unoccupied σ-orbitals $σ_s^*$ and $σ_x^*$.

What happens when these CO_2 orbitals approach the d-orbitals of a transition metal? In the case of an electron-poor metal like iron (Fe [Ar]$3d^6 4s^2$), the d_{z2}-orbitals are unoccupied (Figure 1.29B and Figure 1.32B, middle image), whereas the d_{xy}, d_{xz}, and d_{yz} orbitals are occupied with two electrons (Figure 1.32B, bottom). Upon close contact between metal and CO_2, the occupied d-orbitals transfer

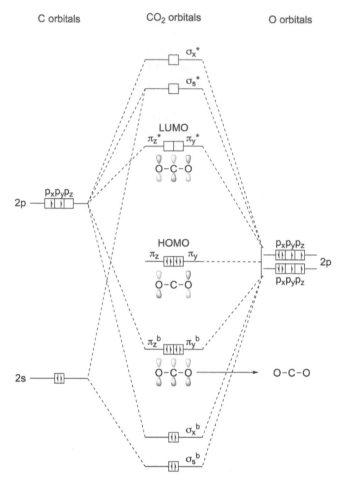

C orbitals   CO₂ orbitals   O orbitals

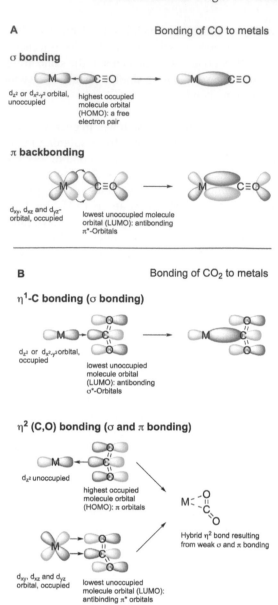

FIGURE 1.31   Correlation diagram of CO₂. Blue and red indicate different signs of the atomic orbitals constituting the molecular orbitals. For the assignment of the orbitals, see text.

A   Bonding of CO to metals

FIGURE 1.32   Bonding of CO (A) and CO₂ (B) bonding to transition metals. Blue and red indicate different signs of the atomic orbitals constituting the molecular orbitals. (A) σ-bonding and π-backbonding activate CO for reaction. (B) η¹-C bonding f.e. in nickel and η² bonding in iron weaken the bonds in CO₂ and activate them for reaction. For the assignment of the orbitals, see the text.

charge to the antibonding LUMO of CO₂ by interaction with its C=O bond. This interaction weakens the C=O bond and, due to the asymmetry of the interaction, bends CO₂. The occupied bonding π_z^b (or π_y^b) orbital transfers electron density to the unoccupied d_{z^2}-orbital of iron (Figure 1.32B, middle). These weak σ- and π-bonds lead to a so-called *hybrid η² bond*. Carbon dioxide is activated by the formation of this bond and is now ready for (re)action.

In electron-rich transition metals like nickel (Ni [Ar]3d⁸4s²), the d_{z^2}-orbital is occupied and can transfer charge to the antibonding LUMO of CO₂ (Figure 1.32B, top). This so-called η¹-C bonding again weakens the CO₂ bonds. Both iron and nickel can activate CO₂.

Carbon monoxide is activated by similar interactions (Figure 1.32A). CO transfers electron density from its HOMO (a free electron pair at the C atom) to the d_{z^2}- or $d_{x^2-y^2}$-orbitals, which are unoccupied in the case of the "electron-poor" transition metals like iron, manganese, chromium, or vanadium (Figure 1.32A, top: σ-*bonding*). In iron, the d_{xy}, d_{xz}, and d_{yz} orbitals are fully occupied and transfer charge to the antibonding LUMO of CO, weakening the CO bond (Figure 1.32A, bottom: π-*backbonding*). These additive σ- and π-contributions lead to very strong bonds of CO to iron and nickel. Examples are nickel carbonyl Ni(CO)₄ and iron carbonyl Fe(CO)₅, both of which are volatile liquids (boiling points 43°C and 103°C, respectively). Here,

the metal in its elemental state is coordinated solely by the carbon of CO. The strong toxicity of CO is based on its effectively irreversible binding to the iron atom of heme in hemoglobin, abolishing the latter's ability to bind and transport O₂.

In "electron-rich" metals like nickel or copper, the d_{z^2}- or $d_{x^2-y^2}$-orbitals are occupied and can transfer electron density to antibonding π*-orbitals of CO in a side-on approach. This interaction weakens the CO bond and makes nickel a particularly good activator of CO.

If H₂ and CO₂ are both present in a gas mixture and bind to a nickel–iron alloy in water, for example, the natural mineral awaruite (Ni₃Fe), which consists of metals in the elemental state and which is formed in serpentinizing hydrothermal vents, they can become activated on the surface and they can react. The main products are formate, acetate, and pyruvate, essential components of microbial metabolism and intermediates as well as the products of the acetyl-CoA pathway of CO_2 fixation. It is remarkable that a simple alloy of transition metals will produce such a clear carbon copy of microbial metabolism. The simplest explanation is that microbial metabolism is a carbon copy of preexisting reactions catalyzed by transition metals alone. We will see in later chapters that, in the biological pathway, Fe and Ni form bonds with H_2 and CO_2 during catalysis.

1.3 LIFE DEVELOPED FROM ABIOTICALLY SYNTHESIZED ORGANIC MOLECULES

Life on Earth is based on organic molecules dissolved in water. The geochemical conditions at the end of the Hadean time must have been suitable to produce sufficient amounts of organics from inorganic compounds—otherwise life would not have had a chance to develop at that time. In the following, we discuss the requirements which the geological sites had to satisfy for the development of organic molecules and primitive life.

1.3.1 LIFE'S EMERGENCE REQUIRED ENERGY, C, H, N, O, CATALYSTS, AND CONCENTRATING MECHANISMS

A few simple criteria must be fulfilled for life to develop—at least for life as we know it, the only kind that we can discuss meaningfully. The second law of thermodynamics states that order in a system (a regulated metabolic network or a biological cell) can be achieved and maintained only if heat is released into the surroundings. Hence, every living system must be fed with a continuous supply of energy to maintain the entropic balance. This criterion applies to all forms of life we know. At the origin of life on Earth, carbon was available as CO_2. Any organics from space that were delivered during accretion went through magma oceans at the moon-forming impact and were converted to CO_2. Any organics that might have been delivered from space after the moon-forming impact either did not survive their own impact or, if they did, were diluted in oceans that were twice as deep as today's.

Many organic reactions proceed only if barriers with appreciable activation energies are surmounted. Often this requires catalysts to lower these barriers and higher temperatures to overcome them. Heat is a ubiquitous and always applicable energy source to overcome thermal barriers of reactions. Light is a possible other energy source, however, chemically only useful for photochemical reactions. The use of visible light for the generation of a steady supply of energy-rich electrons for reduction is difficult and required the evolution of pigments and complex enzymes to evolve as we will see in Chapter 7. In the absence of enzymes, the very first primordial catalysts can only have been inorganic materials. Minerals containing transition metals catalyze many important carbon reactions efficiently and with surprising specificity.

For chemical reactions to occur at some appreciable rate, reactant concentrations must be sufficiently high. This requires the concentration of a limited number of prebiotic reactants in a small volume. The means to achieve that are *thermal diffusion*, *convection*, and *chromatographic separation* (Section 3.2). Pores and microchannels of volcanic rocks have sufficiently small volumes, thermal gradients, and surface activity to concentrate prebiotic reactants for chemical reactions. For example, consider two compounds A and B moving in a cube volume V (Figure 1.33). How much time is needed for molecule A to reach B and react? In a reversible reaction, products can accumulate only if the reaction rate is higher than the decay rate. We estimate 100 seconds for the decay time and require $t = 20$ s for the approach time (the reaction is assumed to be fast compared to the approach: a *diffusion-controlled reaction*).

The average time τ needed for A to move along a path of length d, before it collides with a neighboring molecule and changes direction, can be calculated from the *Einstein–Smoluchowski equation*

$$\tau = \frac{d^2}{2D} \tag{1.18}$$

where D is the diffusion coefficient of molecule A in water. The diffusion coefficient in a liquid can be calculated from the *Stokes–Einstein equation* $D = k_B T/(6\pi\eta R)$. With a viscosity coefficient of water $\eta = 8.9 \times 10^{-4}$ kg m^{-1}s^{-1}, temperature $T = 300$ K, Boltzmann constant $k_B = 1.38 \times 10^{-23}$ JK^{-1}, and $R = 2.28$ Å $= 2.28 \times 10^{-10}$ m (Stokes radius of the amino acid alanine), we obtain $D = 1.09 \times 10^{-9}$ m^{2}s^{-1}.

The cube volume V can be divided into cubic cells of edge length d such that V/d^3 cells match in V. Assuming that A has to pass all cells in V to reach B, we obtain the average time t for a random path A → B

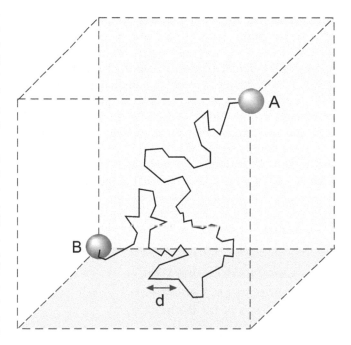

FIGURE 1.33 **Random path of a molecule.** Molecule A moves a path of average length d before it collides with a water molecule in an aqueous medium and changes direction (d is called the mean free path length). Molecules A and B reside in a box of volume V representing a rock pore (drawn dashed).

$$t = \tau \cdot \frac{V}{d^3} = \frac{d^2}{2D} \cdot \frac{V}{d^3} = \frac{V}{2D \cdot d} \qquad (1.19)$$

Molecules A and B react only at a specific orientation, which can be considered by a *steric factor* α. Then we obtain a box volume (pore volume)

$$V = t \cdot 2D \cdot d \cdot \alpha = 4.4 \cdot 10^{-20} \, m^3$$

for $\alpha = 10^{-2}$, $t = 20$ s, $D = 1.09 \times 10^{-9} m^2 s^{-1}$, and $d = 10^{-10}$ $m = 0.1$ nm (estimate of mean free path in liquid water). The edge length of the cube volume V is $(4.4 \times 10^{-20} \, m^3)^{1/3} = 3.52 \times 10^{-7}$ m $= 352$ nm. The rock pore cannot be larger than a few hundred nanometers for reaction time (approach time) to compete with decay time. For comparison, the diameter of a bacterium is in the 1 μm $= 1000$ nm range; however, many bacteria are more rod-like than spherical. The pore opening must be on the order of molecular size such that the residence time in the pore can compete with the approach time. Many pores in volcanic rocks fulfill these requirements.

Such considerations lead quickly to the inevitable conclusion that solid phase catalysis in small volumes was a prerequisite for organic reaction systems to evolve at the origin of life. Reactions by random collisions in a large volume are too slow for reaction sequences to develop. Catalysts introduce specificity into reactions, both for enzymes and for inorganic catalysts. Enzymes in cells are a special case of heterogeneous catalysis in a submicron volume.

These thoughts are even more valid for the evolution of elaborate metabolisms. Chemical networks must be confined to small reactors like pores of solids or micelles to have a chance to develop. The fluid concentrations in the pores must be enriched by thermal gradients (Section 3.2) at low water activity to favor base-catalyzed chemical activation and condensation reactions. These requirements make a *primordial oceanic soup* of organic molecules a highly improbable scenario for the development of life. Molecules produced by lightning and solar radiation in a reduced Earth's atmosphere and by geochemical processes in the Earth's crust will be diluted very fast in the open ocean and not be able to react anymore at a reasonable rate. Chemical enrichment in micropores of volcanic rocks and successive development of a primitive metabolism of simple organic molecules are a much more plausible scenario for chemical evolution. Even more chemical possibilities ensue if the molecules are not just concentrated in pores but also synthesized there.

1.3.2 Organic Molecules Were Synthesized by Geochemical Processes in the Earth's Crust

For the synthesis of organic molecules, a constant source of carbon–hydrogen–nitrogen–oxygen (CHNO) compounds and **energy** in confined volumes (we will call them pores, inorganic compartments, or microcompartments) of catalytically active minerals are key. By **energy** we mean chemical energy, that is, *far from equilibrium* conditions.

For example, we recall that in the reaction of H_2 with CO_2 to organic molecules, the equilibrium lies on the side of reduced carbon compounds. Under physiological conditions, the reaction to acetate is exergonic by -104 kJ/mol:

$$4H_2 + 2HCO_3^- + H^+ \rightarrow CH_3COO^- + 4H_2O \qquad (1.20)$$

while the reaction of H_2, CO_2, and a thiol (the –SH group of coenzyme A, for example) to the level of the energy-rich thioester acetyl CoA is also exergonic by -67 kJ/mol:

$$2CO_2 + 4H_2 + CoASH \rightarrow CH_3COSCoA + 3H_2O \qquad (1.21)$$

Both reactions tend to go forward if (1) suitable catalysts are present and (2) the reactants on the left side of the equation are present but those on the right have not yet accumulated. If the reaction is allowed to go to completion, the products on the right side accumulate according to their equilibrium concentrations, and the forward and reverse reaction rates become equal, that is, the reaction comes to a halt, and it reaches equilibrium. The reactions of living cells do not reach equilibrium, only those of nonliving cells do. Living systems stay far from equilibrium; that is what drives the chemical reaction of life forward. If the products on the right side react further, for example, through Claisen condensation of acetyl moieties in the thioesters or through reaction with other compounds, phosphate, for example, to generate acyl phosphates as it occurs in cells, then at some point a set of reaction products will arise that is at equilibrium and constrained by the stability of the end products of the reactions. No further net reactions will take place; forward and reverse reactions will occur at the same rate.

If, however, the educts H_2 and CO_2 on the left side are permanently supplied, then the products on the right are continuously formed. If these products are simultaneously consumed by downstream reactions and end products removed and if they are waste, then a steady state equilibrium develops. In metabolism, these downstream reactions are sometimes called *pulling reactions*. The steady-state equilibrium is the central overarching principle of how the core physiological chemistry of cells operates (Figure 1.34). At the same time, the products of each reaction are more stable than their educts. Energy-rich compounds react to form increasingly stable products. This is how metabolism works, and it is also very likely a property of the first reaction networks from which cells arose.

Carbon dioxide, which was present in massive amounts in the Hadean ocean (Section 1.1) and carbonates in Earth's crust, are the main sources of carbon. H_2 was continuously available *in situ* from serpentinization. CO_2 must be hydrogenated to organic acids, ketoacids, alcohols, and other organic molecules for chemical networks to develop. Organic molecules and life's building blocks, including amino acids and nucleic acids, can in principle also be synthesized from hydrogen cyanide (HC≡N), acetylene (HC≡CH), or cyanoacetylene (HC≡C–C≡N) instead of CO_2 and H_2 (Section 4.6). Indeed, calculations have shown

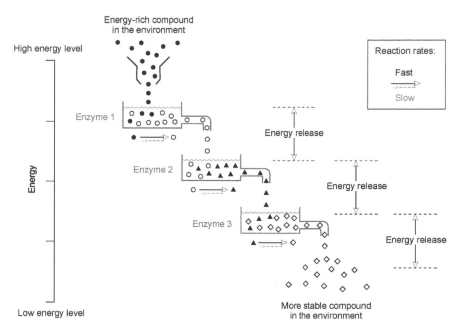

FIGURE 1.34 Steady-state equilibrium in a metabolism-like system. To maintain a steady-state equilibrium, energy-rich substrate (food) has to be continuously supplied and removed by downstream reactions to more stable (energy-poor) final products. If the end products are waste, they must be removed by diffusion or by another member of the food chain; otherwise, the waste accumulates and the final reaction will become very slow until it finally stops, bringing the other upstream reactions to a halt as well. Redrawn from Richter G. (1988) *Stoffwechselphysiologie der Pflanzen*, Thieme, Stuttgart.

that these gases are produced as main components in nitrogen- and carbon-rich magmas at 1200°C. In the Hadean age, these ultra-reduced gases could have erupted in volcanoes or bubbled out in surface hydrothermal pools like the thiocyanate (N≡C–S⁻) species observed in sulfur-rich Yellowstone springs. While the organic chemistry with cyanides is fascinating, efficient, and demanding, there is no hint of a cyanide-based metabolism in any known organism on the Earth, neither in bacteria or archaea nor in higher organisms. If cyanides were used to synthesize organic molecules in the Hadean age, then it was a dead-end path, one that we do not want to follow in this book. In contrast, both carbon dioxide and hydrogen play a major role in the metabolisms of contemporary bacteria and archaea. Some acetogenic bacteria use reaction (1.20) to synthesize ATP. Some methanogenic archaea live on molecular hydrogen and carbon dioxide and synthesize ATP by producing methane in a reaction that under standard conditions is exergonic by −131 kJ/mol.

$$4H_2 + CO_2 \rightarrow CH_4 + 2H_2O \qquad (1.22)$$

Both acetogens and methanogens are strict anaerobes and they use the very ancient acetyl CoA pathway both in energy metabolism and in carbon metabolism (reaction 1.21). Phylogenetic studies that allow us to reconstruct the physiology of the last universal common ancestor of all cells (*LUCA*) point to an anaerobic, thermophilic, CO_2-fixing, and H_2-dependent species that used the chemistry of acetyl CoA pathway, using iron-containing catalysts in a metabolism resembling that of *methanogens* and *acetogens*. Microbes with that kind of metabolism still inhabit the Earth's crust today.

However facile and specific HCN or H_2C_2 chemistry might be for synthesizing the basic building blocks of life, all life we know arises from CO_2, so to explain how the life we know arose, we have to focus on CO_2 chemistry. This raises the question of how the elements H, C, N, O, P, and S enter metabolism in the first place. To answer that question, it might be a good spot to explain how the elements enter metabolism in anaerobic autotrophs. Of course, today, there are many forms of metabolism and most cells live from the chemicals that comprise the dry mass of other cells. But we can ask how the inorganic forms of the elements enter metabolism.

Carbon enters metabolism as CO_2. There are six pathways of CO_2 fixation of which only one occurs both in bacteria and in archaea, the acetyl CoA pathway, which is the most ancient of the six; we will encounter it frequently in this book. Nitrogen enters metabolism as NH_3 via two routes, either via the conversion of the carboxyl group of glutamate into glutamine or via reductive amination of 2-oxoacids, for example, the conversion of 2-oxoglutarate to glutamate. N_2 has to be converted to NH_3 for incorporation; the same is true for nitrates and nitrites. The universally conserved nature of N assimilation into organic compounds via NH_3 indicates that this is how the process occurred in the first cells. H can readily exchange with water, but it enters metabolism mainly through H_2O, NH_3, and H_2 in H_2-dependent organisms. O enters metabolism through any number of routes, via CO_2 and H_2O in anaerobes. S enters metabolism as sulfide via cysteine synthesis. If the side-chain hydroxyl group of serine is activated as *O*-phosphoserine or *O*-acetylserine, then nucleophilic attack by HS⁻, with phosphate or acetate as the leaving group, generates cysteine. Phosphorus enters metabolism as phosphate, mainly during the process of ATP synthesis from ADP.

1.3.3 CARBON-RICH METEORITES CONTAIN MOSTLY UNREACTIVE POLYAROMATIC HYDROCARBONS

Meteorites are another source of organic molecules that in principle could have contributed to the origin of life. During the time 4.1–3.8 Gya, myriads of celestial bodies of different sizes hit the Earth. Today, impacts of solid objects from outer space are much rarer. A meteorite is wreckage from a comet, asteroid, or meteoroid that reaches the surface of a planet or moon. A chondrite is a meteorite that survived the passage through the atmosphere unmodified. Less than 5% of all chondrites contain carbon compounds. These carbonaceous chondrites contain usually about 2% organic matter and 0.3% inorganic carbon.

Much of what we know about the composition of carbonaceous chondrites comes from the Murchison meteorite, which fell to the Earth in 1968 near Murchison, Australia, and remained sufficiently intact to provide about 100 kg of material for chemical analysis. Radiometric dating determines silicon carbide (SiC) particles from the Murchison meteorite to be 7 billion years old, about 2.5 billion years older than our solar system. About 14,000 elemental compositions extending to millions of diverse molecular structures have been detected by high-resolution spectroscopy of probably unaltered interior parts of Murchison fragments. The carbon content of the meteorite is about 2% by weight. The composition of that carbon is given in Table 1.4. Organics from space that reach the Earth today in meteorites are mainly (~94%) polyaromatic hydrocarbons (PAHs), which is the collective name for inert macromolecular material with a chemical structure consisting of huge, honeycomb-like systems of aromatic rings. About 2% of the carbon from carbonaceous chondrites is in the form of monocarboxylic acids; about 1% is CO_2, roughly 0.5% is hydrocarbons, and the remaining 2.5% is distributed across the other compound classes, shown in Table 1.4. Importantly, with the exception of CO_2, CO, and CH_4 the table shows compound classes, not specific compounds, and within each compound class, a myriad of different molecular masses and isomers are found, giving rise to what is called complete structural heterogeneity. The carbon from meteorites is mostly inert and cannot even be used by microbes to fuel fermentations because it is too reduced (see discussion in the following).

We saw earlier in this chapter that in the very early history of the Earth, the moon-forming impact originally generated magma oceans of 2500°C, a temperature that no carbon compound could survive intact. That impact converted all of the carbon on the Earth into CO_2 in the gas phase (atmosphere). We see from Table 1.4 that most of the carbon delivered from meteorites after the moon-forming impact was inert as PAH. If carbon compounds for life did not come from meteorites, where did they come from? There have been (recurrent) suggestions in the literature that any molecule identified in interstellar dust presents a possible source of carbon for life. Life did not, of course, arise in interstellar dust. It arose on Earth. In terms of carbon sources, all the Earth had to offer was CO_2. Carbon dioxide is, however, the best carbon source imaginable for the origin of life, perfect in fact because (1) CO_2 was unquestionably present on the early Earth in abundance, (2) CO_2 is extremely stable in the gas phase and aqueous phase (longevity), but (3) it is highly

TABLE 1.4

Classes of Organic Compounds in the Murchison Meteorite and Their Abundances

Compound	Abundance (ppm) (µg/g)
Macromolecular material (polyaromatic hydrocarbons)	14,500
Carbon dioxide	106
Carbon monoxide	0.06
Methane	0.14
Hydrocarbons: aliphatic	12–35
aromatic	15–28
Acids: monocarboxylic	332
dicarboxylic	25.7
α-hydroxycarboxylic	14.6
Amino acids	60
Alcohols	11
Aldehydes	11
Ketones	16
Sugar-related compounds	~60
Ammonia	19
Amines	8
Urea	25
Basic N-heterocycles (pyridines, quinolines)	0.05–0.5
Pyridine carboxylic acids	>7
Dicarboximides	>50
Pyrimidines (uracil and thymine)	0.06
Purines	1.2
Benzothiophenes	0.3
Sulfonic acids	67
Phosphonic acids	1.5

Source: Data from Sephton M. A. (2002) Organic compounds in carbonaceous meteorites, *Nat. Prod. Rep.* 19: 292–311.

reactive and readily reduced to organic compounds in the presence of H_2 and transition metal catalysts, both of which the early Earth had in ample supply (reactivity), (4) CO_2 is a pure and uniform starting compound for synthesis, meaning that a given catalyst will produce the same kinds of products from CO_2 over and over again (specificity), and (5) CO_2 is the starting point of all biological organic synthesis today (biogenicity).

Clearly, CO_2 is the ideal starting material for life's origin. We do not need to look in interstellar dust for the starting material for origins. We exhale it, CO_2, at every breath. The idea that life arose from CO_2 is called the theory of autotrophic origins, an idea that has a long history dating back well over 100 years. Both Ernst Haeckel in 1903, and Konstantin Mereschkowsky in 1910, made the case for an origin of life from CO_2, but that was long before anyone had the slightest idea how cells, enzymes, or metabolism work. In more recent literature, Günter Wächtershäuser made the case for autotrophic origins. Based on current knowledge about magma oceans in the Earth's early history and the composition of carbon from space, we can be rather certain that the primordial crust, formed from freshly cooled magma, and the primordial oceans that covered it, harbored nothing in the way of an "organic broth" as J.B.S. Haldane termed it in 1929. There were no sequestered, preexisting organic feedstocks that could be used either as an energy supply or as a made-to-order warehouse of biomolecular building blocks to get the life process started. Instead, organic molecules had to be synthesized from inorganic material containing the elements C, H, N, and O, particularly carbon dioxide, hydrogen, and nitrogen gases dissolved in water. This actually makes the origins issue easier, because it casts the problem in a setting of reactants whose existence on the early Earth is certain (H_2, CO_2, and N_2 in water), whose reactions to form organic molecules are exergonic, and whose sequential reactions could, in principle, be very similar to the chemical reactions germane to modern cells with primitive metabolism, as we will see in Chapter 3. But the reactions of those gases require catalysts.

The amounts of organic matter that meteorites and comets can deliver to Earth are remarkable. At an estimated fraction of 19% refractory organic material, the comet Halley with a radius of 5 km could deliver 3×10^{13} kg organic material to the Earth, which is about 10% of the Earth's present biomass. Even if the fraction of organic matter is much smaller, carbonaceous meteorites have certainly contributed to organic material on Earth, to some extent. There are, however, two properties that exclude the direct use of organics from meteorites for biological purposes. First, their major portions are polyaromatic hydrocarbons and aliphatic organic acids, which are too reduced to be fermented. Second, organics from celestial bodies are structurally too heterogeneous to be biodegradable. Consider that fermentation reactions are *disproportionation* reactions in which a carbon substrate of intermediate oxidation state is converted into a more reduced and more oxidized form. For example, glucose converts to ethanol, where the carbon is more reduced than in glucose and to CO_2, where it is more oxidized. Substrates that are already reduced, like lipids, cannot be fermented. There are trace amounts of less reduced compounds like amino acids (D and L forms) that could in principle be fermented. However, the structures of these compounds are completely heterogeneous, which means that all possible isomers for a particular chemical formula can be detected. This, in turn, requires the making of a digestive enzyme for each isomer of each compound— a vast collection of racemases, isomerases, mutases, and the like. The energetic yield (benefit) of trying to ferment trace amounts of individual components present at ppm levels in a complex medium does not come close to outweighing the cost of making the enzymes needed to process the substrate.

There is an additional problem with any contribution of organics from meteorites to the origin of life. We recall the criteria for the evolution of a chemical metabolism, which we discussed above. Biochemical networks can develop only if energy and precursor molecules for organic synthesis are *continuously* delivered to a small, almost closed volume and if the reactions are continuously exergonic, that is, far from equilibrium rather than at equilibrium (where

energy is no longer released). Biologically relevant molecules must be synthesized perpetually, because life is a dynamic equilibrium of synthesis and decay (if we do not eat anymore, we die). Even if its organic compounds remain intact, a meteorite impacting somewhere in the oceans or on land is an isolated, dead body. Its soluble organics will be slowly washed out in the ocean or in terrestrial waters, be diluted there and ultimately decay because there is no renewal of the same compounds. It is a one-off source of organics with a reactivity approximating that of graphite. Organics from meteorites are neither suitable as food for organisms nor as metabolic precursors molecules for chemical evolution. Organics from meteorites might have existed in the Hadean Oceans, but a significant contribution of extraterrestrial organics for the origin of life can be excluded.

We also exclude photosynthesis as the source of the first organic molecules. Photosynthesis is a complex chemical process requiring efficient electron transfer agents which were not yet available at the origin of life. Without these agents or suitable voltage gradients, the photoexcited electrons just recombine with the electron hole generated by photoexcitation and their energy is released as thermal heat. Photoexcitation can be used to generate a proton gradient at the cell wall, but the harvesting of this gradient to chemically stored energy requires a sophisticated molecular machine like the ATP synthase, which was not available at the origin of life. Thermal energy accelerates most chemical reactions and is therefore a natural source of (kinetic) energy to promote reaction rates for synthesis of the first organic molecules. But independent of rates, the most natural and the most likely source of energy at the origin is chemical energy, reactions between educts with an energy state higher than the expected organic products and in far from equilibrium conditions.

1.3.4 HYDROGEN-PRODUCING HYDROTHERMAL SYSTEMS COULD HAVE BEEN SITES FOR ABIOGENESIS

Where did the organic molecules in the Hadean era come from, if not imported by meteorites or generated by HCN chemistry or photochemistry? The most natural scenario of primordial organic synthesis, that is, the scenario with the greatest overall similarity to the far-from-equilibrium metabolic processes underlying life itself, is the reduction of CO_2 with H_2 in submarine hydrothermal fields within the crust. In the following we discuss two kinds of hydrothermal vents: (1) the well-known *black smokers*, first discovered in 1977, the effluent of which comes into close contact with magma at spreading zones and thus emerges with water temperatures exceeding 400°C, usually with an acidic pH, and (2) off-ridge systems like *Lost City*, which were first discovered in 2001, where the water circulating through the vent does not come into contact with magma, being mainly heated by the crustal thermocline instead, and emerges at 70–90°C, alkaline and with large amounts of H_2 from serpentinization. We can safely assume that hydrothermal fields were abundant in the Hadean because this era was geologically much more active than the present Earth.

Submarine vents emit geothermally heated water (*hydrothermal fluid*) and can form chimney-like structures around the crust–ocean interface of the vent. The aqueous fluid comes from seawater that is drawn down into cracks in the Earth's crust to a depth of several kilometers and becomes heated such that it covets back to the sea floor. Black smokers typically reside next to spreading zones where magma is close to, or emerging from, the submarine crust. Magma is molten rock and typically has a temperature on the order of 1200°C. At black smokers, the circulating water comes into close contact with magma. Because of that, the effluent of black smokers can exceed 400°C. At the high pressure of 200–400 bars in 2000–4000 m ocean depth, no boiling occurs, the water stays liquid, and it is superheated. Above 407°C and 298 bar, seawater (with 3.2% average NaCl content) becomes supercritical. The critical point of pure water is 374, 15°C at 221 bars. The deepest-known vent site is at ~5000 m depth in the Cayman Trough of the western Caribbean Sea between Jamaica and the Cayman Islands (see Figure 1.35). It reveals sustained supercritical venting at 407°C and 2.3 wt% NaCl. Supercritical water has the density of liquid water but the viscosity of steam. It can dissolve solids much better than normal water. In close proximity to magma, water may be exposed to such extreme conditions, become supercritical, and dissolve rock minerals more efficiently than normal water. Supercritical water is also less polar and generally has lower water activity (water available for chemical reactions).

In contemporary **black smokers**, the hydrothermal solution is typically acidic and contains sulfides as well as cations of transition metals like iron, copper, manganese, and zinc in addition to volcanic gases like CO_2. The critical point of CO_2 is 74 bars and 31°C. Above 74 bars, CO_2 is supercritical even at high temperatures and dissolves nonpolar substances extremely well. Water, however, is almost immiscible with supercritical CO_2: about 0.006 mol of water dissolves in 1 mol of supercritical CO_2 at 200 bar total pressure and 40°C. Driven by convection, the heated hydrothermal fluid of black smokers pushes upward and flows through a fissure in the crust into the ocean. The hot, acidic volcanic water meets the ~2°C cold seawater and the solved minerals precipitate as a dark solid giving the impression of smoke (Figure 1.35). Solids, primarily pyrite (iron disulfide FeS_2), precipitate at the vent opening and grow several meters per year to roughly cylindrical chimney structures of up to 60 m in height and 180 m in diameter. Some groups of these funnel-like structures extend across areas of up to several kilometers—they are called hydrothermal fields. A hallmark of black smokers is their short lifetime. They form directly over spreading zones, where magma emerges from the mantle to form new crust and pushes oceanic plates apart. As the newly formed crust cools, circulation through the vent subsides. After 10–100 years, black smokers are either crushed under their own weight, plugging the vent openings, or volcanoes erupt and block the vents with lava. Sometimes the vent sources cool down and ebb. However, due to plate tectonics, new crust fissures arise permanently, and new vents develop. The lifetime of individual black smoker-type vents is on the order of decades.

Black Smokers support a 10^4–10^6 higher density of organisms than the typical environment of the abyssal plain on the deep ocean floor. The source of that productivity is

FIGURE 1.35 A black smoker. The vent is from the Von Damm field, adjacent to the Piccard field at the Cayman Rise in the Caribbean. The effluent is pH ~5. For scale, the white tadpole-shaped objects covering the surface of the vent are eyeless shrimps, each ca. 5 cm long, that live from the prokaryotes covering the surface. The vent sits at a depth of 4960 m, and the temperature of vent effluent in the formation is 390–400°C. Information about the geochemistry of the vents at the Cayman Rise in McDermott JM *et al.* (2018) Geochemistry of fluids from Earth's deepest ridge-crest hot-springs: Piccard hydrothermal field, Mid-Cayman Rise. *Geochim. Cosmochim. Acta* 228:95–118. Printed with permission, courtesy of Jeffrey Seewald, WHOI/NSF/ROV Jason 2020 © Woods Hole Oceanographic Institution.

chemosynthetic bacteria that obtain energy from the reaction of reduced sulfur compounds effluent, H_2S and SH^-, with O_2 in seawater. Gutless tube worms up to 2 m tall can live at black smokers and absorb nutrients produced by H_2S-oxidizing bacteria that live as endosymbionts in their tissues. The chemosynthetic bacteria obtain their carbon from CO_2 and are ultimately digested by the worms as food, replacing the function of the worm digestive system, such that the worms have lost the need to maintain a mouth and gut. Instead, they have red plumes that contain hemoglobin. The hemoglobin transports hydrogen sulfide as a growth substrate to the bacteria. The gill tissue of clams living near black smokers also contains such bacterial endosymbionts. Some O_2-dependent methanotrophs feed on methane emitted from the black smokers. Chemosynthetic bacteria are the basis of the black smoker food chain that includes giant tube worms, blind crabs, shrimps, snails, clams, sea stars, and diverse fish. Again, the black smoker food chain is mainly fueled by the reaction of H_2S with O_2. Of course, O_2 is the product of oxygenic photosynthesis, an invention of cyanobacteria that emerged over a billion years after the end of the Hadean. In the Hadean there was no O_2 to supplement food chains at black smokers.

Submarine hydrothermal fields like *Lost City* were first discovered near the mid-Atlantic ridge in 2001. Their underlying geochemistry and chemical composition are different from that of black smokers. Fluids with dissolved calcium, magnesium, and silicate flow out of fissures in the crust and typically precipitate as calcium magnesium carbonate upon contact with CO_2-bearing cold seawater (Figure 1.36). The precipitates form chimneys around the fissures. At Lost City, mantle rocks from greater than 6 km depth are exposed to the seafloor due to magma upwelling and store residual heat from the mantle, which can be harnessed by the penetrating seawater. The mantle rocks contain olivine,

which consumes water by serpentinization—an exergonic (energy-releasing) process that we will encounter in later chapters. The amount of heat liberated is directly proportional to the amount of water consumed. Serpentinization consumes about 300 l of water per m^3 of reactive rock and produces 660,000 kilojoules of heat per m^3 of rock. Without water cooling, serpentinization can raise the rock temperature by about 260°C. Hence it is not only the geothermal thermocline (the increase in temperature of roughly 25°C per kilometer depth in the crust) that drives the hydrothermal system of Lost City but exergonic (energy-releasing) chemical reactions between seawater and mantle rocks.

During serpentinization at Lost City, the rock volume increases by 20–40% because of rock–water interactions, leading to open fractures and seawater migration along grain boundaries to fresh portions of olivine, which allows serpentinization to continue. The heat released by serpentinization and the residual heat from the mantle rocks underlying Lost City ensures the high temperatures of the fluid inclusions, which are necessary for the dissolution of olivine and reaction to serpentine, brucite, and magnetite to continue. The volume increase lowers the rock density from 3.3 to 2.7 g/cm^3 such that the rocks get lighter and more voluminous and lift to greater elevation exposing fresh peridotite. Radiocarbon (^{14}C) dating of the carbonate structures and sediments in Lost City shows an age of at least 30,000 years. There is enough peridotite in the massif below Lost City to drive serpentinization for further hundreds of thousands, if not millions, of years.

Serpentinization leads to a high concentration of hydroxide in the vent fluids (pH 9–11) and concentrations of H_2 up to 15 mM, CH_4 1–2 mM, and very little if any CO_2 at 40–90°C fluid temperature. Biofilms of methanogenic archaea produce CH_4 from H_2 and formate in anoxic regions; regions exposed to O_2 in seawater are rich in methanotrophs that

FIGURE 1.36 A vent at Lost City Hydrothermal field. Left panel: An image of a Lost City vent showing a calcium–magnesium carbonate flange and ca. 70°C warm water emerging from it. The field of vision is roughly 4 m. Printed with permission courtesy of Susan Q. Lang, University of South Carolina/NSF/ROV Jason 2018 © Woods Hole Oceanographic Institution. Right panel. An image of the Beehive structure at Lost City hydrothermal field, warm effluent emerging. Printed with permission of Deborah Kelley, University of Washington and Woods Hole Oceanographic Institute. Information about Lost City hydrothermal field in Kelley D. S. *et al.* (2001) An off-axis hydrothermal vent field near the Mid-Atlantic Ridge at 30°N, *Nature* 412: 145–149.

oxidize methane. Because of serpentinizaiton, H_2 is high in both in the interior of the chimneys (90°C and higher) and at flanges (40–70°C). Furthermore, dense colonies of bacteria with high diversity populate the outside of the chimneys with grey filamentous strands several centimeters in length. Serpentinizing alkaline hydrothermal fields are suitable reactors for primordial organic synthesis, including the synthesis of ammonia and amino acids, as we will see in later sections. Serpentinization can also take place on land, when submarine crust composed of mafic rocks is pushed to the surface (obduction, see chapter 7) and fed by freshwater sources. Several such systems have been studied (see Figure 1.37).

Terrestrial hydrothermal vents can also arise from surface water penetrating into fissures of Earth's crust near magma. Several kinds of terrestrial vents exist. *Geysers* are periodically spouting hot springs known from Yellowstone National Park and from Iceland. The periodic spouting develops from narrowings in the eruption channel. The pressure of the water column above the narrowing hinders the boiling of the geothermally heated water. At temperatures substantially above the normal boiling point, gas bubbles start to form, and the gas expansion forces the overheated water upward in a sudden burst. Geyser periodicity depends on the position and form of the narrowing in the eruption channel. *Fumaroles* are vents in the crust which emit steam and gases such as CO_2, SO_2, HCl, and H_2S. When overheated water pops up from the ground, its pressure drops and steam forms. CO_2-rich fumaroles are called *mofettes*; H_2S-rich fumaroles are called *solfataras*. The ambient air oxidizes H_2S to sulfur and SO_2, which dissolves in the emitted water to sulfurous acid (H_2SO_3). The acid and the water vapor dissolve the soil forming a bubbling sludge vessel. In the Hadean period, of course, in the absence of oxygen in the atmosphere, emitted H_2S would not be oxidized.

Hydrothermal fields similar to Lost City were likely abundant in the Hadean because the chemistry of the crust and mantle were not fundamentally different from today's, and the Hadean Earth was volcanically much more active than today. Arndt and Nisbet estimate that in the Hadean

there was nearly twice as much ocean water covering the surface as today, making the oceans twice as deep. Hence, in the late Hadean, when life began, most of the volcanoes and hydrothermal vents, if not all, were located underwater. Later in Earth's history, a large part of the water in the Hadean oceans was bound in the crust by serpentinization and constantly returned to the mantle by subduction. While vast parts of the surface of the Earth might have been sterilized by meteoritic impacts, the chemical environment within deep-sea hydrothermal vents was probably not affected much, and prebiotic molecules or primitive life could persist there.

1.3.5 Weighing Evidence for the Age of Life

There are two main lines of evidence for the existence of early life on the Earth: isotope (geochemical) and microfossil (palaeontological). The geochemical evidence rests mainly on stable carbon isotope data, ^{13}C, determined from samples taken from formations of known age, dated by U/Pb methods or similar (see Section 1.1.8). The microfossil evidence rests upon the identification of structures that look like organisms from formations of known age.

Is there clear and direct evidence for life on the Earth in the Hadean era (before 4.0 Ga)? The answer at present is no. Evidence for the first life on the Earth is notoriously difficult to identify, mainly because rocks >3.5 Ga that could have preserved such evidence are extremely rare, making the search for life in such rocks a challenge. On top of that, evidence for the most ancient life on Earth is always "in the news" and hence subject to sensationalism and media activity, for the simple reason that everyone wants to know about early life on Earth. It is always a hot topic. This makes it difficult for editors at major journals to make a call on whether or not to publish papers claiming to present evidence for the most ancient life, giving the papers a seal of endorsement from Journal X. The editors have to ponder the merits of a paper submitted for publication that purports to present evidence for very ancient life, and they have to make an

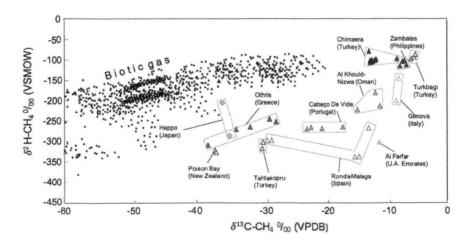

FIGURE 1.37 **Carbon and hydrogen isotope depletion of methane from different sites.** Shown are typical ^{2}H and ^{13}C depletions of biotic methane from methanogenesis or thermogenesis compared to abiotic methane from land-based serpentinization sites. VSMOW: Vienna Standard Mean Ocean Water, a standard for deuterium content. VPDB, versus Pee Dee Belemnite, a geochemical standard for inorganic carbon taken from inorganic carbonate in Cretaceous cephalopod shells of the PeeDee formation South Carolina. Redrawn from Etiope G. (2017) Abiotic methane in continental serpentinization sites: An overview, *Procedia Earth Planet. Sci.* 17: 9–12; Etiope G., Schoell M. (2014) Abiotic gas: Atypical, but not rare, *Elements* 10: 291–296.

informed decision when it comes to the question of what to accept for publication, knowing that what they decline might end up on the cover of Journal Y, with the new record for the most ancient life. Many reports of new evidence for ancient life appear in major multidisciplinary journals. The evidence stays the same though: in rocks older than 3.5 Ga, one tends to find either microfossils or carbon isotope signatures but not both in the same sample.

1.3.5.1 Carbon Isotopes

Evidence from carbon isotope ratios is widely accepted as evidence for life for good reasons. Biotic carbon is depleted in the heavier carbon isotope ^{13}C, because isotopically light CO_2, $^{12}CO_2$, reacts slightly faster in metabolism and therefore preferentially accumulates in biomass during CO_2 fixation. Because ^{13}C is always depleted in biological material, biological carbon is always too light relative to inorganic carbon (abiotic methane can be very light though; see Figure 1.37). The degree of depletion depends upon the pathway that was employed to fix the CO_2. Typical depletions of ^{13}C in organic material are on the order $-5‰$ to $-40‰$. Carbon isotope evidence consistent with the operation of the acetyl-CoA pathway is found in rocks that are 3.8 billion years old and even 3.95 billion years old. However, light carbon isotope evidence for ancient life can also be contested with regard to biogenicity because some abiotic processes can also lead to an enrichment of the lighter carbon isotope, hence the carbon isotope evidence by itself is not always unique to life. Additional isotope signatures, such as deuterium, can be instructive though (Figure 1.37). At the same time, abiotic processes that can generate such light isotopic signatures are geothermal equivalents to Fischer-Tropsch type syntheses, an industrial process for the synthesis of hydrocarbons from H_2 and CO using magnetite (a mineral that occurs in serpentinizing systems) as the catalyst.

Modern hydrothermal systems are still capable of abiotic CO_2 reduction. Lost City produces abiotic methane at about 1 mmol/kg vent effluent and abiotic formate. A number of hydrothermal systems have been found to produce abiotic methane. The abiotic gas is enriched in ^{13}C relative to methane from methanogenesis or from thermogenesis (methane derived from organic carbon at high temperature) and also shows a distinctive fractionation of deuterium, ^2H (Figure 1.37). The figure also shows that a $\delta^{13}C$ value of $-30‰$, though a strong enrichment for ^{12}C, does not by itself indicate a biotic origin, as serpentinization can generate the same carbon isotope signature in methane. Thus, from the standpoint of investigating microbial origin in a serpentinizing geochemical system, an abundance of light carbon isotopes in late Hadean or early Archaean samples indicates either the existence of life or the existence of a process (CO_2 reduction in vents) that could have given rise to life. In the context of the geochemical origins of organic compounds from H_2 and CO_2, both CO_2 fixation processes (biotic and geological) are of interest and, as we will see in later chapters, differ more in grade (their catalysts) than in fundamental nature. In a noteworthy study, Ueno reported carbon isotope data for CH_4 found in methane-bearing fluid inclusions from hydrothermal systems which are generally interpreted as evidence for the existence of methanogens (archaea that use the acetyl-CoA pathway of CO_2 fixation)

at 3.46 Ga. Those fluid inclusions contained a large amount of dissolved CO_2, which is normally lacking in modern hydrothermal effluent, supporting the case for a biogenic origin.

1.3.5.2 Microfossils

The morphology of putatively fossilized microorganisms in radiometrically dated old rocks can be compared with the appearance and mineral assemblage of clearly assigned microfossils in younger rocks, but rocks can form many structures that look like cells without having been formed by life (Figure 1.38). The oldest described samples interpreted as imprints of life forms are micrometer-sized *haematite* (Fe_2O_3) tubes and filaments observed in sedimentary rocks from seafloor hydrothermal vent precipitates in Nuvvuagittuq near Quebec, Canada (Figure 1.38C). Their morphologies resemble microfossil imprints in modern hydrothermal precipitates but can be dated at least 3.77 Gya and possibly 4.28 Gya. Associated with the filaments is isotopically light carbon in carbonate rosettes, together with apatite inclusions and magnetite–hematite granules. The carbon isotope ratio points to a biogenic origin. The assignment was disputed, however, because hematite tubes and filaments with similar morphology and chemical composition as observed in the sedimentary rocks can be produced by the reaction of metal salts like copper sulfate with sodium silicate to plant-like forms. Such abiotic *chemical gardens* are known to fossilize under hydrothermal conditions. Other non-biogenic explanations for the precipitates in Nuvvuagittuq are possible. Unambiguous evidence for life in rocks older than 3.5 Ga is a subjective term because there is always some aspect of evidence for early life that one can contest as ambiguous.

Highly touted evidence of early life on the Earth is mineralized microfossils in 3.465 Ga-old Australian Apex chert rocks (Figure 1.38A) developed from hydrothermal deposits analyzed by Schopf and coworkers. Three-dimensional images of the cell walls and lumina of the microfossils were obtained by Laser Raman Microscopy in a spectral window centered on a vibrational band typical for *kerogen* (insoluble organic matter from degradation of biopolymers in sedimentary rocks). Most images show chains of single cells-like objects, forming filaments, very much like cyanobacteria (blue-green algae) do on the surface of pools. These results are, however, much disputed, as closer inspection of the original fossils suggested that they are not bacteria, as originally reported, but structures that merely look like cells in some cross sections (Figure 1.38C). The alternative explanation is that the structures are derived from serpentinization instead. A problem with evidence for early life is that it has something inherently sensational about it, as everyone ultimately wants to know about our origins, and the older the evidence, the more sensational the claim. Inorganic structures that look very much like living cells can be produced by any number of means. One example is shown in Figure 1.38D. If we ask what is the oldest genuinely undisputed evidence for life, where (almost) everyone agrees that the fossil structures really represent remnants of life, then we can turn to stromatolites ("layer rocks") from the 3.5 Ga-old Dresser formation in Australia and 3.4 Ga fossilized hydrothermal systems from South Africa with distinctive microfossils and distinctive isotopic signatures.

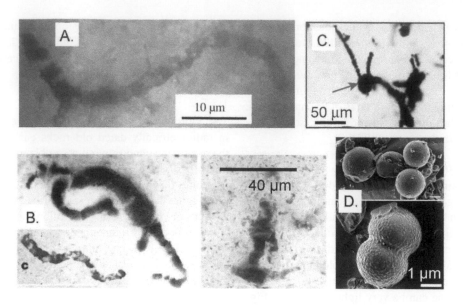

FIGURE 1.38   Microfossils or biomorphs? Microfossils from very ancient rocks are difficult to interpret as unequivocally biogenic. (A) A structure interpreted as a microfossil, possibly of cyanobacteria affinity, from the Apex chert of Australia, 3.45 Ga in age and containing carbon (kerogen) based on Raman microscopy. Reproduced with permission from Schopf J. W. (2006) Fossil evidence of Archaean life. *Phil. Trans. R. Soc. B* 361: 869–885. (B) Structures from the same deposits as in (A), but interpreted as abiogenic, formed by hydrothermal activity, based on morphological data of larger photographic areas and Raman microscopy. Reproduced with permission from Brasier M. D. *et al.* (2002) Questioning the evidence for Earth's oldest fossils, *Nature* 416: 76–81. (C) Hematite filaments in a 3.8 billion-year-old hydrothermal system from the Nuvvuagittuq belt in Canada interpreted as possible mineralized microbial cells. Reprinted with permission from Dodd M. S. *et al.* (2017) Evidence for early life in Earth's oldest hydrothermal vent precipitates, *Nature*, 543: 60–64. (D) Biomorphs, structures produced from a mixture of RNA and quartz in water, and exposed to 200°C and a pressure of 15 bars. From Criouet I. *et al.* (2021) Abiotic formation of organic biomorphs under diagenetic conditions, *Geochem. Persp. Lett.* 16: 40–46.

1.3.6   THE "HYDROLYSIS PROBLEM" AND THE NOTION OF TERRESTRIAL HYDROTHERMAL PONDS

If left in equilibrium with water, nucleic acids and proteins will (eventually) undergo hydrolysis. This observation leads to what is called the "hydrolysis problem"—if proteins and nucleic acids are left in water, how can they accumulate at origins? There have been various proposals aimed at circumventing the "hydrolysis problem" at the origin of life by assuming particular geochemical settings where there is little water. Examples include proposals that life arose in desert settings or terrestrial hydrothermal ponds where periods of no rain lead to dry conditions. How severe is the "hydrolysis problem" as a condition for origins settings? In free solution there is surely a long-term hydrolysis problem, but two issues come into play.

First, biological material is rarely in free solution. It is inherently "sticky" because of the charges that keep it in solution. These charged chemical residues bind to surfaces. Binding can have a strong influence on stability. For example, bound by minerals, protein and DNA can be stable for hundreds, thousands, and even millions of years. Examples include the sequencing of proteins from a 1.7 million-year-old fossil rhinoceros tooth or the sequencing of preserved DNA in fossilized horse tissue 700,000 years of age. These examples make the point that hydrolysis can, in principle, be a problem for nucleic acids, but it is far more severe in free solution than for nucleic acids that are sequestered on a solid phase.

Second, even in free solution, the stability of chemical bonds can be remarkable. Figure 1.39 summarizes findings from work of R. Wolfenden, who studied uncatalyzed

reaction rates for biological reactions for decades. The graph provides an overview of measured rate constants for the uncatalyzed reactions and the extrapolated values of $t_{1/2}$ at 25°C for various hydrolysis or water addition reactions (in water), whereby $t_{1/2}$ is the time required for the reaction to go to 50% completion. The measured rates for phosphodiester bond hydrolysis, for example, are in line with the tendency of DNA to remain stable over long time frames. Note that the rates for the uncatalyzed reactions are often understood as the rates for the nonenzymatic reaction, but enzymes are not the only catalysts in nature. Many inorganic catalysts can accelerate reactions, sometimes by orders of magnitude, though rarely with the same degree of rate enhancement that enzymes afford. The stability of peptide bonds to hydrolysis is noteworthy. The value of $t_{1/2}$ for an internal peptide bond (glycine–glycine) to hydrolyze is about 600 years: after 600 years 50% of the bonds are hydrolyzed; after 1200 years 75% of the bonds are hydrolyzed; after 1800 years 87.5% of the bonds are hydrolyzed; etc. The decay rates in Figure 1.39 are for the reaction in water.

The chemical environment can have a major impact on stability. Under anaerobic conditions, and with the help of humic acids, biological material can be remarkably stable. Moorleichen (peat bog cadavers) are an example. The 2400-year-old Tollund man found in Denmark in 1950 shown in Figure 1.39 was preserved in a hydrated state but under acidic and anoxic conditions. Despite being continuously hydrated during the course of 2400 years, the proteins of Tollund Man clearly did not undergo hydrolysis (the bones were however dissolved by the acidic milieu). The parts of the body made of protein, including individual hairs, which are made of keratin, are still well

A.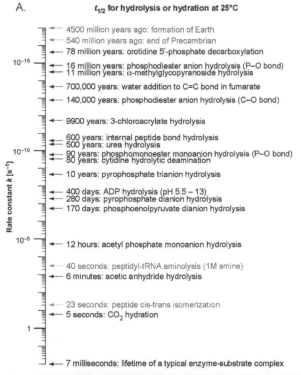

$t_{1/2}$ for hydrolysis or hydration at 25°C

4500 million years ago: formation of Earth
540 million years ago: end of Precambrian
78 million years: orotidine 5'-phosphate decarboxylation
16 million years: phosphodiester anion hydrolysis (P–O bond)
11 million years: α-methylglycopyranoside hydrolysis
700,000 years: water addition to C=C bond in fumarate
140,000 years: phosphodiester anion hydrolysis (C–O bond)
9900 years: 3-chloroacrylate hydrolysis
600 years: internal peptide bond hydrolysis
500 years: urea hydrolysis
90 years: phosphomonoester monoanion hydrolysis (P–O bond)
80 years: cytidine hydrolytic deamination
10 years: pyrophosphate trianion hydrolysis
400 days: ADP hydrolysis (pH 5.5 – 13)
280 days: pyrophosphate dianion hydrolysis
170 days: phosphoenolpyruvate dianion hydrolysis
12 hours: acetyl phosphate monoanion hydrolysis
40 seconds: peptidyl-tRNA aminolysis (1M amine)
6 minutes: acetic anhydride hydrolysis
23 seconds: peptide cis-trans isomerization
5 seconds: CO_2 hydration
7 milliseconds: lifetime of a typical enzyme-substrate complex

Rate constant k [s^{-1}]

FIGURE 1.39 **Biological material can be stable against hydrolysis for thousands of years.** (A) Half times for various hydration and hydrolysis reactions in the absence of enzymes, estimated from measured reaction rate constants. The arrows are not always exactly in place on the scale for reasons of space. Values in gray are not hydration or hydrolysis reactions but are included for comparison. Wolfenden R. (2006) Degrees of difficulty of water-consuming reactions in the absence of enzymes, *Chem. Rev.* 106: 3379–3396 (B) The image is of the Tollund Man, whose well-preserved body was found in a peat moss bog in Denmark in 1950. The body is ^{14}C dated to be 2400 years of age. Reprinted with permission from Silkeborg Museum, Denmark. Photograph Arne Mikkelsen.

preserved, as is his skin, which is made of collagen. At higher magnification, the individual hairs of his beard are distinct.

Tollund Man provides a striking example of the stability of proteins against hydrolysis. The values tabulated by Wolfenden in Figure 1.39 also show that hydrolysis is not so severe that it would force us to look for desert environments to accommodate origins. Recall that cells are ~70% water by dry weight. The chemical reaction of life requires water to go forward. Nonetheless, the 'hydrolysis problem'

is sometimes given so much weight that, to accommodate it, origins theories can be set in sites that avoid water (except when it is absolutely required as a solute). In the extreme, some origins theories assume that origins occurred on dry land, where rainless periods are invoked to reduce water activity to promote the polymerization of monomers (amino acids or nucleotides), followed by a cycle of wetting such that they become extended. As long as activated monomeric precursors are constantly delivered, and as long as no chain termination reactions occur, and as long as no heavy rain takes place, polymers can accumulate. The condition that the site of life's origin never encountered heavy rain is problematic. Anyone who has ever witnessed heavy rain knows that it will wash out everything that is soluble and lying around on the surface.

Despite the risk of one heavy rainfall occurring during the time required for origins, this kind of scenario is popular and commonly referred to as 'wet-dry cycles' in terrestrial ponds and lakes exposed to sunlight. Alternating periods of hydrolysis (high water activity) and condensation (no water activity) are proposed to achieve polymerization. Without question, such well-controlled (rain-free) cycles do deliver promising results for nucleotide synthesis—under laboratory conditions. It is important, however, to consider not only water but also the availability of carbon and energy, the main prerequisites for microbial life.

Do hydrothermal vents have a hydrolysis problem? No. In solid phase–aqueous phase geochemical systems, water activity is constantly low at many sites within the vent system, because serpentinization consumes water, incorporating it into the rock itself. This process was recently shown in some detail for hydrothermal sedimentary layers where pore spaces between volcanic particles can be filled with silica gels, which leads to less free water in the geochemical system. We will encounter this example of water consumption in serpentinizing olivine microcrystals in more detail in Section 2.2.

Do living cells have a hydrolysis problem? No. Microbial cells are typically about 70% water by weight, though most of that water is bound by proteins and solutes. The protein concentration of cytosol is about 400 g/L. In living cells, the hydrolysis problem is countered by thermodynamics and rate. Nucleic acid and peptide polymerization require energy input, these reactions are coupled to the main bioenergetic reaction of the cell such that the rate of polymerization is much faster than the rate of hydrolysis. The net result is growth. Note that the rates of the hydrolysis reactions for some molecules with high energy bonds (phosphoenolpyruvate, acetyl phosphate) are included in Figure 1.39 (compare to Table 1.2 for free energies of hydrolysis). High-energy bonds have a short lifetime, but they are essential for many biological reactions. That means that early metabolic systems necessarily arose in a far-from-equilibrium environment where high-energy bonds could be constantly formed. For polymers to accumulate at origins, the rate of polymerization has to be faster than the rate of hydrolysis, and the rate of monomer synthesis has to be faster than the rate of polymerization.

The key to the chemistry of origins is not the placement of a protein or a nucleic acid in solution in hopes that it will be stable for 1,000,000 years, the solution is to identify a far-from-equilibrium chemical environment where continuous exergonic reactions can thermodynamically drive the synthesis of monomeric components in the presence of catalysts that, as rate accelerators, can substitute for enzymes before genes and proteins arise. We will see in later chapters that H_2-producing hydrothermal vents fulfill such conditions.

The role of sunlight in wet–dry cycles is to provide UV light. UV light is required by origin theories that entail cyanides as starting materials for the synthesis of amino acids and bases. Strong UV light on exposed surfaces will eventually destroy organic molecules (there was no Hadean ozone layer). We will see in later chapters that a constant source of cyanide on the early Earth is not a straightforward issue. A constant source of CO_2 on the early Earth is certain. Hydrothermal vents in the deep sea are protected from UV light, impervious to rain, less affected by destructive meteorite impact than surface environments, and accumulate H_2 to very high partial pressures at the same site where CO_2 and catalysts for chemical reactions are available. Serpentinization even synthesizes awaruite (Ni_3Fe) and magnetite (Fe_3O_4) within vents, both of which are very efficient catalysts for the synthesis of formate, acetate, and pyruvate from H_2 and CO_2, as we will see in later chapters. Serpentinization synthesizes both reactants (H_2) and catalysts for organic reactions in environments where the products can accumulate.

How does the example of Tollund Man in Figure 1.39 relate to origins? If a protein once synthesized in whatever environment can remain stable against hydrolysis for 2000 years, and if the synthesis of a protein in the same environment requires 1000 years, proteins will accumulate. If the synthesis of the protein takes 4000 years, then little or no protein will accumulate in that environment. But if the synthesis of one protein takes 4000 years, the problems on the way to generating microbial cells are going to be insurmountable anyway. A typical microbial cell contains on the order of 2–5,000,000 proteins. The millions of proteins required to form a cell take too long to synthesize at a rate of 4000 years per protein (see the rate scale in Figure 1.39).

The same reasoning applies to nucleic acids. The point is this: the problem at origins is not the rate of decay of organic compounds or polymers; the problem is their rate of synthesis. That places the focus on geochemically stable sites for origins with continuously abundant chemical energy (as opposed to high energy photons), abundant reactive starting compounds (like H_2 and CO_2), and abundant catalysts and mechanisms to retain the products of organic synthesis so that they can react further to form more complex molecules within a time frame that meets the constraints set forth by Earth's history (see Figure 1.7). Life arose after the appearance of liquid water 4.3 Ga and was in existence probably by 3.8 Ga (light methane) and, at the very latest, by the time that stromatolites (fossil microbial colonies) with a clear ^{13}C signature appear (3.4 Ga). That puts constraints on the rates at which reactions need to take place for origins to have occurred.

1.3.7 SEVERAL FORMS OF WATER-SOLUBLE PHOSPHATES AND SULFIDES EXIST

Most phosphorus-containing minerals do not dissolve in water—they are chemically almost inert at normal temperatures and neutral pH. Indeed, most of the phosphorous in the Hadean period was in the form of water-insoluble *apatite*

(calcium phosphate). Only water-soluble phosphates, however, can react chemically. One possible route to phosphate availability was presented by a laboratory experiment which simulated phosphate rocks meeting water at magmatic temperatures. In the experiment, water-soluble phosphates and polyphosphates were produced from phosphate rocks or calcium phosphate at temperatures of ~1300°C in the presence of water (Figure 1.38). P_4O_{10} vapor was analyzed to be the main discharge of the hot phosphate rock. The vapor was rapidly cooled in a condenser and directed into cold water to simulate the eruption of volcanic gases at high pressures and temperatures into the air at near-ambient conditions. Similar sudden cooling may take place in the interior of a hydrothermal vent when phosphate minerals in volcanic rocks come into contact with intruding cold seawater.

Polyphosphates are remarkably stable against hydrolysis at <100°C and neutral pH. Hence, pyrophosphate and the larger cyclic polyphosphates may have enough lifetime in water to activate organic molecules by phosphate transfer. Probably reduced phosphorous compounds like phosphites, HPO_3^{2-}, were also produced in the magma-simulating experiment but could not be analyzed unambiguously. Phosphites are more water soluble than phosphates and a potential source of energy-rich electrons for redox reactions. The midpoint potential of the HPO_4^{2-}/HPO_3^{2-} redox couple under standard physiological conditions is –690 mV, which is very low but not as low as the midpoint potential of alkaline serpentinizing vents, where midpoint potentials can reach –700 to –900 mV in present-day vents because of the high pH and the large amounts of H_2 that serpentinization generates. Recent investigations by M. Pasek suggest the possibility that the low midpoint potential of serpentinizing systems could have mobilized phosphate by converting it to the more soluble phosphite, which itself is a powerful reductant that readily reduces sulfite to sulfide in the presence of suitable catalysts, in a reaction that generates phosphate. Furthermore, recent work by Mao and colleagues in the laboratory of Bernhard Schink identified an enzyme that phosphorylates AMP with phosphite, whereby phosphite is oxidized to phosphate during the energy-releasing reaction. This reaction opens up vast new possibilites in the field of early phosphorylation chemistry at origins.

1.3.8 HIGH ENERGY CARBON BONDS REACT WITH SULFUR AND PHOSPHATE

Phosphorus, like sulfur, and in contrast to C, N, and O, has a third electron shell: [Neon]$3s^23p^3$. Although its 3d orbitals are unfilled, its 3s and 3p orbitals can hybridize with the d orbitals. Therefore, bonds between P and O in phosphates HO–PO have a rather long bond length of 1.66 Å (compared to C–O and N–O lengths of 1.43 and 1.44 Å, respectively) and are energy rich. The long and rather weak P–O bonds make the P atom in phosphates easily accessible to nucleophilic attack and susceptible to hydrolysis. Therefore, phosphate is an excellent leaving group and is often involved in reactions where substrate activation by free energy change is required.

Consider the synthesis of ATP by phosphate transfer from acetyl phosphate to ADP (Figure 1.25) as an example. In modern organisms, ATP activates molecules for further reaction, adding specificity to metabolic reaction sequences. At the origin of life, inorganic reactive forms of phosphorous like pyrophosphate and metaphosphates and reduced forms like phosphite or could have adopted that role in principle. There are, however, little or no hints that inorganic phosphates activate organic molecules in modern organisms. There is evidence for the role of polyphosphates as energy reservoirs but not as activating agents. Note also the low free enthalpy of pyrophosphate hydrolysis (Table 1.2), making pyrophosphate unable to fuel ADP phosphorylation. What about energy-rich reduced phosphorus compounds like phosphite then? About 1% of bacteria harness phosphite as an electron donor and some use it as a source of ATP synthesis in addition, as Mao *et al.* recently showed.

How was phosphate activated in early metabolic evolution, if polyphosphates or similar were not used? The comparison with modern organisms and ancient bacterial life forms points to phosphite and acyl phosphates as early activating agents. Formyl phosphate ($H–C(=O)–OPO_3^{2-}$) occurs in the fixation of CO_2 in the acetyl-CoA pathway of some bacteria; acetyl phosphate ($CH_3–C(=O)–OPO_3^{2-}$) conserves the energy of the acetyl-CoA pathway and transfers phosphate to ADP forming ATP; and various acyl phosphates conserve energy during fermentation or as γ-glutamyl phosphate in nitrogen assimilation. Succinyl phosphate is used for ATP synthesis during the Krebs cycle in human mitochondria. Acetyl phosphate concentrations can reach 3 mM in *E. coli* cells. Let us take the synthesis of acetyl phosphate as an example. In Section 3.1 we will consider the transition-metal-catalyzed synthesis of acetyl fragments from CO_2 and H_2 and their reaction with SH^- to get thioacetic acid ($CH_3(C=O)SH$). This *activated acetic acid* has enough free enthalpy of hydrolysis to generate acetyl phosphate from inorganic phosphate (P_i), whereas a corresponding

thioester like acetyl-CoA does not based on free energies under standard conditions (Table 1.2). It was shown experimentally that acetyl phosphate can be readily synthesized in water from thioacetate but not from the thioester methyl thioacetate ($CH_3(C=O)SCH_3$) under conditions tested so far. The experiment also showed that acetyl phosphate can phosphorylate ADP to ATP in water at 50°C. Importantly, those phosphorylation experiments were performed without enzymes, under prebiotic conditions. With regard to the thioester experiment we stress here, however, that, in many growing cells, acetyl phosphate is constantly made from acetyl-CoA and P_i. The reaction goes forward because the reactant concentration is kept high and the product concentration is kept low, such that the free reaction enthalpy remains negative (see Equation 1.12).

Hence, in early metabolism, phosphate was likely activated by energy from *carbon redox chemistry* as a high-energy P–O bond in an organic molecule, not in a geochemically activated inorganic polyphosphate. The energy in substrate-level phosphorylation stems from activated carbon reacting with phosphate, not from activated *environmental* phosphate species such as pyrophosphate, polyphosphates, or phosphides (Figure 1.40) reacting with unreactive carbon species. Phosphite occurs in serpentinized rocks and can be used by bacteria for ATP synthesis, it is an interesting source of phosphorous, energy and electrons in one, as the findings of Mao *et al.* show. This does not mean that polyphosphates do not play a role in metabolism. For example, both translation and DNA replication are driven forward by the hydrolysis of pyrophosphate, yet the energy does not come from environmental PP_i but from ATP synthesized by the cell instead.

* * *

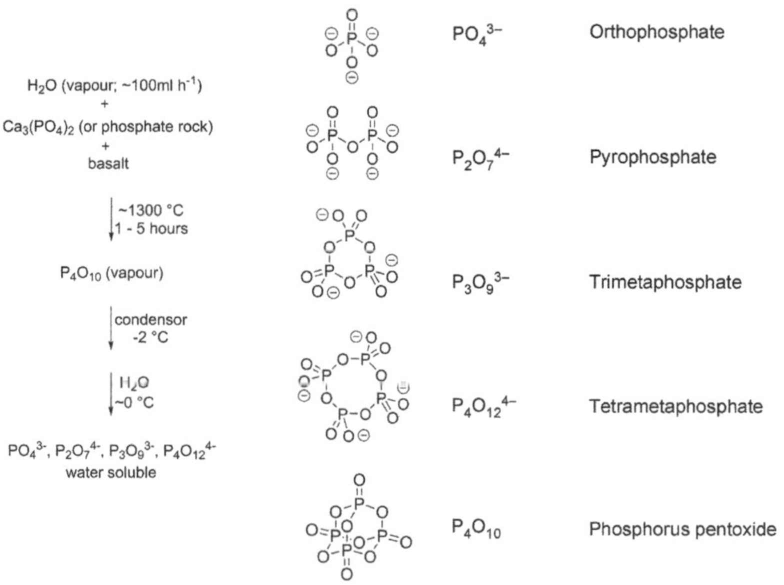

FIGURE 1.40 **Forms of phosphate.** Production of water-soluble phosphates, pyrophosphate, and cyclic polyphosphates by heating water, phosphate rock, and basalt to 1300°C. The polyphosphates arise from the partial hydrolysis of P_4O_{10} vapor. Basalt melts at ~1200°C and dissolves calcium phosphate and phosphate rock, which melt at >> 1300°C. For P_4O_{10} to form, dissolved phosphate ions must collide with each other, which can occur only in the melt not in the solid. Hydrolysis of polyphosphates can occur at 100°C for 1 hour with 0.5 M H_2SO_4. Data from Yamagata Y. *et al.* (1991) Volcanic production of polyphosphates and its relevance to prebiotic evolution, *Nature* 352: 516–519.

1.4   CHAPTER SUMMARY

Section 1.1 Some basic knowledge about chemistry, physics, and geology is necessary to understand the origin of organic molecules on the early Earth. Stars, planets, and chemical elements set the stage for life. Stars develop from the gravitational collapse of interstellar clouds of hydrogen, helium, and other ionized gases which developed after the Big Bang. The compression in the center of the nebulae heats them up until nuclear fusion of hydrogen into helium begins and stars form. Heavier material from the explosion of burned-out stars can accrete to planets which orbit the stars. The Earth was formed this way about 4.54 Ga. Water was brought to Earth during accretion and from comets probably not as bulk ice but bound in hydrated silicate minerals. Volcanic eruptions and large impacts later released such bound water into the Earth's atmosphere. Due to the heat-releasing accretion of heavy material, the early Earth was molten and differentiated by gravity into a dense core of iron and nickel (density roughly 8 g/cm^3) and a mantle of lighter magma (density roughly 4–5 g/cm^3) containing elements like silicon, aluminum, magnesium, oxygen, nitrogen, and sulfur. Soon after the origin of the Earth, the moon-forming impact transformed carbon and nitrogen chemically bound in mantle material as gases into the atmosphere, mainly as water vapor, carbon dioxide, nitrogen gas, and sulfur gases. Heat radiation cooled the outer magma of the mantle and formed a thin crust floating upon the mantle. Atmospheric water vapor condensed to oceans and left behind a hot, dense atmosphere of N_2 and mainly CO_2, which gradually dissolved in ocean water. Subduction of colliding tectonic plates into the mantle, volcanic activity, and formation of hydrothermal vents from fractures on the seafloor were probably common at the end of the Hadean period at 4 Ga. Radiometric dating indicates that prebiotic organic synthesis started 4.2–4 billion years ago. Mantle-derived rocks contained inexhaustible quantities of ferrous iron, which can reduce hot water from hydrothermal vents to H_2, a process called *serpentinization*. H_2 is the ubiquitous reductant for the fixation of CO_2 to organic molecules.

Section 1.2 To reach an energetically more stable state, carbon, oxygen, and nitrogen in organic molecules populate their outer shells with eight electrons (*octet rule*). The atoms in an organic molecule share electrons with their binding partners to obey this octet rule, which explains the number of covalent chemical bonds in a stable organic molecule. In this way, the octet rule places energetic guardrails along the paths governing the synthesis of organic molecules. Noncovalent bonds stem from electrostatic interactions, hydrogen bonding, dipole, and induced dipole interactions. They are weaker than covalent bonds but are often important for processes determining chemical organization and origins. Acid–base reactions of organic and inorganic molecules, redox processes, and nucleophilic substitutions or additions are central in the chemical networks of early life. A chemical reaction must release free energy to occur spontaneously. This free reaction energy can be determined from the equilibrium constant of the reaction or from the free heat of the formation of reactants and products. The rate of a chemical reaction can be measured by following the concentration of reactants and products with time. The

reaction rate indicates how fast a given process actually proceeds. Transition metals such as iron and nickel on mineral surfaces and in solution can activate molecules and catalyze (accelerate) chemical reactions.

Section 1.3 Life emerged on Earth at sites with sustainable energy supply; sources of C, H, N, and O; solid-state mineral catalysis; and concentration enrichment in micropores. The organic molecules of life were synthesized by geochemical processes in the Earth's crust. Another widely discussed source of bioorganic molecules is organic material from meteorites. However, such material becomes diluted in the oceans after meteorite impact; almost all of it is organized as inert polyaromatic hydrocarbons and that which is not inert is chemically too diverse and reduced to serve as food or metabolic precursors for chemical evolution. Serpentinizing ultramafic rocks in deep-sea hydrothermal vents and forearc regions are more probable sites for abiogenesis. Water-soluble phosphates and sulfides react with high-energy organic molecules to very reactive composite compounds.

*   *   *

PROBLEMS FOR CHAPTER 1

1. *Formation of the Earth.* Where did the solid material and the water on the Earth come from?
2. *Earth's early atmosphere.* How did the moon-forming impact influence the composition of the Earth's atmosphere?
3. *Radiometric age determination.* Calculate the age of a very ancient piece of rock with a ^{207}Pb to ^{235}U ratio of 10:1.
4. *Noncovalent interactions.* What noncovalent interactions are important for biomolecules?
5. *Functional chemical groups.* What are the most important functional groups of biological molecules? Write the chemical structures of the corresponding moieties.
6. *Protonation states.* Calculate the concentration of phosphoric acid in its different protonation states in water at pH 4. Hint: See pKa values in Figure 1.20.
7. *Water activity.* Calculate the water activity for a water pressure of 10 mbar above an aqueous salt solution at 25°C/1 bar total pressure. Determine the mole fraction of water and salt in the solution for an activity coefficient of 1. Hint: Use eq. 1.7 and 1.8 for the calculation. Search the water vapor pressure at 25°C/1 bar in the internet.
8. *Free reaction enthalpy and equilibrium constant.* Calculate the free enthalpy for a chemical reaction with an equilibrium constant of 20 at 25°C. Hint: Use eq. 1.13 for the calculation.
9. *Free reaction enthalpy and free enthalpies of formation.* Calculate the free enthalpy for the reaction of phosphoenolpyruvate (−1185.46) with ADP (−1428.93) to ATP (−2292.61) and pyruvate (−352.40) and compare with the difference of the corresponding free enthalpies of hydrolysis from Table 1.2. The values in parentheses are the

standard free enthalpies of formation in kJ/mol at zero ionic strength taken from the *Handbook of Biochemistry and Molecular Biology*, CRC Press. 5th Edition 2018, pp. 572–575.

10. *Adsorption processes.* What is the difference between chemisorption and physisorption?

11. *Transition metal catalysts.* Why are many transition metals good catalysts in chemical reactions?

12. *Molecular diffusion.* Calculate the diffusion time of alanine in a rock pore of $50 \times 50 \times 100$ nm filled with water: Hint: Use eq. 1.19 and data there.

13. *Organics in meteorites.* Give arguments for and against a significant contribution of meteorite material to the origin of organic molecules on the Earth.

14. *Lost City hydrothermal field.* What are the typical temperature, pH, total pressure, and H_2 concentration in the Lost City hydrothermal vent?

2 Origin of Organic Molecules

2.1 CARBON DIOXIDE REACTS WITH H_2 TO GENERATE ORGANIC COMPOUNDS

Molecules that contain carbon are called organic molecules. Carbon can form linear or branched chains or rings with other carbon atoms. Carbon can also form single bonds, double bonds, or triple bonds with other carbon atoms. In natural settings, it can form bonds with hydrogen, oxygen, nitrogen, sulfur, and metals such that millions of different organic molecules exist. Some carbon-containing compounds, such as carbides, carbonates, cyanides, CO, and CO_2 are regarded as inorganic. Generally, all molecules that contain carbon and hydrogen are considered as organic. Because silicon (Si) resides below carbon in the periodic system, it is sometimes speculated that there might be forms of life based on Si instead of C. This can be excluded, however, because, under the conditions that existed during the Earth's formation and in natural environments, Si does not readily form bonds with other Si atoms. Instead, it makes very strong bonds with oxygen atoms, forming extremely stable silicates such as quartz, which erodes to sand over eons. The ability of C to make metastable bonds with other C atoms sets it apart.

Today, all life we know arises from CO_2 and, when life is over, everything ultimately ends up as sediment on the ocean floor. Sediment on the ocean floor is sooner or later (after about 300 million years at the latest, the age of the oldest submarine crust) subducted back into the mantle, and there the carbon is converted back to CO_2 that can enter the biosphere once more through volcanic eruptions. Life starts with CO_2 and ends as CO_2; hence, we need to focus on CO_2 chemistry to understand the origin of life. At origins, H_2 was the main environmentally available reductant with a sufficiently negative redox potential to reduce CO_2. To get acquainted with the chemistry of primordial organic synthesis, we have to consider the specifics of CO_2 reduction with activated hydrogen or with energy-rich electrons to obtain organic molecules. On the early Earth, H_2 was not the only available reductant capable of reducing CO_2, because some native metals such as elemental iron (Fe^0) have a sufficiently negative redox potential to reduce CO_2 and there were surely at least some native metals available in the primordial crust in the wake of differentiation (the formation of the Earth's core). However, under anaerobic conditions, Fe^0 reacts with water to form H_2. In an aqueous environment as was required for origins, native metals that could reduce CO_2 would first reduce H_2O to H_2 before reducing CO_2 to organic compounds because, as we will see in this chapter, H_2O ($E_0' = -414$ mV) is a better electron acceptor than CO_2 ($E_0' = -434$ mV). Organic molecules are the prerequisite for life on Earth. An old German saying has it that 'the devil is in the details'. If we want to place constraints on CO_2 reduction, we have to delve into the details of the reactions of H_2 and CO_2.

2.1.1 CO_2 IS THE MAIN PREBIOTIC SOURCE OF CARBON; IT REACTS WITH H_2 ON METAL CATALYSTS

Carbon dioxide (O=C=O) is a linear molecule with strong polar bonds between carbon and oxygen (C=O bond distance 1.16 Å). The polar bonds point in opposite directions ($O^{\delta-}-C^{\delta 2+}-O^{\delta-}$); hence, the net electric dipole is zero. The two back-to-back dipoles, however, form a large quadrupole moment such that CO_2 is attracted by a polar solvent or a solid surface. A quadrupole moment can be represented as two dipoles oriented antiparallel; hence, it consists of four poles and is polar but has no net electric dipole.

Compared to other atmospheric gases like N_2, O_2, and H_2, CO_2 is highly soluble in water. For example, according to the ideal gas law at 25°C (298 K) and $p = 1$ bar CO_2 pressure, the concentration of CO_2 in the gas phase is $c = p/RT = 1$ bar$/(0.0831 \times 1 \cdot$ bar$/$K$/$mol $\cdot 298$ K$) = 0.040$ mol/l. According to Henry's law, the amount of dissolved gas in a liquid is proportional to the (partial) pressure of the gas above the liquid. For CO_2, the proportionality factor (Henry's law constant k_H) is $k_H = 29.4$ l·bar/mol $= p_{CO2}/[CO_2(aq)]$. Hence $[CO_2(aq)] = 1$ bar$/29.4$ l·bar/mol $= 0.034$ mol/l at 25°C and 1 bar CO_2, making it very soluble in water. The abbreviation M denotes mol/l. Once in water, CO_2 forms the acid H_2CO_3 (carbonic acid):

$$H_2O + CO_2 \rightleftharpoons H_2CO_3$$

The equilibrium constant $K_1 = [H_2CO_3]/[CO_2(aq)] = 1.7 \times 10^{-3}$ (1.2×10^{-3} in seawater) is low; hence, most CO_2 is not transformed into carbonic acid but remains as physically dissolved (solvated) CO_2 in water without affecting the pH if no base is added. Assuming CO_2 (aq) $= 0.034$ mol/l at 25°C and 1 bar CO_2, we obtain $[H_2CO_3] = K_1 \cdot [CO_2 (aq)] = 1.7 \times 10^{-3} \cdot 0.034$ mol/l $= 5.8 \times 10^{-5}$ mol/l. The equilibrium for the hydration of CO_2 to H_2CO_3 is reached quite slowly without a catalyst, however. The rate constants are 0.039 s^{-1} for 2 and 23 s^{-1} for the reverse reaction $H_2CO_3 \rightarrow CO_2 + H_2O$. H_2CO_3 is a weak acid

$$H_2CO_3 \rightleftharpoons H^+ + HCO_3^-$$

with an equilibrium constant $K_2 = [HCO_3^-] [H^+]/[H_2CO_3] = 2.5 \times 10^{-4}$ mol/l. Hence $[HCO_3^-] = [H^+] = (2.5 \times 10^{-4}$ mol/l $\times 5.8 \times 10^{-5}$ mol/l$)^{0.5} = 1.2 \times 10^{-4}$ mol/l and pH $= -\log [H^+] = 3.92$ at 1 bar CO_2. For normal atmospheric pressure conditions $p_{CO2} = 3.5 \times 10^{-4}$ bar and pH $= 5.65$. Hydrogen carbonate HCO_3^- dissociates into the carbonate ion CO_3^{2-}:

$$HCO_3^- \rightleftharpoons H^+ + CO_3^{2-}$$

DOI: 10.1201/9781003378617-2

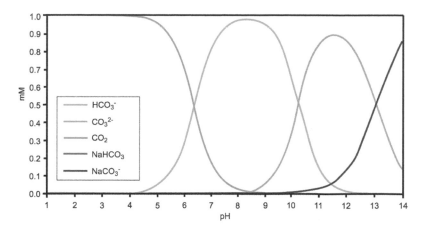

FIGURE 2.1 Bicarbonate and carbonate concentrations for 1 mM CO$_2$ in water versus pH. The proton concentration is altered to pH < 4.68 (pH at 1 mM Dissolved Inorganic Carbon DIC:ΣCO$_2$) by adding HCl and to pH > 4.68 by adding NaOH. At very high pH, Na$^+$ forms aqueous complexes NaHCO$_3$ and NaCO$_3$$^-$. Data from AQION software at www.aqion.de (free download at aqion.de). For solubility of CO$_2$ at different temperatures and pressures, see also Diamond L.W. *et al.* in the literature section.

with an equilibrium constant K_3 = [CO$_3$$^{2-}$]·[H$^+$]/ [HCO$_3$$^-$] = 4.7 × 10^{-11} mol/l. Hence [CO$_3$$^{2-}$] = [H$^+$] = (4.7 × 10^{-11} mol/l · 1.2 × 10^{-4} mol/l)$^{0.5}$ = 7.5 × 10^{-8} mol/l. The CO$_3$$^{2-}$ concentration is therefore negligible with respect to the HCO$_3$$^-$ concentration, and H$^+$ from this reaction does not contribute noticeably to the pH. For normal atmospheric conditions, the CO$_2$ pressure is 3.5 × 10^{-4} bar, [CO$_2$ (aq)] = 1.22 × 10^{-5} M, [H$_2$CO$_3$] = 2.07 × 10^{-8} M, and [HCO$_3$$^-$] = 2.27 × 10^{-6} M and pH = 5.65. At 10 bar CO$_2$, [CO$_2$ (aq)] = 0.34 M, [H$_2$CO$_3$] = 5.8 × 10^{-4} M, and [HCO$_3$$^-$] = 3.8 × 10^{-4} M and pH = 3.4. That is, at high CO$_2$ pressures, [H$_2$CO$_3$] > [HCO$_3$$^-$].

How do carbonate concentrations change with pH and CO$_2$ pressure? Of course, addition of a base increases the concentration of hydrogen carbonate (bicarbonate) and carbonate. At higher pH, mole and charge balance have to be considered, because concentration changes due to consecutive dissociation cannot be neglected anymore. The coupled equations can be solved numerically. Figure 2.1 shows an example for the calculation of the HCO$_3$$^-$ and CO$_3$$^{2-}$ concentrations for CO$_2$ dissolved in water at different pH. At pH 8–9, hydrogen carbonate is the prevailing species. In alkaline hydrothermal systems like Lost City, there are always regions of that pH in the gradient between Lost City's effluent at pH 9–11 and seawater, which was slightly acidic in the Hadean era. We will see later that hydrogen carbonate is the main reactive species for CO$_2$ reduction in water.

Hydrogen gas is about 40 times less soluble in water than carbon dioxide: 0.8 mM (1.6 mg/l) H$_2$ at 25°C and 1 atm H$_2$ gas pressure compared to 34 mM (1.5 g/l) carbon dioxide at 25°C and 1 atm CO$_2$. For other common gases, the situation is similar. About 1.3 mM O$_2$ and 0.6 mM N$_2$ dissolve in water at 25°C and 1 atm gas.

The situation is quite different, however, for the affinity of these gasses to metals. On the surface of iron and other transition metals, H$_2$, O$_2$, and N$_2$ adsorb and 'dissolve' (physisorption/chemisorption) readily. Hence, transition metals concentrate these gases from the aqueous phase onto the solid phase, remove them from equilibrium in the aqueous phase and activate them (see chemisorption in Section 1.2) for chemical reaction through orbital interactions.

We recall that the moon-forming impact transformed carbon-containing rocks to gaseous CO$_2$, which, after cooling of the crust and condensation of the water vapor, remained at high concentration in the atmosphere (Section 1.1). With time, atmospheric CO$_2$ dissolved in the oceans as hydrated CO$_2$, H$_2$CO$_3$, HCO$_3$$^-$, and CO$_3$$^{2-}$ (carbonate). Carbonate has a low molar solubility product with divalent ions of the alkaline earth metals (group 2 of the periodic system) like Mg^{2+} and Ca^{2+}. Massive amounts of CO$_2$ from the moon-forming impact did not persist over long time scales in the atmosphere or the ocean. Dissolved CO$_2$ precipitated as carbonates that sedimented and were ultimately transferred to the mantle by subduction. The CO$_2$ that remained in the ocean was furthermore chemically inert, it had to be activated on transition metal surfaces to react to organic molecules (Section 1.2). While gaseous CO$_2$ is retained by the Earth's gravity or becomes dissolved in the Earth's water, H$_2$ dissolves very poorly in water and can escape the Earth's gravitational pull by diffusion into space. Where did the H$_2$ needed for CO$_2$ reduction come from? It came from the process of serpentinization.

2.2 SERPENTINIZATION GENERATES H$_2$ FROM REACTIONS OF ROCKS WITH WATER

2.2.1 SERPENTINIZATION REACTIONS IN HYDROTHERMAL SYSTEMS GENERATE H$_2$

Serpentinization is a geochemical process that occurs when iron- and magnesium-rich rocks (*mafic* or *ultramafic* rocks) of the upper mantle interact with seawater that is drawn by gravity into cracks in the crust, allowing the circulating water to penetrate the crust to a depth of several kilometers. *Mafic* rocks have low silica content and high magnesium (*ma*) and iron (ferrum: *fe*) content. *Ultramafic* rocks have very low silicate content and very high Mg^{2+} and Fe^{2+} contents. Because of the geothermal gradient in the crust (today roughly 25°C to 30°C per kilometer of depth) and because of heat released during the exergonic reaction process of serpentinization, the circulating water is heated at depth to roughly 100°C to 200°C and returns to the surface via convection, forming a

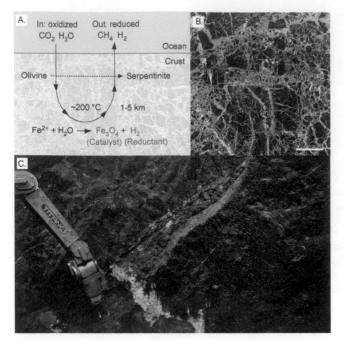

FIGURE 2.2 Serpentinization. (A) Schematic representation of the serpentinization process in a hydrothermal field. Catalysts produced by serpentinization can include Fe^0, Ni^0, $FeNi_3$, and Fe_3O_4, depending on the composition of the hydrothermal fluid. Oxidized species like CO_2, H_2O, and SO_2 can be reduced depending on H_2 pressure, temperature, catalyst, and fluid composition. The depth of circulation is on the order of 1 to 5 km. (B) A section of serpentinite from a storefront in London (Image courtesy John F. Allen, University College London). Scale bar 10 cm. The white quartz veins indicate where water flow and rock water interactions took place. (C) A closeup of native and actively reacting serpentinite directly exposed to seawater on the eastern flank of Lost City. Image kindly provided by Susan Q. Lang, University of South Carolina with the help of funding from NSF/ROV. Image taken on a 2018 expedition to Lost City. White carbonate veins indicate routes of fluid flow. The team was able to sample effluent coming directly out of the rock. Printed with permission courtesy of Susan Q. Lang, University of South Carolina/NSF/ROV Jason 2018 © Woods Hole Oceanographic Institution.

hydrothermal current (Figure 2.2). The fluid that emerges from a hydrothermal vent is called effluent. The mineral product of serpentinization is serpentinite, a familiar building material used by stonemasons for centuries.

The main gas phase product of serpentinization is molecular hydrogen (H_2), resulting from the reduction of protons from water with electrons from Fe(II) minerals. The amount of H_2 generated by serpentinization depends upon temperature and the rock-to-water ratio (the ratio of the mass of rock to the mass of fluid phase in contact with rock in that volume). For example, at 100°C, serpentinization starting with harzburgite, an ultramafic olivine-containing rock, generates about 0.9 moles of H_2 per kg rock at a rock:water ratio of 5 (excess of rock over water), but that value increases to about 130 moles of H_2 per kg rock as the rock:water ratio decreases to 0.1 (excess of water over rock). H_2 is a light gas; 130 moles of H_2 correspond to 260 g or a volume of 2.9 m³ of H_2 gas at 1 bar (22.4 l/mol). Harzburgite has a density of roughly 3 kg/l such that 1 kg of pure harzburgite (0.33 l) can generate up to 8800 times its volume in H_2 gas during serpentinization. The hydrogen atoms are extracted

from water; the oxygen atoms from water end up in the rock as iron oxides (Fe_xO_y). The addition of oxygen and water (as hydroxides; see Figure 2.3) to the host rock increases the mass and volume of the rock as the process proceeds.

Hydrogen from serpentinization reacts with carbon dioxide in the hydrothermal fluid or, with carbonates in host rocks, to hydrocarbons and other organic molecules if mineral catalysts arising from serpentinization are present to catalyze the reactions. One of the best-studied serpentinization sites is the Lost City hydrothermal field. The effluent of Lost City contains about 1.5 mM isotopically heavy (^{12}C depleted, abiotic) methane, about 100 μM isotopically heavy (abiotic) formate along with about 20 μM acetate that is isotopically lighter and more likely of biotic origin. Many serpentinizing sites contain effluent that is rich in H_2, almost devoid of CO_2, bicarbonate, or carbonate (together called dissolved inorganic carbon or DIC) but rich in formate instead. Formate, HCOOH, is the simplest product of H_2-dependent CO_2 reduction and is readily formed both in nature and in the laboratory from H_2 and CO_2 in the presence of transition metal catalysts.

During serpentinization, hydrogen is continuously replenished in pores of olivine. It adsorbs and accumulates on the iron-containing products of serpentinization. Hydrogen's low solubility in water therefore poses no disadvantage for hydrogenation reactions to occur. The chemisorption of H_2 onto iron minerals sequesters the reductant and activates it for further reactions. Hydrogen and methane from serpentinization can also become sequestered in pores of submarine rock as gas and persist there on geological time scales.

Recent findings by Shang and colleagues reveal that serpentinization can generate vast amounts of ammonia in laboratory experiments via the reaction $2Fe^{2+} + 2H_2O \rightarrow 2Fe^{3+} + H_2 + 2OH^-$; $3H_2 + N_2 \rightarrow 2NH_3$, whereby up to 200 μmol NH_3 was obtained per gram of peridotite after 30 days in laboratory scale serpentinization experiments. In metabolism, ammonia is the substrate for the synthesis of amino acids from α-ketoacids via reductive amination.

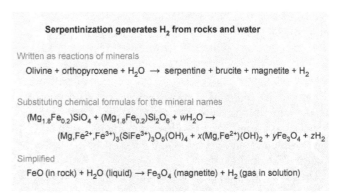

FIGURE 2.3 Serpentinization described as a series of reactions. For explanation, see text. Reactions from Sleep N. H. *et al.* (2011) Serpentinite and the dawn of life, *Philos. Trans. R. Soc. B* 366: 2857–2869; McCollom T. M. (2016) Abiotic methane formation during experimental serpentinization of olivine, *Proc. Natl. Acad. Sci. USA* 113: 13965–13970; Preiner M. *et al.* (2018) Serpentinization: Connecting geochemistry, ancient metabolism and industrial hydrogenation, *Life* 8: 41.

2.2.2 Olivine Is a Very Common and Widespread Component of the Earth's Crust

Olivine is an iron(II) magnesium silicate that reacts with hot pressurized water to hydrogen, the iron oxide *magnetite*, the magnesium hydroxide *brucite*, and the hydrous magnesium silicate *serpentine*. The Earth's mantle consists mainly of *peridotite*, which is composed mostly of *pyroxene* (chain silicate forming rocks) and *olivine*. Olivine is an iron-magnesium silicate, a solid solution between the magnesium silicate *forsterite* (Mg_2SiO_4) and the iron silicate *fayalite* (Fe_2SiO_4). Below 400°C forsterite dissolves in water:

$$Mg_2SiO_4 + 4\,H^+ \rightarrow 2\,Mg^{2+} + SiO_2\,(aq) + 2\,H_2O$$

At sufficiently high concentrations of dissolved species, *serpentinite* ($Mg_3Si_2O_5(OH)_4$), which is a synonym for serpentine, and *brucite* ($Mg(OH)_2$) nucleate and precipitate:

$$3\,Mg^{2+} + 2\,SiO_2\,(aq) + 5\,H_2O \rightarrow Mg_3Si_2O_5\,(OH)_4 \downarrow +6\,H^+$$

$$Mg^{2+} + 2\,H_2O \rightarrow Mg\,(OH)_2^- + 2\,H^+$$

Both reactions consume water and produce H^+, which promotes the dissolution of forsterite. Fayalite reacts with hydrothermal water to magnetite (Fe_3O_4) and H_2.

$$3\,Fe_2SiO_4 + 2\,H_2O \rightarrow 2\,Fe_3O_4 \downarrow + 3\,SiO_2\,(aq) + 2\,H_2$$

This reaction also consumes water. Although magnetite is a minor component of the serpentinization process, it is the main product of Fe^{2+} oxidation. Most of the water consumed by rock during serpentinization is consumed through the reactions to serpentine and brucite. The equilibrium pH of the hydrothermal fluid is nearly neutral at temperatures near 300°C but increases to about pH 11 at 50°C in serpentinizing sites like Lost City, because the solubility of brucite increases at lower temperatures, releasing dissolved Mg^{2+} and OH^- ions. The reactions of serpentinization are summarized in Figure 2.3.

2.2.3 Serpentinization Can Be Directly Observed in Real Time by Laser Raman Microscopy

A recent *in situ* serpentinization experiment by Lamadrid and coworkers provides important mechanistic insights into the interactions between olivine and hot pressurized salt water at hydrothermal conditions (280°C and 500 bar). The concentrations of all serpentinization products were monitored in the micropores of olivine by Laser Raman microscopy (Figure 2.4).

The formation of serpentine minerals, brucite ($Mg(OH)_2$), magnetite (Fe_3O_4), and H_2 consumes water, which is incorporated into the reaction products of serpentinization. As the water content in the pores decreases, the concentrations of salts and dissolved minerals increase. Ultimately, the pores are filled with a highly concentrated, 'crowded'

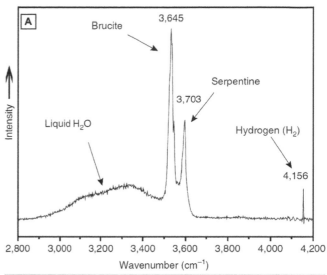

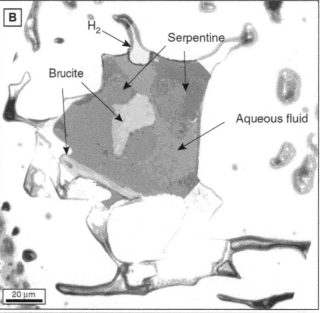

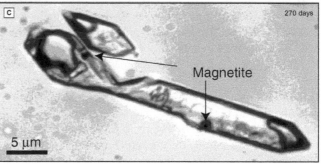

FIGURE 2.4 Fine-scale analysis of serpentinization products. The aqueous inclusions contain NaCl and $MgCl_2$ at molar ratios like in seawater (Na/Mg: 8/1) and total salinities of 1–10 wt%. (A) Laser Raman spectrum of the characteristic O–H bands of crystalline serpentine and brucite, the broad symmetric stretch vibration of liquid water, and the H_2 stretch vibration in the vapor bubble inside the olivine crystal. (B) Raman Map of the saltwater inclusion inside the olivine crystal: serpentine (red), brucite (green), aqueous solution (blue), and H_2 (yellow). (C) Photomicrograph of the saltwater inclusion inside the olivine crystal. After 15 days at 280°C, crystals of brucite and serpentine appear to be visible; after 45 days, the fluid inclusion is full of crystals and after 270 days magnetite crystals are clearly discernible. Modified with permission from Lamadrid H. M. *et al.* (2016) Effect of water activity on rates of serpentinization of olivine, *Nat. Commun.* 8: 1–9.

hydrothermal fluid with low water activity (most water is bound to salt ions), hydrogen gas, and catalytically active magnetite (and iron sulfides depending on the composition of the hydrothermal fluid).

2.2.4 GEOCHEMICAL HOMEOSTASIS: THE RATE OF SERPENTINIZATION IS REGULATED BY WATER SALINITY

Lamadrid and coworkers measured the salt concentration inside olivine pores during serpentinization by determining the freezing point depression ΔT_f of water as the function of salinity and reaction time. T_f was obtained from the melting temperature of the last ice crystal in olivine during serpentinization. ΔT_f of an ideal (highly diluted) solution of salt in water is

$$\Delta T_f = K_f I \ b$$

where K_f is the kryoscopic constant 1.853 K kg/mol, I is the number of ion particles per molecule of solute, for example, $I = 2$ for NaCl, $I = 3$ for $MgCl_2$, and b is the salt molality in mol salt/kg water. By comparing the initial weighted sodium chloride molality with the salt molality measured by the freezing point depression ΔT_f after a defined time of serpentinization (increased due to the consumption of water), they obtained the amount of water lost (in kilograms) at constant moles of salt. The rate of serpentinization was quantified by the amount of water removed from the aqueous solution as a function of time. Their main finding was that the *serpentinization rates decrease rapidly with increasing salinity (decreasing water activity)*. The serpentinization process has a salinity-dependent self-regulating mechanism, analogous to self-regulating mechanisms in biological systems.

At high salinity, serpentinization and H_2 synthesis stop until fresh water refills olivine pores. In the micro inclusions of olivine they investigated, serpentinization stops as soon as the water activity gets too low (meaning the salinity gets too high). New seawater with lower salinity must diffuse into the system to restart serpentinization. This scenario of fluctuating high and low water activity resembles wet–dry cycles known to be well suited for condensation reactions like polymerization of amino acids to peptides. Self-regulation of micropore salinity during serpentinization resembles ion homeostasis (control of ion concentration) in biological cells in many respects.

2.2.5 H₂ FROM SERPENTINIZATION REDUCES Nɪ²⁺ AND FE²⁺ IONS AND S INSIDE HYDRATED OLIVINE PORES

The mineral content in serpentinizing systems varies strongly with the environment surrounding a given system. If enough H_2 accumulates in surroundings that bear Ni^{2+}-containing compounds, native NiFe alloys such as awaruite (Ni_3Fe) can form.

$$FeO + 3 \ NiO + 4 \ H_2 \rightarrow Ni_3Fe + 4 \ H_2O$$

In fossilized vents formed during serpentinization, awaruite is observed as small grains. Likewise, iron sulfides Fe_3S_4 and/or FeS_2 or nickel sulfide NiS will be formed in sulfurous hydrothermal solutions. In the following, we will see that awaruite Ni_3Fe, greigite (Fe_3S_4), and magnetite (Fe_3O_4) are excellent transition metal catalysts for the fixation of carbon dioxide.

2.3 REDUCTION OF CO₂ WITH H₂ IS FACILE WITH TRANSITION METAL CATALYSTS

In the absence of catalysts, H_2 does not readily react with CO_2 at all. In the presence of suitable catalysts, the reaction is facile. This is because H_2 by itself is quite inert, but on the surface of transition metals, H_2 is readily activated. It dissociates into hydrogen atoms (H·) bound to the surface or into a hydride ion (H⁻) and a proton (H⁺). These hydrogen species are, in contrast to H_2 gas, very reactive. This process of binding, dissociation, and covalent bond formation to catalyst surfaces (chemisorption) is how transition metals activate H_2.

In chemical industry, magnetite (Fe_3O_4) is employed on a large scale as the catalyst for a variety of important chemical processes in which H_2 is used as the reductant, including the Haber–Bosch process, the name for the industrial process of H_2-dependent reduction of N_2 to NH_3 at high pressure and temperature. It is also a catalyst for Fischer–Tropsch synthesis, the name for the industrial process of H_2-dependent reduction of CO to hydrocarbons. In both processes, the reactions take place on the surface of a catalyst, not (or only to a very minor extent) in the aqueous or gaseous phase. Natural magnetite from serpentinization is therefore a clear candidate for prebiotic catalysis of carbon fixation and possibly also for primordial N_2 fixation using H_2.

Magnetite is also synthesized by serpentinization, the same process that generates H_2. Serpentinizing systems are an ideal site for organic synthesis because both the catalysts for CO_2 activation, Fe_3O_4 and Ni_3Fe, and the reductant, H_2, are synthesized in the same place (in the crust) and by the same process: interactions of rock with water. In Hadean serpentinizing systems, CO_2 was present not only in the seawater that was driving serpentinization but additionally as carbonates that had precipitated onto the vents of hydrothermal systems. Laboratory experiments, performed under simulated hydrothermal vent conditions with dissolved gasses and a catalyst submersed in an aqueous phase, demonstrated the ease with which CO_2 is reduced by H_2 to biologically relevant compounds.

2.3.1 TRANSITION METALS CONVERT H₂ + CO₂ TO PYRUVATE, THE CENTRAL MOLECULE OF METABOLISM

Experiments by Martina Preiner and colleagues simulating hydrothermal vent conditions show that the hydrothermal minerals awaruite (Ni_3Fe), greigite (Fe_3S_4), and magnetite (Fe_3O_4) catalyze the reduction of CO_2 with H_2 under alkaline aqueous conditions. That in itself is important for geochemical origins. Even more striking, however, is the product spectrum. Using only 25 bar total pressure of H_2 and CO_2 (ratio of the partial pressures 40:60) and incubation at

100°C for 16 hours, typical products in those experiments were formate (100 mM), acetate (100 µM), pyruvate (10 µm), methanol (100 µM), and traces of methane. At 20°C, only formate was detected, while at 60°C acetate also accumulates. At 2 bar pressure of H_2 and CO_2 and 100°C, formate and acetate were synthesized when using greigite (Fe_3S_4) as a catalyst. In its atomic structure, the mineral greigite resembles the iron–sulfur clusters of many modern enzymes. Notably, the reaction product spectrum that Preiner *et al.* observed is virtually identical to the products of the acetyl-CoA pathway of bacteria and archaea (Figure 2.5).

Of particular importance, formate, acetate, pyruvate, methanol, and methane were the *only* major products observed, although in some experiments ethanol was also found, which probably derived from pyruvate breakdown. There is a nearly one-to-one correspondence between the products of the biological pathway, which requires roughly ten enzymes and ten cofactors for the conversion of H_2 and CO_2 to pyruvate (as we will see in Chapter 3), and the

products of abiotic synthesis from H_2 and CO_2 using only hydrothermal minerals as catalysts. Stated another way, the catalytic function of the 10 enzymes in the acetyl-CoA pathway plus the roughly 100 enzymes needed to synthesize its 10 cofactors (127 enzymes total as calculated by Mrnjavac and colleagues) can be performed by a simple piece of metal, awaruite, or the simple iron oxide magnetite, which are both formed in hydrothermal vents during serpentinization.

This strongly suggests that, in early chemical evolution, the reaction sequence from H_2 and CO_2 to acetate and pyruvate, which is the backbone of carbon and energy metabolism in primitive anaerobic bacteria and archaea that live from H_2 and CO_2, occurred readily, at rapid rates, and with considerable product specificity long before the origin of genes or enzymes. The yields for acetate and pyruvate observed are furthermore in the range of concentrations, 10–100 µM, observed for typical metabolites of central metabolism in modern cells. Most metabolites in exponentially growing *E. coli* are typically in the 1–1000 µM range. Beyazay and

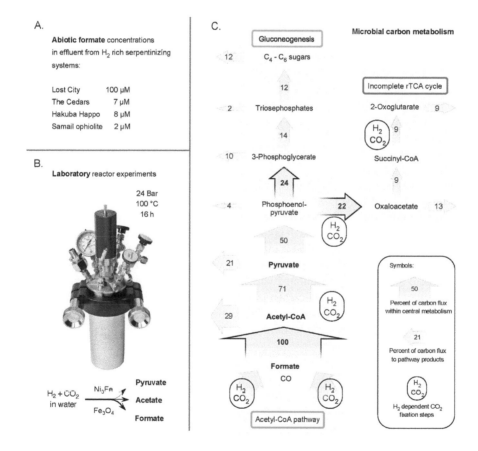

FIGURE 2.5 Similar reactions of H_2 and CO_2 in serpentinizing systems, laboratory reactors, and microbial carbon metabolism. (A) The effluent of modern, H_2-rich serpentinizing systems typically contains formate. Lost City effluent is particularly rich in formate (100 µM) (Lang S. Q., Brazelton W. J. (2020) Habitability of the marine serpentinite subsurface: A case study of the Lost City hydrothermal field, *Philos. Trans. R. Soc. A* 378: 20180429.). Other serpentinizing sites synthesize enough formate to support growth (Martin W. F. (2022) Narrowing gaps between Earth and life, *Proc. Natl Acad. Sci USA* 119: e2216017119.). (B) In laboratory experiments, CO_2 is readily reduced to formate >> acetate > pyruvate overnight in water using only Ni_3Fe or Fe_3O_4 as a catalyst. The same reaction set also takes place at 20°C. (Preiner M. *et al.* (2020) A hydrogen-dependent geochemical analogue of primordial carbon and energy metabolism, *Nat. Ecol. Evol.* 4: 534–542; Beyazay T. *et al.* (2023) Ambient temperature CO_2 fixation to pyruvate and subsequently to citramalate over iron and nickel nanoparticles, *Nature Comms.* https://doi.org/10.1038/s41467-023-36088-w). (C) A general map of carbon metabolism in microbes that use the acetyl-CoA pathway. The arrows indicate the flux of carbon through the steps shown. Modified from (Fuchs G., Stupperich E. (1978) Evidence for an incomplete reductive carboxylic acid cycle in *Methanobacterium thermoautotrophicum*, *Arch. Microbiol.* 118: 121–125; Stupperich E., Fuchs G. (1984) Autotrophic synthesis of activated acetic acid from two CO_2 in *Methanobacterium thermoautotrophicum*: I. Properties of *in vitro* system, *Arch. Microbiol.* 139: 8–13; and Fuchs G. (2011) Alternative pathways of carbon dioxide fixation: Insights into the early evolution of life? *Annu. Rev. Microbiol.* 65: 631–658).

colleagues obtained 200 μM pyruvate in experiments using Ni as the catalyst, equal to the cytosolic concentration of pyruvate that Furdui and Ragsdale meausured in acetogens growing on H_2 and CO_2. This particular abiotic reaction sequence, $H_2 + CO_2 \rightarrow$ formate, acetate, and pyruvate, is remarkably biological in character. As we will see in Chapter 3, roughly 90% of the carbon atoms in autotrophs that use the acetyl-CoA pathway for CO_2 fixation enter metabolism through that reaction sequence to pyruvate, and an additional 10% are incorporated in CO_2-fixating reactions of the (incomplete) reverse citric acid cycle that generate oxalacetate and α-ketoglutarate.

This in turn suggests that the abiotic reaction could have laid down a pattern of natural chemical reactions from which the backbone of autotrophic metabolism arose. Moreover, given a continuous supply of H_2 and CO_2, the reaction could, in principle, have run continuously for thousands, tens of thousands, or even hundreds of thousands of years in the far from equilibrium conditions of a serpentinizing hydrothermal vent—constantly producing important biological precursors and intermediates of metabolism. Because neither water nor CO_2 nor rock was in short supply on the early Earth, and because Lost City has been active for 30,000 years at least, a continuous and natural far from equilibrium geochemical reaction that synthesizes biological precursors from H_2 and CO_2 provides a starting point for chemical evolution that meshes 1:1 with the starting point of microbial metabolism. This suggests that there is something very natural about the most primitive form of central carbon metabolism, if we investigate the right conditions (hydrothermal conditions), the right reactants (H_2 and CO_2), and the right catalysts (iron minerals that form naturally in serpentinizing hydrothermal vents). Many modern hydrothermal systems synthesize enough formate to support the growth of microorganisms that are typically dependent upon H_2 and CO_2 for growth because they can convert formate into H_2 and CO_2. Note that carbon in the metabolism of primitive anaerobic autotrophs (Figure 2.5) is mainly assimilated via the acetyl-CoA pathway: roughly half of the fixed carbon is diverted to other syntheses at acetyl-CoA and pyruvate, and the remaining half is divided across reactions of gluconeogenesis and the incomplete reverse citric acid cycle.

Abiotic pyruvate and acetate have so far not been detected at Lost City, most likely because today's CO_2 concentrations are orders of magnitude lower than they were 4 billion years ago, leading to low product levels, and because the Earth's crust, including hydrothermal vents, is populated up to several kilometers deep with microbes that would readily consume such useful products of organic synthesis. It is estimated that roughly as much microbial biomass exists below the surface as all biomass above the surface, whereby most of that biomass below the surface is concentrated in sediment, although a substantial portion is present in crust.

A similar product spectrum of formate, acetate, and pyruvate was found in the reaction of native iron with CO_2 in water, whereby iron reacts with water to generate H_2, which likely served as the reductant. Though the exact mechanism of the mineral-catalyzed pathway is not known, the intermediate products that were not experimentally detected were certainly chemisorbed on the catalyst. In the biological pathway, methanol can be used as a substrate; it is not a

product of the reaction from H_2 and CO_2, but all reactions in the pathway are reversible.

Importantly, the reactions shown in Figure 2.5 did not take place at all in the absence of catalysts, indicating that both reactants had to be adsorbed onto the catalyst for the reaction to take place (see Section 3.1). In an earlier report by Horita and Berndt using Ni_3Fe as the catalyst, H_2 and CO_2 were almost completely converted to CH_4 in the course of weeks at 200–400°C and roughly 500 bar, but those conditions, though they simulate conditions of methane synthesis reactions that occur in the crust today, are too harsh for the accumulation of energy-rich (metastable) compounds such as pyruvate. Viewed from the standpoint of metabolism (Figure 2.5C), a central challenge of origins research is to find conditions under which organic molecules arise and react along routes that are similar to those in metabolism, but using only inorganic catalysts instead of enzymes. In this sense, the facile synthesis of pyruvate under sustainable hydrothermal conditions is a step in the right direction and the catalysts are decisive.

The enzymes that activate H_2 today, hydrogenases, all have transition metals at their active sites and the transition metals are not just there to stabilize protein structure, they physically perform the splitting of the H_2 molecule into two protons and two electrons via chemisorption (the formation of M–H bonds). These hydrogenase active sites are shown in the right-hand panel of Figure 2.6, while the left-hand panel of Figure 2.6 shows the inorganic structures that likely served as the catalytically active primordial precursors of the enzyme active sites: the naturally formed inorganic active sites that the enzymes came to replace during evolution.

In the acetyl-CoA pathway of CO_2 fixation, which we will encounter in more detail in Chapter 3, involves several key reactions in which carbon forms covalent bonds with active-site metals during the enzymatic reaction mechanism. The enzymes that activate CO_2, that transfer the methyl group for acetate formation, or that generate the C—C bond in acetate are metalloenzymes, with transition metals at their active sites. In the acetyl-CoA pathway alone, Fe, Ni, and Co form covalent bonds with reaction intermediates (Figure 2.7).

This commonality of chemisorption in catalysis shared between hydrothermal minerals and ancient enzymes is most readily understood as a kind of continuity in chemical evolution: Before there were enzymes, the transition metals in mineral form performed the catalysis for H_2 activation and CO_2 reduction. As genes and enzymes arose, the transition metals were incorporated into proteins *via* cysteine thiols. In the acetyl-CoA pathway, sulfur atoms do not perform any catalytic steps (though sulfur does act as a substrate in one step, see Chapter 3). The catalysis is always provided by electrons in the *d*-orbitals of transition metals, the sulfur atoms are just a means of holding the metals in place with the tools available to proteins (cysteine residues).

The role of the metals as opposed to the role of sulfur is something to keep in mind more generally, because there is a long (and fascinating) literature about the role of iron sulfides in early chemical evolution. The idea that FeS clusters are ancient goes back to biochemical literature from the 1960s (Fritz Lipman's papers, for example) and the discovery of ferredoxin, the first FeS protein characterized.

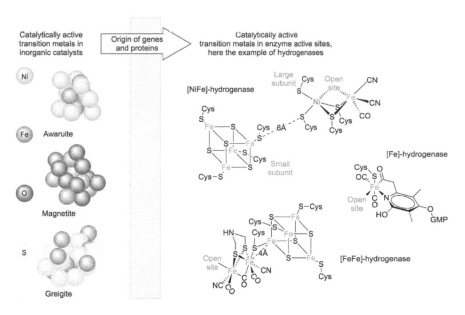

FIGURE 2.6   Possible evolutionary relationship linking minerals to FeS clusters in hydrogenases. All three minerals shown can catalyze H_2 activation (the dissociation of H_2 into protons and electrons) during abiotic CO_2 fixation. Once proteins arose in evolution, the ligands of the polypeptide chains allowed the catalytic properties of the transition metals to be preserved by the coordination of the metals into functional prosthetic groups of proteins, such as FeS and FeNiS clusters. The most common transition metal ligands in proteins are sulfur and cysteine thiols, however in [Fe]-hydrogenase nitrogen and carbon coordinate Fe as well. In the molybdenum cofactor MoCo (not shown), molybdenum is coordinated by two thiol groups of the cofactor rather than by the thiol groups of a protein. Hydrogenase active sites redrawn from Thauer R. K. *et al.* (2010) Hydrogenases from methanogenic archaea, nickel, a novel cofactor, and H_2 storage, *Annu. Rev. Biochem.* 79: 507–536.The dithiomethylamine ligand near the open site of FeFe hydrogenase is unique to this enzyme (Betz J. N. *et al.* (2015) [FeFe]-Hydrogenase maturation: Insights into the role HydE plays in dithiomethylamine biosynthesis, *Biochemistry* 54: 1807–1818). The minerals in the left panel are hand sketched on the basis of publicly available crystal structure data from www.mindat.org, an outreach activity of the Hudson Institute of Mineralogy, with irregularities deliberately introduced into the crystal lattice to convey the message that minerals in highly ordered crystal form were likely not required for inorganic catalytic activity at origins. Effective heterogeneous catalysts used in industry or the laboratory rarely, if ever, have a well-defined crystal structure and the catalytically active regions of the catalysts are rarely, if ever, known.

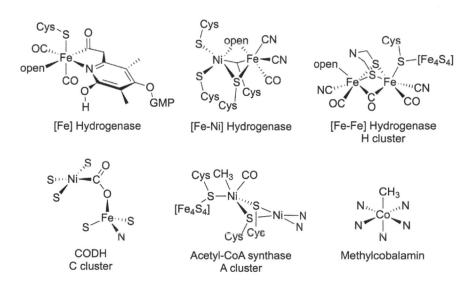

FIGURE 2.7   Transition metals in active sites of hydrogenases and enzymes of the acetyl-CoA pathway. Redrawn from Martin W. F. (2019) Carbon metal bonds, rare and primordial in metabolism. *Trends Biochem. Sci.* 44: 807–818. The CN^- and CO ligands of [Fe-Fe] hydrogenase stem from the enzymatic breakdown of tyrosine into cresol during hydrogenase maturation (Sheperd E. M. *et al.* (2021) HydG, the 'dangler' iron, and catalytic production of free CO and CN–: Implications for [FeFe]-hydrogenase maturation. *Dalton Trans.* 50: 10405–10422). The structures of the active sites for enzymes of the acetyl-CoA pathway derive in large part from work by Holger Dobbeck and Steve Ragsdale (see Gregg C. M. *et al.* (2016) AcsF catalyzes the ATP-dependent insertion of nickel into the Ni, Ni-[4Fe4S] cluster of Acetyl-CoA synthase, *J. Biol. Chem.* 291: 18129–18138; or Svetlitchnaia T. *et al.* (2006) Structural insights into methyltransfer reactions of a corrinoid iron-sulfur protein involved in acetyl-CoA synthesis. *Proc. Natl Acad. Sci. USA* 103: 14331–14336; Can M. *et al.* (2014) Structure, function, and mechanism of the nickel metalloenzymes, CO dehydrogenase, and acetyl-CoA synthase, *Chem. Rev.* 114: 4149–4174).

Wächtershauser developed the idea of FeS catalysis into the more general theory of an 'iron–sulfur world', coopting Walter Gilbert's metaphor of an 'RNA world' for metabolism. But with the help of crystal structures of active sites for enzymes of the acetyl-CoA pathway provided by Holger Dobbeck and colleagues as well as Steve Ragsdale and colleagues, we can see today that catalysis in the pathway is provided by the metals.

To underscore that point, one important message of Preiner's experiment is that iron–sulfur minerals are not required for catalysis, because awaruite (Ni_3Fe, pure metal) and magnetite (Fe_3O_4) also activate H_2 and CO_2 for reaction and generate products. This indicates that the incorporation of transition metals into proteins did not require direct FeS precursors. Rather, the coordination of Fe, Ni, and other transition metals as sulfides reflects constraints concerning the number of ways that metabolism can hold transition metals in place as heterogeneous catalysts. Pure metal clusters as in awaruite are so far unknown in biology. Magnetite is an iron oxide. A number of enzymes do harbor Fe–O bonds, usually involving pairs of Fe atoms coordinated in diiron centers; these diiron proteins are however typically O_2-utilizing, O_2-detoxifying, or O_2-activating enzymes. Molecular oxygen is a product of cyanobacterial photosynthesis (see Chapter 7) and did not exist on the early Earth in amounts that could have impacted chemical evolution and therefore can be excluded as a reactant at the origin of life. Therefore, there is no evidence for a role for diiron proteins in primitive *enzymes*, even though a primitive catalytic role for magnetite itself seems likely. Iron–sulfur centers probably reflect biology's way of getting transition metal catalysis into proteins more than they reflect a direct requirement for sulfide itself in prebiotic synthesis.

Once proteins had arisen in evolution (more on how that might have happened later), the catalytic properties of Fe and Ni incorporated into proteins as FeS and FeNiS centers, carried over from mineral precursors to protein-bound catalysts. This made the catalysts portable and offered evolutionary opportunities to adjust the way in which the metals are coordinated, thereby modulating their catalytic properties. For example, the Fe–Fe and Fe–Ni hydrogenases transfer single electrons through FeS clusters in the protein to FeS clusters in ferredoxins, whereas the [Fe] hydrogenase at the right in Figure 2.6 and left in Figure 2.7 transfers the electron pair of H_2 to an organic cofactor, F_{420}. Transition metal catalysis in minerals and in enzymes likely stand in a progenitor-descendant relationship in which the mineral-bound metals are ancient; once genes and proteins arose (a huge leap that we will deal with in later chapters), the transition metals, catalytically active by nature, were merely incorporated into a new catalytic site. There they could subsequently be fine-tuned by the protein environment to render reaction rate, regulation, and substrate as well as product specificity adjustable through natural variation and natural selection.

2.3.2 Molecular Nitrogen, N_2, Can Be Reduced with H_2 to NH_3 by Transition Metal Catalysts

Another problem of prebiotic synthesis is the fixation of molecular nitrogen to ammonia as a precursor of organic

nitrogen compounds. Iron sulfides from a freshly precipitated mixture of $FeSO_4$ and Na_2S in water were used as catalysts to synthesize ammonia from molecular nitrogen and hydrogen at 1 atm N_2 and 70–80°C by Dörr et al. in the group of Wolfgang Weigand. Clusters or nanoparticles of different Fe–S-moieties were probably produced from the fresh FeS precipitate and some of them catalyzed the ammonia synthesis. Commercial iron sulfide or aged iron sulfide did not generate ammonia. Hydrogen was obtained by the exergonic reaction

$$FeS(s) + H_2S(aq) \rightarrow FeS_2(s) + H_2(g)$$

where H_2S was generated by the reaction of FeS with sulfuric acid and s, aq, and g indicate solid, aqueous, and gas, respectively. The nascent H_2 is probably attached to the iron sulfide nanoparticles as Fe–H and reacts in this activated form with chemisorbed nitrogen by successive hydrogenation. The reaction yields for this FeS-catalyzed N_2 fixation were low (3 mmol NH_3 at 3 mol iron sulfide) and had to be carried out at large scale for two weeks to obtain enough ammonia for analysis. However, as the authors pointed out, iron sulfides and H_2S were abundant on the primordial Earth and their reaction with molecular nitrogen may well have contributed to the global ammonia budget in the Hadean era.

Nickel and iron metals and alloys are somewhat more effective catalysts for reducing molecular nitrogen with hydrogen to ammonia. At 200°C and 55 bars H_2/N_2 pressure, up to 2.5% yield of nitrogen reduction was obtained with elemental iron (Fe^0) as an effective catalyst; however, the quantity of background ammonia in the catalyst revealed by an argon blank test was large. Reduction of nitrite (NO_2^-) and nitrate (NO_3^-) to ammonia was complete (100% yield) at these experimental conditions; however, these reactions are not kinetically challenging, whereas N_2 reduction is. The reduction of N_2 with H_2 is exergonic, but the activation energy of this reaction is very high because the strong N≡N triple bond is very difficult to activate. In the industrial Haber–Bosch process, the reaction has to be performed at roughly 300°C and 500 bars to achieve good yields (on the order of 10% to 14% conversion of N_2 to NH_3 per passage across the catalyst). The high temperature and the large scale of ammonia production correspond to an enormous global energy input of roughly 1–2% of the world's annual energy production. For over 100 years, magnetite has been used as the catalyst for the Haber–Bosch reaction. A more cost-effective catalyst has not been found.

Effective fixation of dinitrogen to ammonia has traditionally been difficult, even under hydrothermal conditions and high H_2/N_2 pressures. In biological systems, there is only one entry point of N_2 into the nitrogen cycle: the enzyme nitrogenase. There are three variants of nitrogenase enzymes that are all related and contain Mo, V, or Fe coordinated by S and inhibited by O_2 at their active site cluster (Figure 2.8). In the biological reaction, the electrons are donated by one-electron carrier called ferredoxins with a midpoint potential on the order of −500 mV, slightly more negative than H_2 at physiological conditions. The active site of the enzyme is unique among all enzymes known not only by virtue of the Fe_7S_9

cluster but also due to an active site carbide carbon atom. The existence of this carbide is of interest because, in the Haber–Bosch process, nitrides are formed on the catalyst that may play an important role in the reaction. In the Fischer–Tropsch reaction, carbides are formed on the catalyst.

Nitrogenase is a remarkable enzyme. It is the only enzyme known to possess a carbide atom and C–N intermediates may play a role here too. It is the only molybdenum-dependent enzyme that does not utilize Mo ligated by the pterin molybdenum cofactor MoCo. The enzymatic reaction requires eight electrons from low-potential reduced ferredoxin and involves the generation of one molecule of H_2 for every molecule of N_2 reduced:

$$N_2 + 8\,H^+ + 16\,MgATP + 8\,e^-$$
$$\rightarrow 2NH_3 + H_2 + 16\,MgADP + 16\,P_i.$$

The hydrolysis of two ATP per e^- drives strong conformational changes in nitrogenase subunits with reduced FeS clusters that lower their midpoint potential. This conformationally induced generation and release of steric tension is similar to the process of an archer drawing and releasing the bow, for which reason the nitrogenase subunit has been designated as an 'archerase'. The nitrogenase active site catalyzes a number of reactions in addition to N_2 fixation, including hydrocarbon synthesis from CO. The fact that biological systems have not yet devised an alternative catalyst to nitrogenase for reducing N_2 indicates the existence of very strict mechanistic constraints on the reaction in terms of substrate–catalyst orbital interaction. Given the low-pressure reactions observed by Dörr et al. as well as the high temperature (300–800°C) and high pressure (ca. 1 GPa) necessary for the Fe^0-dependent N_2 reduction observed by Brandes et al., the nature of the nitrogenase catalyst and the highly reducing conditions in the ancient crust around serpentinizing systems, it seems likely that primordial geochemical systems were able to reduce N_2 to NH_3, probably to an extent that it was not rate limiting for N incorporation into organic compounds. Note however that when it comes to organic syntheses, local concentrations of activated and reactive N species in serpentinizing systems were likely far more important than the accumulation of stable N species in the ocean or the atmosphere. It is also possible that reactive N species were not in solution but restricted to the surfaces of catalysts instead.

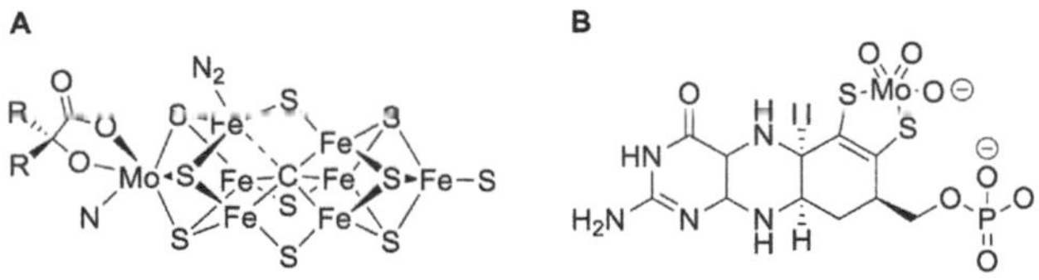

Experiments show that the fixation of CO_2 and N_2 with H_2 under hydrothermal conditions using naturally occurring catalysts from hydrothermal systems works in the laboratory. By inference, the fixation of CO_2 and N_2 with H_2 took place in the primordial crust during serpentinization with the help of transition metal catalysts that can chemisorb the unreactive gasses to activate them. The primordial ocean was about twice as deep as now because the crust and mantle have bound about one ocean volume of water in the last 4 billion years (see Section 1.3). The CO_2 concentration of the primordial atmosphere was possibly 100–1000 times higher than now, and huge amounts of CO_2 were, therefore, dissolved in the ocean. Convective water currents through the dry, olivine-containing, crust led to large-scale rock–water interactions and serpentinization.

2.4 CARBON AND NITROGEN FIXATION DEPEND ON TRANSFER OF HIGH-ENERGY ELECTRONS

2.4.1 ELECTRONS CAN BE SHARED BETWEEN ATOMS (COVALENT BONDS) OR TRANSFERRED (REDOX REACTIONS)

At its heart, life is chemistry. Chemistry is, in turn, about the **sharing** or **transfer** of electrons between atoms and molecules. This is a subtle point but one worth underscoring. Biochemical and geochemical reactions never alter the nucleus of an atom; they involve the redistribution of electrons between and among atomic nuclei. Reactions that distribute electrons in such a way as to undergo **sharing** by two atoms create chemical bonds. As we saw in Chapter 1.2, chemical bonds arise when electron pairs come to occupy orbitals that connect two nuclei in a covalent bond. Carbon's ability to form covalent bonds with so many different kinds of atoms, from hydrogen (H), with one proton, to nickel (Ni) with 28 protons, is one reason that it sits at the center of life's chemistry.

Tungsten (W) is the heaviest atom known that has a functional role in biology. It occurs in tungsten-dependent formate dehydrogenases (FDH) among others. Many FDH enzymes are Mo-dependent and use molybdopterin (Figure 2.8) as a cofactor. In some organisms, W can substitute for Mo and is coordinated by the molybdenum cofactor (Figure 2.8) to form tungstopterin. Mo-dependent FDH catalyzes the initial reaction in the acetyl-CoA pathway:

$$CO_2 + 2H^+ + 2e^- \rightleftharpoons HCOO^- + H^+$$

In contrast to the carbon metal bonds in Figure 2.7, current models for the FDH reaction mechanism do not suggest covalent bonding between the carbon atom in CO_2 and Mo (or W). Instead, CO_2 binds the active site via a Mo–O–C=O (by analogy, a W–O–C=O) intermediate. In general, C–W bonds are extremely stable. Tungsten carbide, CW, is one of the hardest substances known, almost as hard as diamond and with a density double that of iron. This is likely the reason that C–W bonds are not formed during the reaction. In the Mo-dependent FDH reaction mechanism, an electron pair

FIGURE 2.8 Molybdenum in the active site of nitrogenase. (A) The transition metal sulfide cluster with the proposed binding site for N_2 (active site). (B) The molybdenum cofactor molybdopterin. Nitrogenase is exceptional not only in the reaction it catalyzes but also in that it is the only Mo-containing enzyme that does not use the molybdenum cofactor MoCo (a pterin) to coordinate Mo. (A) Redrawn from Anderson J. S. et al. (2013) Catalytic conversion of nitrogen to ammonia by an iron model complex, *Nature* 501: 84–87.

leaves Mo^{+4} via a hydride donated by a Mo-bound sulfhydryl, to form a C–H bond with CO_2. The carbon atom in CO_2 is thereby reduced, changing the oxidation state from C^{+4} to C^{+2} because of the negative charges contributed by the two electrons of the hydride. The Mo atom undergoes oxidation from Mo^{+4} to Mo^{+6}, which it does readily: Mo (like W) is a good two-electron donor in biological reactions. Formate leaves the active center of the enzyme, taking the two electrons with it and leaving Mo as Mo^{6+}. Before the enzyme can complete another round of catalysis, the two electrons that were extracted from Mo^{4+} to yield Mo^{6+} have to be replaced. They are transferred from reduced FeS centers within the protein to regenerate the Mo^{4+} initial state (with the Mo–S–H hydride donor) so that the next round of catalysis can proceed. The electrons in the FeS centers have to be replenished in turn, which requires a source of electrons in the cytosol, which can be either one-electron carrier like ferredoxin or two-electron carriers like NADH. These carriers have to obtain their electrons from the environment, for example, from H_2 via hydrogenases in modern cells or via metal hydrides in prebiotic systems. The source of H_2 in serpentinizing systems is, as we have seen, reactions of rocks and water.

The process just described is a process of **electron transfer** from H_2 (leaving protons in solution as H_3O^+) to FeS centers to Mo to CO_2, reducing the latter to yield formate. In electron transfer reactions, electrons are not shared, they are transferred from one atom to another. The electrons have to come from somewhere, and they have to go somewhere. In modern biological systems and at life's origin, they come from an environmentally available reductant, an electron donor (such as H_2), which becomes oxidized, and are transferred to an oxidant, an electron acceptor (such as CO_2), which becomes reduced. Electron transfer reactions are therefore called reduction–oxidation reactions or **redox** reactions.

2.4.2   Redox Reactions Are Essential to Life

Redox reactions are integral to the chemistry of life. They are the drivers that push all other processes forward. The central energy-releasing reactions that allow all other reactions in metabolism to take place are redox reactions. For example, humans need oxygen to support ATP synthesis in mitochondria. In the human respiratory chain, O_2 is the terminal acceptor for electrons from amino acids, sugars, and fats in our diet. This generates the steady supply of ATP that keeps us alive; our ATP synthesis (humans synthesize about a bodyweight of ATP per day) depends on the redox reaction of organic foodstuffs and O_2. That redox reaction releases energy that the cell harnesses to stay alive. All organisms have some form of redox reaction at the core of their energy metabolism. It is a rule to which there are no exceptions. This rule also applies to parasites that do not have any ATP synthesis of their own. They still depend upon redox reactions and ATP synthesis, because they use the ATP that their host cells synthesize from redox reactions. That does not give them independence from redox reactions; it just makes them dependent upon redox reactions that other cells catalyze and harness.

The transfer of electrons from a donor to an acceptor releases energy. We will make acquaintance in this chapter with the Nernst equation, which relates the tendency of donor–acceptor pairs to transfer electrons to concentrations of reactants and products and thus to the change of free energy ΔG of the reaction. It is difficult to generalize in biology because there are always exceptions. But, generally speaking, the changes of free energy that cells harness for life are founded in the energy that is released during the transfer of electrons from donors to acceptors. The intrinsic dependence of the life process upon redox reactions impacts our views concerning the nature of origins problem. Origins is not just a process of chemical synthesis; it is chemical synthesis coupled to exergonic redox chemistry.

2.4.3   The Electron Energy in Atoms and Molecules Is Measured as Redox Potential

Hydrogen is a source of electrons; it is also the smallest molecule in nature, making it a highly useful and mobile source of electrons because, under all conditions on Earth, it is always a gas. On the modern Earth serpentinization generates roughly 3.7 cubic kilometers of hydrogen gas every year. On the early Earth H_2 production was greater. At the modern rate, serpentinization generates an amount of H_2 corresponding roughly to the volume of the atmosphere every billion years. But it does not accumulate because it is too light and diffuses into space. The hydrogenation of carbon dioxide is the key step of the primordial transition from inorganic to organic matter. Hydrogen gas can reduce CO_2 to organic molecules, particularly carboxylic acids, if the reaction is catalyzed by transition metals such as iron or nickel and the temperature and hydrogen pressure are high enough. The hydrogen atom consists of a proton, H^+, which is ubiquitous in water, and an electron e^-

$$H^+ + OH^- \rightleftharpoons H_2O$$

$$H^+ + e^- \rightleftharpoons 0.5H_2$$

Of course, electrons from sources other than H_2 can reduce CO_2 in water (a source of available protons) *if the electron energy is high enough to run the reduction in an exergonic reaction*. In chemistry, electrons are transferred between atoms or molecules in a *reduction–oxidation (redox)* reaction. The *redox potential* of an electron transfer reaction (sometimes called the *reduction potential* or *oxidation–reduction potential*) is a measure of the electron energy and can be determined in an electrochemical cell (Figure 2.9).

The apparatus to measure the redox potential of a donor–acceptor pair consists of a sample half-cell and a standard reference half-cell. Electrodes immersed in the cells are connected by a wire to a voltmeter. As an example, consider a standard sample solution of 1 M Fe^{2+} and 1M Fe^{3+} and a standard reference solution of 1M H^+ in equilibrium with 1 atm H_2 gas at 25°C (standard conditions). Then

$$Fe^{3+} + e^- \rightleftharpoons Fe^{2+}$$

$$2H^+ + 2e^- \rightleftharpoons H_2$$

If the reaction proceeds in the direction

$$\tfrac{1}{2} H_2 + Fe^{3+} \rightarrow Fe^{2+} + H^+$$

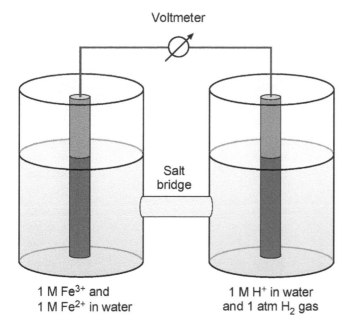

Voltmeter

Salt
bridge

1 M Fe³⁺ and
1 M Fe²⁺ in water

1 M H⁺ in water
and 1 atm H₂ gas

FIGURE 2.9 The electrochemical cell. Electrodes connected by a wire and a voltmeter measure the electron potential difference between a sample redox couple (here Fe²⁺/Fe³⁺) and the H₂/2H+ reference couple at standard conditions (also called a standard hydrogen electrode, or SHE). A salt bridge allows the flow of charge to keep electrical continuity between the half-cells but keeps their content otherwise separate.

then electrons flow from the standard reference solution to the sample solution and the sample-cell electrode is taken to be positive relative to the reference electrode. The voltmeter measures +0.77 V as the potential difference between the half-cells for the Fe²⁺/Fe³⁺ reaction. By convention, *the reduction potential of the 2H+/H₂ couple is defined to be 0 volts at standard conditions*. Then the standard redox potential E_0 of Fe²⁺/Fe³⁺ is +0.77 V.

To better imagine and memorize the meaning of a redox potential, consider the reverse of the process discussed above: an external voltage is applied to the electrochemical cell. Then the voltage necessary to detach the electron from its atomic or molecular carrier is a measure of the electron energy. Strongly bound electrons have low energy and can be detached only by a sufficiently positive voltage. Electrons that are weakly bound to their atomic or molecular carrier can be detached more readily by a smaller positive voltage or by a negative voltage if they can be detached more easily than from H₂. Hence, a smaller (or more negative) redox potential means higher electron energy and thus greater capability to reduce inorganic carbon to organic molecules. Spontaneous electron flow proceeds only from a carrier of stronger electron donating ability (a small or *negative* redox potential) to a carrier of stronger electron accepting ability (a more *positive* redox potential). This is like a stone rolling spontaneously only *down* a hill (from higher altitude to lower altitude). Gravity pulls the stone to Earth, and the tendency of atoms to sequester or donate electrons determines their flow. When electrons flow energetically downhill, they flow from more negative to more positive potentials. What gravity is to mass, redox chemistry is to electrons.

2.4.4   REDOX POTENTIAL DEPENDS ON CONCENTRATION, TEMPERATURE, AND ACIDITY

The concentrations of different redox-active chemicals, temperature, and pH value vary considerably in different hydrothermal habitats. These parameters can, and do, exert an influence on redox potentials. The *Nernst equation* relates the redox potential E_0 to the concentration of reactants and products of a half-reaction at equilibrium by

$$E = E_0 - \left(RT/n_e F \right) 2.303 \log \left(a_{Red}/a_{Ox} \right)$$

in which E_0 is the standard redox potential in volts, R is the *gas constant* (8.314 J/mol/K), T is the temperature in Kelvin, n_e is the number of electrons transferred, F is the *Faraday constant* (96,484.3 J/mol/V), and a_{Red} and a_{Ox} are the effective concentrations (also called 'activities') of reductants and oxidants. By convention, reduction potentials refer to half-reactions written as reductions: Ox + ne^- → Red, and therefore a_{Red}/a_{Ox} is the equilibrium constant of the reduction reaction.

For the hydrogen reference electrode, we obtain log a_{Red}/a_{Ox} = log $a_{H2}/(a_{H+}·a_{H+})$ = log 1/(1·1) = 0 and therefore E = 0 mV. In most textbooks of biology, standard redox potentials are given at pH 7 rather than at pH 0. At pH 7, 1 atm H₂ and 25°C, the redox potential of hydrogen from 2H⁺ + 2e⁻ ⇌ H₂ with [H⁺] = 10⁻⁷ M at pH 7 is:

$$E = E_0 - 29.6 \log 1/\left(10^{-7} \times 10^{-7} \right) mV$$

$$= 0 \, mV - 29.6 \cdot 14 \, mV = -414 \, mV$$

Serpentinizing hydrothermal vents have very different H₂, pH values, and temperatures depending on the specific site. The hydrogen redox potential (measure of electron energy of hydrogen) varies to a large extent (Table 2.1).

Consider the low values of the redox potential (high electron energy; strong reducing power) of H₂ at high pH (alkaline milieu), high temperature, and high H₂ partial pressure in Table 2.1. This makes H₂ a powerful reductant at alkaline hydrothermal conditions within a serpentinization site. Hydrothermal vents of serpentinizing systems contain about 10 to 50 mmol H₂/l fluid (10–50 mM).

For obtaining the redox potential we must calculate the pressure of hydrogen above the fluid because the Nernst equation relates to pressure in the gas phase above the electrode for a reacting gas, not to the concentration of the gas in the solution of the electrochemical cell. By using Henry's law $p = Hc$ with Henry's constant $H = 1300$ l·atm/mol for H₂ gas solved in water, c as concentration of H₂ solved in water, and p as pressure of H₂ above water, we obtain p = 1300 atm/M·15 mM = 19.5 atm, which corresponds at 90°C to a redox potential of −695 mV at pH 9 and −839 mV at pH 11. Even for very low H₂ concentrations of 0.25 mmol/kg seawater (0.33 atm H₂ gas above the sample) measured at Lost City's vents at 40°C and pH 9, the redox potential is −544 mV. The Henry constant is for 1 atm external (air column) pressure and fresh water, but at high water column pressure of, for example, 100 bar in 1000 m deep sea the Henry constant is probably larger such that a lower H₂ partial pressure is sufficient to achieve the 15 mM concentration in

TABLE 2.1

Some Midpoint Potentials E for $H_2 \rightarrow 2H^+ + 2e^-$

H_2 (atm)	pH	Temperature (°C)	E (mV)
10	10	100/200	−777/−1128
1	10	100/200	−740/−940
0.1	10	100/200	−703/−893
10	9	100/200	−703/−893
1	9	100/200	−666/−846
0.1	9	100/200	−629/−799
10	8	100/200	−629/−799
1	8	100/200	−592/−752
0.1	8	100/200	−555/−705
10	7	100/200	−555/−705
1	7	100/200	−518/−657
0.1	7	100/200	−481/−610

Note: Note the large effect of pH on E. This is because at alkaline pH, the reaction of H^+ with OH^- serves as a pulling reaction.

TABLE 2.2

Some Standard Midpoint Potentials

Donor	Acceptor	n_e	E_0' (V)
Fe^0	$Fe(OH)_2$	2	−0.88
Acetaldehyde	Acetate	2	−0.60
Fe^0	Fe^{2+}	2	−0.44
HCOOH	$CO_2 + 2H^+$	2	−0.43
H_2	$2H^+$	2	−0.41
HS^-	S^0	2	−0.27
Ni^0	Ni^{2+}	2	−0.25
Ethanol	Acetaldehyde	2	−0.20
HS^-	HSO_3^-	6	−0.12
Fe^{2+}	Fe^{3+}	1	0.77
H_2O	$\frac{1}{2} O_2 + 2H^+$	2	0.82

Note: In this table E_0' is the redox potential at pH 7, 1 atm H_2, 25°C, and n_e is the number of transferred electrons. The electron donor is also called the reductant; the electron acceptor is also called the oxidant. E_0' refers to the half-cell reaction written as oxidant + $ne^- \rightarrow$ reductant.

water; salinity decreases the Henry coefficient. Serpentinizing alkaline hydrothermal vents like those in Lost City (pH 9–11, up to 90°C effluent temperature and 15 mM H_2 partial hydrogen concentration in the fluid at 90°C) have especially negative redox potentials.

Methane-forming microbes known as *methanogens* stop oxidizing H_2 at about 1–10 Pa (10^{-5} to 10^{-4} bar) partial H_2 pressure in the gas phase above the microbe culture and grow at $p_{H2} > 10^{-4}$ bar. Hence more than enough H_2 is available as a growth substrate for methanogens at a serpentinization site. High-temperature tolerant (*hyperthermophilic*) methanogens are very much at home in serpentinizing deep-sea hydrothermal vents.

The redox potentials of many prebiotically important reduction–oxidation couples are known at pH 7 and shown for some selected examples in Table 2.2. A more comprehensive list is given in Appendix 1.

The comparison between Tables 2.1 and 2.2 shows that hydrogen at pH 7 and modest 25°C ($E_0' = -0.414$ V) cannot reduce CO_2 to HCOOH (−0.43 V), whereas the reaction *is* possible under hydrothermal conditions at alkaline pH, because the midpoint potential of the H_2 oxidizing reaction becomes sufficiently negative to permit CO_2 reduction. But, even at very negative potentials, a catalyst is necessary to activate CO_2 for reduction at biologically relevant rates.

At suitable hydrothermal conditions, H_2 from serpentinization can reduce Fe^{2+} and Ni^{2+} to elemental Fe^0 and Ni^0. For example, at 1 atm H_2, pH 10 and 200°C the redox potential of H_2 is −940 mV (Table 2.1) and thus more negative than that of $Fe(OH)_2$ (Table 2.2). At pH 7 and lower, the iron and nickel ions can dissolve in water. Then their redox potentials are more positive (Table 2.2: −440 and −250 mV), and they can be reduced to the elemental state

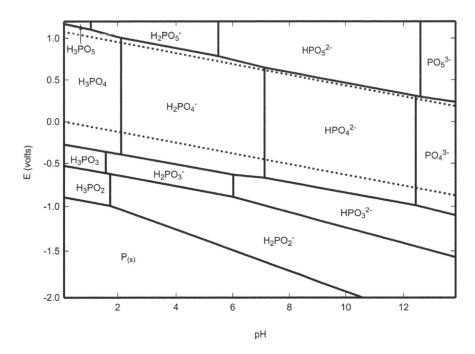

FIGURE 2.10 **Phase diagram for aqueous phosphorous for redox potential E (volts) *vs.* pH.** The more reduced forms H_2PO^{2-} with $P1^+$ are unstable in water. See text. Redrawn from data in Pasek M. A. (2008) Rethinking early Earth phosphorus geochemistry, *Proc. Natl Acad. Sci. USA* 105: 853–858.

even at mild temperatures and hydrogen pressures. The metallic mineral *awaruite* $FeNi_3$ is formed that way at serpentinization sites. Awaruite is a very effective catalyst for reduction of carbon dioxide to various organic molecules as discussed in Section 2.3.

The strongly reducing conditions at serpentinizing sites can, in principle, lead to unusual oxidation states of elements in a chemical compound. For example, the phosphorus in phosphoric acid H_3PO_4 has the oxidation state +5, written P^{5+}, P(V), or P^{V+}. At pH ≥ 7, the acid loses one or two protons yielding dihydrogenphosphate ($H_2PO_4^-$) and hydrogenphosphate (HPO_4^{2-}). These ions can, in principle, be reduced with hydrogen of $E < -0.8$ V to the phosphite ion HPO_3^{2-} having a P^{3+} oxidation state. The redox potentials of the phosphorous acids are highly pH dependent. This is shown in Figure 2.10, which is a phase diagram for P in water as a function of pH and midpoint potential E. The dark lines in the phase diagram delineate conditions in which the form of P indicated will tend to be thermodynamically stable at the pH and redox potentials shown. For example, at pH 5 and $E = +500$ mV, P should exist as dihydrogen phosphate ($H_2PO_4^-$). But at pH 10 and -900 mV, P should exist as the more reduced form of hydrogenphosphite (HPO_3^{2-}).

Phosphite ions can, in principle, reduce and activate organic compounds at serpentinization sites, leading to unusual chemistry that has hitherto been little explored. Whether phosphite played a role in geochemical reactions that gave rise to life is so far unknown. The reactions of central carbon and energy metabolism in most microbes only involve phosphate. Some microbes obtain energy and electrons from phosphite, however, as work by Bernhard Schink has shown. It is possible that the highly reducing conditions of hydrothermal vents helped to leach phosphate

from host rocks by reducing it to the more soluble, more reactive, and more strongly reducing but less stable, phosphite ion.

2.4.5 THE FREE ENTHALPY OF A REDOX REACTION CAN BE CALCULATED FROM THE REDOX POTENTIALS

The change in free enthalpy of a redox reaction $\Delta G^{0'}$ at pH 7, 1 atm gas pressure, and 25°C is related to the change in reduction potential at pH 7 ΔE_0' by

$$\Delta G^{0'} = -n_e F \Delta E_0{}'$$

where $\Delta G^{0'}$ is expressed in kilojoules per mol (kJ/mol) and ΔE_0' is in volts (V).

Using this formula, the free enthalpy change of a redox reaction can be readily calculated from the redox potentials of the reactants. For example, consider the reduction of carbon dioxide with hydrogen to formic acid (formate ion at pH 7):

$$CO_2\left(aq\right) + H_2 \rightarrow HCOOH\left(HCOO^- + H^+\right) \quad (A)$$

Recall that, by convention, reduction potentials refer to half-reactions written as reductions: oxidant $+ ne^- \rightarrow$ reductant. Hence, with reduction potentials from Table 2.2

$$CO_2\left(aq\right) + 2H^+ + 2e^- \rightarrow HCOOH \quad E_0{}' = -0.432 \text{ V} \quad (B)$$

$$2H^+ + 2e^- \rightarrow H_2 \qquad E_0{}' = -0.414 \text{ V} \quad (C)$$

Reaction A can be obtained by adding reaction B and the reverse of reaction C. To reverse reaction C to $H_2 \rightarrow 2H^+ + 2e^-$, the sign of E_0' must be changed: $E_0' = +0.414$ V. For reaction B, the free reaction enthalpy is:

$$\Delta G^{0'} = -2 \cdot 96.485 \, \text{kJ mol}^{-1} \text{V}^{-1} \cdot -0.432 \text{V} = +83.36 \, \text{kJ mol}^{-1}$$

Likewise, we obtain for the reverse of reaction C:

$$\Delta G^{0'} = -2 \cdot 96.485 \, \text{kJ mol}^{-1} \text{V}^{-1} \cdot +0.414 \text{ V} = -79.89 \, \text{kJ mol}^{-1}$$

Hence the free enthalpy of reaction A at pH 7 is given by:

$$\Delta G^{0'} = \Delta G^{0'} \left(\text{for reaction B}\right) + \Delta G^{0'} \left(\text{for reverse of reaction C}\right)$$

$$= +83.36 \, \text{kJ mol}^{-1} + \left(-79.89 \, \text{kJ mol}^{-1}\right) = +3.47 \, \text{kJ mol}^{-1}$$

Reaction A is slightly endergonic according to this calculation. H_2 has to be activated by metals or by an enzymatic process called electron bifurcation as described in the following for the reaction to proceed.

We see from Equations (B) and (C) that the midpoint potential of the H_2–proton couple is more positive than that of the CO_2–formate couple or, in other words, that the reaction is endergonic. To reduce CO_2 under physiological conditions, the electrons would effectively have to flow uphill, against the thermodynamic imperative. To circumvent this problem, cells reduce CO_2 with electrons from a donor with a more negative midpoint potential than the CO_2–formate couple. In biological systems, the physiological donors of electrons for CO_2 reduction are **ferredoxins**.

Ferredoxins are small proteins having on the order of 50 to 100 amino acids. They possess either one or two 4Fe4S iron–sulfur clusters. Iron–sulfur clusters (FeS clusters) are inorganic **prosthetic groups** in proteins. A prosthetic group is a functional group of a protein that is incorporated into the protein after its synthesis on the ribosome. Prosthetic groups can be covalently or non-covalently bound to the protein. The ferredoxins that cells use for CO_2 reduction typically have clusters that possess 4 Fe and 4 S atoms each, whereby the Fe atoms are additionally coordinated by covalent bonds

to the thiol sulfur from the side chain of cysteine residues in the peptide chain, as shown in Figure 2.11. Ferredoxins have one function in cells: they transport electrons, one at a time. Ferredoxins are one-electron carriers, as opposed to NAD(P)H, the most common two-electron carrier.

The oxidation state of the iron atoms in the oxidized state of the 4Fe4S cluster is typically such that two atoms are formally Fe^{2+} and two are formally Fe^{3+}. But just as in the case of the conjugated double bonds of the benzene ring, the electrons are delocalized, each Fe atom having an average oxidation state of $Fe^{2.5+}$. This allows the cluster to readily accept one additional electron, giving the cluster in the reduced form of ferredoxin (usually abbreviated as Fd^- or Fd_{red}) an overall increase in negative charge relative to the oxidized cluster in oxidized ferredoxin (typically written as Fd or Fd_{ox}). The resulting average oxidation state of iron in the reduced ferredoxin is $Fe^{2.25+}$. The redox reaction is readily reversible (Figure 2.11).

In living anaerobic cells, ferredoxin typically has a concentration in the cytosol on the order of 80–400 μM. That is a very high concentration for a protein; it lies in the same range as intermediates of the citric acid cycle, such as succinate. That high concentration reflects ferredoxin's role as a metabolite (effectively soluble single electrons) and reflects the importance of ferredoxins in the metabolism of anaerobes. Ferredoxins are not enzymes; nor are they catalysts; they are one-electron carriers, cofactors made of protein. In living anaerobes, most of the ferredoxin in the cell is reduced, giving the cytosol an overall low midpoint potential. There are the other electron carriers in cells. In anaerobes, **flavodoxins** are common, small proteins that use **flavins** (pterin-derived redox cofactors) instead of FeS centers for one-electron transfers. Flavodoxins can have similar midpoint potentials as ferredoxins and often substitute for ferredoxins under growth conditions where iron is limiting. Other common one-electron carriers are cytochromes, proteins that contain a single iron atom, coordinated in a tetrapyrrole, which can accept or donate an electron, with the iron atom alternating between the Fe^{3+} and Fe^{2+} oxidation states. Ferredoxins are generally regarded as more ancient than flavodoxins, because the electron-carrying prosthetic group is inorganic (FeS) rather than organic (flavin), and both are generally regarded as more ancient than cytochromes. Some acetogens and methanogens that live from H_2 and CO_2 have no cytochromes at all, which likely reflects an ancient ancestral state.

FIGURE 2.11 Structure of a ferredoxin, a protein essential to electron transfer. Ferredoxin from *Clostridium pasteurianum*, a ferredoxin with two 4Fe4S clusters (left; PDB 1CLF). The redox reaction in a 4Fe4S cluster (right). The electron is not associated with one specific Fe atom in the cluster but is delocalized across all four. The cysteine sulfur ligands are indicated. From Bertini I. *et al.* (1995) Solution structure of the oxidized 2[4Fe-4S] ferredoxin from *Clostridium pasteurianum*, *Eur. J. Biochem.* 232: 192–205.

The ferredoxins that are used for CO_2 fixation are often called low-potential ferredoxins because their midpoint potential in the cell is on the order of −500 mV (there are also high-potential ferredoxins or high-potential iron–sulfur proteins with a midpoint potential on the order of +300 mV). As Georg Fuchs has pointed out, this very low midpoint potential of ca. −500 mV is required for several key reactions of anaerobic autotrophs—including CO_2 reduction to methyl groups, CO_2-dependent pyruvate synthesis, and CO_2-dependent 2-oxoglutarate synthesis—to go forward. The midpoint potential of low-potential reduced ferredoxin is sufficiently negative to allow the electrons to flow energetically downhill to CO_2.

The question of how cells generate low-potential reduced ferredoxin (−500 mV) with electrons from the more positive electron donor H_2 (−414 mV) was a longstanding puzzle at the basis of anaerobe bioenergetics. The puzzle was only recently solved by the team of Wolfgang Buckel and Rolf Thauer, who discovered the process known as flavin-based electron bifurcation or electron bifurcation for short. In *electron bifurcation*, one electron from H_2 flows uphill to generate the low-potential reduced ferredoxin, while the other electron is transferred downhill to NAD^+, for example, which has a midpoint potential of −320 mV in

the physiological conditions that exist in the cell. Because one reaction is strongly downhill in terms of thermodynamics, it can permit the uphill reaction to take place. This is an example of energetic coupling in metabolism. The uphill electrons have enough energy to reduce Fd, and Fd_{red} can reduce carbon dioxide to formate, but the reaction has to be coupled to another energy-releasing reaction, for example, acetate synthesis in acetogens or methane synthesis in methanogens so that the whole process can go forward.

At alkaline hydrothermal sites, however, H_2 has a sufficiently negative redox potential to reduce hydrated carbon dioxide to formate—electron bifurcation is thermodynamically not necessary. The effect of pH on the midpoint potential of the reaction $H_2 \rightleftharpoons 2H^+ + 2e^-$ is substantial (Table 2.1). This is because at alkaline pH, OH^- ions scavenge the protons generated from H_2 oxidation, shifting the equilibrium of the reaction $H_2 \rightleftharpoons 2H^+ + 2e^-$ to the right. The more alkaline, the more negative the midpoint potential of the hydrogen oxidation reaction. We can observe this principle in action in the experiments of Preiner *et al.* that we discussed in Section 2.3. They were able to reduce CO_2 to formate, acetate, pyruvate, and methane using H_2 but without ferredoxin, without electron bifurcation, without any protein for that matter, and without ATP, phosphate, thioesters, ion

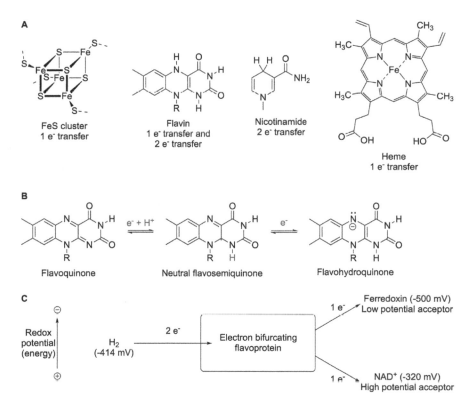

FIGURE 2.12 **Biological electron carriers.** (A) Prosthetic groups that can donate single electrons in one-electron reactions or electron pairs (hydride) in two-electron reactions. FeS clusters, flavins, nicotinamide, heme. (B) Flavins can participate in one-electron reactions and two-electron reactions because the flavin semiquinone can accept or transfer single electrons or electron pairs. Flavins are derived in biosynthesis from pterins, which are ancient cofactors and the functional moiety of tetrahydrofolate and tetrahydromethanopterin, the carriers of C1 units in the acetyl-CoA pathway. Hemes are derived from late intermediates of cobalamin biosynthesis. Cobalamin is required in the acetyl-CoA pathway as a methyl transfer cofactor; heme is not. Like O_2-dependent reactions, heme-dependent reactions were probably not present at the origin of life; they are the result of later evolutionary processes in free-living cells. By contrast, FeS clusters, flavins, and pterins were involved in the chemical reactions of the most primitive life forms. (C) Electron bifurcation utilizes the ability of flavins to transfer electrons in pairs or in one-electron transfer. As one example, the electron pair in H_2 can be split, or bifurcated, with one electron transferred to a high-potential acceptor such as NAD+ and the other to a low-potential acceptor such as ferredoxin. Redrawn from Buckel W., Thauer R. K. (2018) Flavin-based electron bifurcation, a new mechanism of biological energy coupling, *Chem. Rev.* 118: 3862–3886.

gradients, or anything else that cells use as energy currency. They performed their reactions with 10 bar H_2 at 100°C and pH 9, which corresponds to 8 mM H_2 solved in water very similar to the conditions at Lost City. Under those conditions, the midpoint potential of the H_2 oxidation reaction is roughly −700 mV, far more negative than reduced ferredoxin; hence, no electron bifurcation was needed. However, the reaction did require transition metals as catalysts, in the elemental form, as oxides, or as sulfides. The energy in those reactions is the pure chemical energy of redox reactions. No light was required, UV or visible, no supporting high-energy compounds had to be added, no phosphorus in any form, no bolide impact, and no wet–dry cycles were needed. H_2 and CO_2, when combined, are themselves a high-energy mixture under hydrothermal conditions (high pH and transition metal catalysts). Energy is released in the reaction of H_2 with CO_2 during the synthesis of formate, acetate, and pyruvate. It is both electrochemical energy and Gibbs free energy, two quantities which are two sides of the same coin and interconverted by the relation between free enthalpy and cell potential. Without a catalyst, the reaction of H_2 with CO_2 will effectively not take place: it proceeds at such a slow rate as to be irrelevant to biological processes.

2.5 CHAPTER SUMMARY

Section 2.1 All molecules that contain carbon and hydrogen are considered organic. CO_2 dissolved in water is the main prebiotic source of carbon. It has to be reduced with activated H_2 to obtain organic molecules. The moon-forming impact transformed carbon-containing rocks to gaseous CO_2 in the atmosphere. Atmospheric CO_2 dissolved gradually in the oceans and precipitated there mainly as insoluble magnesium-, calcium- or iron carbonates that sedimented and were ultimately transferred to the mantle by subduction. The CO_2 that remained in the ocean as hydrated CO_2, H_2CO_3, HCO_3^-, and CO_3^{2-} was chemically inert and had to be activated on transition metal surfaces to react with activated H_2 to organic molecules.

Section 2.2 Olivine-containing rocks in the upper mantle have very low silicate and very high Mg^{2+} and Fe^{2+} content and are called ultramafic (from *ma*gnesium and *f*errum). The reduced iron ions in ultramafic rocks react with hot seawater circulating in 1–5 km deep fissures of the crust to hydrogen, iron oxides like magnetite (Fe_3O_4), brucite ($Mg(OH)_2$), and the mineral serpentine. This process of serpentinization has been going on since the first water condensed on the Earth 4 billion years and is still ongoing today. Roughly one ocean volume of water is bound in the crust and mantle, meaning that the primordial oceans were roughly twice as deep as today's.

Section 2.3 Catalysts produced by serpentinization can include Fe^0, Ni^0, $FeNi_3$, Fe_3S_4, and Fe_3O_4, depending on the composition of the hydrothermal fluid and the reduction power of H_2. One of the best-investigated serpentinization sites is the Lost City hydrothermal field located on the Atlantis Massif in the Atlantic Ocean. Its effluent contains large concentrations of isotopically heavy (^{12}C depleted, abiotic) molecule methane. The process of serpentinization with the generation of hydrogen, serpentine, brucite,

and magnetite can be followed in real time in micropores of olivine by Laser Raman Microscopy. The investigations showed that the serpentinization rates in the pores decrease rapidly with increasing salinity (decreasing water activity), and new seawater with lower salinity has to diffuse into the system by self-regulation to restart serpentinization. Hence high and low water activity fluctuates in serpentinizing olivine micropores which resemble wet–dry cycles known to be well suited for condensation reactions like polymerization of amino acids to peptides. The gas H_2 by itself is quite inert. On the surface of transition metals, however, H_2 is readily activated for reaction: it dissociates into hydrogen atoms (H·) bound to the surface or into a hydride ion (H⁻) and a proton (H⁺). Transition metal compounds produced by serpentinization like magnetite (Fe_3O_4), awaruite ($FeNi_3$) and greigite (Fe_3S_4) activate hydrogen and CO_2, which then react to energy-rich carboxylic acids that are central to metabolism and N_2 to NH_3 in laboratory experiments. Once proteins arose in evolution, ligands of the polypeptide chains like cysteine thiols coordinated these catalytically active transition metal compounds into functional prosthetic groups of proteins, such as FeS and FeNiS clusters in ferredoxins, hydrogenases, and the CO-generating and CO-methylating enzyme carbon monoxide dehydrogenase acetyl-CoA synthase.

Section 2.4 Hydrogen reduces CO_2 to organic compounds and N_2 to ammonia by transfer of its electrons only if the electrons have sufficient energy to allow hydrogenation occur spontaneously in an exergonic reaction. According to the Nernst equation, the electron energy of hydrogen is very high (its redox potential is very negative) at the high pH (alkaline milieu), high temperature, and high H_2 partial pressure of a serpentinizing alkaline hydrothermal system like Lost City. Hydrogen at pH 7 and 25°C has a redox potential of −0.414 V and cannot reduce CO_2 to HCOOH (−0.43 V), whereas this is possible under hydrothermal conditions at alkaline pH 8 to pH 10, where the midpoint potential of H_2 can reach −0.5 to −0.9 V. To reduce CO_2 with electrons from H_2, cells need a complicated process called electron bifurcation, which increases the energy of one of two electrons at the expense of the energy of the other electron to allow more endergonic reduction reactions to occur. This is not necessary for electrons of H_2 at alkaline serpentinization sites. At suitable hydrothermal conditions, H_2 from serpentinization can even reduce Fe^{2+} and Ni^{2+} to elemental Fe^0 and Ni^0. Newer findings suggest that serpentinization can even reduce phosphate ions (P^{5+}) to phosphite ions (P^{3+}). Transition metal catalysts are necessary to drive CO_2 reduction with H_2 at biologically relevant rates.

* * *

PROBLEMS FOR CHAPTER 2

1. *Carbon dioxide in water.* Calculate the concentration of CO_2, H_2CO_3, HCO_3^-, and the pH in water at 5 bar carbon dioxide pressure above the water solution. Hint:Use equations in chapter 2.1.1.

2. *Serpentinization.* What is the reducing agent in the serpentinization reaction and what are the reaction products?

3. *Water activity during serpentinization.* How do salt concentration and water activity change during serpentinization?

4. *Transition metal catalysis.* How do transition metals catalyze hydrogenation reactions?

5. *Carbon dioxide fixation.* What is the name of the only exergonic pathway of biological CO_2 fixation that operates with H_2 and CO_2, and what are its products?

6. *Redox potential.* Calculate the redox potential of H_2 at 80°C, 0.5 atm H_2 gas above the water solution, and pH 10.

7. *Redox potential and free reaction enthalpy of acetaldehyde oxidation.* Calculate the free reaction enthalpy and the redox potential for the reaction CH_3CHO (20.83) + H_2O (−157.28) → CH_3COOH (−249.46) + $2H^+$ + $2e^-$ at standard conditions and compare them with the redox potential in Table 2.2. The values in parentheses are the standard free enthalpies of formation in kJ/mol at zero ionic strength taken from the *Handbook of Biochemistry and Molecular Biology*, CRC Press. 5th Ed. 2018, pp. 572–575. Hint: Use the relation between the free reaction enthalpy and the redox potential in section 2.4.5.

8. *Redox potential and free reaction enthalpy of pyruvate reduction.* Write the reaction equation for the reduction of pyruvate (−352.40) to lactate (−316.94) and calculate the free reaction enthalpy and the redox potential at standard conditions; compare with the redox potential in Table A1.1. The values in parentheses are the standard free enthalpies of formation in kJ/mol at zero ionic strength taken from the *Handbook of Biochemistry and Molecular Biology*, CRC Press. 5th Ed. 2018, pp. 572–575.

9. *The pH dependence of H_2 oxidation catalyzed by metals.* When hydrogen gas in water meets native iron surfaces, Fe(0), it chemisorbs. The H–H bond becomes homolytically cleaved and the two resulting hydrogen atoms undergo covalent bonding with iron atoms on the metal surface (two separate Fe–H bonds form). Those H atoms can diffuse on the surface in two dimensions. The midpoint potential of the iron-catalyzed hydrogen oxidation reaction becomes increasingly negative in the pH range 8 to 11, the range found in serpentiniziung vents. Serpentinzing vents also contain Fe(0). A) Write a chemical reaction that could help to explain why hydrogen becomes a stronger reductant at high pH. B) Can you propose a schematic mechanism for that reaction on interface of the iron surface with the aqueous phase? (Assume the presence of a soluble one electron acceptor like ferredoxin.)

10. *Compression energy.* Calculate the change of free compression enthalpy of an ideal gas from above surface to 1000 m and 5000 m water depth and to 5000 m sediment depth (Hint: $\Delta G = RT \ln p/p_0$; pressure increase 100 bar/km water depth, 250 bar/km sediment depth. P_0 is the atmospheric pressure and p is the hydrostatic pressure at the corresponding water or sediment depth). Calculate the free compression enthalpy for CO_2 at 1000 m water depth (Hint: CO_2 is not an ideal gas and has a fugacity coefficient of 0.58 at 100 bar, such that $p' = 0.58 \cdot p$ and $\Delta G = RT \ln p'/p_0$, see Kuhn H., Försterling H.-D. (1999) *Principles of Physical Chemistry*, Wiley Publisher, Table 20.1).

11. *Biological electron carriers.* Name the main biological electron carriers and designate them as one-electron or two-electron carriers.

12. *N_2 fixation by rocks.* Extract the essential information from the following paper and summarize it critically: Shang X. *et al.* (2023) Formation of ammonia through serpentinization in the Hadean Eon, *Sci. Bull.* 68(11): 1109–1112.

13. *Redox potentials at modern vents.* Extract the essential information from the following paper and summarize it critically: Yamamoto M. *et al.* (2018) Deep-sea hydrothermal fields as natural power plants, *ChemElectroChem* 5: 2162–2166.

3 Primordial Reaction Networks and Energy Metabolism

3.1 TRANSITION METALS BIND AND ACTIVATE MOLECULES TO METAL BOUND FRAGMENTS

In Chapters 1 and 2, we saw that CO_2 was the starting material for organic synthesis at the origin of metabolism and life. In today's environment, CO_2 is still the starting material for all modern ecosystems because autotrophs supply the growth substrates for all heterotrophs. Microbiologists generally describe the growth characteristics of microbes (their trophic mode, from Greek *trophe*, nourishment) in terms of the sources of **energy** for ATP synthesis, **electrons** for redox reactions, and **carbon** for biosynthesis. **Phototrophs** synthesize ATP by harnessing the energy of photons (light) with the help of pigments like chlorophyll. **Chemotrophs** synthesize ATP by harnessing the energy of light-independent chemical reactions. **Lithotrophs** obtain their electrons from inorganic donors such as H_2. **Organotrophs** obtain their electrons from reduced organic compounds such as sugars, fats, or amino acids. **Autotrophs** obtain their carbon for biosynthesis from CO_2. **Heterotrophs** are dependent upon a source of reduced organic compounds for biosyntheses. Acetogens and methanogens satisfy their carbon and energy needs from chemical energy in the reduction of CO_2 with H_2, they are chemolithoautotrophs. *E. coli*, humans, and yeast are chemoorganoheterotrophs. Plants are photolithoautotrophs. Modern metabolism starts from CO_2. Ancient metabolism did as well, so it is instructive to look at the modern pathways of CO_2 fixation.

3.1.1 THERE ARE SEVEN MODERN CO$_2$ FIXATION PATHWAYS

There are six different CO_2 fixation pathways known among modern microbes, seven including the reductive glycine pathway recently described by Alfons Stams and colleagues, which is an ammonia-dependent modification of the acetyl-CoA pathway. The **acetyl-CoA pathway** is the most ancient among them. It is used by a number of diverse strictly anaerobic bacteria and archaea. It is the only exergonic pathway of biological CO_2 fixation known:

$$2\,CO_2 + 4\,H_2 + CoASH \rightarrow CH_3COSCoA + 3\,H_2O$$

$$\Delta G^{0'} = -59 \text{ kJ/mol}$$

The acetyl-CoA pathway is H_2 dependent and sometimes also called the Wood–Ljungdahl pathway for the work of Lars Ljungdahl and Harland Wood on the enzymology of the pathway in clostridia (bacteria). It turned out that methanogens (archaea) use a pathway to synthesize methyl groups

from H_2 and CO_2, that is, in terms of the intermediates and reaction mechanisms, chemically quite similar to the acetyl-CoA pathway of bacteria, but the archaeal and bacterial pathways are fundamentally different in terms of cofactors and enzymes. The archaeal pathway was elucidated through work on methanogens, mainly by Ralph Wolfe in Urbana, Illinois, who elucidated the novel cofactors involved in the pathway, and the group of Rolf Thauer, Marburg, who characterized and crystallized many of the archaeal enzymes. The archaeal version of the pathway is chemically similar but enzymatically distinct from bacterial pathways, suggesting that the basic chemistry of H_2-dependent CO_2 fixation is more ancient than the enzymes that catalyze the reactions in microbes.

This basic thought, that many (possibly most) of the chemical reactions of ancient microbial metabolism existed before enzymes existed, is a recurrent common theme of early metabolic evolution. It is linked to the idea that when enzymes finally emerged on the scene of early evolution, they did not invent reactions that otherwise did not exist in an abiotic environment, they merely accelerated organic reactions that tend to occur anyway, provided that suitable catalysts exist. In that sense, the role of enzymes was not to invent biochemical pathways, but to replace inorganic catalysts that got the pathways going to begin with. Metabolism starts with reduced carbon, hence CO_2-fixing pathways are integral to understanding origins. A comparison of some features of the six ammonia-independent pathways of CO_2 fixation is given in Table 3.1.

In H_2-dependent acetogens (bacteria) and H_2-dependent methanogens (archaea), the acetyl-CoA pathway performs a dual role. It not only supplies reduced carbon for biosynthesis, it is also integral to the central energy-conserving reaction of the cell. In acetogens energy is obtained via acetate formation from H_2 and CO_2 (acetogenesis)

$$4H_2 + 2CO_2 \rightarrow CH_3COOH + 2H_2O$$

$$\Delta G^{0'} = -95 \text{ kJ/mol}$$

In methanogens, the methanogenic reaction proceeds to a more reduced end product, methane

$$4H_2 + CO_2 \rightarrow CH_4 + 2H_2O$$

$$\Delta G^{0'} = -131 \text{ kJ/mol}$$

and is slightly more exergonic per mol H_2 consumed. Both the acetogenic and the methanogenic pathways involve the coupling of CO_2 reduction to the generation of an ion

DOI: 10.1201/9781003378617-3

gradient that is harnessed by an ATPase. This allows both acetogens to synthesize substoichiometric amounts of ATP relative to the end product. This entails ion gradients (for the mechanisms involved see Chapter 6). Under optimal growth conditions, acetogens obtain about 0.3 ATP per acetate, methanogens about 0.5 ATP per methane. Moreover, the amount of CO_2 that flows through the cell as acetate or methane exceeds the amount of carbon that is fixed as cell mass by about 24:1 (acetogens) or 20:1 (methanogens). That is, for every 12 molecules of the C2 unit acetate that is produced and excreted for the purpose of ATP synthesis in acetogens, only one C atom ends up as cell mass. We will return to that mass balance shortly.

The acetyl-CoA pathway can be used by cells both to fix CO_2 and to harness energy in the form of ATP. That cannot be said for any other CO_2-fixing or energy-conserving pathway. Among autotrophs, the other 5 CO_2-fixing pathways operate at ATP expense, and that ATP has to be supplied by some other form of energy metabolism. In acetogens and methanogens, carbon metabolism (CO_2 reduction) and energy metabolism (ATP synthesis) are united in one and the same pathway. That makes the acetyl-CoA pathway unique in both carbon and energy metabolism. It is also the only pathway of biological CO_2 fixation known that can be catalyzed by a mineral or piece of metal in the laboratory instead of proteins and cofactors. Georg Fuchs was the first strong proponent of the idea that the acetyl-CoA pathway is ancient:

The total synthesis of acetyl CoA fulfills most of the criteria postulated for an ancient pathway. Its distribution in only distantly related anaerobes (Archaebacteria and Eubacteria) . . . and its unusual biochemistry are noteworthy. It requires the lowest amount of ATP. It is a versatile one-carbon and two-carbon assimilation path.

(Fuchs and Stupperich, 1985, pp. 245–246)

Fuchs subsequently discovered three of the six known pathways of CO_2 fixation pathways in Table 3.1: the dicarboxylate/4-hydroxybutyrate cycle, the 3-hydroxypropionate/4-hydroxybutyrate cycle, and the 3-hydroxypropionate bicycle.

The acetyl-CoA pathway is (still) the only exergonic pathway of CO_2 fixation that has been discovered so far. It is a linear pathway starting from H_2 and CO_2, there is no cycle involved, a unique attribute. Formate and CO are the only intermediates of the acetyl-CoA pathway that occur without covalent bonds to enzymes or cofactors. Other pathways have more free intermediates and involve larger numbers of intermediates in general, often soluble as CoA esters. A larger number of free intermediates means that side reactions become more probable, requiring more enzymatic help for specificity. This is reflected in a larger number of enzymes required in the cycles relative to the linear acetyl-CoA pathway. In the reactions from CO_2 to acetyl-CoA or pyruvate, there are no chiral centers in any of the fixed carbon atoms or the pathway intermediates. By contrast, the intermediates of the cycles have many chiral centers (Table 3.1).

TABLE 3.1
Comparison of CO_2-Fixing Pathways

Pathway Name	Pathway Properties						
	Linear or Cyclic?	ATP[b] per Pyruvate	Free Inter-mediates[c]	Chiral Centers[d]	Bacteria and Archaea[e]	O_2 Tolerant[f]	Non-enzymatic Version[g]
Acetyl-CoA	linear	−0.5	2	0	Both	No	Yes
Reverse TCA	cycle	1	9	3	Bacteria	No	No
Dicarbox/4-HB[a]	cycle	5	8	2	Archaea	No	No
3-HP/4-HB[a]	cycle	9	4	2	Archaea	Yes	No
3-HP[a] bi-cycle	bi-cycle	7	3	6	Bacteria	Yes	No
Calvin cycle	cycle	7	13	18	Bacteria	Yes	No

Information summarized from Fuchs, G. (2011) Alternative pathways of carbon dioxide fixation: Insights into the early evolution of life? *Annu. Rev. Microbiol.* 65: 631–658; Mall A. *et al.* (2018) Reversibility of citrate synthase allows autotrophic growth of a thermophilic bacterium, *Science* 359: 563–567; Berg I. A. *et al.* (2010) Autotrophic carbon fixation in archaea, *Nat. Rev. Microbiol.* 8: 447–460. [a] Dicarbox, dicarboxylate; HB, hydroxybutyrate; HP, hydroxypropionate. [b] Number of ATP expended per pyruvate synthesizes via the pathway. The acetyl-CoA pathway using H_2 as reductant is exergonic under standard physiological conditions by = −59 kJ/mol and involves ion gradient formation for ATP synthesis in acetogens and methanogens. There is not enough energy released in the pathway to permit CO_2 fixation and energy conservation by SLP using H_2 as reductant under physiological conditions, but there is enough energy to permit simultaneous CO_2 fixation and ion pumping that can be used for ATP synthesis via rotor stator ATPases (see Chapter 6). [c] Number of pathway intermediates that are covalently bound neither to enzymes nor cofactors. Intermediates in the acetyl-CoA pathway are mostly enzyme bound. Intermediates in the hydroxybutyrate and hydroxypropionate pathways are usually covalently bound as CoA esters. [d] Number of chiral centers in pathway intermediates (excluding cofactors). Formate and CO are the only free intermediates in the acetyl-CoA pathway. [e] Distribution of the pathway among prokaryotes. [f] Refers to the participation of enzymes in the pathway that are readily inactivated or inhibited by O_2, these are typically enzymes that are ferredoxin dependent or contain redox-active FeS centers. [g] Refers to the question of whether pyruvate can be obtained from H_2 and CO_2 via the reactions of the pathway but without enzymes in a laboratory experiment. See Preiner M. *et al.* (2020) A hydrogen dependent geochemical analogue of primordial carbon and energy metabolism, *Nat. Ecol. Evol.* 4: 534–542.

The acetyl-CoA pathway is linear, not cyclic, and is the only dedicated CO_2 fixation route known to occur both in archaea and in bacteria. Many of the enzymes involved are oxygen sensitive, this has to do with a large number of redox-active transition metal clusters in the pathway. For example, the first enzyme of the archaeal version of the acetyl-CoA pathway, formylmethanofuran dehydrogenase, contains 46 individual 4Fe4S clusters and four catalytically active W atoms. The enzyme is very oxygen sensitive. The Calvin cycle, the 3-hydroxypropionate/4-hydroxybutyrate cycle, and the 3-hydroxypropionate bicycle have no O_2 sensitive enzymes and occur in O_2 tolerant organisms; the involvement of catalytic centers that are restricted to anaerobic environments is generally seen as an ancient trait. Importantly, the products of the acetyl-CoA pathway up to pyruvate can be obtained from H_2 and CO_2 without enzymes. No other pathway of CO_2 fixation has been shown to display a similar level of enzymatic independence.

3.1.2   THE ACETYL-CoA PATHWAY IS THE ONLY EXERGONIC PATHWAY OF CO_2 FIXATION

Common to the archaeal and the bacterial versions of the acetyl-CoA pathway are the chemical conversions shown in Figure 3.1. In carbon metabolism, the pathway forms acetyl CoA from two molecules of CO_2. In all organisms that use the pathway, acetyl-CoA synthesis is followed by one more incorporation of CO_2 to form pyruvate. The acetyl-CoA pathway requires reduced ferredoxins with a sufficiently low midpoint potential, that is, with sufficiently energetic electrons, for conversion of CO_2 to CO. In cells that live from H_2, the reduced ferredoxins are supplied by a process called electron bifurcation (see Chapter 6). It is instructive to compare the microbial and hydrothermal pathways for the formation of activated acetyl in some detail.

At first sight, the microbial pathway appears daunting in the details (Figure 3.1). Ten enzymes and ten cofactors or prosthetic groups (cofactors or FeS clusters) are involved in both the bacterial and the archaeal pathways. When confronted with such a complex biochemical pathway, the critical observer, in particular geochemists or prebiotic chemists, might react with utter disbelief to the proposal that such a pathway might be ancient. In comparative biochemistry, it is extremely helpful to hold on to a basic tenet of enzymology: Enzymes do not catalyze reactions that otherwise could not occur. That is, enzymes do not make impossible reactions possible, they merely accelerate reactions that tend to occur anyway by orienting reactants to one another in space so as to lower the activation energy. The question of whether or not reactions can take place in terms of thermodynamics is decisive at origins, and if reactions are exergonic one can consider which simple and naturally occurring catalysts might have accelerated biochemical reactions before the advent of enzymes. That said, the acetyl-CoA pathway involves three steps: methyl synthesis, acetyl synthesis, and pyruvate synthesis.

The methyl synthesis routes in bacteria and archaea involve six enzymes each which are not homologous at the level of amino acid sequences, though the formate synthesizing subunits do share recognizable structural similarity. This is a somewhat unusual circumstance. For most biochemical pathways of central metabolism, the enzymes of

pathways tend to be homologous across all organisms in which the pathway occurs. The methyl synthesizing branch of the acetyl-CoA pathway occurs in two versions. The bacterial methyl synthesis route involves NAD(P)H and ATP, the archaeal does not, and the corresponding bacterial and archaeal enzymes are not related at the level of amino acid sequence comparisons. Other differences in cofactor requirements are shown in Figure 3.1. The formyl to methyl conversions occur with the carbon substrate covalently on a pterin cofactor, tetrahydrofolate (H_4F) in the bacterial route, and tetrahydromethanopterin (H_4MPT) in the archaeal route. Methyl synthesis from CO_2 requires three reduction steps of two electrons each.

Formation of the methyl group from formic acid in acetogens and methanogens is outlined in Figure 3.2A and B in more detail. A cyclic intermediate is formed (methenyl-H_4F and methenyl-H_4MPT), which is subsequently hydrogenated to the methylene and methyl group. The synthesis of CO from CO_2 and reduced ferredoxin at carbon monoxide dehydrogenase (CODH) is outlined in Figure 3.2C. The reduced state of the active site is indicated in the figure as a formal elemental oxidation state for Ni^0 for convenience, although the electrons are delocalized in the cluster. The active site nickel binds the carbon of CO_2, while the iron atom ligated by Cys^{295} binds one oxygen atom of CO_2. C–O bond cleavage leaves a carbonyl moiety on the active site nickel. CO then exits the active site, taking the electron pair by which it is bound to the nickel, such that the active site is oxidized, indicated as Ni^{2+}, and needs to be re-reduced by ferredoxin to complete the catalytic cycle.

At the active site of acetyl-CoA synthase (ACS), CO from the CODH reaction is bound to an active site nickel, the carbonylation step in Figure 3.3. The methyl group from the methyl synthesis pathway is transferred from the pterin to the cobalt atom of cobalamin, the cofactor in a protein called corrinoid iron-sulfur protein (CoFeSP), indicated in Figure 3.1. The cobalt-bound methyl group is transferred by CoFeS to the same nickel atom where the CO is bound at the active site of acetyl-CoA synthase. This is a rare example of a metal-to-metal methyl transfer reaction.

In some microbes, CODH and ACS are organized as a single multisubunit enzyme called bifunctional CODH/ACS. The bifunctional enzyme performs two functions, reduction of CO_2 to CO and condensation of CO with the methyl group via carbonyl insertion to generate a nickel-bound acetyl moiety, which is removed from the enzyme via thiolysis to generate the thioester acetyl-CoA. These steps are outlined in Figure 3.3.

As we saw in Figure 2.5, in microbes that use the pathway for carbon assimilation, about 30% of the acetyl groups in acetyl CoA are diverted to biosynthesis using C2 units (synthesis of lipids for example), while the majority, about 70%, are routed to further CO_2 incorporation at the step catalyzed by the enzyme pyruvate synthase. Pyruvate can rightly be seen as a product of the acetyl-CoA pathway. From the first contact of CO_2 with the enzymes and cofactors involved, formate and CO are the only free intermediates of the acetyl-CoA pathway. In the archaeal pathway formate is not even released to the cytosol, it is channeled within the formylmethanofuran dehydrogenase enzyme to

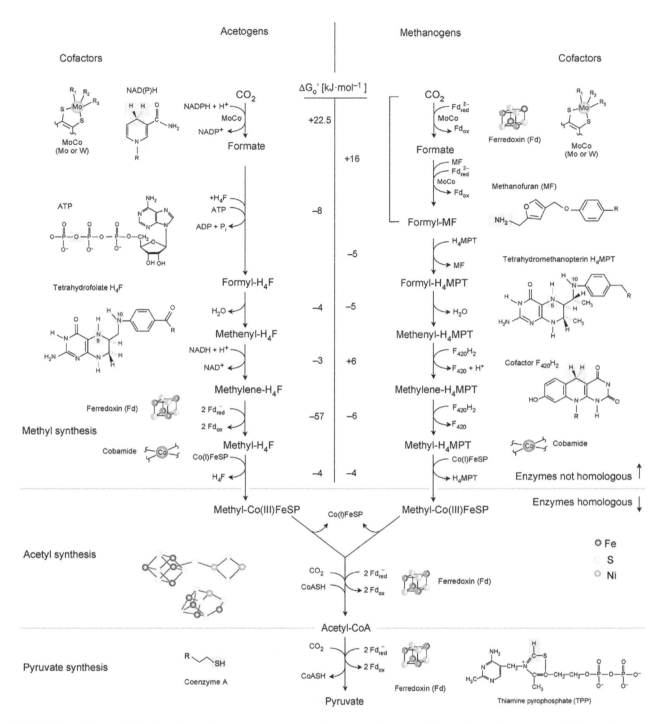

FIGURE 3.1 The acetyl-CoA pathway in acetogens and methanogens. Active moieties of the cofactors are highlighted in green. Modified from Sousa F. L., Martin W. F. (2014) Biochemical fossils of the ancient transition from geoenergetics to bioenergetics in prokaryotic one carbon compound metabolism, *Biochim. Biophys. Acta* 1837: 964–981; Fuchs G. (2011) Alternative pathways of carbon dioxide fixation: Insights into the early evolution of life? *Annu. Rev. Microbiol.* 65: 631–658. The CO_2 to formyl-MF reaction is catalyzed by one bifunctional enzyme: Wagner T. *et al.* (2016) The methanogenic CO_2 reducing-and-fixing enzyme is bifunctional and contains 46 [4Fe-4S] clusters, *Science* 354: 114–117.

condense with the primary amino group of methanofuran (formyl-MF in Figure 3.2B). A similar phenomenon occurs in the reaction catalyzed by bifunctional CODH/ACS, in that CO generated from CO_2 does not leave the enzyme, it travels through a channel in the polypeptide to the nickel atom in the ACS active site. This tendency of the intermediates to be covalently bound rather than free in solution can be seen as a relic of metabolic origin on solid-state catalysts.

The transformations of the carbon atom in CO_2 from formate to acetate all take place either covalently bound to

the enzyme or covalently bound to a cofactor, hence we can represent the pathway as in Figure 3.4 where an analogous covalent binding of the substrate carbon to the surface was observed, because only formate, methanol, acetate, and pyruvate were detected. This sets the acetyl-CoA pathway apart from other CO_2-fixing pathways, which can involve up to 13 different free intermediates (Table 3.1).

The acetyl-CoA pathway can be compared with a chemically similar hydrothermal pathway based on laboratory experiments (Figure 3.5). While the synthesis of the

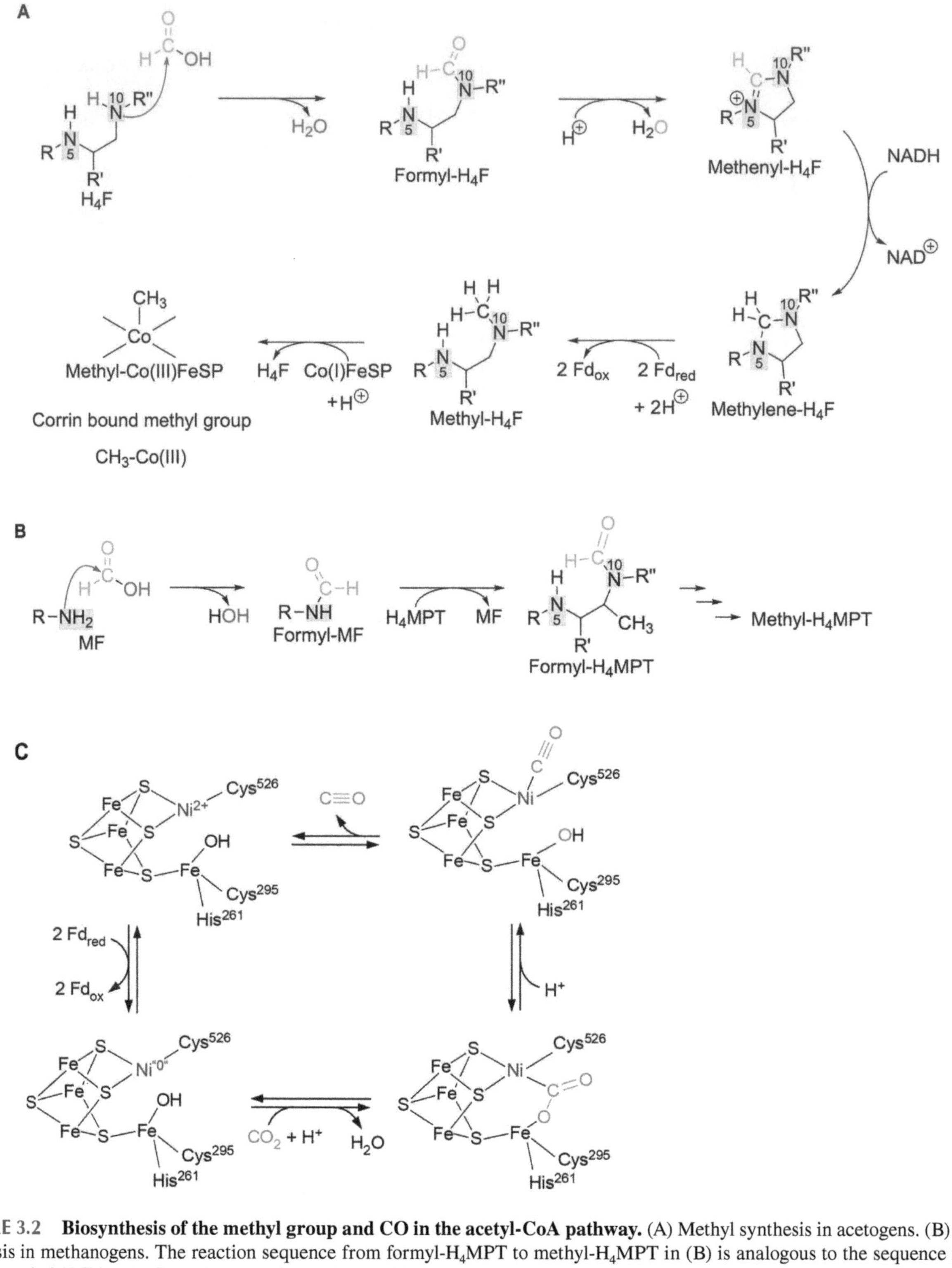

FIGURE 3.2 Biosynthesis of the methyl group and CO in the acetyl-CoA pathway. (A) Methyl synthesis in acetogens. (B) Methyl synthesis in methanogens. The reaction sequence from formyl-H$_4$MPT to methyl-H$_4$MPT in (B) is analogous to the sequence formyl-H$_4$F to methyl-H$_4$F in (A). Data from Maden B. E. H. (2000) Tetrahydrofolate and tetrahydromethanopterin compared: functionally distinct carriers in C1 metabolism, *Biochem. J.* 350: 609–629; and Wagner T. *et al.* (2016) The methanogenic CO$_2$ reducing-and-fixing enzyme is bifunctional and contains 46 [4Fe-4S] clusters, *Science* 354: 114–117. (C) Proposed mechanism for CO synthesis from CO$_2$ and reduced ferredoxin in carbon monoxide dehydrogenase. Redrawn after Ragsdale S. W. (2009) Nickel-based enzyme systems, *J Biol Chem.* 284: 18571–18575; and Can M. *et al.* (2014) Structure, function, and mechanism of the nickel metalloenzymes, CO dehydrogenase, and acetyl-CoA synthase, *Chem. Rev.* 114: 4149–4174.

thioester acetyl CoA from H$_2$ and CO$_2$ is exergonic, the subsequent reaction of the thioester to acetyl phosphate is slightly endergonic under standard physiological conditions (1 M reactant concentrations). The free energy of hydrolysis of acetyl-CoA is −31.4 kJ/mol which is not negative enough to react to acetyl phosphate with a hydrolysis energy of −42.3 kJ/mol (see Table 1.3, phosphate transfer) under standard conditions. Nonetheless, the reaction readily takes place in cells that use the acetyl-CoA pathway. The reason is that the conditions in cells differ from standard conditions under which changes in free energy are typically calculated. Like many reactions in metabolism that are (slightly)

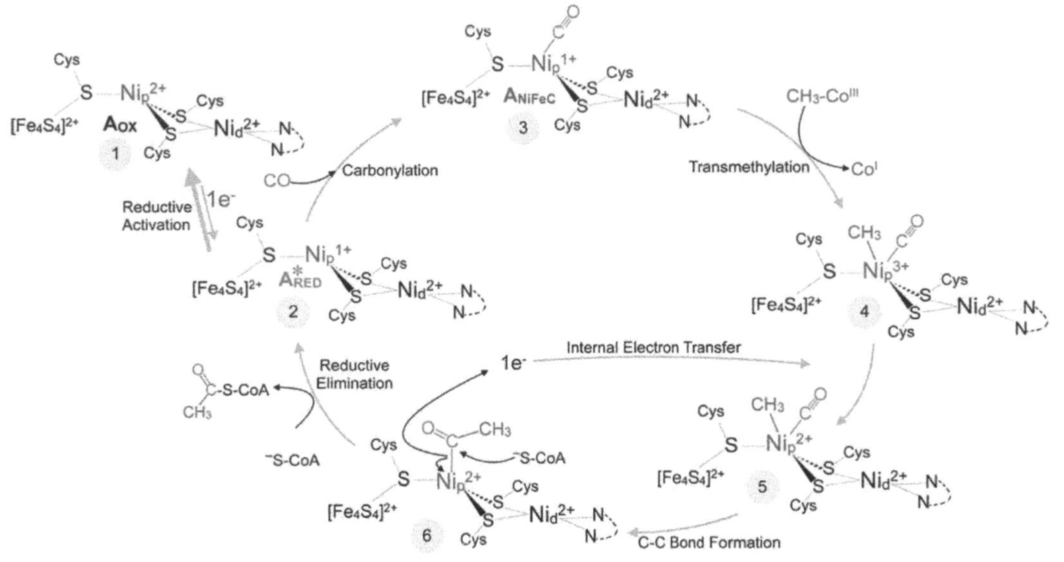

FIGURE 3.3 **Formation of an acetyl moiety in the acetyl-CoA pathway.** The reaction mechanism of acetyl-CoA synthase (ACS) proposed by Stephen Ragsdale and colleagues. The C-C bond is formed by condensation of CO with the methyl group via carbonyl insertion on a nickel atom at the ACS active site nickel. CH_3–Co^{III} is a methyl CoFeS protein. The methyl group is synthesized on pterins and transferred to CoFeS (see Figures 3.1 and 3.2). The electron pair in the C–Co bond remains associated with Co, generating Co^I, and increasing the oxidation state of the active site nickel, Ni_p. The nickel-bound acetyl moiety is removed from the enzyme via thiolysis to generate the thioester acetyl-CoA. The nickel-bound acetyl group is removed from the enzyme via thiolysis by the thiol form of coenzyme A, CoASH. A_{ox} indicates an oxidized state of the active site. Ni_p and Ni_d refer to the nickel atoms proximal and distal to the FeS cluster. The arrow 'internal electron transfer' posits a yet unidentified docking site (probably a FeS cluster) that can sequester an electron from one catalytic cycle for use in the next. An alternative mechanism involves a zero-valent state for the active site nickel, Ni_p: Lindahl P. A. (2004) Acetyl-coenzyme A synthase: the case for a Ni $^{(0)}$-based mechanism of catalysis, *J. Biol. Inorg. Chem.* 9: 516–524. Reproduced with permission from: Can M. *et al.* (2017) X-ray absorption spectroscopy reveals an organometallic Ni–C bond in the CO-treated form of acetyl-CoA synthase, *Biochemistry* 56: 1248–1260.

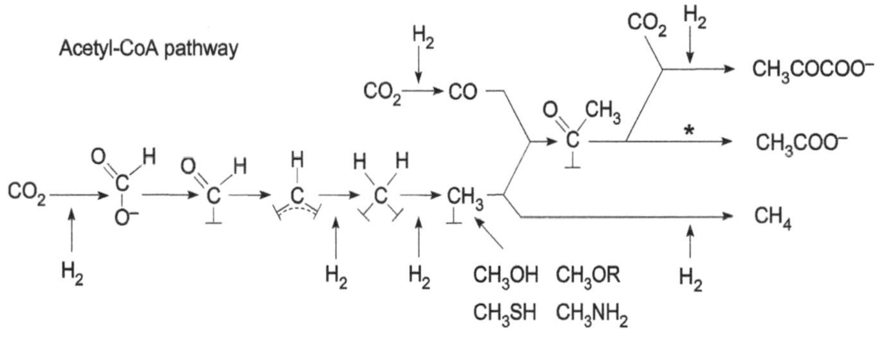

FIGURE 3.4 **Acetyl-CoA pathway reactions occur as covalently bound intermediates.** The carbon intermediates are bound to a cofactor or enzyme in bacteria and archaea, whereas in laboratory experiments under hydrothermal conditions using Ni_3Fe as a catalyst, the intermediates are covalently bound to metal atoms. '⊥' denotes covalent bonding. In the biological pathway, H_2 is converted by hydrogenases to reduced ferredoxin. In the laboratory experiments (and at hydrothermal vents), H_2 is converted by the catalysts to 2 H^+ and 2 e– under alkaline conditions. In the biological pathway, the ligands representing '⊥' are, from left to right, N, N, N, Co, Ni (and S) atoms and methyl donors (lower right) can be incorporated. In the Ni_3Fe catalyzed pathway, the ligands are either Ni or Fe and methanol can be produced. Redrawn from Preiner M. *et al.* (2020) A hydrogen dependent geochemical analogue of primordial carbon and energy metabolism, *Nat. Ecol. Evol.* 4: 534–542.

endergonic under standard conditions, the reaction can go forward in the cell because reactants do not have 1M concentrations in cytosol and subsequent pulling reactions keep product concentrations low in steady state equilibrium. For example, the synthesis of acetyl phosphate from acetyl CoA and phosphate is central in acetogen metabolism and takes place about 60 times for every CO_2 that is incorporated into acetogen cell mass, although $\Delta G^{0\prime}$ for the reaction is slightly endergonic. Synthesis of acetyl phosphate from acetyl CoA

or similar thioester has not been demonstrated in nonenzymatic laboratory reactions, although older work reported the synthesis of pyrophosphate from acetyl CoA and phosphate, from which the synthesis of acetyl phosphate as an intermediate was inferred but not shown. The efficient synthesis of acyl phosphates from thioesters without enzymes has not yet been reported, suggesting that the right inorganic catalysts, or the right reactions, or both, have not yet been identified.

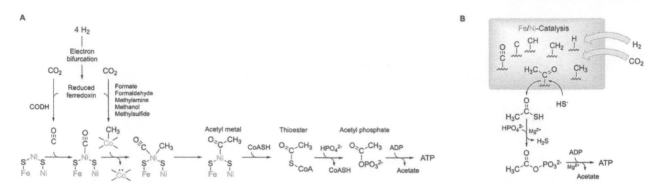

FIGURE 3.5 **The acetyl-CoA pathway and a possible inorganic analogue.** (A) Simplified scheme of the microbial acetyl-CoA pathway from ferredoxin to ATP. Microbes employ a mechanism called electron bifurcation to generate reduced ferredoxin from H_2 (more on that in Chapter 7). If the reactions are performed in the presence of H_2 under alkaline conditions, electron bifurcation is not needed, H_2 can reduce CO_2 directly in the presence of suitable catalysts. Redrawn from Martin W. F., Thauer R. K. (2017) Energy in ancient metabolism, *Cell* 168: 953–955; and Sousa F. L. *et al.* (2018) Native metals, electron bifurcation, and CO_2 reduction in early biochemical evolution, *Curr. Opin. Microbiol.* 43: 77–83. The methyl groups for the acetyl moiety (CH_3CO) of acetyl CoA are obtained from hydrogenation of formate, formaldehyde, or methanol or from the reduction of CO_2 with H_2. CO is obtained from enzymatic reduction of CO_2. The cofactors for CO_2 reduction and synthesis of the acetyl group contain the transition metals Fe (iron), Ni (nickel), and Co (cobalt). (B) Hypothetical scheme of the reaction of H_2, CO_2, and SH– to thioacetic acid $CH_3C=OSH$ on a transition metal surface without enzymes. Thioacetate reacts with inorganic phosphate ions to acetyl phosphate and via substrate-level phosphorylation to ATP. The reaction of thioacetate with phosphate to acetyl phosphate is exergonic (Table 1.3) and occurs without enzymes under mild hydrothermal conditions (Whicher A. *et al.* (2018) Acetyl phosphate as a primordial energy currency at the origin of life, *Orig. Life Evol. Biosph.* 48: 159–179). The reaction of methyl thioacetate (thioester) with phosphate to generate acyl phosphates is slightly endergonic under standard physiological conditions (1 M reactant concentrations) and has not been demonstrated in laboratory experiments so far (see text).

3.1.3 CO_2 REDUCTION WITH H_2 ON TRANSITION METAL CATALYSTS IN WATER GENERATES ACETYL GROUPS

The nonenzymatic generation of C–C bonds and acetyl groups from H_2 and CO_2 under hydrothermal conditions is central to metabolic origin. Martina Preiner and colleagues (2020) obtained the carboxylic acids formate, acetate, and pyruvate as products from the reaction of H_2 with CO_2 catalyzed by awaruite $FeNi_3$, greigite Fe_3S_4, and magnetite Fe_3O_4 (Section 2.3). Product concentrations up to 200 mM formate, 100 µM acetate, and 10 µM pyruvate were obtained at 10 bar H_2, 15 bar CO_2, 1 mmol metal atoms in Ni_3Fe, 100°C, and 16 hours reaction time. The experiments indicate a build-up of acetyl groups on various transition metal surfaces typical for serpentinization sites.

Joseph Moran and coworkers showed that the zero-valent (native) transition metals molybdenum (Mo), nickel (Ni), cobalt (Co), manganese (Mn), and tungsten (W) donate electrons and catalyze carbon fixation to acetyl groups as well (Figure 3.6) without the addition of exogenous H_2, although H_2 was formed from reactions of water with the native metals under their reaction conditions. Note that the reaction of H_2 with CO_2 catalyzed by awaruite in the experiment of Preiner *et al.* produced five orders of magnitude more formate than reactions reported by Hudson *et al.* that were driven by a pH gradient and two orders of magnitude more formate than the reactions of Varma *et al.* using native metals as the catalyst and the reductant but without added H_2 (Figure 3.6).

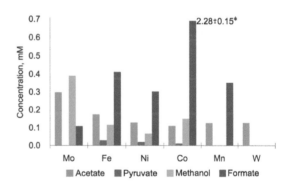

FIGURE 3.6 **Product formation from CO_2 using transition metals as catalysts and electron donors.** Experimental conditions: 100°C, 35 bar CO_2, 1 M KCl in H_2O, pH = 7, 16 h reaction time, maximum formate concentration 2.28 mM (not to scale). The surface-bound formyl, methyl, acetyl, or pyruvyl species must be cleaved from the surface with KOH at the end of the reaction to obtain solved formate, methanol, acetate, or pyruvate for analysis. Therefore, at prebiotic conditions, the pH value of the hydrothermal solution must have increased to very alkaline conditions by some means to obtain the free products of carbon fixation. Redrawn from Varma S. J. *et al.* (2018) Native iron reduces CO_2 to intermediates and end-products of the acetyl CoA pathway, *Nat. Ecol. Evol.* 2: 1019–1024.

3.1.4 THIOACETATE REACTS WITH PHOSPHATE TO ACETYL PHOSPHATE, WHICH CAN PHOSPHORYLATE ADP

What happens if compounds with reduced sulfur are added to the CO_2/H_2 mixture? Heinen and Lauwers showed in an early groundbreaking experiment that iron sulfide, hydrogen sulfide, and CO_2 react in water under mild but very acidic anaerobic conditions to nanomole quantities of methanethiol (CH_3SH) and carbonyl sulfide (COS).

$$FeS + H_2S \rightarrow FeS_2 + H_2$$

$$CO_2 + H_2S + 3\,H_2 \rightarrow CH_3SH + 2\,H_2O$$

$$CO_2 + H_2S \rightarrow COS + H_2O$$

Iron-sulfur clusters probably catalyzed the reactions. Both COS and CH_3SH are useful compounds for prebiotic metabolic reactions. COS catalyzes the oligomerization of amino acids to peptides (Section 4.2); CH_3SH is a precursor of thioacetate and methyl thioacetate. Huber and Wächtershäuser reported the synthesis of μmol quantities of the thioester CH_3COSCH_3 (methyl thioacetate) from CH_3SH and 1 bar CO at 100°C and pH 1.7, using freshly coprecipitated FeS and NiS as a catalyst.

$$2\,CH_3SH + CO \rightarrow [CH_3COSH]$$
$$+ CH_3SH \rightarrow CH_3(CO)SCH_3 + H_2S$$

Thioacetic acid (CH_3COSH) is probably an intermediate of this reaction, but it was not detected. The thioester hydrolyzes in 20 hours to acetic acid and methanethiol excluding, in their view, the accumulation of thioesters as a primordial energy reservoir. However, had they chosen conditions under which thioesters are constantly synthesized, they might have reached a different conclusion concerning thioesters. Martin and Russell made the case for the role of acetyl phosphate as the primordial energy currency before ATP. Whicher and coworkers demonstrated that thioacetic acid reacts with inorganic phosphate under Mg^{2+}-catalysis and ambient conditions within minutes to acetyl phosphate

$$CH_3(CO)SH + HPO_4^{2-} \rightarrow CH_3(CO)OPO_3^{2-} + H_2S$$

whereas the thioester CH_3COSCH_3 does not. Again, acetyl phosphate easily hydrolyzes and therefore would have to be constantly resynthesized if it is to be suitable as a primordial energy reservoir. However, Kitani and coworkers have shown that, in the presence of soluble Fe(III) ions, acetyl phosphate readily transfers its phosphate group to adenosine diphosphate (ADP) to obtain ATP

$$CH_3(CO)OPO_3^{2-} + ADP \rightarrow ATP + acetate$$

The final product of this substrate phosphorylation is ATP, the universal energy currency of life (Figure 3.5B). Hence it is possible, in principle, that acetyl phosphate could have coupled exergonic carbon chemistry to the phosphorylation of ADP at biochemical origins. However, acetyl phosphate synthesis using thioacetate as a precursor under realistic hydrothermal vent conditions remains an open question. Another possible candidate for substrate-level phosphorylation of ADP to ATP is formyl phosphate, which releases

about 5 kJ/mol more hydrolysis energy than the thioester formyl CoA. A more recently recognized mechanism of substrate level phosphorylation is the phosphite dependent reaction discovered by Mao and colleagues in the laboratory of Bernhard Schink, in which phosphite serves both as a reductant and a phosphorylating agent:

$$AMP + NAD^+ + phosphite \rightarrow NADH + ADP$$

Reactions of phosphate under strongly reducing hydrothermal conditions where phosphite might be formed have not yet been reported. Reactions of thioesters or thiols with organic acids and phosphate have also not yet been investigated in the presence of H_2 and minerals that catalyze the synthesis of pyruvate. While thioacetate has been shown to react with phosphate to form acetyl phosphate, the synthesis of thioacetate itself from H_2, CO_2, and H_2S has not yet been reported. Thioacetate is not a compound of modern metabolism. The idea that thioacetate might be important in early metabolism traces to Wächtershäuser, whose iron-sulfur world hypothesis is interesting, but generally contains too much sulfur and too many organosulfur compounds relative to modern metabolism.

Based upon its widespread occurrence in microbial metabolism, its chemical and biological proximity to the acetyl-CoA pathway, its chemical simplicity relative to ADP and ATP, and its high free energy of hydrolysis ($\Delta G^{0'} = -43$ kJ/mol), acetyl phosphate probably played an important role in early energy metabolism before the emergence of ATP. However, it does not activate the amino acid glycine for polymerization to peptides or AMP for polymerization to nucleic acids. Instead, acetyl phosphate efficiently acetylates the unprotected amino group of glycine to form N-acetyl glycine as Whicher et al. showed. AMP only forms stacks of monomers in the presence of acetyl phosphate. Efficient polymerization probably requires protection of the amino group (Chapter 4.2) and more elaborate catalysis than just coordination by Mg^{2+}. Moran and coworkers showed that metaphosphates like P_4O_{10} (the hydrolysis of which is exothermic by -177 kJ/mol) and strongly reduced phosphorous compounds like FeP are very efficient activating agents to drive energy-demanding reactions such as sugar and nucleotide synthesis. However, whether iron phosphide (P^{3-}) reactions were relevant for prebiotic synthesis is questionable. The main form of phosphorus on the early Earth was phosphate. Yet phosphite, which occurs in geothermal samples, and which is utilized as an energy and electron source by many microbes, should form under the highly reducing conditions of serpentinizing hydrothermal vents. Morton and coworkers showed that reduced phosphorus compounds are present in the rust that forms on iron and in other environmental samples. Schink and colleagues showed that phosphite will phosphorylate AMP to ADP in the presence of NAD^+ and an enzyme.

The results of experiments with metal-catalyzed, H_2-dependent CO_2 reduction reactions support the idea that primordial reaction networks preceded the evolution of polymeric enzymes or ribozymes. Acetyl phosphate is currently the best candidate for a primordial phosphate-based energy currency. It is however noteworthy that in the biological synthesis of acetyl phosphate, the high-energy mixed anhydride bond is formed by the reaction of an energy-rich carbonyl group with an inert phosphate residue, that is, the energy in the anhydride bond of acetyl phosphate stems from the energy released during reactions of H_2-dependent CO_2 reduction, not from sulfur and not from phosphate.

To summarize these findings concerning the source and currency of chemical energy at origins, it is prudent to keep one thought in mind. The source of metabolic energy in methanogen metabolism, acetogen metabolism, and the synthesis of acetyl phosphate from H_2, CO_2, and phosphate in the acetyl-CoA pathway is the highly exergonic reaction of H_2 with CO_2, which involves the synthesis of reactive carbonyls as intermediates that can react with inert phosphate, not reactions of inert carbon with chemically reactive phosphorous compounds.

3.1.5 CATALYSIS IS ESSENTIAL TO INCREASE REACTION RATES

The rate of most chemical reactions in cells increases with temperature because the kinetic energy of reactants increases, increasing the probability of a molecular collision that results in bond formation. On average, the increase in reaction rate for biological reactions is roughly a factor of 2 for every increase in temperature of 10°C. A reaction with that increase is said to have a Q_{10} of 2. All reactions that are catalyzed by enzymes today can occur spontaneously without an enzyme or other catalyst. The spontaneous reaction rate can however be very low. Wolfenden studied many rates of uncatalyzed biochemical reactions and found that they vary across 20 orders of magnitude. Importantly, he also found a general principle behind enzymatic catalysis across different reactions: *Enzymes increase the rates of reaction such that the roughly 1000 reactions that make up the core metabolism of a cell all tend to take place at approximately the same rate* within an order of magnitude. This allows the complex, ramified chemical reaction network of the cell to go forward.

In a complex network of 1000 reactions, it is imaginable, for example, that one essential reaction might suddenly operate at a rate that is five to six orders of magnitude slower than all others. That is exactly the case when *E. coli* (or other modern microbe) undergoes a simple mutation that leads to a massively defective mutant enzyme. If the reaction catalyzed by the enzyme is essential, for example, to provide an essential metabolite like an amino acid or a cofactor, one extremely slow reaction among 1000 is enough to bring the growth (life) process to a halt. In metabolic networks, reactions need to operate at roughly the same rate. That is the job of enzymes.

Another important insight from Wolfenden concerns the effect of temperature at origins. At high temperatures in the stability range for biochemical compounds (temperatures near 100°C), reactions proceed more rapidly than at lower temperatures. At high temperatures, the increase in reaction rate afforded by enzymes is not as pronounced as at lower temperatures. In theories for hydrothermal origins, that means that reaction rates during a hot start were high and that with decreasing temperature, the effects of catalysts—inorganic catalysts, organic cofactors, and finally enzymes—on

reaction rate enhancement became increasingly pronounced. In the absence of enzymes, the very slow rate of most biologically relevant chemical reactions is an argument in favor of chemical origins in hot environments. By hot we mean ~100°C, but not ~300°C where biomolecules become quite unstable and degrade.

Catalysis does not change the equilibrium constant of a reaction, it just changes the rate at which equilibrium is reached (Section 1.2). Cofactors can have a similar effect on the reaction rate as an enzyme has. An instructive example is provided by a decarboxylase. The enzyme requires pyridoxal phosphate (PLP) as the cofactor. The rate of reaction is increased 10^{18}-fold by the enzyme together with the cofactor. The cofactor alone accelerates the reaction by a factor of 10^{10}, the remaining 10^8-fold rate acceleration is provided by the enzyme. In the 1950s, Metzler and coworkers reported that in the absence of enzymes, most amino acids are efficiently converted by pyridoxal to the corresponding α-keto acids over a broad pH range at 100°C in the presence of Fe(II), Fe(III), or alum $(NH_4Al(SO_4)_2)$ as catalysts and that the reverse reaction works well too.

Temperature and catalysts affect reaction rates. The effect of energetic coupling via high-energy bonds is different. Transfer of a phosphate group from a phosphorylated energy-rich organic molecule (substrate phosphorylation) is a simple means to obtain activated monomers for chemical reactions. Provided that activation energy barriers for the reaction are not too high, this kind of 'cold activation' enables slightly unfavorable primordial reactions to occur at reasonable rates by coupling the reaction to free energy release involving hydrolysis of the high energy bond, even in low-temperature habitats, where thermolabile monomers and polymers are stable. ATP synthesis by substrate-level phosphorylation can take place without enzymes. In chemical evolution, substrate-level phosphorylation preceded enzymatic ATP synthesis driven by a proton gradient at the cell membrane (chemiosmosis; see Chapter 7).

3.1.6 METAL CATALYSIS ON SURFACES GENERATES PRODUCTS OF NONENZYMATIC CARBON FIXATION

There exists a vast chemical literature on transition metal catalyzed organic reactions, reactions that were not performed in an origins context. From that literature, enough is known that we can suggest a likely mechanism for the observed chemical products of CO_2 hydrogenation. We can assume that CO_2 and H_2 undergo *dissociative attachment* on the transition metal surface and the resulting activated CO, C, and H fragments successively build up methyl and acetyl groups. In the presence of sulfide ions, the acetyl groups can react to thioacetate and thioester. Thioacetate reacts in the aqueous phase with inorganic phosphate to acetyl phosphate, which can nonenzymatically phosphorylate ADP to ATP (Figure 3.5B). Possible intermediates of the surface reactions of H_2, CO_2, N_2, H_2S, and phosphate ions are shown in Figure 3.7. We stress that 'one-pot' reactions involving H_2, CO_2, N_2, H_2S, and phosphate on Ni_3Fe or Fe_3O_4 catalysts have not yet been reported, but it can be expected that they will generate a diverse product spectrum, although it can also be expected that the catalyst surface might rapidly

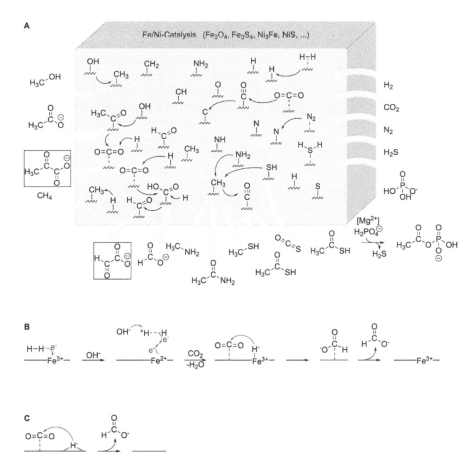

FIGURE 3.7 **Surface catalyzed hydrogenation of CO$_2$ and N$_2$.** (A) Possible intermediates and products of carbon and nitrogen fixation on transition metal surfaces. The most important possible products are boxed. Formate HCOO$^-$, acetate CH$_3$COO$^-$, methanol CH$_3$OH, pyruvate CH$_3$COCOO$^-$ and CH$_4$ were observed by Preiner *et al.*, CH$_3$SH, COS and the thioester CH$_3$COSCH$_3$ by other research groups. See text. (B) Hydride mechanism for synthesis of formate from hydrogen carbonate HCO$_3^-$ with Fe$_3$O$_4$ and Fe$_3$S$_4$ catalysts. (C) Hydride mechanism for synthesis of formate with Ni$_3$Fe.

be 'poisoned' by various tightly binding products. Whether those products will include biologically relevant compounds is not yet known.

Formate is the carbon hydrogenation product with the highest yield under most reaction conditions investigated so far. The close-to-equilibrium reaction requires the transfer of two electrons to proceed. A plausible mechanism for the formation of the two-electron donor *hydride* H$^-$ is shown in Figure 3.7B and C (*ionic hydrogenation*). H$_2$ approaches the Fe$_3$O$_4$ (or Fe$_3$S$_4$) surface and reduces Fe^{3+} to Fe^{2+} (Figure 3.7B). The generated H$_2^+$ is unstable and decomposes to H$^+$ (assisted by OH$^-$ at pH > 7) and to a hydrogen atom (H·) which picks up an electron from Fe^{2+} to become a hydride ion (H$^-$). Hydrated CO$_2$ (HCO$_3^-$; see Section 2.1) physisorbs on the magnetite surface and reacts with the hydride to HCOO$^-$, which is displaced from the surface by the addition of OH$^-$. Experiments with minerals containing merely ferrous iron (FeO and FeS) give far lower yields of CO$_2$ fixation products than Fe$_3$O$_4$ or Fe$_3$S$_4$.

What about catalysis by awaruite, Ni$_3$Fe? As Ni$_3$Fe consists of zero-valent metals, H$_2$ can dissociate on the metal surface to H atoms which diffuse into the awaruite solid where they can capture mobile electrons from the conduction band of the metal alloy to form H$^-$. Carbon dioxide is probably reduced by hydride ions on the awaruite surface to formic acid. The catalytic activity of Ni$_3$Fe to generate products germane to microbial physiology in high yields underscores the obvious: the transition metals are the catalytically active agents (*d* electrons), sulfur and oxygen merely hold them in place, as in enzymes that use Fe^{2+} in a catalytic role.

3.1.7 TRANSITION METAL CATALYZED REACTIONS GENERATE UNIVERSAL METABOLIC PRECURSORS

Muchowska, Varma, and Moran showed that ferrous ions (Fe^{2+}) dissolved in acidic water promote the synthesis of three of five metabolic precursors considered to be universal to biochemistry: oxaloacetate, succinate, and α-ketoglutarate. The remaining precursors acetate and pyruvate can be obtained from H$_2$ and CO$_2$ with iron-containing catalysts as already discussed. They started their experiments from aqueous glyoxylate and pyruvate and obtained hydroxyketoglutarate by aldol condensation using Fe^{2+} ions for catalysis (Figure 3.8; for the mechanism of aldol condensation by nucleophilic addition see Figure 1.22). Hydroxyketoglutarate dehydrated to oxopentenedioate, was reduced to ketoglutarate and decarboxylated oxidatively to succinate. In a parallel pathway, oxopentenedioate hydrated and then oxidatively decarboxylated to malate and oxidized to oxaloacetate.

Those reduction, oxidation, and decarboxylation reactions occurred all in the same vessel. Obviously, iron ions shuttle between the oxidation states Fe^{2+} and Fe^{3+}

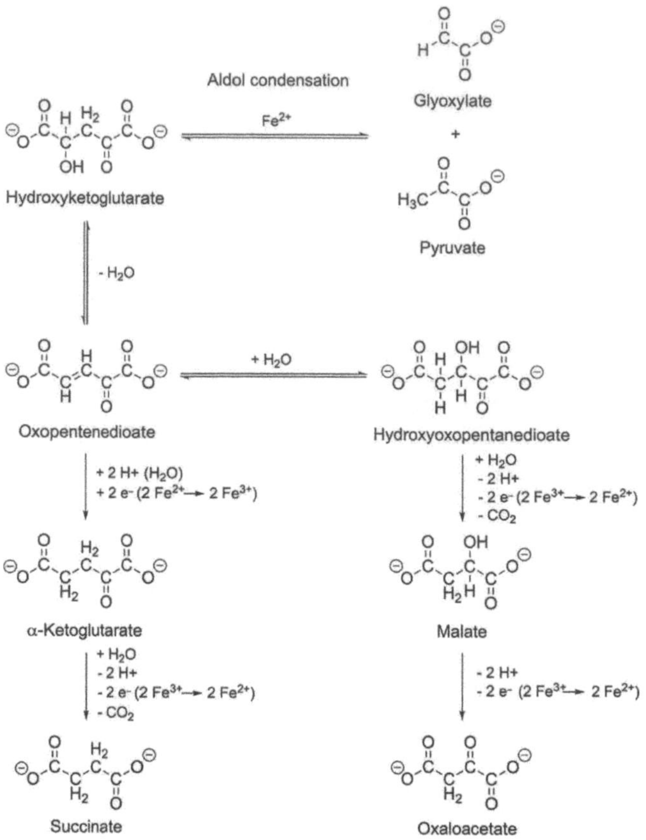

FIGURE 3.8   Network of Fe²⁺ catalyzed reactions of pyruvate and glyoxylate at 70°C. Selected reactions of the reverse tricarboxylic acid (TCA) cycle obtained nonenzymatically by transition metal catalysis are shown. Source substrates are glyoxylate and pyruvate obtained from CO_2 and H_2 under hydrothermal conditions.

during reduction and oxidation, and indeed, addition of Fe^{3+} enhanced the efficiency of the oxidation processes. Consider also that the dehydration reactions occurred in an aqueous solvent. This points to lowered water activity near the Fe^{2+} catalyst, probably due to binding of water to the ferrous ion. These reactions were performed using homogeneous catalysis, that is, dissolved Fe^{2+} and Fe^{3+} in solution, as opposed to heterogeneous catalysis (solid state catalysts) used in the synthesis of pyruvate from H_2 and CO_2. In a hydrothermal system, both kinds of catalysis are possible. By reductive amination of glyoxylate, pyruvate, oxaloacetate, and ketoglutarate with hydroxylamine and Fe^0, Moran and coworkers generated four biological amino acids. Hydroxylamine is an extremely aggressive reagent however, a property not common in biochemical reactions. Alternative routes for the incorporation of nitrogen in organic molecules are discussed in Section 4.1.

Some of the reactions in Figure 3.8 involve segments of the biologically important *tricarboxylic acid (TCA) cycle* (also called the *citric acid* or *Krebs cycle*). The cycle is a set of reversible reactions. In human cells, the TCA cycle operates in the oxidative direction, breaking down acetyl CoA into carbon dioxide, water, and electrons. The energy released during oxidation is conserved as GTP (equivalent to ATP) and NADH (*catabolic* breakdown of biomolecules). In the reductive direction, the reverse TCA cycle is a pathway of CO_2 fixation; it leads to the synthesis of acetyl CoA

from reduced ferredoxin and CO_2. Moran and coworkers showed that 6 of the 11 reactions of the reverse TCA cycle (rTCA; *anabolic* network with buildup of biomolecules) occur nonenzymatically in an acidic, metal-rich, reducing (Fe^0 or Ni^0, Zn^{2+}, Cr^{3+}) medium; the observed pathways are indicated with yellow in Figure 3.9. They proposed a coupling of the acetyl-CoA pathway and the rTCA cycle in primordial metabolism.

However, in cells that use the acetyl-CoA pathway for carbon and energy metabolism, the TCA cycle is usually incomplete, such that citrate is not formed. There are two routes that close the rTCA cycle by forming citrate, as work by Ivan Berg has shown. Cells that use the acetyl-CoA pathway do not simultaneously use the cyclic (complete) form of the rTCA cycle as a CO_2 fixation pathway, which the scheme of Figure 3.9 would seem to suggest. In organisms that use the acetyl-CoA pathway, the rTCA cycle is typically incomplete, missing the steps in the cycle that generate acetyl CoA and oxaloacetate. In these organisms, the rTCA cycle operates as a linear pathway, like the acetyl-CoA pathway itself, and forms a U-shaped or 'horseshoe' pathway that leads to linear accumulation of essential metabolites. This is shown in Figure 3.10. The incomplete rTCA cycle starts with oxaloacetate that comes from the incorporation of CO_2 by phosphoenolpyruvate and incorporates one more CO_2 at the step leading to 2-oxoglutarate (α-ketoglutarate). This sequence of CO_2 incorporations provides C2 units (acetyl CoA), C3 units (pyruvate), C4 units (oxalacetate), and C5 units (2-oxoglutarate) from which all of the carbon atoms in amino acids stem. Amino acids are not only the building blocks of proteins; they are also the starting point for the synthesis of the bases of nucleic acids, requiring nitrogen incorporation, a topic to which we will return shortly. The terms α-ketoglutarate and 2-oxoglutarate are synonymous and interchangeable, which applies to all α-ketoacids and 2-oxoacids. The α-ketoacid designations are older. In modern enzyme nomenclature, the 2-oxoacid designations are used.

Based on observations from thermodynamics and comparative physiology, Fuchs has made a clear case for the antiquity of the acetyl-CoA pathway since the early 1980s. Nonetheless, a large number of publications, usually founded in observations other than biology or thermodynamics, have favored the rTCA cycle as the starting point of metabolism. For many years it was thought that the rTCA cycle operates in archaea, which turned out not to be the case (Table 3.1). To date, the acetyl-CoA pathway is the only core pathway of CO_2 fixation that is found in both bacteria and archaea.

Harold Morowitz and Günter Wächtershäuser championed the idea of an autocatalytic nature of the rTCA cycle. **Autocatalytic** is a term used to designate reactions that can be catalyzed by one of the reaction products. The rTCA cycle is not autocatalytic in the strict sense, because it falls short of this definition. However, the rTCA cycle can be presented in such a way as to make it seem that it produces more of its own constituents (red arrow in Figure 3.9). When it operates, it fixes two molecules of CO_2 to an acetyl unit in acetyl-CoA that is exported for other syntheses. Furthermore, yet not bearing upon the catalytic criterion of autocatalysis, its intermediates are depleted to provide carbon backbones for amino acids, so it does not actually make more of itself as an autocatalytic cycle, it makes C2, C4, and

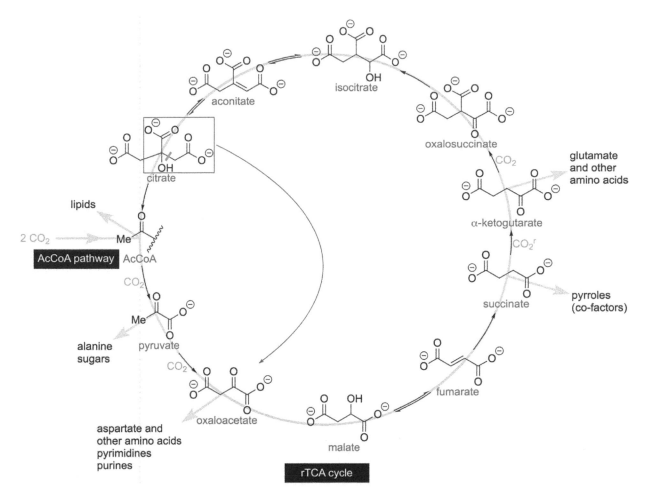

FIGURE 3.9 Hypothetical reaction network connecting the acetyl-CoA pathway to the rTCA cycle. The two sequences marked yellow in the rTCA cycle were observed experimentally using Fe^0 (Ni^0), Zn^{2+}, and Cr^{3+} as catalysts. The observed sequences encompass C=O and C=C hydrogenation (reduction) and hydration and dehydration in 6 of the 11 reactions of the reverse TCA cycle. The slightly endergonic dehydrations are probably driven by coordination to Zn^{2+} and pushed forward by the subsequent irreversible reduction step using Fe^0. CO_2^r indicates reductive carboxylation ($CO_2 + 2H^+ + 2 e^- + ATP$), α-ketoglutarate carboxylation is ATP dependent in metabolism. The retro-aldol cleavage of citrate is shown as a green dash. The two reaction sequences marked yellow were observed using Fe^0 (Ni^0), Zn^{2+}, and Cr^{3+} as catalysts. The cleavage product oxaloacetate feeds back in the rTCA cycle to enhance the reaction sequences (red arrow). Potential side reactions like off-cycle reductions were not observed. Most of the precursor molecules to amino acids, purines, pyrimidines, purines, pyrroles, and lipids could be obtained from H_2 and CO_2 and a source of activated nitrogen if the two pathways would couple. Adapted from Muchowska K. B. *et al.* (2017) Metals promote sequences of the reverse Krebs cycle, *Nat. Ecol. Evol.* 1: 1716–1721. Although some bacteria can use succinate for tetrapyrrole biosynthesis, as shown in the figure, the primordial pathway of tetrapyrrole biosynthesis starts from glutamyl-tRNA^{Glu} and proceeds via glutamate-1-semialdehyde (Dailey H. A. *et al.* (2017) Prokaryotic heme biosynthesis: Multiple pathways to a common essential product, *Microbiol. Mol. Biol. Rev.* 81: e00048–16).

C5 carbon units, the latter two of which need to be replaced by anaplerotic (replenishing) reactions that allow all reactions of the cycle to keep pace.

In cells that use the acetyl-CoA pathway, the rTCA cycle is typically incomplete, operating as a horseshoe (Figure 3.10) without the synthesis of citrate. Starting from the acetyl-CoA pathway and the horseshoe rTCA cycle as ancient and natural (nonenzymatic) CO_2 incorporation pathways, the connections to amino acid synthesis and nucleotide synthesis become very short and direct. Wächtershäuser and colleagues demonstrated reductive aminations of α-ketoacids with NH_3 using FeS as the reductant. Muchowska, Moran, and colleagues used hydroxylamine to generate amino acids from 2-oxoacids (α-ketoacids) using Fe^0 as the reductant. In both cases, the reaction is reductive amination of a keto group to an amino group (Figure 3.11). In biological systems, the main reduction amination reaction is the incorporation of NH_3 into 2-oxoglutarate to form glutamate using electrons from NADPH or ferredoxin. As the name 'reductive amination' indicates, the reaction requires the addition not only of an amino group but also the addition of 2 electrons. In the studies using FeS or Fe^0 as the reductant, H_2 was probably formed and served as the electron donor in both reactions, being activated by transition metals.

Reductive amination of the first 2-oxoacids that arise from CO_2 incorporation via the acetyl-CoA pathway and the incomplete horseshoe TCA cycle—pyruvate, oxalacetate, 2-oxoglutarate, and glyoxylate—give rise to alanine, aspartate, glutamate, and glycine. These are starting points not only for other amino acid biosyntheses but also starting points for the synthesis of nucleobases. In biological chemistry, the bases of nucleic acids are formed from amino acids, one carbon intermediates of the acetyl-CoA pathway, NH_3, and CO_2. This is shown in Figure 3.12B. The route

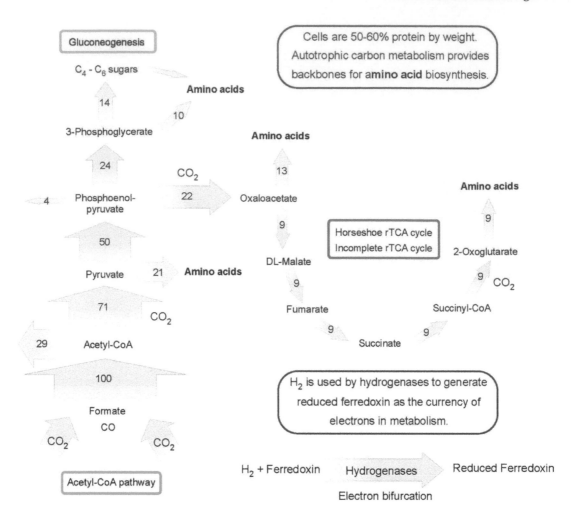

FIGURE 3.10 **The acetyl-CoA pathway feeds the rTCA cycle and amino acid synthesis.** As mentioned in Figure 2.5, carbon metabolism in microbes that use the acetyl-CoA pathway branches into gluconeogenesis and the incomplete rTCA pathway. The flux of carbon (in%) is indicated in the arrows as in Fuchs G. (2011) Alternative pathways of carbon dioxide fixation: Insights into the early evolution of life? *Annu. Rev. Microbiol.* 65: 631–658. The main message of the figure is to once again underscore that the rTCA cycle in organisms that use the acetyl-CoA pathway is primarily a source of 2-oxoacids for amino acid synthesis. Modified from: Fuchs G., Stupperich E. (1978) Evidence for an incomplete reductive carboxylic acid cycle in *Methanobacterium thermoautotrophicum*, *Arch. Microbiol.* 118: 121–125; Stupperich E., Fuchs G. (1984) Autotrophic synthesis of activated acetic acid from two CO_2 in *Methanobacterium thermoautotrophicum*: I. Properties of in vitro system, *Arch. Microbiol.* 139: 8–13; Martin and Russell 2007 On the origin of biochemistry at an alkaline hydrothermal vent. *Philos. Trans. R. Soc. Lond. B* 362: 1887–1925.

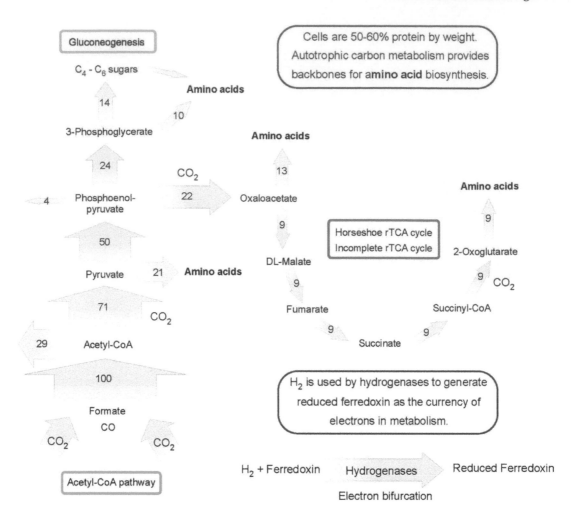

FIGURE 3.11 **Reductive amination.** The conversion of keto groups to amino groups requires the addition of NH_3 and H_2 as $2H^+$ and $2e^-$ or as H^- and H^+. Water is formed in the reaction. See Section 4.1 for details of amino acid synthesis. Recent findings reveal that serpentinization, discussed in Chapter 1, has recently been shown to generate vast amounts of ammonia in laboratory experiments via the reaction $2Fe^{2+} + 2H_2O \rightarrow 2Fe^{3+} + H_2 + 2OH^-$; $3H_2 + N_2 \rightarrow 2NH_3$, whereby up to 200 μmol NH_3 was obtained per gram of peridotite after 30d: Shang X., Hunag R., Sun W. (2023) Formation of ammonia through serpentinization in the Hadean Eon, *Sci. Bull.* 68: 1109–1112. Ammonia, generated by serpentinization, can be used as a substrate for reductive aminations of 10 different alpha-keto acids (see Figure 3.10) in laboratory experiments using H_2 as the reductant and Ni^0 as the catalyst (Kaur H. et al. (2024) A prebiotic Krebs cycle analogue generates amino acids with H_2 and NH_3 over nickel. CHEM 10 doi.org/10.1016/j.chempr.2024.02.001). In this way, serpentinization can contribute to primordial amino acid synthesis.

from the acetyl-CoA pathway to amino acids and nucleobases is short, direct, and involves naturally arising intermediates.

Is it reasonable to assume that pterins like tetrahydrofolate, which participate both in bacterial purine synthesis and in the acetyl-CoA pathway (Figure 3.1), could have existed in prebiotic evolution before enzymes? There are three things to consider. First, the essential contribution of the pterin in the reaction is simply to donate a reactive C1 unit. This function could have been fulfilled, in principle, by activated C1 units on a transition metal catalyst surface as Preiner *et al.* showed. The C1 unit is the irreplaceable unit of function, the pterin, and the metal catalysts are interchangeable. Second, the archaeal purine pathway appears to use formyl phosphate rather than the pterin-bound C1 unit, this is a simpler reaction than that involving the pterin. Third, the synthesis of pterins is admittedly complicated in modern biology, but pterins can be obtained in surprisingly simple reactions by heating amino acids under conditions of very low water activity (Figure 3.13).

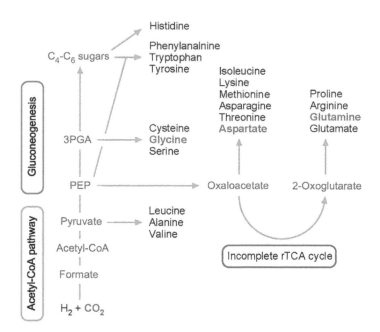

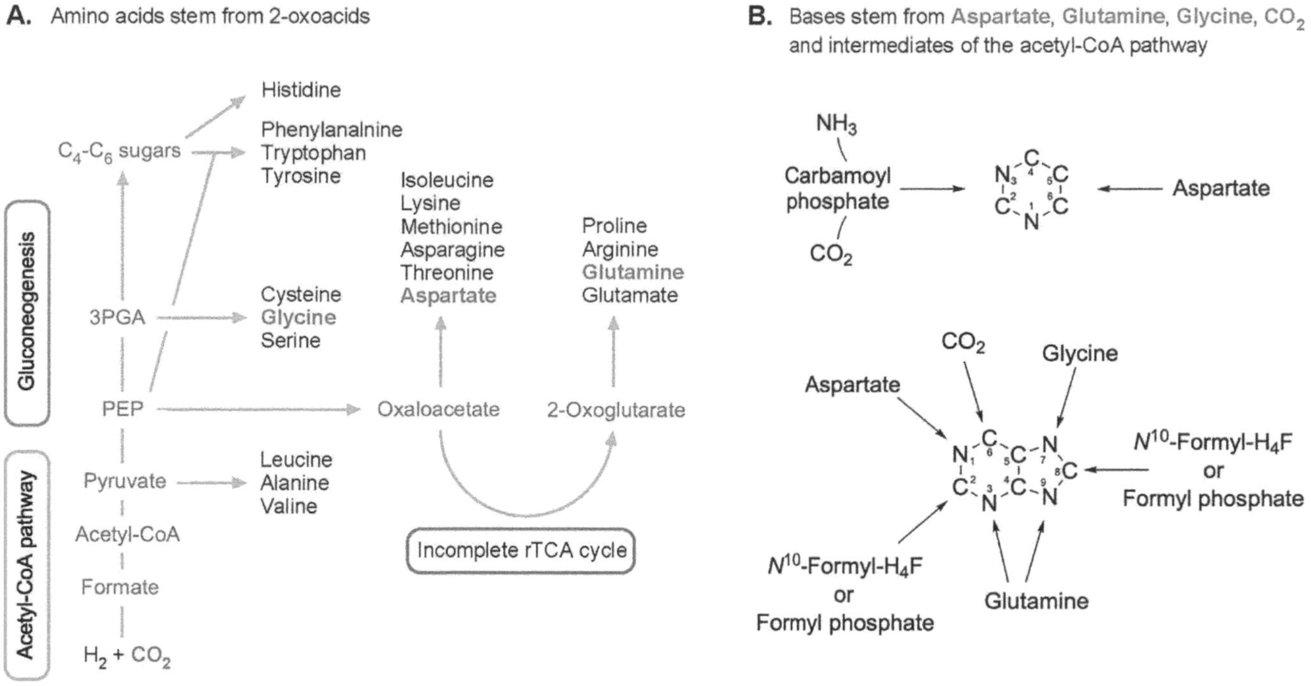

A. Amino acids stem from 2-oxoacids

B. Bases stem from Aspartate, Glutamine, Glycine, CO_2 and intermediates of the acetyl-CoA pathway

FIGURE 3.12 **Amino acid biosynthesis and the origin of the atoms in bases in metabolism.** (A) An abbreviated form of Figure 3.10 shows the biosynthetic families of the 20 amino acids. See Section 4.1 for more details on amino acid synthesis and nucleotide synthesis. Modified from Martin W. F. (2020) Older than genes: The acetyl-CoA pathway and origins. *Front. Microbiol.* 11: 817. Amino acids shown in boldface type are precursors of purine and pyrimidine biosynthesis. (B) The origin of the atoms in purines and pyrimidines. Some methanogens use formyl phosphate instead of formyl tetrahydrofolate. Modified from Stryer (1975) *Biochemistry*, Freemann, San Francisco, CA, p. 949.

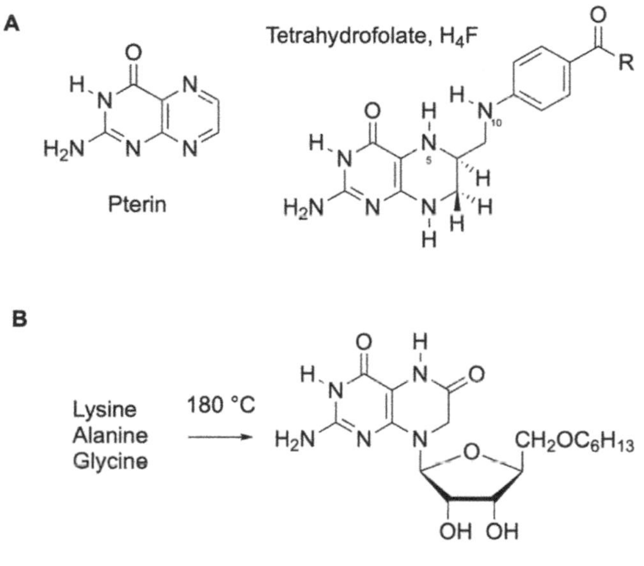

A

Tetrahydrofolate, H_4F

Pterin

B

Lysine
Alanine $\xrightarrow{180\,°C}$
Glycine

FIGURE 3.13 **Pterins.** (A) Pterin and tetrahydrofolate (H_4F). (B) Abiotic formation of a pterin from simply heating amino acids in an anhydrous state. The structure shown is compound 7 in Heinz B. *et al.* (1979) Thermische Erzeugung von Pteridinen und Flavinen aus Aminosauregemischen. *Angew. Chemie* 91: 510–511. Note the *N*-glycosidic bond.

Metabolism as in Figure 3.1 (the archaeal and bacterial acetyl-CoA pathway) shows us the reactions that are used in the life process. Nonenzymatic reactions as in Figure 3.4 or 3.9 show us the reactions that unfold spontaneously if we start with the right starting compounds using hydrothermal conditions and, crucially, suitable catalysts

that occur naturally in hydrothermal systems. There is a growing realization that the degree of correspondence between the chemical reactions of central metabolism and their nonenzymatic counterparts is much higher than most might have dared to expect even 20 years ago. Is that shocking or surprising? Not if we recall the basics: Enzymes do not catalyze reactions that would otherwise be impossible, they just accelerate reactions that tend to occur anyway.

Thermodynamics dictates the direction of reactions, and kinetics determines rates. Does that mean that life (cell mass) is in a lower energy state than the components from which it is made? Under hydrothermal conditions, yes. This might seem surprising at first sight. In anaerobes living from H_2, CO_2, and NH_3, the changes in free energy for the synthesis of cell mass—the synthesis of the components of cell mass, not the growth, division, or movement process—are negative under hydrothermal vent conditions at 50°C (Table 3.2).

In other words, the synthesis of cells under hydrothermal vent conditions is exergonic, but the synthesis of cells in modern seawater is not an exergonic process. The thermodynamic estimates in Table 3.2 show that at 25°C in slightly acidic Hadean seawater, there is not enough reducing power in the form of electrons to synthesize organic compounds from HCO_3^-. Table 2.2 shows that electrons from H_2 lack sufficient energy to synthesize for example HCOOH from HCO_3^- (redox potentials of −0.41 and −0.43 V, respectively). In the mixing zone of cool seawater and hot alkaline hydrothermal vent effluent, however, the redox potential of H_2 is sufficiently negative to synthesize organic compounds

TABLE 3.2

Thermodynamics of Cell Mass Synthesis under Hydrothermal Conditions

Cell Constituent	Amount (mg per Gram of Cells)	ΔG_R (J per Gram of Cells)				
		25°C	50°C	75°C	100°C	125°C
Fatty acids	80	−75	−347	−330	−298	−230
Nucleotides	249	405	214	247	295	380
Saccharides	47	38	−27	−22	−13	8
Amino acids	631	117	−841	−754	−603	−305
Amines	17	14	−15	−13	−10	−3
Total	1024	500	−1016	−873	−628	−150

Note: The calculations are based on the mixing of cool, relatively oxidized Hadean Seawater (redox midpoint potential E_h −300 mV at 25°C, pH 6.5) with hot, reduced alkaline hydrothermal vent effluent (140°C, pH 9, E_h −700 mV) in endmember fluids at 250 bar. The values in the column for 25°C are for Hadean seawater without mixing; the different temperatures result from different mixing ratios of the cool seawater with the hot hydrothermal effluent. From Amend J. P., McCollom T. M. (2009) Energetics of biomolecule synthesis on early earth. In *Chemical Evolution II: From the Origins of Life to Modern Society.* Zaikowski, L. *et al.* eds., American Chemical Society, Washington, DC, pp. 63–94.

(Table 2.1), leading to negative Gibbs energies for cell mass production. At moderately elevated temperatures, for example, 50°C, the energy contributions from HCO_3^-, H^+ (alkaline pH), and electrons are optimized, resulting in more negative values of ΔG_R (−1016 J/g of cells in Table 3.2). At higher temperatures, for example, 125°C, reducing power is very high (Table 2.1); however, this effect on ΔG_R is partially offset by the low HCO_3^- concentrations in nearly pure alkaline hydrothermal vent fluids (removal of HCO_3^- by precipitation as $MgCO_3$ and $CaCO_3$) and low H^+ concentration at high pH (H^+ is necessary for hydrogenation of HCO_3^- to an organic compound).

The first cells had to be autotrophic because organic food was not available, although organic catalysts, which are not the same as foodstuffs, likely were available, at least in some form. The first cells developed a metabolism to build up cell mass (proteins and others) from inorganics by an overall exergonic process which is only possible under conditions with sufficient reducing power, not in an oxidizing environment. In the presence of H_2 and the absence of oxidants, which capture electrons, the synthesis of cell mass is exergonic under the strongly reducing conditions of hydrothermal environments. This is a very strong argument in favor of a hydrogen-dependent origin of life. The chemistry of the system works over geological time scales in favor of the synthesis of the building blocks of life arising from the elements.

There are versions of hydrothermal theories founded on the idea that methane oxidation preceded CO_2 reduction in evolution as a source of chemical energy. That theory is problematic, however, because the C−H bond of methane is extremely stable, reminiscent of Teflon (C−F bonds), requiring the assumed presence of very strong oxidants like O_2 or NO to break such bonds nonenzymatically. But if such strong oxidants are around to open up the C−H bond of methane, the accumulation of the components of cells is no longer thermodynamically favored and any synthesized organics are oxidized to CO_2. In the absence of oxygen, CO_2 is far more reactive than methane, which is why CO_2 reacts so well on transition metal catalysts while

CH_4 does not. Reducing environments are conducive to the preservation of biological material whereas oxidizing environments are not. We will see in Section 3.2 that hydrothermal systems also provide natural mechanisms to concentrate reaction products around their site of synthesis so that more complex chemical compounds including polymers can arise. The most significant property of hydrothermal systems in an origins context, however, is their highly reducing environment, which drives the synthesis of pyruvate from H_2 and CO_2 in the laboratory and the synthesis of TCA or rTCA cycle intermediates from pyruvate.

If the nonenzymatic analog of the acetyl-CoA pathway (Figure 3.5) links via pyruvate to the linear horseshoe variant of the rTCA cycle (Figure 3.10), then, with a source of NH_3, all of the precursor molecules of the building blocks of proteins, RNA, lipids and cofactors, including cobalamin can, in principle, be generated, because primitive anaerobes do so by that route. Additionally, various α-keto acids and aldehydes of the rTCA cycle react to thioesters in the presence of Fe^{2+}, thiol, and oxidant. Hence a coupled acetyl CoA—horseshoe rTCA reaction hub would provide a rich set of precursor molecules for biomolecular building blocks and phosphorylating potential for energy metabolism.

A prebiotic connection between the two central segments of carbon metabolism is not improbable because there is a common set of catalysts (Fe^0, Zn^{2+}, Cr^{3+}) and reactants (hydrated CO_2 and H_2 or e^- plus H^+ from Fe^{2+} and H_2O). The chemical conditions are quite different for the two networks, however, using the catalysts studied so far, especially the pH value. While the nonenzymatic analog of the acetyl-CoA pathway runs optimally in alkaline conditions, Zn^{2+} and Cr^{3+} promote dehydration and hydration in the rTCA cycle most efficiently under strongly acidic conditions, where metal cations become hydrated and form polynuclear species that can complex the rTCA acid intermediates. In contrast, under alkaline conditions, many transition metal ions react to hydroxides that do not bind carboxylic acids well. Coordination of the carboxylic

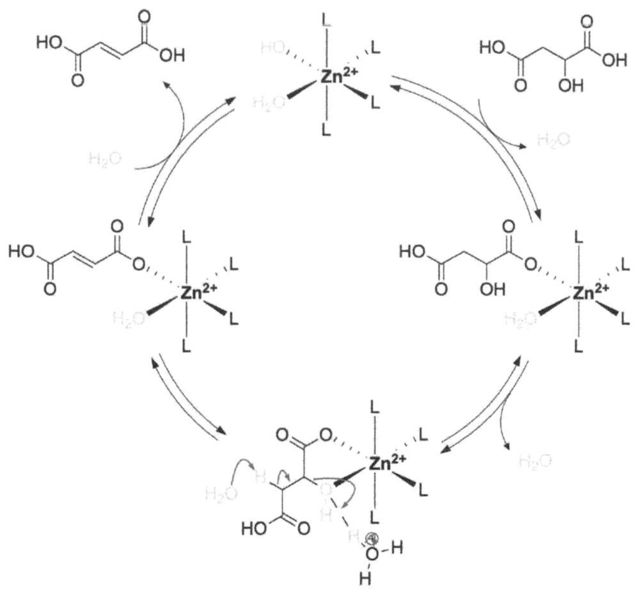

FIGURE 3.14 Possible mechanism for Zn²⁺ promoted dehydration of malate to fumarate. Hydration (mechanism not shown here) is catalyzed best by Cr^{3+} where the water ligands just have the right position for hydration of the ketoacids. Redrawn from Muchowska K. B. *et al.* (2017) Metals promote sequences of the reverse Krebs cycle, *Nat. Ecol. Evol.* 1: 1716–1721.

acid with the metal cation, however, is necessary for dehydration because bulk water activity is lowered inside polynuclear metal clusters (Zn^{2+} catalysis in Figure 3.14). There is still much to explore in the realm of nonenzymatic catalysis.

How probable is a unified acetyl CoA–rTCA reaction network at a serpentinizing field like Lost City? Recall that the equilibrium pH of the Lost City's hydrothermal fluid is nearly neutral at temperatures near 300°C but increases to about pH 11 at 50°C, because the solubility of brucite increases at lower temperatures, releasing dissolved Mg^{2+} and OH^- ions. Serpentinization generates H^+ which promotes the dissolution of the forsterite component of olivine (Section 2.2) and hence enhances serpentinization by feedback. There is a strong proton gradient between the acidic serpentinization site and the low-temperature alkaline effluent. This leaves the possibility that the surface reactions of the acetyl-CoA pathway and an incomplete rTCA network could take place at different serpentinization sites coupled by the transport of pyruvate, for example, or other free intermediates. Fundamental transport and accumulation processes are discussed in the next section. Nonetheless, the two observed reaction sequences of the rTCA cycle shown in Figure 3.9 proceeded also under the rather mildly acidic conditions of CO_2-saturated water. Hence carbon fixation and sequences of the rTCA cycle are potentially compatible at the same site.

3.1.8 ORGANIC CATALYSTS: THE FUNCTIONS OF COFACTORS

A recurrent theme in early biochemical evolution is that catalysis by cofactors alone preceded catalysis by enzymes assisted by cofactors. This makes sense because there are

many examples known where a cofactor alone can provide a marked increase in reaction rate. One of the most instructive examples is the case of a pyridoxal phosphate (PLP) dependent transaminase reaction investigated by Zabinsky and Toney. The PLP-dependent enzyme accelerates the rate of the uncatalyzed reaction by ~10^{18} fold. They found that at pH 8, the addition of free PLP alone catalyzes 2-aminoisobutyrate decarboxylation by ~10^{10}-fold, with the enzyme contributing an additional ~10^8-fold in the presence of the cofactor. The PLP cofactor alone is the main source of catalysis, but the enzyme helps of course, in particular with regard to substrate specificity. Cofactors are essential to modern metabolism and they were likely essential to primordial metabolism as well. In Figure 3.15 we have drawn the structures of most of the cofactors used by microbial metabolism. The active moieties of the cofactors are highlighted in green. What functional roles do the cofactors have?

We start with **nicotinamide adenine dinucleotide** (**NAD⁺**, vitamin B_3), a redox cofactor that serves as a hydride donor or acceptor in 2-electron transfer reactions. We have seen that metal minerals such as Ni_3Fe can also catalyze 2-electron transfer reactions, but Ni_3Fe cannot be incorporated into proteins at an active site, whereas NAD^+ can. In prokaryotic metabolism, NAD^+ is synthesized from aspartate via quinolinic acid, a substituted pyridine.

Corrins include vitamin B_{12}, the most complex cofactor. They belong to the class of **tetrapyrroles**, which differ in their properties according to the central coordinated metal and are derived in metabolism from succinyl-CoA and glycine. Corrins coordinate cobalt. Acetogens and methanogens require the corrin B_{12} for an essential step in the acetyl-CoA pathway: the transfer of the methyl group from corrinoid Co^{2+} to a nickel atom in the active site of CODH (Figure 3.15). This rare metal-to-metal methyl transfer reaction is catalyzed in cells by a protein called CoFeS (corrinoid iron-sulfur protein) but can also be catalyzed by Ni_3Fe, yet with the same caveat as in the case of NAD^+, namely, the native metal Ni_3Fe complex cannot be incorporated into protein. Other tetrapyrroles include F_{430}, which coordinates Ni^{2+} and is essential for the last step of methanogenesis: methyl CoM reduction to generate methane, a reaction that has a radical mechanism. Tetrapyrroles that coordinate Fe^{2+} are typically called heme. In siroheme, a special form of heme specific to sulfite and nitrite reduction, Fe^{2+} also coordinates a 4Fe4S cluster.

S-adenosyl methionine (SAM) is an essential cofactor, best known for its role in methyl transferase reactions. In conjunction with a **4Fe-4S** cluster in **radical SAM enzymes**, it can also function as a radical-generating cofactor. The addition of an electron from ferredoxin via the FeS cluster yields methionine and the 5′-deoxyadenosyl radical. As reviewed by Joan Broderick, SAM is one of the functionally most diverse cofactors, it is involved in radical generation, sulfur insertions, radical-dependent mutase reactions, thiamine biosynthesis, molybdenum cofactor (MoCo) biosynthesis, tRNA methylations, methylthiolations, metal cluster synthesis in nitrogenase maturation, and tetrapyrrole modifications.

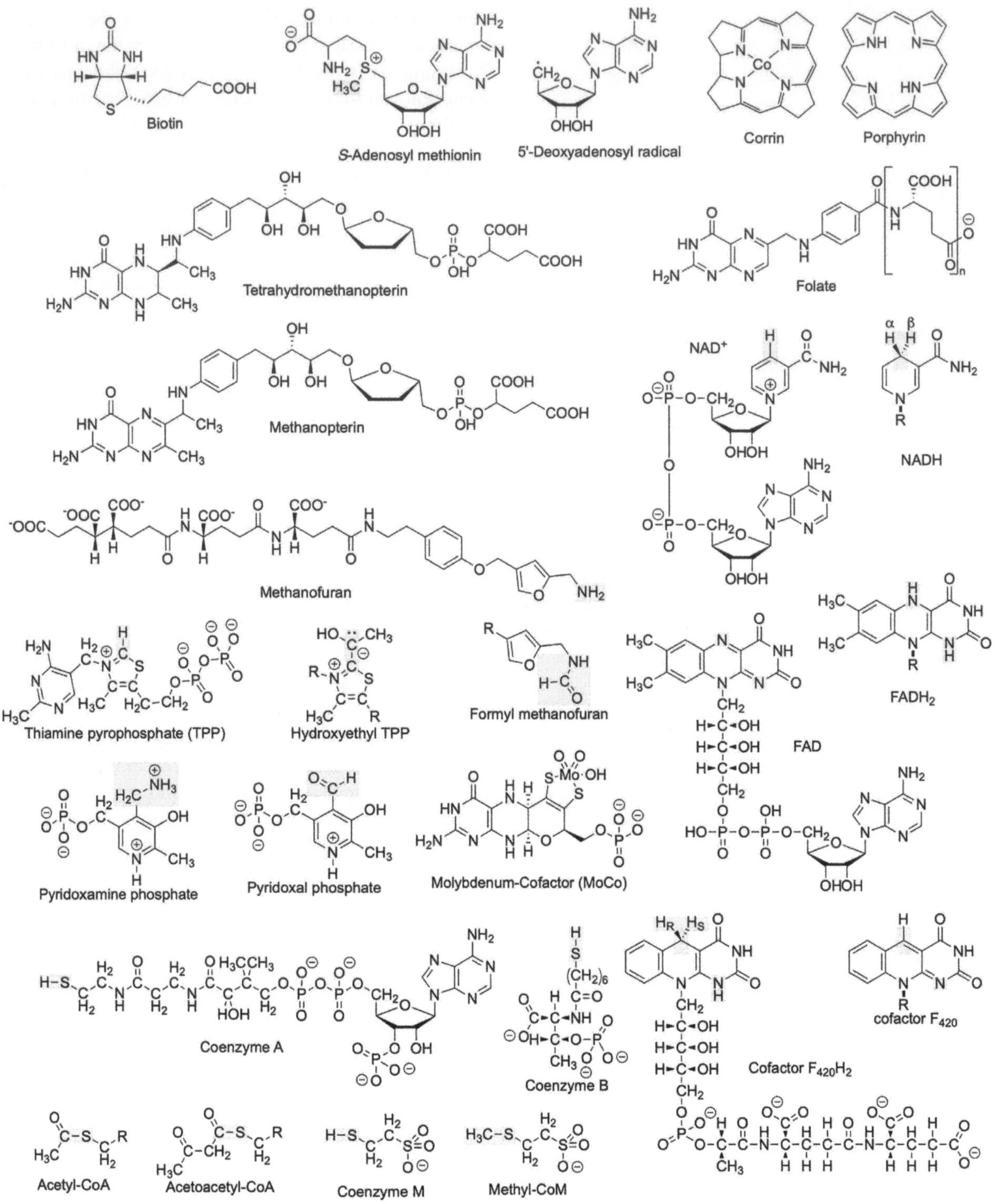

FIGURE 3.15 Cofactors have no chiral centers at their functional moieties. The active moieties of the individual cofactors are highlighted in green. The roles of the various cofactors in biochemical reactions are outlined in the text.

Pyridoxal phosphate (PLP, vitamin B_6) is another extremely versatile cofactor. In biosynthesis, it is derived from the condensation of phosphothreonine with xylulose in a highly exergonic reaction involving cyclization and water eliminations leading to aromaticity ($\Delta G^{0'} = -383$ kJ/mol). Mainly known for its role in transamination (amino acid synthesis), PLP is also involved in elimination reactions, racemization reactions, and aldol condensations. The main active moiety of PLP is its free aldehyde group. Aldehydes are highly reactive and therefore generally rare in metabolism. PLP is one of the most commonly employed cofactors in metabolism. In transamination reactions, PLP toggles between the aldehyde and the amine (pyridoxamine, PNP) state during amino transfer. Transamination reactions can be catalyzed by Fe^{2+}.

Flavin adenine dinucleotide (FAD, vitamin B_2) is both a hydride carrier in two-electron reactions, like NAD$^+$ but in addition can function as a transducer of one electron to two electron transfers because of the metastable nature of the flavin semiquinone radical (see Figure 2.12). Flavins belong to the class of pterins, in biosynthesis, they are derived from GTP via the incorporation of two carbon atoms from ribose (the 1' and 2' positions) into the guanosine ring under formate elimination. Flavins are essential to flavin-based electron bifurcation, which is essential for microbes that reduce CO_2 with electrons from H_2. The function of electron bifurcation (CO_2 reduction with electrons from H_2) can be supplied by Ni_3Fe under hydrothermal conditions. **Methanofuran (MF)** is an archaeal-specific cofactor involved in the acetyl-CoA pathway during the reduction of CO_2 to formate and binding as the cofactor bound N-formyl group in formyl-MF, which transfers the formyl group to tetrahydromethanopterin. Like F_{430}, the role of MF in metabolism appears to be restricted to methanogenesis. **Factor F_{420}** is, like FAD, a pterin, however, one N atom of FAD is replaced by C in F_{420}. Because of this difference, it does not perform one-electron reactions. It is involved in hydride transfer (two-electron reactions), like NAD^+. As Shima and team have shown, methanogens have an additional cofactor that has a special place in physiology, **iron guanylylpyrinidol (FeGP)**. It is the cofactor of the [Fe]-only hydrogenase Hmd of methanogens, which transfers electrons from H_2 directly to the reaction substrate, methenyl-H_4MPT, to generate methylene-H_4MPT during the reaction catalyzed by Hmd. All other hydrogenases transfer their electrons from H_2 to FeS clusters, either in the enzyme itself or in ferredoxin. The FeGP cofactor of methanogens is also unique in that its catalytically active Fe atom is coordinated by three carbonyl C atoms, one S atom, one aromatic N (in the pyridine ring), and one O (in H_2O) as ligands.

Thiamine pyrophosphate (TPP, vitamin B_1) is typically involved in the transfer or metabolism of C2 units. It is synthesized from tyrosine, deoxyxylulose-5-phosphate, cysteine, and intermediate of purine biosynthesis. TPP is widely involved in C2 transfers in sugar metabolism. It is also a key cofactor in the enzymatic pyruvate synthase reaction that generates pyruvate in the acetyl-CoA pathway via condensation of acetyl-CoA with CO_2 and H_2-dependent reduction. That function, H_2-dependent pyruvate synthesis from CO_2 and an acetyl moiety, can be replaced by Ni_3Fe. The active moiety of **coenzyme A (CoA)** is a thiol group that forms thioesters as the activated forms of organic acids. The thioester bond has roughly the same free energy of hydrolysis as ATP, $\Delta G^{0'} = -32$ kJ/mol, giving it a high group transfer potential for acyl residues, sufficient to generate acetyl phosphate. The hydrides in the methyl group of acetyl CoA are in α−position to the carbonyl and hence slightly acidic, promoting generation of the carbanion, which can perform nucleophilic attack or undergo Claisen condensations, forming β keto acids like acetoacetyl-CoA. In metabolism, the pantothenate (vitamin B_5) arm of CoA is derived from amino acids. **Coenzyme M (CoM)** and **coenzyme B (CoB)** are specific to methanogenesis. Both have a thiol as the active moiety. They are involved in the terminal step of methane synthesis. CoM forms methyl CoM as the substrate for methane formation during methanogenesis. Methane can also be formed from CO_2 and H_2 using Fe_3O_4 or Fe_3S_4 as a catalyst. **Biotin (vitamin B_7)** is a cofactor of carboxylation reactions. In a typical biotin-dependent reaction, CO_2 is phosphorylated by ATP to yield carboxyphosphate, which reacts with biotin at the amino group to form carboxybiotin

for the substrate carboxylation reaction.

Tetrahydrofolate (H_4F, vitamin B_9) and its archaeal counterpart **tetrahydromethanopterin (H_4MPT)** are pterins, derived from GTP as outlined above for FAD. They are C1 carriers. In biosyntheses, biotin has the role of donating carboxy groups, while H_4F and H_4MPT donate more reduced C1 units. The synthesis of these C1 units is outlined in Figure 3.15, where the methyl synthesis branch of the acetyl-CoA pathway is shown. H_4F and H_4MPT form the backbone of the acetyl-CoA pathway, holding formyl (−CHO), methenyl (−CH=), methylene (−CH_2−), and methyl (−CH_3) groups in place, bonded to N^5 and N^{10} of the cofactor (see Figure 3.2). Again, the synthesis of the methyl groups from H_2 and CO_2 can be performed by Ni_3Fe, but Ni_3Fe cannot be incorporated into proteins as a prosthetic group.

A very notable aspect of the cofactors is the complete absence of chiral centers at their active moieties. This suggests that catalysis involving cofactors could have preceded the origin of chirality in biochemical evolution.

3.2 SEPARATION, ACCUMULATION, AND AUTOCATALYSIS GENERATE REACTION NETWORKS

Returning to abiotic CO_2 fixation experiments using Ni_3Fe or native metals, we recall that although the product spectrum corresponds nearly 1:1 to biochemical products, the yields in the acetyl-CoA pathway simulation and related rTCA carbon fixation experiments are generally quite low. For example, only $\approx 1\%$ of the available electrons from Fe^0 were channeled to organic products pointing to the predominant formation of gaseous products like CO and H_2 and/or surface coverage with inert iron hydroxides and iron carbonates. High pressures of H_2 can increase the yield of organic products substantially, up to 10 μM pyruvate in the typical experiments described in Section 3.1. Under some conditions, transition metal catalysts can generate 200 μM pyruvate from H_2 and CO_2 as Beyazay and colleagues have shown; that is the physiological concentration of pyruvate that accumulates in the cytosol of acetogens growing on H_2 and CO_2, as measured by Furdui and Ragsdale. As in metabolism, consecutive reactions of products to more complex carboxylic acids and from there to amino acids and nucleotides, and as metabolism fans out like a river delta, each path will lead to even smaller concentrations. Therefore, products such as amino acids or bases must somehow accumulate to reach concentrations necessary for polymerization.

Hydrothermal fluids are transported through volcanic or other hydrothermal pores, channels, and fissures by pressure gradients (*bulk transport*, for example, vent effluent) and temperature gradients (*convection*), and on a small length scale by temperature-dependent diffusion (called *thermodiffusion* or *thermophoresis*) and concentration-dependent diffusion (called *diffusion*) (see Figure 3.16). The group of Dieter Braun pioneered the use of these techniques for the investigation of primordial evolution. Dissolved molecules can chromatographically separate on the surfaces of rocks and accumulate in cooler, lower-lying zones of inorganic pores and fissures by convection and thermodiffusion.

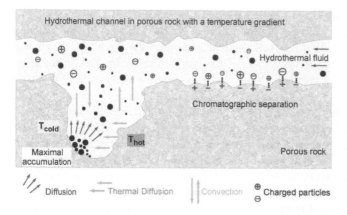

Diffusion   ← Thermal Diffusion   ‖ Convection   ⊕ Charged particles ⊖

FIGURE 3.16 **Molecular separation and accumulation of hydrothermal fluids in volcanic rocks.** Different kinds of interaction between rock surface and hydrothermal fluid like electrostatic attraction and repulsion (+ - and + +, - -), hydrogen bonding, hydrophobic effects, and van der Waals forces lead to chromatographic separation. Convection (green arrows), thermal diffusion (yellow arrows), and counter balancing concentration-dependent diffusion (red arrows) lead to the accumulation of molecules in lower-lying parts of low-temperature zones of the hydrothermal channels and pores.

3.2.1 Molecules Accumulate in Mineral Pores and Channels by Convection and Thermodiffusion

Gases or liquids expand upon heating such that their density (weight per volume) decreases. Then, driven by gravity (displacement by cooler, more dense gases or liquids), they flow upward and with them dissolved molecules and heat (*convection*). For example, warm air moves up and sinks upon cooling. Water vapor moves with the air and condenses to water droplets in the coolest part of a room. Sailors know that currents in open waters can often be traced back to temperature differences as well. Furthermore, molecules diffuse from a region of high molecular concentration to a region of lower concentration, a process called *diffusion*. Diffusion results from molecular motion, which is driven by random collisions. However, on average, more molecules will move out from crowded places than from emptier places: This generates a uniform distribution of molecules, which has greater entropy than a non-uniform distribution of the same molecules.

Molecular motion can also be driven by a thermal gradient. Molecules diffuse from warmer to colder places because collisions with species from the warmer side have a higher energy per collision owing to the higher average particle velocity. This migration of particles induced by a temperature gradient is called *thermodiffusion* (also named *thermophoresis* or *Soret effect*).

In a binary (two-component) fluid mixture, the total mass flux J (mol/m²/s) of ordinary diffusion and thermodiffusion in direction x is given by

$$J_x = -D\frac{\partial c}{\partial x} - c(1-c)D_T\frac{\partial T}{\partial x}$$

where $\partial c/\partial x$ is the change of concentration c (mol/m³) along direction x (concentration gradient), D is the mass diffusion or Fickian coefficient (m²/s), D_T is the thermodiffusion coefficient (K⁻¹ m² s⁻¹) and $\partial T/\partial x$ is the temperature gradient. If $D_T > 0$ (as for most molecules dissolved in water at ambient temperatures), thermodiffusion transports mass from high to low temperature. If the temperature gradient is time-independent (does not change with time), then a steady state $J_x = 0$ is reached when the mass fluxes from thermodiffusion and (opposite directed) diffusion just cancel at some concentration distribution. For $J_x = 0$ we obtain

$$\frac{\partial c}{\partial x} = -c(1-c)\frac{D_T}{D}\frac{\partial T}{\partial x}$$

where $S_T = D_T/D$ is the Soret coefficient. A larger Soret coefficient indicates a larger concentration increase for a given temperature decrease. The corresponding three-dimensional equation including terms for convection (not shown here) can be numerically integrated, or an approximate analytical solution can be used to obtain the stationary concentration c as a function of x, y, and z (see Baaske *et al.*, 2007).

The accumulation of formamide, a typical product expected from the fixation of CO_2 and N_2, can serve as an example to illustrate. Formamide is used by chemists to synthesize nucleotides (Section 4.6). Figure 3.17 shows the accumulation c_{max}/c_0 of formamide as obtained by Niether, Wiegand, and coworkers by solving the diffusion equations with a measured Soret coefficient. They used an initial concentration (weight fraction) $c_0 = 10^{-5}$ kg formamide/1 kg water (molar concentration 0.22 mM) and square pores of different length-to-width ('aspect') ratios in their calculations. An aspect ratio r of 100 corresponds to 10 mm pore length and 100 µm pore width, for example. In their calculations, the mean temperature T_{mean} was 75°C (Figure 3.17) and the temperature difference between hot and cold vertical walls was 60°C. At $r = 150$, accumulation is ≈10⁵, and maximal formamide concentration at the bottom of the pore on the cold side is close to the pure component. At lower formamide concentrations and higher temperature differences accumulation can reach 10⁹ (curves not shown here). A time-dependent study showed that at $c_0 = 10^{-5}$ mol/l FA, $r = 156$, and $T_{mean} = 45°C$, the saturation plateau is reached 45 days after the start of the temperature gradient. At $c_0 = 10^{-7}$ FA it takes 90 days.

For comparison, Wiegand and coworkers also calculated the accumulation of nucleotides in hydrothermal pores. The solubility of nucleotides in water is limited to a concentration of about 35 wt% (saturation concentration), which was reached at an accumulation of ≈10⁵ at $c_0 = 10^{-5}$ nucleotide weight fraction and $r ≈ 25$ (red line in Figure 3.17).

Accumulation depends on the Soret coefficient S_T which increases with increasing formamide concentration in water up to a factor of 10 at 10°C and a factor of 2 at 61°C. This is probably due to the hydrogen bonding of formamide. Formamide can form up to five hydrogen bonds to other formamide molecules, water only 3.5–4. Therefore, formamide forms aggregates more effectively than water

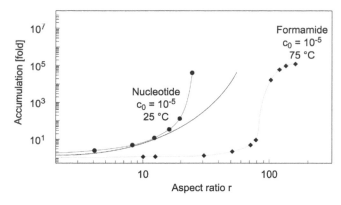

FIGURE 3.17 Theoretical accumulation of formamide (blue curve) and a nucleotide (red and black curve) in pores of different aspect ratios. The black line has been calculated from the analytical solution for accumulation of nucleotides and the red and blue lines were obtained by numerical integration. The temperature difference was 60°C, the initial weight fraction was 10^{-5} kg formamide (FA) in 1 kg water, and the mean temperature was 75°C. Formamide and water are freely miscible. At the plateau, the formamide concentration is close to the pure component. At lower formamide concentrations and higher temperatures, accumulation can reach 109 (curves not shown here). The solubility of nucleotides is limited to a concentration of 35 wt% (last data point on the red line) and hence accumulation is limited to this concentration. Accumulation increases with the Soret coefficient, which is typically 10^{-3} K^{-1} for FA. Redrawn from Niether D. *et al.* (2016) Accumulation of formamide in hydrothermal pores to form prebiotic nucleobases, *Proc. Natl Acad. Sci USA* 113: 4272–4277.

and has a much higher boiling point (T_{boil} = 210°C). Concentration diffusion of large formamide aggregates is inefficient (small D) such that S_T gets large and accumulation high. Generally, D decreases and S_T increases with the size of the respective molecule, aggregate or polymer. This is an important point for the accumulation of biopolymers in a small spot at the bottom of hydrothermal pores on the cold side. We will see in Section 4.7 that a 10^{15}-fold increase of concentration (!) can be reached for double-stranded, 1000 base pairs long DNA in 33 minutes by convection, thermodiffusion, and diffusion in pores of modest aspect ratio 10 at 30 K temperature difference—a very remarkable concentrating mechanism.

3.2.2 COMPLEX SOLUTIONS CAN BE CHROMATOGRAPHICALLY SEPARATED ON MINERAL SURFACES AND PORES

Chromatography is used to separate components of a mixture dissolved in a fluid (*mobile phase*). The mobile phase flows or is driven through a *stationary phase*, for example, a packed column, a sheet of porous paper, or a thin layer of adsorbents on a solid support. Depending on the strength of interaction with the stationary phase, the molecules in the fluid move a smaller or larger distance over a specific time period. For example, a polar substance bonds stronger to polar cellulose paper than a nonpolar one and therefore does not travel as far. Historically, chromatography (Greek *chroma*: 'color' and *graphein*: 'to write') was used to separate colored plant pigments.

Porous inorganic deposits with clefts, channels, pores, and extended surfaces from different minerals or covered by molecular layers from abiotic hydrothermal synthesis potentially present a rich variety of different stationary phases. Depending on pressure or temperature gradient, the hydrothermal fluid can move through rock columns and capillaries filled with small packing particles to achieve very efficient separation (similar to high-pressure liquid chromatography: HPLC). Organic molecules can be bound to the surface and bind molecules in the hydrothermal mobile phase specifically (affinity chromatography; Section 4.8). Some biomolecules bind efficiently to Zn, Cu, or Fe ions immobilized on the mineral surface (metal affinity chromatography) and are retained by electrostatic interaction (Figure 3.16).

Ions in the mobile phase can exchange with ions in the stationary phase and get reversibly bound (ion exchange chromatography). For example, a negatively charged nucleotide can bind to positively charged groups of the stationary phase and displace the negative counterions there. Ultrasmall hydrothermal pores and flow bottlenecks can retain molecules according to their size. Smaller molecules can enter the pores or larger ones stick at bottlenecks such that they are removed from the mobile phase. The average trapping (residence) time depends upon the size of the molecule (including the hydration shell) and other properties like charge and hydrogen bonding capability, for example. Molecules that are larger than the average pore size are excluded and flow with low or little retention (size exclusion chromatography).

If the stationary phase is hydrophobic, for example, with 'tar'-like polymers coating the pores of volcanic rocks (Section 4.3), then hydrophilic substances elute first and hydrophobic molecules or nonpolar side groups of polymers are retained and elute later or only after a change of fluid composition or pressure (*hydrophobic interaction chromatography*). Separation can be improved by a second column (volcanic pore or channel) with different physical and chemical properties or by a change of fluid composition during elution.

3.2.3 MASS TRANSPORT AND SURFACE INTERACTIONS REMOVE PRODUCTS AND COUNTERACT BACK REACTIONS

Chromatographic separation is not only important for the accumulation of pure substances for further reactions, it can also prevent reverse reactions by separating the reaction products from the reactants. Consider an equilibrium reaction $A + B \rightleftarrows C + D$. When equilibrium is reached, forward and reverse reactions are equally fast such that there is no net-production of C and D anymore. If, however, C and/or D are removed, for example by hydrothermal flow, while A and B are trapped, then A and B can produce further C and D until a new equilibrium is reached. Life ceases at chemical equilibrium. Physical or chemical removal of reaction products is imperative to keep the chemical reaction of life far from equilibrium.

3.2.4 Some Reaction Products Can Catalyze Their Own Synthesis

Physical processes like chromatographic separation and thermal-induced accumulation are not the only means to enhance selectivity and reactivity. Chemical enrichment can add to physical enrichment. Generally speaking, a chemical reaction is called *autocatalytic* if one of the reaction products catalyzes the reaction leading to is own formation, or if it increases reactant concentrations in a network, a sequence of interconnected reactions (see legend of Figure 3.18). DNA replication is an example of an autocatalytic biochemical reaction. Probably the simplest autocatalytic reaction is

$$A + B \rightleftarrows 2B$$

$$\frac{d[A]}{dt} = -k_{\rightarrow}[A][B] + k_{\leftarrow}[B][B]$$

$$\frac{d[B]}{dt} = +2k_{\rightarrow}[A][B] - k_{\leftarrow}[B][B]$$

The kinetic equations are nonlinear because the second terms on the right vary with the square of concentration ($[B]^2$). Integration leads to *sigmoid* product concentration versus time [B(t)] curve. The reaction starts slowly because initially there is little 'catalyst' *B* available (*induction period*). Then the concentration curve increases more rapidly because more *B* is available. Finally, the curve flattens because reactant concentration [*A*] decreases. If a kinetic measurement shows a sigmoid concentration versus time behavior, then the reaction can be autocatalytic. For example, an acid-catalyzed hydrolysis of an ester produces carboxylic acid and alcohol and the acid product catalyzes the reaction (proton autocatalysis). A sigmoid product concentration versus time curve would indicate this autocatalysis. An example of a simple autocatalytic network is the predator–prey relationships of the *Lotka–Volterra equation*, in which *A*≡Food, *X*≡Rabbit, *Y*≡Fox, *E*≡Dead fox

$$A + X \rightarrow 2X$$

$$X + Y \rightarrow 2Y$$

$$Y \rightarrow E$$

FIGURE 3.18 A reflexively autocatalytic food generated network (RAF) tracing to the common ancestor of acetogens and methanogens. The procedure to obtain the network was as follows. A cell's metabolism is expressed as reactions and reactants. A starting food set is provided and an algorithm determines which reactions of metabolism are generated from the food set to form a RAF. For each identified reaction, each reactant is connected to each product by a grey line. The largest RAF (maxRAF) is identified. As an example: Using a metabolic map corresponding to the generic metabolism of an anaerobe, an inorganic food set containing H_2O, H_2, H^+, CO_2, CO, P_i, sulfate, bicarbonate, PP_i, S, H_2S, NH_3, N_2 and all required metals and metal clusters is provided. This generates a minute maxRAF with eight reactions linking ammonia, carbon, and sulfide transformations. The addition of formate, methanol, acetate, and pyruvate to the inorganic food set doubles the maxRAF size to 16 reactions. Adding organic cofactors that occur in the network to the food set (see Figure 3.15) expands the maxRAF from 16 to 1335 reactions spanning 25% of the starting anaerobic network. In the specific case of this figure, that procedure was repeated independently for the metabolism of the acetogen *Moorella thermoacetica* (Ace) and the methanogen *Methanococcus maripaludis* (Met). This generates maxRAFs that contain 394 and 209 reactions for Ace and Met, respectively. These two separately generated maxRAFs overlap in a connected network harboring 172 reactions and 175 metabolites. That network, shown in the figure, is a RAF and can be seen as a RAF tracing to the metabolism of LUCA. Reproduced with permission from Xavier J. C. *et al.* (2020) Autocatalytic chemical networks at the origin of metabolism, *Proc. R. Soc. B: Biol. Sci.* 287: 20192377.

with reverse reactions neglected at large concentrations of A. If rabbit X (molecule X) has enough food (molecule A), then it breeds more rabbits ($2X$). If fox Y eats rabbit X, then Y can breed more Y ($2Y$). When all rabbits are eaten up, the foxes will die ($Y \rightarrow E$). Integration of the rate equations

$$\frac{d[X]}{dt} = 2k_1[A][X] - k_2[X][Y]$$

$$\frac{d[Y]}{dt} = 2k_2[X][Y] - k_3[Y]$$

gives oscillating concentrations of prey and predator (molecules X and Y) with a temporal shift between $X(t)$ and $Y(t)$: When prey population $X(t)$ goes down, then somewhat later predator population $Y(t)$ goes down too, in which case prey population recovers and somewhat later predator population increases, too, and so on. Indeed, several temporally (or spatially) oscillating chemical reactions driven by autocatalysis have been observed experimentally.

The concept of autocatalysis can be extended to a network of chemical reactions. If the reaction products increase the rates in the network by feedback sufficiently to make it self-sustaining (given enough input of energy and food molecules), then the network is called autocatalytic (legend of Figure 3.18). For example, the reactions

$$A + B \xrightarrow{\text{catalyst } E} C + D$$

$$C + B \xrightarrow{\text{catalyst } D} E + A$$

constitute an autocatalytic set if there are molecules B freely available from the surroundings ('food') and only a few 'starting' molecules A around.

Now consider what happens if just a small amount of A flows from one hydrothermal pore into a neighboring hydrothermal pore which is exposed to an influx of 'food' B as well. Then the whole *autocatalytic set* with all reactions included is reproduced in this neighboring pore. One is tempted to liken this behavior to molecular or cellular replication but it is only an effect of reproducibility not stable inheritance. Nonetheless, molecular autocatalytic sets are attractive concepts for understanding the origin of life before the origin of special replicators like RNA evolved (which require complex organic synthesis and continuous, specific phosphorylation). This is because autocatalytic networks capture the natural tendency of organic molecules to *react in the same manner to generate the same products given the same conditions*. If the products formed feed back catalytically into the synthesis of more of the components of the network, and if the same starting compounds (food) are provided, available components are organized into more of the same products and an autocatalytically self-amplifying network emerges. Remember, however, that disorder of a system plus disorder of its surroundings must increase with time (Section 1.1). Autocatalytic metabolic networks increase the (temporal or spatial) order in a pore (or cell) and are therefore only stable by releasing more disorder (heat) to the surroundings while requiring a steady influx of reactive food molecules. Autocatalytic networks can also

generate endproducts. Thermodynamically stable, kinetically unreactive, and catalytically inert compounds like methane represent endproducts of autocatalytic networks.

Increased molecular variety and complexity leads to a growing number of catalyzed reactions and enhanced probability of a self-sustained chemical network. Autocatalytic networks can grow in size, provided that there is a constant supply of food molecules to support the reaction system. In a hydrothermal context, H_2 and CO_2 would be the initial food source from which higher molecular weight reactants would ensue, probably much along the lines of core microbial metabolism. This is a recurrent theme in our approach to the origins problem: Enzymes, once they arose, did not invent core metabolic pathways, rather they just accelerated preexisting reactions that tended to occur anyway. That is, core metabolism can be viewed as having arisen through a process in which a core set of reactions took place spontaneously, but accelerated by non-enzymatic catalysts provided by the environment. The inorganic catalysts were eventually replaced by cofactors and enzymes that accelerated the reactions and channeled them into greater educt and adduct specificity. By that measure, metabolism would have necessarily evolved from CO_2 to the tips of biosynthetic networks, or from the bottom left of Figure 3.20 to the upper right, toward the ribosome that synthesizes the genetically encoded catalysts (enzymes).

Throughout this chapter we have made the case that reactions very similar to those in the metabolism of primordial cells could have taken place before the appearance of enzymes. We have however not yet discussed the issue of how genetically encoded enzymes arose. That concerns the origin of the genetic code, a very difficult problem to which we will turn in subsequent chapters.

3.2.5 AUTOCATALYTIC REACTION NETWORKS LEAD TO REPRODUCIBLE MOLECULAR ORGANIZATION PROCESSES

The evolutionary transition from H_2 and CO_2 to cells cannot have taken place all at once. There had to be intermediate stages of chemical organization. In the search for intermediate stages, autocatalytic networks are a useful concept. In the original definition, autocatalytic sets simply generate more of the members of the set, without reference to catalysis at individual reactions or steps. Many cofactors are involved in steps of their own biosynthesis, for example, thiamine and PLP. This is also a form of autocatalysis. Of special interest for the study of metabolic evolution is a class of mathematical objects investigated by Mike Steel and his colleagues called reflexively autocatalytic food-generated networks, or RAFs for short. RAFs have the property that each reaction in the network is catalyzed by a molecule from within the network and all molecules can be produced by the network itself starting from a defined set of food molecules. In a geochemical context, the food molecules would be compounds provided by geochemical reactions. Using computational tools, RAFs can be detected in the metabolic networks of real cells.

Recent work identified a RAF contained in the reactions common to an acetogen and a methanogen, tracing

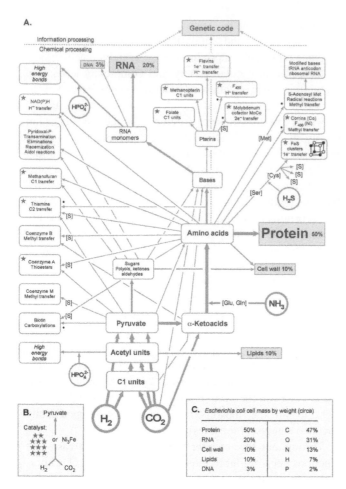

FIGURE 3.19 A general map of metabolic pathways to cofactors and cell mass. (A) The main connections between acetyl, pyruvate, and horseshoe TCA cycle intermediates and base, amino acid, and cofactor biosynthesis. The purpose of the figure is to show where the main carbon units in cofactors and cell mass stem, not to present a detailed map of reactions. The cofactors shown are the main ones required by acetogens and methanogens that lack cytochromes. For example, the guanylylpyridinol cofactor of the Fe hydrogenase of methanogens is not shown, nor is the generation of pyruvoyl enzymes that are involved in CoA biosynthesis. CoM, CoB, methanofuran, methanopterin, and F430 are not required by acetogens. The 11 cofactors labeled with a star are required for the synthesis of pyruvate from H_2 and CO_2, in acetogens, methanogens, or both. [S]: sulfur. (B) The 11 cofactors and their corresponding enzymes can be replaced by a piece of metal, Ni_3Fe, that occurs in serpentinizing hydrothermal vents. (C) Components of cell mass of *Escherichia coli* by weight. From Wimmer J. L. E. *et al.* (2021) The autotrophic core: An ancient network of 404 reactions converts H_2, CO_2, and NH_3 into amino acids, bases, and cofactors, *Microorganisms* 9: 458.

very deep into the origin of metabolism (Figure 3.18). The RAF comprised a network harboring 172 reactions and 175 metabolites involved in core metabolism and the synthesis of basic building blocks like amino acids, bases, and cofactors. Autocatalytic networks are objects of molecular self-organization in the sense that they have the capacity to grow in size and complexity as units of biochemical evolution. Compounds generated from the food set become part of the network, hence autocatalytic networks can start small and grow, in principle to a size approaching the complexity of metabolic networks of modern cells, as long as the compounds in the network feed into the synthesis of other compounds. A surprising result was that only very little rate enhancement (roughly 10% increase) needs to be provided by individual compounds of the network for autocatalytic networks to emerge, at least in computer simulations. RAFs and other kinds of autocatalytic networks provide underexplored tools to model biochemical evolution.

The RAF in Figure 3.18 reveals an interesting property: A food set that contains cofactors generates a large network that contains amino acids and bases, whereas a food set that contains bases and amino acids does not generate the cofactors. This indicates a more ancient status for cofactors (catalysts) over nucleic acids and proteins in the early evolution of metabolism, which is another way of saying that organic catalysis in small molecule reaction networks preceded both RNA and protein in evolution, which makes sense.

A noteworthy aspect of the RAF in Figure 3.18 is that water is the most frequent reaction partner. This is a general property of metabolism. Metabolic networks have been constructed for a number of microorganisms. They can be accessed at the Kyoto Encyclopedia of Genes and Genomes, a uniquely useful resource for the comparative study of metabolism. For investigation with computers, metabolic networks are typically coded as a set of chemical reactions, one per line, and the sum of those reactions describes the metabolic map. With simple searching tools, one can ascertain the number of times that different reactants occur. That water is the most common reactant in metabolism makes sense, because cells are 80% water by fresh weight and water is the solvent for the molecules of life, except lipid soluble compounds like lipids themselves or hydrophobic compounds like quiniones. One of the most important effects that enzymes afford in catalysis is to exclude water from the active site of reactions so that it does not interfere with the reaction mechanism. At the same time, in proteins that have been crystallized for the purpose of determining their structure, about 40% of the mass in the protein structure comes from bound H_2O molecules. We will deal with water in more detail in Section 4.2.

3.3 SCALING UP TO THE REACTION NETWORKS OF A CELL

So far, we have looked at the synthesis of acetate and pyruvate, further CO_2 incorporation in the synthesis of some simple building blocks like TCA cycle intermediates, the synthesis of the simplest amino acids, the components of which bases are made, and some properties of autocatalytic networks. Is the synthesis of components sufficient to get life started? No, of course not. Life is a complex chemical reaction, an energy-releasing reaction, but if we look across the breadth of microbial cells, it has a very conserved metabolic core.

In thermodynamic terms, the main energy investment that cells make is not in the synthesis of their monomeric components but in the synthesis of ATP that is required to make all the reactions that are endergonic go forward. Peptide bond formation is such an endergonic process. The main energy investment that a cell undergoes is protein synthesis.

3.3.1 Cells Are About 50–60% Protein by Weight

Using *E. coli* as a typical cell, about 75% of the ATP expended by a cell for biosynthetic purposes is expended for protein synthesis (Table 3.3). Each peptide bond costs four ATP (two at amino acid activation on tRNA and two at moving the ribosome by one codon along the mRNA, more on this in Chapter 6), and some ATP must be expended to synthesize amino acids or import them from the surrounding medium. About 12% of the energy budget goes to making RNA, and about 3% goes to making DNA. The rest goes for lipids and other things.

In the following we want to get a feel for the scale of the problem at the origin of metabolism. This is not asking how all of the reactions of metabolism arose, and in exactly which order but rather which reactions had to be there in the common ancestors of the first archaeal and the first bacterial cells and what might have come later. The goal is to see how carbon and energy are allocated across pathways to produce cell mass in an exergonic reaction from H_2 and CO_2.

In Table 3.3 we see the overall biosynthetic energy expenditures in terms of ATP that are required to synthesize a cell. What prerequisites must be fulfilled for those reactions to take place? The basic requirements are a source of carbon (CO_2), electrons (H_2), energy (H_2 and CO_2), nitrogen, phosphate, sulfur, trace elements, plus cofactors, and enzymes as catalysts. Figure 3.19 shows the relationships between the starting compounds for autotrophic growth, the basic components of metabolism (acetyl units, pyruvate, amino acids, sugars), the main end products of biosynthesis (protein, nucleic acids, lipids, cell wall), and the cofactors required by primitive

cells. If we count FeS as a cofactor, which it is, Figure 3.19 provides a general overview of roughly how many and what kinds of cofactors need to be synthesized in primitive cells.

The circumstance that all five pterin cofactor families—folate, methanopterin, flavins, F_{420}, and MoCo—are involved in the basal H_2–CO_2 to pyruvate conversion is of interest. This indicates a very ancient nature of pterin cofactors, possibly as ancient intermediates between metal-catalyzed methyl synthesis and enzymatic methyl synthesis. If we exclude the many modified bases in tRNA that allow the genetic code to function as it does (they can be numerous and can require specific enzymes each), we can ask how many enzymes are required to make the 16 cofactors, 20 amino acids, and the main nucleic acid bases. It turns out that this requires only about 400 enzymes. Four hundred reactions to get to the core components of metabolism complex enough to run a cell from the elements is a very finite number.

The problem of metabolic origin is not as severe as a typical 'Biochemical Pathways' map hanging on many university department walls might suggest. The synthesis of acetyl and pyruvate from H_2 and CO_2 works well without enzymes. Many transamination reactions generating amino acids from 2-oxoacids work well without enzymes, provided that pyridoxal phosphate (PLP) is present. The synthesis of pterins from amino acids works well without enzymes (Figure 3.13). The challenge of obtaining a rich mixture of compounds that could bring forth the main catalysts and substance of cells from inorganic components is not insurmountable. Enzymes do not make impossible reactions possible, they just accelerate reactions that tend to occur anyway, and they add some specificity in the process, channelling general reaction classes into specific reaction pathways. The reactions

TABLE 3.3
Biosynthetic ATP Requirement per Gram of *E. coli* for Growth on Glucose, NH_4^+, and Inorganic Salts

Molecule Synthesis	Amount (g/g cells)	ATP per Monomer	ATP (mol·10^4)	% of ATP
Polysaccharide	0.17	2	21	6.1
Lipid	0.09	1	1	0.3
DNA	0.03			3.2
dNMP formation		4	9	
Polymerization		2	2	
Protein	0.52			59.1
Amino acids			14	
Polymerization		4	191	
RNA	0.16			16.4
NMP formation		3	34	
Polymerization		2	9	
mRNA turnover			14	
Membrane transport				14.9
Ammonium ions			42	
Potassium ions			2	
Phosphate			8	
Total ATP required			347	100

Source: From Stouthamer A. H. (1973) A theoretical study on the amount of ATP required for synthesis of microbial cell material, *Antonie van Leeuwenhoek* 39: 545–565 and Harold F. M. (1986) *The Vital Force: A Study of Bioenergetics*, W. H. Freeman, New York, NY.

that tend to occur between H_2 and CO_2 on transition metal catalysts bring forth the backbone of microbial carbon and energy metabolism.

3.3.2 A BALANCE BETWEEN CARBON AND ENERGY IN THE CHEMICAL REACTION OF LIFE

How does the overall energy-harnessing reaction of a cell compare with the amount of ATP required for the synthesis of components? To get an estimate we need to scale up the reactions to the size of a cell and set the amount of ATP the cell synthesizes in relation to the carbon flux. Figure 3.20 summarizes metabolism in a modern acetogen as a model for an ancient, anaerobic, H_2-dependent, autotrophic cell. The starting point is a study by Harold Drake and colleagues (Daniel *et al.*, 1990) in which they quantified the carbon flux through the cell as carbon flux into cell mass and acetate. For *Clostridium thermoaceticum*, they found that during growth on H_2 and CO_2, approximately 0.1 mol of carbon accumulates as cell mass for each 2.4 mol of CO_2 consumed (large gray vertical arrow at the left). Note however that about 10% of the carbon in their experiment was not recovered, hence the values in the following for Figure 3.20 calculations remain approximate.

If we start with 2500 atoms of carbon in CO_2, approximately 2400 of them are converted to acetate for energy metabolism and approximately 100 of them go to cell mass (carbon metabolism). Using information from growth yields, very similar numbers (but ca. 1:20 rather than 1:24) can be obtained for methanogens growing on H_2 and CO_2. The biosynthetic fate of the 100 carbons that go to cell mass is given by Fuchs (2011), who provided a summary of carbon distribution in an idealized primordial metabolism based on the acetyl-CoA pathway (numbers next to the arrows in the scheme). These acetyl moieties go to C2 metabolism or are extended by further CO_2 incorporations to pyruvate, phosphoenolpyruvate (PEP), 3-phosphoglycerate (PGA), oxaloacetate, 2-oxoglutarate, and sugars. A check with a good biochemistry textbook will reveal that those are exactly the compounds from which the syntheses of amino acids start. The amino acids are used to make protein, which comprises 50% to 60% of the cell's mass.

Peptide bonds cost 4 ATP each. The energy for peptide bond synthesis comes from acetogenesis, which delivers approximately 0.27 or, rounded, 0.3 ATP per acetate, as Volker Müller and colleagues (2018) worked out. For the 1200 acetate produced, that yields approximately 360 ATP, which, if we consult Stouthamer (1978) regarding the rough distribution of energy costs across the cell, is enough to make approximately 52 peptide bonds, with a smaller amount of ATP being required for RNA, DNA, and other things. Only 60% of cell carbon goes to protein (cells are roughly 50% carbon in total and 30% carbon in protein by dry weight), such that 60% of the 100 carbon atoms assimilated per 1200 acetate, or 60 carbon atoms, can be directed toward peptide synthesis. But an average amino acid has somewhat less than five carbons, so there is enough energy to make 52 peptide bonds but only enough carbon to make about 15 amino acids which can polymerize to peptides (conservatively assuming an average amino acid size of four

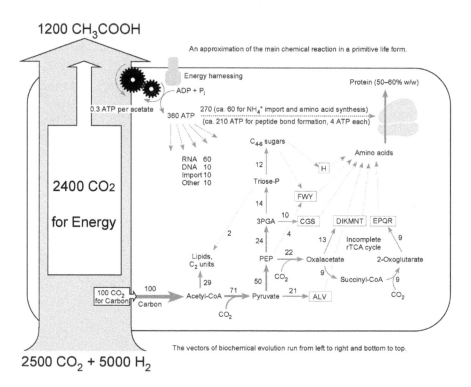

FIGURE 3.20 The flux of carbon and energy through an idealized acetogen as a model for a primitive cell. See text. Numbers above the arrows at right indicate the approximate flux of carbon atoms to biosynthetic routes. PEP, phosphoenolpyruvate; 3PGA, 3-phosphoglycerate; Triose-P, glyceraldehyde-3-phosphate and dihydroxyacetone phosphate. ALV, CGS, FWY, H, DIKMNT, and EPQR indicate the one-letter abbreviations for amino acids: A alanine, L leucine, V valine, C cysteine, G glycine, S serine, F phenylalanine, W tryptophan, Y tyrosine, H histidine, D aspartic acid, I isoleucine, K lysine, M methionine, N asparagine, T threonine, E glutamic acid, P proline, Q glutamine, R arginine. From Martin W. F. (2020) Older than genes: The acetyl-CoA pathway and origins, *Front. Microbiol.* 11: 817.

carbons). In this rough calculation, the available energy for peptide synthesis exceeds the available carbon for peptide synthesis by a factor of approximately three. Can that be right? If correct, where does the excess energy go?

A three-fold excess of energy relative to protein cell mass seems odd, but we have not considered maintenance energy or ATP spilling, which can be substantial. Maintenance energy is still not fully accounted for in bioenergetics. It can be seen as the cost of staying alive but without growing or as an energy expense that does not lead to cell mass. Stouthamer showed that the theoretical maximum yield, Y_{ATP}^{max}, for cell mass of *E. coli* (on the order of 28 g per mol ATP synthesized) is approximately three times the measured value (approximately 10 g of cells per mol ATP synthesized). That is, *Escherichia coli* cells apparently synthesize approximately three times more ATP than they require for biomass synthesis. Similar values can be measured for other cells. The 3:1 excess of energy production to mass accumulation that we see in Figure 3.20 is normal for cells. The efficiency of ATP utilization in living cells is very often approximately three-fold lower in terms of cell mass accumulation than would be predicted from standard biosynthetic costs. This is accounted for by processes such as maintenance energy, futile cycling, ATP spilling, and uncoupling, processes that consume ATP or diminish ATP synthesis efficiency, with no yield in terms of growth or cell mass. The theoretical maximum yield in terms of net cell mass increase per ATP, Y_{ATP}^{max}, is always higher than the observed value in studies of modern cells; it is also higher in Figure 3.20. However, part of the three-fold excess of energy production relative to cell mass comes from a discrepancy between the theoretical maximum ATP yield and the *de facto* ATP yield during growth. For example, when *E. coli* has a very rich medium and O_2 to power its respiratory chain, it grows in conditions where it can obtain more than 20 ATP per glucose. But under such optimal conditions, *E. coli* enters what is called overflow metabolism, switching its ATP synthesis to acetate-producing fermentations that yield about 5 ATP per glucose. The actual ATP yield is often much lower than the theoretical maximum. Part of the discrepancy between the expected maximum growth yield, Y_{ATP}^{max}, and the measured growth yield stems from the expectation that *E. coli* will synthesize ~20 ATP per glucose when it has the opportunity (instead it switches to acetate production at ~5 ATP per glucose even in the presence of oxygen).

How many genes are required to make the complete chemical reaction of such a cell work? The smallest genomes of free-living autotrophs are found among the methanogens (archaea) and comprise about 1300 protein-coding genes. If we assume that the first primitive cells had not yet developed a full set of highly specific enzymes for all biosynthetic steps, but instead perhaps got by with some enzymes having a broader substrate specificity for a given reaction class where today specialized and dedicated gene families exist, we might venture an estimate that for the first cells to operate in the free-living state, about 1000 protein-coding genes would be required. About 400 of those would be dedicated to the basic synthesis of required cofactors, amino acids, bases, and the like (Figure 3.19). The remainder would be dedicated to lipids, RNA maturation, tRNA modification, the ribosome, protein synthesis, the cell wall, division, membrane importers, gene regulation, and the like.

Does this present a paradox? We just said that roughly 1000 proteins encoded by genes might be required to support the first free-living cells, with roughly 400 of those dedicated to the synthesis of the basic building blocks that the cell requires (cofactors, bases, amino acids). So where, the critical reader might ask, did the required components come from before the origin of genes and enzymes? Put another way, where did the amino acids come from that were used to synthesize the first amino acid biosynthetic (proto)enzymes? The answer is a fundamental component of all theories for origins: there had to be initial, sustained, and relatively specific input of chemical building blocks from the environment (serpentinizing geochemical systems in our view). In theories that have life arising on the Earth's surface, the input comes from space, but as we saw in an earlier chapter, organics from space are not suitable starting material for biochemical origins. The recent experiments by Moran, Preiner, and Muchowska as well as older experiments by Wächtershäuser and colleagues that we have outlined in this chapter indicate that there is a basic tendency of carbon, hydrogen, oxygen, and nitrogen to organize into the components of core carbon metabolism in the acetyl-CoA pathway and the incomplete rTCA cycle, provided that the right catalysts and the right starting compounds are available: H_2, CO_2, H_2O and activated N species. There is something very natural about the core chemistry of life. The reactions tend to occur all by themselves given the right catalysts. That does not solve the problem of origins, but it does solve the problem of CO_2 conversion into biologically relevant organics, the origin of metabolism, which even just a few years ago seemed an almost insurmountable hurdle in laboratory simulations. Adding hydrogen in the presence of the right catalysts induces the first steps of autotrophic metabolism to unfold naturally.

A subtle but important message of Figure 3.20 is that in modern, highly refined, fine-tuned cells that have been optimized over billions of years of evolution and natural selection, only 4% of the amount of carbon that flows through the cell in the acetyl-CoA pathway remains in the cell as cell mass, 96% is excreted as the waste product of a main energy releasing reaction, acetate from H_2 and CO_2 in the case of acetogens. The main bioenergetic reaction that synthesizes ATP energetically drives the full set of metabolic reactions, about 1300 reactions in a minimal autotroph, forward. These enzymatic reactions have great specificity, but still only 4% of the carbon is deposited as cell mass. At the origin of metabolism, before there were enzymes, it was probably not 4% nor even 1% nor 10^{-3} nor 10^{-4} of the carbon that was directed toward the constructive process of synthesizing what would become a cell. It was almost certainly less, but how much less we do not know, perhaps 10^{-6} of the CO_2 entering the system, perhaps less. The point is this: Cell mass arises as a minor byproduct of the main energy-releasing reaction of metabolism. Any ideas that entertain the accumulation of molecules that are replicated, whether protein, RNA, or other, require the coexistence, in time and space, of a main supporting (nonenzymatic at the outset) energy-releasing reaction that is quantitatively dominant by orders of magnitude in terms of carbon and energy flux. That main reaction is essential to provide the carbon units and thermodynamic impetus needed to energetically finance any other side reaction that could lead to metastable building blocks of life.

Whatever organic reactions took place at origins, they were the byproduct of a main exergonic reaction of CO_2. Figure 3.20 provides a rough conceptual target of what state of complexity those reactions need to attain in order to sustain the replication of something like a cell. Life is a side product of a main energy-releasing reaction. That also had to be true at origins, for thermodynamic reasons. That energy-releasing reaction also had to be stable over geological timescales and ideally should produce molecules relevant to life in large amounts. The reactions of H_2 and CO_2 at hydrothermal vents meet these demands.

Inspection of Figure 3.20 reveals that the evolutionary progression of events has to start at the lower left, where acetate and pyruvate are made, proceeding to the upper right, where the ribosome is sketched. The RNA in the ribosome is made of bases, the bases are derived from amino acids, and the amino acids are derived from 2-oxoacids in the horseshoe rTCA, which in turn starts with oxalacetate from pyruvate out of the acetyl-CoA pathway. We have not yet discussed the mechanism of how acetogens actually obtain net ATP from acetate synthesis. That will come in Chapter 7. Nor have we addressed ribosome synthesis (Chapter 7). Yet overall, the scheme in Figure 3.20 is useful because it summarizes the chemical reaction of life by depicting the metabolism of a primitive cell. Obviously, there are a large number of hurdles to be taken in the transition from pyruvate synthesis in a hydrothermal vent to a functioning cell. We are not suggesting that the transition was easy, and under no circumstances was it inevitable. We are, however, suggesting that it was possible.

3.3.3 RECONSTRUCTING THE LAST UNIVERSAL COMMON ANCESTOR LUCA FROM GENOMES

Approaching origins from the biological and physiological perspective is necessarily a top-down endeavor. One considers the organisms alive today and tries to identify homologies: properties that, by inference, were present in a common ancestor. This general approach is called the **comparative method**, it allows one to construct ancestral states and understand evolutionary trajectories. The entire field of evolutionary biology was built on the comparative method, specifically on comparative morphology. For vertebrates, the comparative method works well. One can trace the origin of limbs, lungs, wings, and fingers across the vertebrate tree and, more importantly, one can *construct* that tree from those traits. In prokaryotes, the comparative method has limitations, because homologous properties can either be the result of vertical inheritance from a common ancestor, or they can result from lateral gene transfer (LGT), the dispersal of genes without regard for taxonomic boundaries.

To illustrate LGT briefly, there are three main mechanisms by which prokaryotes undergo genetic recombination. Recombination in prokaryotes is never reciprocal, it is unidirectional from donor to recipient. Conjugation is the transfer of genes by plasmids (small circular double-stranded DNA), it is the source and mechanism of antibiotic resistance spread. Transduction is the transfer of genes by phage (bacterium infecting virus). Transformation is the uptake of naked DNA from the environment. In **conjugation**, donor and recipient cells, which need not be of the same phylum, come into physical contact via pili (hair-like protein appendage on the bacterial surface), and a copy of the plasmid is transferred from donor to recipient. Even complex traits, including photosynthesis, can be spread among diverse prokaryotes by individual plasmids of over 50 kb in length that contain all genes needed for reaction center biogenesis, chlorophyll biosynthesis, and carotenoid biosynthesis, as Henner Brinkmann and colleagues have shown. In **transduction**, phages incorporate DNA from the cell in which they were replicated into their genome and are released into the environment, where they persist until they find a new host to infect. Phages exist in vast numbers. In the ocean, for example, phages are roughly 10 to 100 times more abundant than prokaryotes. When they meet a new host cell, which can, in principle, be from any taxonomic group, they can infect, integrate, and introduce the new DNA segment (up to about 20 kb) into the genome. In **transformation**, prokaryotes just take up naked, 'free' DNA from the environment, this is also called natural competence. Some prokaryotes possess special membrane receptors that actively introduce DNA into the cell. DNA in the environment comes from dead cells, for example, cells lysed by phages. In some sediment environments, up to 90% of the DNA is extracellular, not packaged in cells.

By those mechanisms, DNA in prokaryotes is always on the move and always has been for the last 4 billion years. That severely impacts efforts to reconstruct the last universal ancestor of all cells from genome data, because it decouples the evolution of traits (genes) from vertical descent common ancestors. In prokaryotes, the comparative method can still be applied to infer the nature of the first life forms, but after the application of some additional filters. This is shown in Figure 3.21. The traditional method for inferring the properties of LUCA has been to look for genes that are present in all genomes, these should have been present in LUCA too. If we do that using a sample of roughly 2000 genomes, we find that about 30 genes are present in all genomes. These encode ribosomal proteins, components involved in translation, and a couple of ATP synthase subunits. That tells us that LUCA had ribosomes and used the universal genetic code, which we have known for about 50 years, and that it had the rotor-stator ATP synthase, which places the principle of chemiosmotic energy harnessing in LUCA (see Figure 3.21). But what about genes that are present in some bacteria and in some archaea but not universally present in all bacteria and archaea? For 2000 genomes there are about 11,000 such genes, corresponding to 10 times more genes than LUCA needed and four to five times more genes than an average modern genome. Those 11,000 could be the result of vertical inheritance from LUCA or the result of LGT between the domains of Archaea, Bacteria, and Eukarya much later in evolution. Such genes need not have anything to do with LUCA. How to decide what is ancient? Weiss *et al.* developed a simple test: One looks for genes that are present in both bacteria and archaea, but that are also present in at least **two** phyla thereof, and that preserve domain monophyly in phylogenetic trees. For unicellular

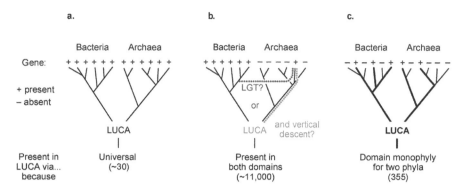

FIGURE 3.21 Tracing genes to LUCA despite lateral gene transfer. Three ways to infer genes present in LUCA. The gene presence is indicated with a plus sign and the absence with a minus sign. (A) If a gene is found universally in both domains, regardless of their tree, then it likely traces to LUCA. About 30 fulfill this criterion. (B) Another way to trace genes to LUCA is to say that a gene found in just one of the phyla of archaea or bacteria and in several phyla of the other domain was present in LUCA. However, thousands of these genes will have been transferred between bacteria and archaea by LGT so they were not necessarily present in LUCA. (C) Genes present in only one bacterial or archaeal phylum could easily be the result of LGT and are removed. But the presence of two phyla per domain while preserving domain monophyly yields good candidates to have been present in LUCA. Such phylogenies would only result from LGT under very specific and restrictive conditions. They require exactly one transdomain transfer followed by either (1) one additional transdomain LGT from the same donor lineage to a different recipient phylum or (2) retention during phylum divergence in the recipient domain. In addition to either criteria there is an additional, highly restrictive criterion: No further transdomain LGTs occurred during all of evolution because subsequent transdomain LGT would violate domain monophyly for the gene. Reproduced with permission from Weiss M. C. *et al.* (2018) The last universal common ancestor between ancient Earth chemistry and the onset of genetics, *PLoS Genet.* 14: e1007518.

organisms, phyla are the biological taxon below domain, for example, proteobacteria or cyanobacteria. The criterion 'present in two phyla' of bacteria places the origin of the gene deep in bacterial phylogeny. Hence if a gene is present in at least two different phyla, and bacteria are monophyletic in the tree in question, then such a gene tree topology (Figure 3.21C) is unlikely (though not impossible) to result from LGT. Domain monophyly means that the respective gene in the different phyla of the domain is from a common ancestor and in no phylum from transdomain LGT. Adding that simple filter sorts 97% of the 11,000 genes into the LGT bin and leaves only 355 in the LUCA bin. This is shown in Figure 3.21C.

The finding that 97% of genes tend to reflect LGT between prokaryotic domains might seem high but that number squares off reasonably well with other estimates based on comparisons of more closely related species. The 355 genes that trace to LUCA not only trace the ribosome and the code to LUCA, but they also provide insights into the physiology and habitat of LUCA that fit remarkably well with the predictions of hydrothermal origins.

This is summarized in Figure 3.22. The 355 genes uncover an autotrophic lifestyle with carbon and energy metabolism based on the acetyl-CoA pathway. This indicates that LUCA could use substrate-level phosphorylation but was dependent upon a geochemical supply of methyl groups, the synthesis of which under hydrothermal conditions we now know to be facile in the presence of H_2, CO_2, and Ni_3Fe. LUCA was an anaerobe, as long predicted by microbiologists. Its metabolism was replete with O_2-sensitive enzymes. These include proteins rich in O_2-sensitive iron-sulfur (FeS) clusters and enzymes that entail the generation of O_2-sensitive radicals via S-adenosylmethionine (SAM) in their reaction mechanisms. LUCA's environment was rich in sulfur. Thioesters, SAM, proteins rich in FeS,

and iron–nickel–sulfur (FeNiS) clusters, sulfurtransferases, and thioredoxins were part of its repertoire, as were hydrogenases that could channel electrons from environmental H_2 to reduce ferredoxin, which is the main currency of reducing power (electrons) in anaerobes.

Based on the phylogenetic reconstruction, LUCA lived from gases: H_2, CO_2, N_2, and H_2S. For carbon assimilation, LUCA used the acetyl-CoA pathway. For N_2 assimilation, it possessed nitrogenase subunits. It possessed subunits of the rotor-stator ATP synthetase, suggesting that it could harness naturally preexisting ion gradients, but it lacked proteins involved in generating ion gradients via chemiosmotic coupling. It was able to convert H^+ gradients into Na^+ gradients with the help of an H^+/Na^+ antiporter, a strategy that is widespread among anaerobes today that live in acetate-rich environments because acetate is an uncoupler for protons. What do we mean by uncoupler? An uncoupler dissipates the ion gradients that cells generate to harness energy. Acetic acid, for example, is a weak acid, acetate can readily be protonated. Because acetate is charged (−1), it does not traverse membranes freely. When acetate is protonated, it is uncharged and unpolar and thus can readily traverse membranes in the protonated state. This process will dissipate proton gradients (uncoupling). Acetate cannot bind Na^+ ions strongly enough to become an uncharged molecule because Na^+ is too tightly hydrated by water. Na^+ gradients are therefore more stable in the presence of acetate.

LUCA had a vast repertoire of modified tRNAs, which are part of the genetic code. Transfer RNA requires modified bases for proper interaction with mRNA (codon–anticodon wobble base pairing) and with rRNA in the ribosome during translation. Several of LUCA's tRNA modifying enzymes are methyltransferases (many SAM dependent), a reminder that before the genetic code arose, the four main RNA bases could hardly have been in great supply in pure

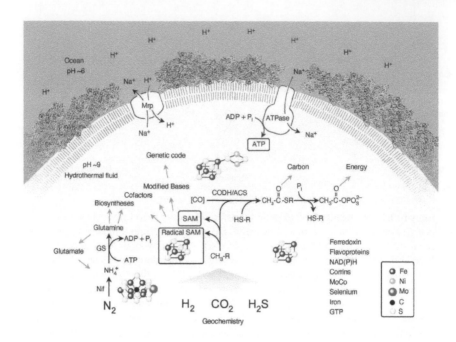

FIGURE 3.22   The physiology of LUCA reconstructed from genomes. Summary of the main interactions of LUCA with its environment, reprinted with permission from Weiss M. C. *et al.* (2016) The physiology and habitat of the last universal common ancestor, *Nature Microbiol.* 1: 16116. The components listed at the lower right are present in LUCA. The figure does not make a statement regarding the source of CO in primordial metabolism, symbolized by [CO]. LUCA indisputably possessed genes because it had a genetic code. Transition metal clusters are symbolized. CH_3-R, methyl groups; CODH/ACS, carbon monoxide dehydrogenase/acetyl-CoA synthase; GS, glutamine synthetase; HS-R, organic thiols; LUCA, last universal common ancestor; Mrp, MrP type Na^+/H^+ antiporter; Nif, nitrogenase subunit; SAM, S-adenosylmethionine. LUCA did not have a cell wall or a lipid membrane but was instead caged in inorganic pores which were possibly coated with hydrophobic organics as indicated with short gray lines in the figure.

form because there were no genes or enzymes, only chemical reactions. Spontaneous synthesis of bases in a real early Earth environment like a hydrothermal vent is likely to generate many side products, including chemically modified bases. There are 28 modified bases, mainly occurring in tRNA, that are shared by bacteria and archaea. The modifications are chemically simple, such as the introduction of methyl groups or sulfur and occasionally of acetyl groups. These are, notably, the same chemical moieties as are germane to the hydrothermal origin theory. Amino acid recognition by tRNA was possible or might even have required these modified bases (Section 5.2).

A repertoire of only 355 genes is not enough for a free-living lifestyle. Genes for cell wall synthesis as well as lipid synthesis were lacking. In that sense, LUCA as reconstructed from vertically inherited genes was only half alive in that it could assimilate carbon, harness energy, maintain genes, and synthesize proteins but was not endowed with a free-living lifestyle. Instead, it was stuck within the confines of the inorganic geochemical surroundings in which its chemical constituents and its simpler molecular precursors arose. A crucial step was the transition to the free-living state. This required lipid synthesis (unrelated pathways in archaea and bacteria), cell wall synthesis (unrelated pathways in bacteria and archaea), cell division mechanisms (unrelated in bacteria and archaea), and ion pumping mechanisms, which fundamentally differ in acetogens and methanogens, the likely founding lineages of the bacteria and archaea respectively. This suggests that the first bacteria and archaea are the result of independent transitions to the free-living state.

The concept of a LUCA that was half-alive might seem unfamiliar, but there are only three possibilities: LUCA was a bacterium, an archaeon, or more primitive than either. Genome data favor the last option. This suggests that genes can provide a glimpse into evolution before the origin of free-living cells, an exciting insight in its own right. As with metabolism, where the whole network cannot have arisen all at once, there had to be intermediate stages along the path from organized and, at some point, genetically instructed chemical reactions to free-living cells. A LUCA that was half-alive, evolving within inorganic compartments, and still tied to geochemical processes via a kind of molecular umbilical cord, would be a natural intermediate in that transition.

3.3.4   No RNA World?

Some readers will notice that our rendering of biochemical origins so far does not entail the very familiar concept of self-replicating RNA or even RNA that is replicated by other molecules. That is in contrast to a substantial segment of mainstream origins literature that is replete with, or exclusively focused on, narratives of how self-replicating RNA can evolve in laboratory experiments. Such experiments start from highly pure and energetically activated commercially available monomers that are polymerized in the laboratory by some means. We will consider some possible reactions of RNA in later chapters. We will also consider the origin of genomes in later chapters, where RNA will be discussed as the first genetic material, the precursor to DNA.

In the present chapter, however, we emphasize that there are no reactions in growing, resting, or dying prokaryotic cells that involve self-replicating RNA. None. RNA does not self-replicate in real cells. Instead, RNA folds into stable conformations like proteins do. Its main function is to promote the specific condensation of amino acids via peptide bonds to form proteins via rRNA-tRNA-mRNA interactions.

The idea of exponential growth among self-replicating RNA molecules traces to laboratory experiments by Manfred Eigen and team in the 1970s using an enzyme obtained from a virus called Qβ replicase, which in the presence of NTPs as substrate could replicate both strands of an RNA template (the + and the – strand), causing exponential accumulation of both strands. This was a kind of PCR (polymerase chain reaction) but long before PCR had been discovered. Their work showed exponential growth of molecules and, obviously, intense selection for the fastest RNA replicators.

This notion of selection among molecules, as opposed to cells, for replication efficiency revolutionized concepts about primordial evolution. It also spawned a niche of theoretical population genetics that aims to characterize the properties of replicating RNA molecules in an origins context, based on laboratory experiments using selection among RNAs (Section 4.8). This has gradually led to a widespread notion that the origin of life was somehow a fast process, initially involving exponential replication among molecules (not cells) right from the outset. This is where the heart of the RNA world concept resides. It is certainly possible to apply principles of genetics and natural selection to

molecules instead of cells, but the RNA world concept is not founded either in observations from cells or in observations from early Earth habitats. For that reason, the RNA world concept does not interface readily either with microbiology or with geochemistry. The RNA world is founded in laboratory experiments by Manfred Eigen involving commercially available chemicals and a purified RNA dependent RNA polymerase enzyme. It is an enticing idea, but it might be wrong. It provides a realm to theorize about early evolutionary processes, but it neither directly addresses nor explains any aspect of the biology of cells.

In contrast to the concept of replicating RNAs, there are a number of catalytically active RNAs in cells. These include the RNA components of RNaseP, which cleaves RNA and is involved in tRNA processing, and the peptidyl transferase reaction in the ribosome. The peptide condensing reaction of rRNA in the ribosome is the most important function that RNA fulfills today. It is possible that it has always been that way.

3.3.5 All Roads Start with Pyruvate

In this chapter we have highlighted the importance of pyruvate because (1) it is spontaneously synthesized from H_2 and CO_2 on transition metal catalysts, and because (2) it has an extremely central position in biosynthetic metabolism. Much like the saying 'all roads lead to Rome', all pathways start with pyruvate. This can be seen in Figures 3.19 or 3.20. Another way to underscore the central position of pyruvate in the metabolism of cells is shown in Figure 3.23, where

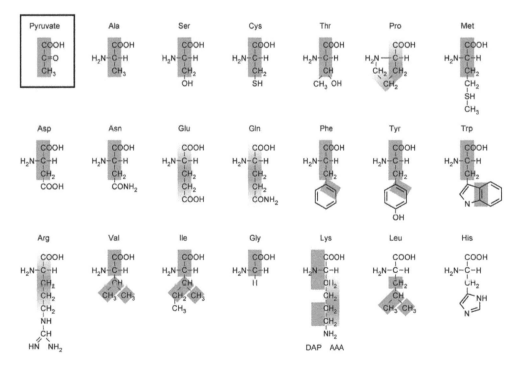

FIGURE 3.23 **Amino acids carry the imprint of metabolic origins from pyruvate.** Carbon atoms that stem from pyruvate in the pathway are indicated with red shading. Lighter shading indicates that only a portion of the C atoms at that position stem directly from pyruvate (see text). DAP: diaminopimelate pathway. AAA alpha amino adipate pathway. The DAP pathway to lysine is typical of bacteria and proceeds via DAP as an intermediate. DAP and lysine are common crosslinker components of peptidoglycan. The AAA pathway is used by some Gram-positive bacteria and archaea. Some methanogenic archaea synthesize a cell wall polymer pseudomureine that contains lysine as a crosslinker, see Albers S. V., Meyer B. H. (2011) The archaeal cell envelope, *Nat. Rev. Microbiol.* 9: 414–426. The carbon atoms derived from pyruvate are compiled from the section on amino acid synthesis in Lengeler J. W., Drews G., Schlegel, H.G. (1999) *Biology of the Prokaryotes.* Georg Thieme Verlag, Stuttgart.

the carbon atoms in the amino acids that stem directly from pyruvate in amino acid biosynthetic pathways are indicated.

Lighter shading in Figure 3.23 indicates that either carboxylate of succinate can be activated as succinyl-CoA and carboxylated to 2-oxoglutarate, such that the identity of the carboxylate carbon in pyruvate is lost in amino acids of the glutamate family (glutamate, glutamine, proline, and arginine). In biosynthetic pathways, about 75% of the carbon atoms in amino acids stem directly from pyruvate. This is a biochemical relict, it clearly reflects the imprint of pyruvate on metabolic origin. Pyruvate gives rise to amino acids, which are, in turn, the starting point for the synthesis of cofactors and bases (Figure 3.19). Why pyruvate? What made pyruvate the starting point for the synthesis and, later in evolution, biosynthesis and a central hub of metabolism, a position that it has conserved from primitive anaerobic prokaryotes all the way to energy metabolism in mammalian mitochondria? The answer is probably twofold: function and opportunity.

Functional moieties of pyruvate render it reactive, both with itself and with other molecules: a carbonyl group at C2 that carries a partial positive charge and is prone to nucleophilic attack, hydrogen atoms at C3 that are in α-position relative to the carbonyl and are hence slightly acidic generating carbanion intermediates at C3, and a carboxyl group at C1 that can leave as CO_2. This combination makes pyruvate prone to participate in Claisen condensations (see Figure 1.22), which generate larger carbon backbones. Many of the reactions of pyruvate in amino acid biosyntheses involve exactly these reactive moieties in addition to reactions of pyruvate with acetyl-CoA, which also undergoes Claisen condensations by virtue of its carbonyl and α-hydrogens. The map in Figure 3.20 shows that in H_2-dependent autotrophs, all of the carbon entering metabolism does so as acetyl groups and pyruvate. They are reactive forms of carbon, which is exactly what is needed for prebiotic synthesis.

Pyruvate had the opportunity to serve as a biosynthetic starting material because it forms spontaneously from H_2 and CO_2 on transition metal catalysts under the conditions of serpentinizing hydrothermal systems (see Section 3.1.3). In addition, the kinds of mineral catalysts that promote pyruvate formation are themselves formed in serpentinizing hydrothermal systems. Furthermore, serpentinization has been going on since there was water on the Earth because it is itself an exergonic process. As long as there is water in the ocean there is no way to turn it off. The same applies to pyruvate formation. That would generate the backbone of a primordial carbon metabolism consisting of 2-oxoacids, with the incorporation of nitrogen occurring via reductive amination of 2-oxoacids, as it occurs in metabolism. This general view of chemistry at origins is called metabolism first.

* * *

3.4   CHAPTER SUMMARY

Section 3.1 Several linear and cyclic reaction networks fix CO_2 in cells to generate organic molecules for metabolism. The acetyl-CoA pathway is the most ancient pathway of CO_2 fixation and the only exergonic pathway of CO_2 fixation. It generates, in combination with the reverse

'horseshoe' tricarboxylic acid (TCA) cycle, diverse organic acids as precursors of amino acids and nucleobases. The organic products of the acetyl-CoA pathway are generated with high specificity from H_2 and CO_2 in laboratory experiments using transition metals as catalysts.

Transition metal catalysts including the hydrothermal minerals Ni_3Fe and Fe_3O_4 dissociate H_2 and CO_2 and bind the fragments to the metal atoms on the surface. The fragments are activated by these metal bonds and react with each other to new molecules. For example, H_2 and CO_2 form acetate and pyruvate in the laboratory in the presence of transition metal catalysts. Thioacetic acid reacts with inorganic phosphate to acetyl phosphate which reacts with ADP by phosphate transfer (substrate level phosphorylation) to ATP, the 'energy currency' of cells. Six of the 11 reactions of the reverse TCA cycle were observed experimentally in an acidic, metal-rich, reducing medium using Fe^0 or Ni^0, Zn^{2+}, and Cr^{3+} (but no enzymes) as catalysts, and alpha-ketoacids as reactants. Carboxylation however was not observed. Reductive amination of alpha-ketoacids to diverse amino acids was observed when activated ammonia compounds and Fe^0 as electron donor were used. Calculations of the thermodynamics of cell mass synthesis show that the synthesis of amino acids, amines, saccharides, nucleotides, and fatty acids from scratch is generally exergonic under the strongly reducing conditions of hydrothermal vents. This is a very strong argument in favor of a hydrothermal origin of life. Without catalysis however primordial reaction networks cannot develop and life would not exist. Transition metals and later small organic cofactors seem to have preceded catalysis by enzymes assisted by cofactors.

Section 3.2 Hydrothermal fluids are transported through volcanic pores, channels, and fissures by pressure, temperature, and concentration gradients. Dissolved molecules can chromatographically separate on the surfaces of rocks and between layers of mineral inclusions and accumulate in cooler, lower-lying zones. The molecular flux induced by these convection and diffusion processes can be calculated by numerical integration of the corresponding differential equations. The calculations show that nucleotides accumulate for example in rock pores of typical length-to-width ratios of 30 and temperature differences of 60°C to levels approaching saturation (10^5 accumulation). Without such separation and accumulation effects, it would not be possible to obtain precursor concentrations necessary for targeted biochemical synthesis. This physical enrichment can be enhanced by chemical enrichment if one of the reaction products catalyzes a reaction leading to its own formation (autocatalysis). The concept of autocatalysis can be extended to a network of chemical reactions. If the reaction products increase the rates in the network by feedback sufficiently to make it self-sustaining (given enough input of energy and food molecules), then the network is called autocatalytic. As an example, a reflexively autocatalytic food-generated network (RAF) tracing to the common ancestor of acetogens and methanogens is given.

Section 3.3 The autotrophic core of metabolism in a primordial cell converts H_2, CO_2, and NH_3 into amino acids, bases, and cofactors and consists of only 404 reactions to generate the main chemical components of a cell from the elements. In addition, the 11 cofactors required for the synthesis of pyruvate from H_2 and CO_2 and their corresponding enzymes can be replaced by a piece of metal, Ni_3Fe, that occurs in serpentinizing hydrothermal vents. The flux of carbon and energy through an idealized acetogen as a model for a primitive cell is compared with the flux required for the synthesis of its components and exceeds it by a factor of approximately three (similar to the situation in *E. coli* cells). The excess energy can be attributed to overflow metabolism, the energy necessary for the maintenance of the cell, futile cycling, ATP spilling, and ATP uncoupling. The last universal common ancestor LUCA can be partially reconstructed from physiology and from genomes even in the presence of lateral gene transfer (LGT). Genes that are present in both bacteria and archaea can reflect either common descent from LUCA or transfer between bacteria and archaea. Gene trees can distinguish these cases, indicating that LUCA lived from the gases H_2, CO_2, NH_3, and H_2S, used the acetyl-CoA pathway, subunits of the rotor-stator ATP synthetase, and could harness naturally preexisting ion gradients but lacked proteins involved in generating ion gradients via chemiosmotic coupling. Pyruvate plays a crucial role in the origin and structure of metabolism. It is reactive because of its central carbonyl group, it is the source of three-fourths of all carbon atoms in amino acids and it arises spontaneously from H_2 and CO_2 in the presence of natural transition metal catalysts.

* * *

PROBLEMS FOR CHAPTER 3

1. *CO$_2$-fixing pathways.* Name at least three CO_2-fixing pathways, the reactants, and the class of product molecules.
2. *Acetyl-CoA pathway.* What intermediates are formed in the acetyl-CoA pathway from CO_2 to a methyl group ($-CH_3$) in bacteria and archaea?
3. *Nonenzymatic pathway to ATP.* Describe a possible pathway without enzymes from transition metal surface bound H_2, CO_2, $SH-$, and HPO_4^{2-} as reactants to ATP as the final product.
4. *Aldol condensation.* Write the aldol condensation of glyoxylate and pyruvate to alpha-ketoglutarate and indicate the mechanism in your reaction equation.
5. *Reductive amination.* Describe the pathway from alpha-ketoacids to amino acids in metabolism.
6. *Nucleobase synthesis.* From what compounds are the purine nucleobases formed in metabolism?
7. *Energetics of cell synthesis.* Why is the synthesis of cells under hydrothermal vent conditions, but not in modern seawater, exergonic?
8. *Organic cofactors.* Name at least six important organic cofactors and specify their function.
9. *Accumulation of molecules.* By what processes can molecules accumulate in hydrothermal pores? What is the common nature of these processes?
10. *Accumulation in rock pores.* What factors govern molecular accumulation in rock pores?
11. *Autocatalytic chemical networks.* How is a reflexively autocatalytic food-generated network (a RAF) defined?
12. *Energy balance in cells.* Why does the available energy for peptide synthesis exceed the available carbon for peptide synthesis by a factor of approximately three in many biological cells?
13. *The physiology of LUCA.* What does phylogenetic reconstruction tell us about the lifestyle and physiology of LUCA (the Last Universal Common Ancestor)?

4 Prebiotic Synthesis of Monomers and Polymers

4.1 AMINO ACIDS WERE SYNTHESIZED BY REDUCTIVE AMINATION OF α-KETOACIDS

4.1.1 Intermediates of the TCA Cycle React with Activated NH_3 and H_2 to Amino Acids

We saw in Section 3.1 that most intermediates of the acetyl-CoA pathway and tricarboxylic acid cycle can be synthesized without enzymes by the transition metal catalyzed reaction of CO_2 with H_2. Also, molecular nitrogen can be hydrogenated to ammonia and other reduced nitrogen compounds if transition metals catalyze the reaction.

In a recent experiment, the groups of Joseph Moran and Harun Tüysüz showed that α-ketoacids from the pathways displayed in Figure 3.8 react with ammonium on nickel catalysts in the presence of H_2 as reductant to generate biological amino acids. The reactions took place under very mild, natural conditions (Figure 4.1): in water at room temperature and with H_2 partial pressures that are observed in serpentinizing systems today. The amino acid glycine was obtained from glyoxylate, aspartate from oxaloacetate, and glutamate from 2-oxoglutarate,

Ni⁰ was bound on a silicate–Al₂O₃ support

Conversion: 6% to 50%

Glyoxylate ⟶ Glycine
Pyruvate ⟶ Alanine
Oxalacetate ⟶ Aspartate
2-Oxoglutarate ⟶ Glutamate

2-Oxoisocaproate ⟶ Leucine
2-Oxoisovalerate ⟶ Valine
2-Oxomethylvalerate ⟶ Isoleucine

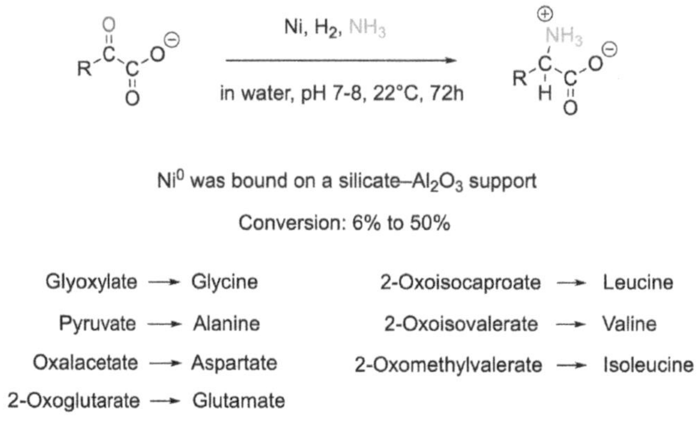

FIGURE 4.1 **Reductive amination of ketoacids with H_2 and ammonia over nickel.** The scheme is an excerpt of results from recent work by the groups of Joseph Moran and Harun Tüysüz: Kaur H. et al. (2024) A prebiotic Krebs cycle analog generates amino acids with H_2 and NH_3 over nickel. Chem (https://doi.org/10.1016/j.chempr.2024.02.001). Earlier work by the Moran group had shown that iron effeciently catalyzes reductive aminations using hydroxylamine: Muchowska K. B. *et al.* (2019) Synthesis and breakdown of the universal precursors to biological metabolism promoted by ferrous iron, *Nature* 569: 104–107. Biological reductive amination starts with ammonia instead of hydroxylamine. Ammonia can be generated by serpentinization (Shang X. *et al.* (2023) Formation of ammonia through serpentinization in the Hadean Eon, *Sci. Bull.* 68: 1109–1112) and can be used as a substrate for reductive aminations of alpha keto acids under hydrothermal conditions using H_2 as the reductant and Ni⁰ as the catalyst (Kaur H. et al. (2024) A prebiotic Krebs cycle analogue generates amino acids with H_2 and NH_3 over nickel. CHEM 10 doi.org/10.1016/j.chempr.2024.02.001).

but valine, leucine and isoleucine were also obtained from the corresponding 2-oxoacids. The yields ranged between 6 and 50% conversion. As the main side products, the corresponding 2-hydroxyacids were observed (for example lactate instead of alanine). Today, the reductive amination of α-ketoglutarate to glutamate is the main entry point of nitrogen into matabolism and the biosphere. The nickel-hydrogen-ammonium synthesis of amino acids by Kaur et al. is a natural and efficient prebiotic amino acid synthesis route. It uses reactants and catalysts generated by serpentinization and it closely mirrors the biological process, because in the biosynthesis of the 20 amino acids shown in Figure 3.23, amination of the corresponding 2-oxoacid is almost always the last biosynthetic step.

In a similar experiment, the amino acids alanine, glutamate, phenylalanine, and tyrosine were obtained by reductive amination of the respective α-ketoacids with ammonium salts and freshly precipitated FeS or $Fe(OH)_2$ as electron donors and catalysts.

Experiments by Barge *et al.* showed that the alanine yield from reductive amination of pyruvic acid is maximal at alkaline conditions and a (prebiotically unrealistic) 1:1 $Fe^{2+}:Fe^{3+}$ ratio with catalyzing/electron donating iron oxyhydroxides. Purely ferric (Fe^{3+}) hydroxides did not lead to any reaction (no electron donor), whereas purely ferrous (Fe^{2+}) hydroxides reduced pyruvate to lactate (CH_3CHO-$HCOO^-$) without amination, probably because reduction of the keto group was faster than amination under their conditions.

4.2 PEPTIDES WERE SYNTHESIZED BY CONDENSATION OF ACTIVATED AMINO ACIDS

4.2.1 Activated Amino Acids Condense in Water to Short Peptides of Random Sequence

Peptide bonds –CO–NH– are formed when the carboxyl group of an amino acid reacts with the amino group of another amino acid releasing water (*condensation reaction* or *dehydration synthesis;* Figure 4.2A).

Hydrolysis (addition of water) can break the peptide bond –CO–NH– and release 8–16 kJ/mol free enthalpy depending on the amino acids involved. However, without catalysis, peptides are kinetically stable for many years in water at pH 7 and ambient temperature (Figure 4.2B). The reverse reaction, peptide bond formation, is endergonic by 8–16 kJ/mol, accordingly, and the peptide yield is low in neutral aqueous solution at ambient temperature ($K_{Gly-Gly} \approx 2$–3×10^{-3} M^{-1} for Gly–Gly, corresponding to 2–3 mM Gly–Gly for 1 M glycine, for example). The equilibrium constant for peptide formation increases with temperature, but even at 200–250°C, peptide equilibrium concentrations are negligible.

DOI: 10.1201/9781003378617-4

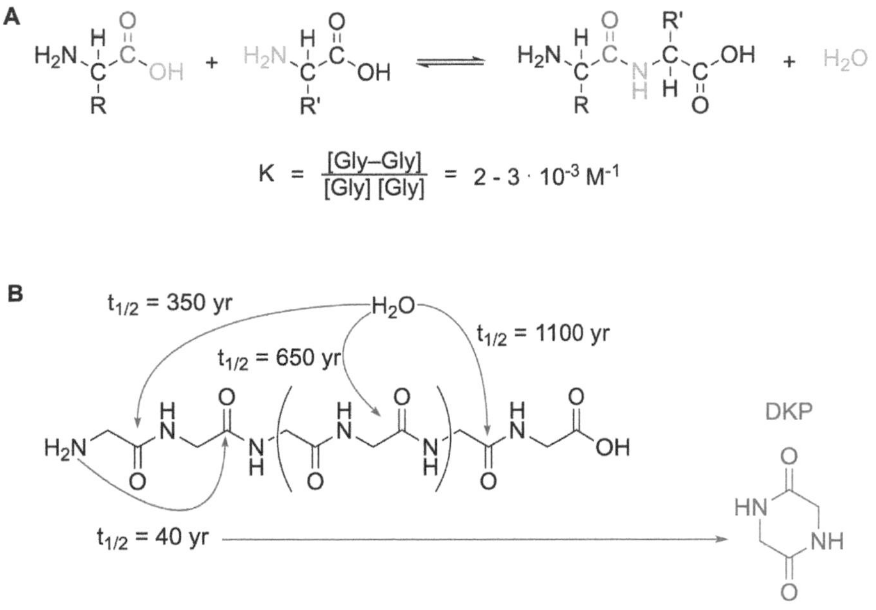

FIGURE 4.2 Polymerization of glycine and hydrolysis of oligoglycine. (A) Formal description of amino acid dimerization (for glycine, R=H) and equilibrium constant for dimerization of glycine. (B) Spontaneous hydrolysis and intramolecular cleavage of the peptide bonds in oligoglycines at 25°C in neutral water. The intramolecular cleavage is usually the fastest process and leads to diketopiperazine (DKP, drawn red). In the reverse reaction, DKP elongates oligoglycines. Redrawn from Radzicka A., Wolfenden R. (1996) Rates of uncatalyzed peptide bond hydrolysis in neutral solution and the transition state affinities of proteases, *J. Am. Chem. Soc.* 118: 6105–6109.

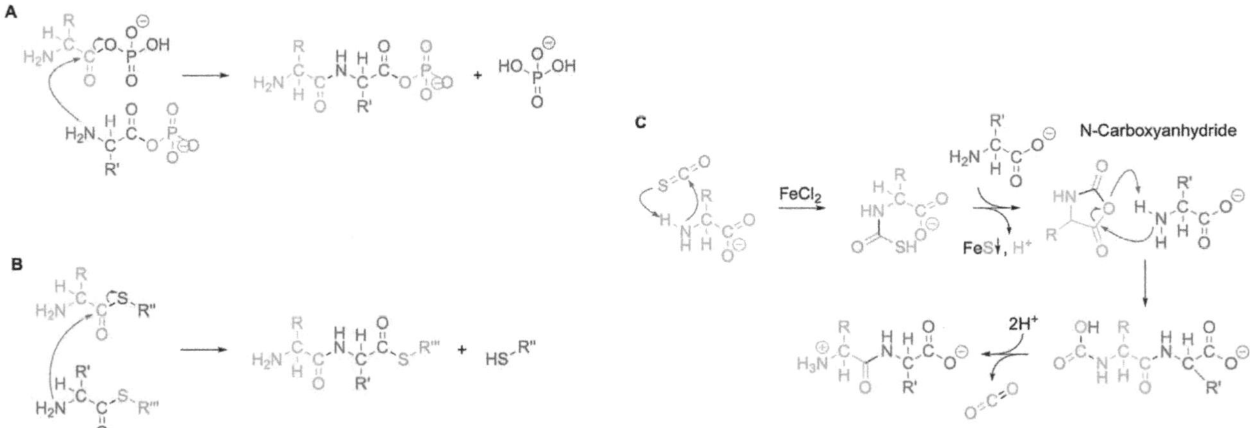

FIGURE 4.3 Condensation of activated amino acids to peptides. (A) Condensation of phosphorylated amino acids by nucleophilic substitution. The amino group must be protected to exclude its reaction with the activator (see text). (B) Condensation of thiolated amino acids by nucleophilic substitution. The amino group must be protected to exclude its acetylation by the thiol-activated amino acid. (C) Condensation of COS-activated amino acids by *N*-carboxyanhydride formation. (D) Condensation of bicarbonate-activated amino acid esters by *N*-carboxyanhydride formation.

For efficient polymerization, the amino acids must be activated. Phosphorylation, thiolation, and anhydride formation with COS, HCO_3^-, or other dehydrating agents are possible means of activation (Figure 4.3). Before activation, however, the reactive amino group must be protected to exclude its reaction with the activator. For example, the activator acetyl phosphate did not promote polymerization of glycine as shown in Figure 4.3A, but acetylated the amino group of glycine. Activation with thiols (Figure 4.3B) leads to the same problem: activation experiments with thioacetic acid merely led to acetylation of the amino group. In automated laboratory peptide synthesis, the amino group is blocked by bulky substituents before activation and condensation. In prebiotic synthesis, the positively charged amino group $-NH_3^+$ might have been blocked by interaction with negatively charged surface groups like silicates, for example, however, this possibility has not experimentally been tested yet.

The activator carbonyl sulfide (O=C=S; Figure 4.3C) reacts selectively with the amino group and not with the carboxyl group or non-amino side chain groups; hence, protection of the amino acid functional groups is not required in most cases. The amino acids are activated for oligomerization by forming a N-carboxy anhydride. Initially, NH adds to the C=S double bond of COS. Then the sulfide group is eliminated by a reaction with Fe^{2+} to precipitate as FeS↓ and a reactive N-carboxyanhydride (NCA) is formed. Another amino acid adds to the anhydride and generates a dipeptide after decarboxylation. Peptide formation by activation of amino acids with CO and H_2S on (Ni,Fe)S surfaces probably runs via carboxy anhydrides too. Wächtershäuser and coworkers observed peptide synthesis with anabolic and catabolic segments if peptide formation was driven by CO and catalyzed by (Ni,Fe)S.

The mechanism of the COS-driven polymerization of glycine on a pyrite (FeS_2) surface was investigated by *ab initio* molecular dynamics simulations (Figure 4.4). The

high-level calculations of Marx and coworkers showed in atomic detail how the amino acid carbonyl group is activated by N-carboxy anhydride formation and how the amino group of another amino acid nucleophilically attacks the activated carbonyl group to form the glycine dipeptide. The calculations were performed at the temperature and pressure conditions ($T = 500K$ and $p \approx 20$ MPa) of hydrothermal vents.

4.2.1.1 Acid Phosphoanhydrides

Amino acid esters can also be activated by prebiotically abundant bicarbonate ions to carbamate followed by cyclization to N-carboxyanhydride (Figure 4.3D). The N-carboxyanhydride then reacts readily with another amino acid and decarboxylates to a dipeptide (Figure 4.3D). Oligoleucine with an average length of 10 was precipitated in 77% yield from L-leucine ester with bicarbonate as a dehydrating (activating) agent. Several other dehydrating agents like NH_2CONH_2, HNCO, HCN, RCN, and HC≡CH have been investigated or proposed, but their prebiotic significance is unclear (Section 4.6).

4.2.2 ACTIVATED PEPTIDES ARE LESS SENSITIVE TO HYDROLYSIS THAN ACTIVATED AMINO ACIDS

Several experiments showed that amino acids activated by adenylation with ATP hydrolyze fast at alkaline conditions. Probably, the charge neutralization of phosphate and amino groups (blue arrow in Figure 4.5A) facilitates the nucleophilic attack of OH- on the carbonyl group (red arrow in Figure 4.5A). Adenylated peptides hydrolyze much more slowly because the separated charges do not neutralize each other such that the attacking OH^- is repelled by O^- from the phosphate group. At acidic pH only hydrolysis, not polymerization, takes place.

If all activated amino acids hydrolyze before polymerization takes place, no peptides or proteins will be formed. At what conditions can polymerization compete with hydrolysis of the adenylated amino acid at alkaline conditions? Experimentally, the rate constant for the hydrolysis of the adenylated amino acid alanine A_1P is $k_h = 0.46$ [min⁻¹] (Figure 4.5B). The polymerization constants of adenylated alanine polypeptides at pH 10 are $k_i \approx 25$ [l/mol/min] with all k_i approximately equal $k_1 = k_2 \cdots = k_i$ for the growth of a polyalanine chain in a homogeneous aqueous solution. Then,

$$R' = \frac{\text{Rate}_{\text{polymerisation}}}{\text{Rate}_{\text{hydrolysis}}} = \frac{k_i[A_1P][A_1]}{k_h[A_1P]} = \frac{k_i[A_1]}{k_h}$$

For $k_i[A_1] \gg k_h$, polymerization outpaces hydrolysis. This applies to $[A_1] = 100$ mM, $k_i[A_1] = 2.5$ [min⁻¹] $\gg$ $k_h = 0.46$ [min⁻¹], for example. At such concentrations, A_1 (or some A_i) reacts with A_1P, before A_1P can hydrolyze much.

4.2.3 WATER ACTIVITY IN SALT SOLUTIONS AFFECTS PEPTIDE STRUCTURE

Once simple prebiotic peptides are formed, be it either polypeptides consisting of just one amino acid or peptides

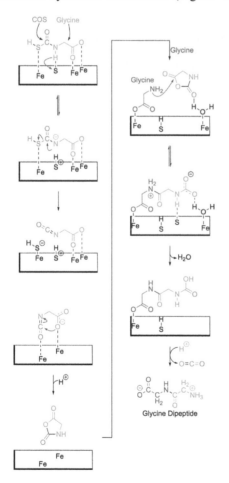

FIGURE 4.4 Proposed mechanism of glycine dipeptide formation with COS at a water-pyrite interface. Surface atoms of Fe and S and sulfur vacancies ('defects') polarize the NH, C=O, and C–S bonds leading to the isocyanate N=C=O as intermediate. From here, ring-closure to the N-carboxy anhydride takes place. Water bound to iron on the FeS_2 surface polarizes the carbonyl group further and prepares for the attack of the next amino acid on the carboxy anhydride. For further assignments, see text. Redrawn from Schreiner E. *et al.* (2011) Peptide synthesis in aqueous environments: The role of extreme conditions and pyrite mineral surfaces on formation and hydrolysis of peptides, *J. Am. Chem. Soc.* 133: 8216–8226.

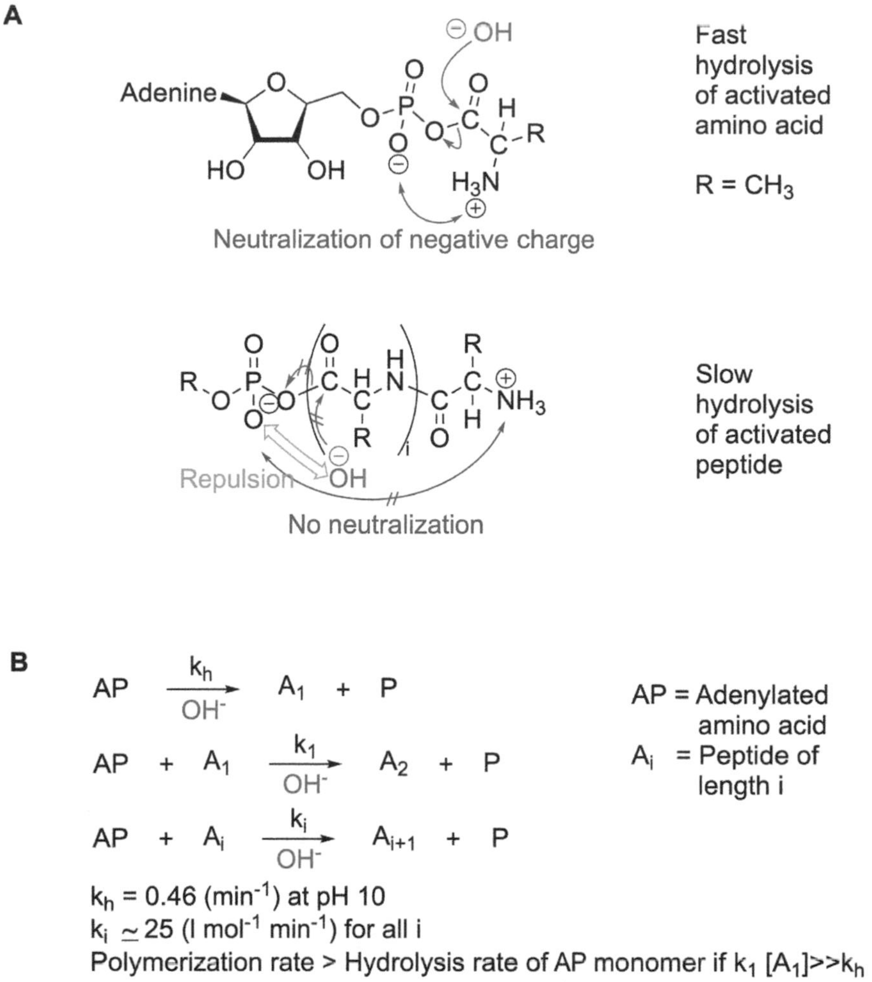

A

Neutralization of negative charge

Fast
hydrolysis
of activated
amino acid

$R = CH_3$

Slow
hydrolysis
of activated
peptide

No neutralization

B

$$AP \xrightarrow[OH^-]{k_h} A_1 + P$$

$$AP + A_1 \xrightarrow[OH^-]{k_1} A_2 + P$$

$$AP + A_i \xrightarrow[OH^-]{k_i} A_{i+1} + P$$

AP = Adenylated
amino acid

A_i = Peptide of
length i

$k_h = 0.46 \ (min^{-1})$ at pH 10
$k_i \simeq 25 \ (l \ mol^{-1} \ min^{-1})$ for all i
Polymerization rate > Hydrolysis rate of AP monomer if $k_1 [A_1] >> k_h$

FIGURE 4.5 **Kinetics of polymerization and hydrolysis of activated amino acids.** (A) Mechanism of OH⁻ catalyzed hydrolysis of adenylated alanine ($R=CH_3$) and polyalanine. (B) Kinetics of OH⁻ catalyzed hydrolysis and polymerization of adenylated amino acids. At pH 1–6 only hydrolysis takes place with a rate constant k_h which is nearly independent of ionic strength, extraneous salt, and small amounts of free alanine A1 at pH 4. At alkaline pH, adenylated alanine polymerizes. Rate constants are from Lewinsohn *et al.* (1967).

with two or three different amino acids, what structures can emerge? Can such simple peptides already function as enzymes to catalyze chemical reactions?

Salt can have interesting effects on the three-dimensional structure of peptides and proteins under abiotic conditions. André Brack showed in 1993 that alternating homochiral leucine-lysine polypeptides (poly Leu–Lys) fold as random coils in pure water. Recall that the positive charges of the lysine amino groups repel each other ruling out formation of an ordered structure. But in salt the charges are sufficiently screened by counterions to allow the formation of thermostable and hydrolysis-stable bilayers of β-sheets (Figure 4.6).

The hydrophilic side chains of Lys ($R=-(CH_2)_4NH_3^+$) in the β-sheets point outward into the salt solution, whereas the hydrophobic side chains of Leu (R = isobutyl group) point inward. With 50% 2-propanol added to the salt solution α-helices are formed. Propanol probably disrupts some of the NH···O=C hydrogen bonds of the β-sheets by forming H bonds of its own and by hydrophobic propyl-isobutyl interaction. The resulting single strands fold to α-helices which are stabilized by hydrogen bonds between the C=O of the *i*-th and the NH of the (*i* + 4)-th amino acid. Heating

transforms the thermolabile α-helices to the thermostable β-sheets.

Longo and coworkers showed that the structures of larger proteins from ten prebiotically plausible amino acids are also influenced by high salt concentrations. The 'prebiotic' set contains only acidic or simple hydrophobic amino acids, no amino acids with basic or aromatic side groups. Reproducible folding to trefoil beta-sheet arrangements was obtained such that the negatively charged side groups of the acidic amino acids point outwards and are stabilized by salt cations. With simple hydrophobic amino acids such as Leu, Ile (isoleucine), and Val (valine) pointing inside, hydrophobic core/pocket structures with protecting hydrophilic outer layers are formed. Such proteins are *halophilic* and *thermostable*. These 'prebiotic' proteins naturally fold into the structures that concentrate substrates in their protected interior without informational instruction provided by nucleic acids as templates. Their reproducible folding is merely directed by the nature of peptide bonds, the properties of amino acid side chains, water activity, and salt concentration. Amino acids with nucleophilic side chains and side chains able to bind to transition metal complexes like in cysteine can make these halophilic proteins enzymatically active.

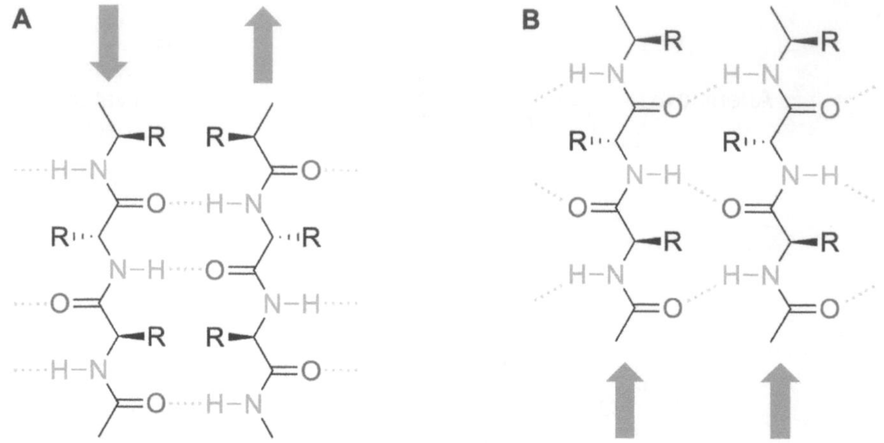

FIGURE 4.6 **Schematic depiction of peptide β-sheets.** (A) Antiparallel β-sheets. (B) Parallel β-sheets. The amino acid side chains R can be hydrophilic (charged or uncharged) or hydrophobic. They are located alternately above and below the peptide strand which is only possible if the amino acid monomers are homochiral. For more discussion of chirality see text in Section 4.2.4.

4.2.4 HOMOCHIRAL AMINO ACIDS FORM STABLE α-HELICES AND β-SHEETS IN SALT SOLUTION

Chiral amino acids are necessary for the formation of α-helices and β-sheets in proteins because of the strictly regular location of the side chains in these structures. Amino acids are *chiral* (from Greek *kheir*, hand) because their central carbon atom binds tetrahedrally to four different groups (except glycine). Two mirror-image structures exist for amino acids which are called L- and D-isomers depending on whether they rotate plane-polarized light counterclockwise (L) or clockwise (D) (Figure 4.7).

In origin of life studies, the existence of L-amino acids in proteins synthesized on ribosomes is called the homochirality problem. It has long been known that cells use L-amino acids in proteins and D-sugars in nucleic acids. There are two approaches to explaining the origin of homochirality. The traditional approach is that chemical processes before the origin of metabolism favored the accumulation of one or the other enantiomeric monomer (amino acids, sugars) before the origin of polymers. In that view, primordial polymers were then synthesized from a mixture with a preexisting enantiomeric excess. Experiments indicate that L-amino acids are somewhat better soluble in water, whereas racemic mixtures (D/L = 1:1) crystallize more easily. The small difference in solubility might have influenced L-amino acid frequency in polymers with time.

The other alternative to the origin of homochirality is that the synthesis of neither enantiomer was favored, but over the course of time one enantiomer was preferentially incorporated into polymers. Small preferences can tip the scale and a few calculations illustrate the salient problem. For example, consider the possibility that a racemic mixture of small peptides might accumulate. A racemic mixture contains both the handed compound and its mirror image in equal amounts. Enantiopure polymers can however not be attained from random, non-stereospecific polymerization steps because of the large number of molecules required. For example, a dipeptide can form from $2^2 = 4$ different combinations of amino acids with D and L handedness: DD, DL, LD, LL; only DD and LL form a racemate of enantiopure

strands. A tripeptide gives $2^3 = 8$ different combinations: DDD, DDL, DLD, DLL, LDD, LDL, LLD, LLL with DDD and LLL as a racemate of enantiopure strands.

A synthetic mixture of a modern peptide only 28 amino acids long, consists of 2^{28} or 2.7×10^8 different polymers of L- and D-amino acids, if that is, they all have exactly the same amino acid sequence, and if all peptide bonds

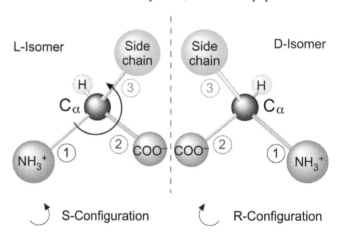

FIGURE 4.7 **L- and D-isomers of amino acids.** The dashed vertical line indicates a mirror. Proteins synthesized on ribosomes consist of L-amino acids in the S-configuration (from Latin: sinister, left) with substituents (1), (2), and (3) arranged counterclockwise according to decreasing priority (atomic number). The D-isomers are in the R-configuration (from Latin: rectus, rechts) with clockwise arrangement of the substituents. The substituent with the lowest priority (the H atom) points away from the viewer. L- and D-isomers cannot be superimposed upon each other by rotation but they can be chemically interconverted by mutases or by spontaneous decay processes; the latter is a method for estimating the age or degree of decay of biological samples, for example, Csapó J. *et al.* (1994) Age determination based on amino acid racemization: A new possibility, *Amino Acids* 7: 317–325. Note that the designations D and L designate the optical activity of the compound, not its absolute configuration (R and S). If a dissolved compound rotates the plane of polarized light to the left, it is called the levorotatory (L-) isomer, if it rotates the plane of polarized light to the right it is called the dextrorotatory (D-) isomer. Compounds that are mirror images of one another, as in the figure, are called enantiomers.

involve the proper amino and carboxyl moieties (not the side chains), and only two of them form a peptide with all amino acids having the same handedness. If we, however, assume a mixture of peptides 28 amino acids long, with all possible random sequences of the 20 proteogenic amino acids, the enantiopure polymers become an unattainable synthetic feat, and their yield would be extremely small. The mixture would have to consist of $2.7 \times 10^8 \cdot 20^{28} = 7.2 \times 10^{44}$ peptide molecules. Since an average amino acid has a molecular mass of 110 D, the racemate of our 28 amino acid peptide with molecular mass ca. 3080 g/mol (not taking loss of water during condensation into account for this rough calculation) corresponds to about 3.7×10^{18} tons of peptide. That is more than the mass of all carbon on (and inside) the Earth, which is about 1.8×10^{18} tons. For comparison, there are about 1.4×10^{18} tons of water in all of the oceans. Those numbers are for a short peptide, the length of a small ferredoxin. We can do a similar calculation for a nucleic acid of 50 bases in length. The ribose D and L combinations correspond to 1.1×10^{15} combinations for poly A. If we multiply that by 4^{50} or 1.3×10^{30} (the number of 50mer sequences for four bases) we get 1.4×10^{45} molecules each with a mass of roughly 16,500 D, that weigh in total 3.8×10^{25} g, or 3.8×10^{19} tons, also more than the mass of all carbon on Earth.

The point of the foregoing calculations is to drive home the message that there is not enough carbon on Earth to get a racemate of enantiopure polymers of any complex mixture of biologically relevant molecules. That means that even for random incorporation of D or L monomers at every step, short, handed oligomers and polymers with some optical activity were an unavoidable consequence of prebiotic polymerization reactions, because there is just not enough carbon on Earth to allow enantiopure polymers to arise, in particular if we start to think about a specific locality for that synthesis (Earth has a surface area of roughly 5×10^{14} m^2 and life must have developed at specific places of that area). In Chapter 3 we saw that cofactors always lack chiral carbon atoms in their active moieties, suggesting that in prebiotic chemical evolution, chirality arose after primitive organic catalysts had already formed.

Physical processes may have been an issue to enrich enantiomers. Racemic asparagine for example forms nearly enantiopure crystals from solution. Whether L- or D-asparagine enriches in the crystal seems, however, to be a random event once the crystal nucleus of a single configuration is formed, probably for thermodynamic reasons. Up to 100% enantiomeric excess in such crystals has been reported. Recrystallization of a mixture of racemic asparagine with other racemic amino acids can lead to crystals of these amino acids with an excess of the same configuration as the nearly enantiopure asparagine crystal. Hence, spontaneous and effective racemate splitting is possible by crystallization even in the case of a nonstereospecific primordial synthesis. It would be interesting to investigate if crystallization at gas bubbles in pores (Figure 4.23) leads to racemate splitting of an amino acid mixture. In the case of molecular accumulation in hydrothermal pores, as explained later in this chapter, an oversaturated amino acid solution will crystallize spontaneously at cooler places in the pores leading to racemate splitting. From this standpoint, the homochirality

problem appears to be much less severe than one might think, possibly even trivial in some respects, because various physicochemical processes can lead to enantiomeric excess starting from racemic mixtures.

Chemical reactions are another possibility to enrich enantiomers. If a randomly synthesized handed molecule has any kind of catalytic activity to promote the synthesis of an excess of one monomeric enantiomer over the other, or perhaps more importantly, the polymerization of one enantiomer over the other, then the next generation of handed catalysts will arise and the scales will quickly tip, perhaps irreversibly, in the direction of homochirality. The key thought in that line of reasoning, though, is that catalytic activity can be exerted by a component within the system, whereby the synthesis of the catalyst was attained with the help of a previously synthesized component of the system. This is a property of autocatalytic networks, which we discussed in Chapter 3 and as shown in Figure 4.8.

In this context we can think of possible enrichment cycles of simple catalytic activities afforded by the extraordinary stability of homochiral β-sheets to temperature and hydrolysis. These cycles involve several serial rounds of handed catalysts promoting the synthesis of other handed catalysts. The L-amino acids that we see in proteins today would be the end of the line in any such progression. One possibility is that the amino acid mixture from which amino acids were incorporated at the origin of translation was a more or less equimolar mixture of D and L enantiomers. That would be the case if inorganic or organic catalysts with no stereospecifity were generating amino acids from 2-oxoacids in a hydrothermal setting. Since the mechanism of peptidyl transferase reaction involves nucleophilic attack of a carbonyl by a primary amine, the primordial peptidyl transferase active site of the primordial protoribosome would be able to polymerize either D-amino acids or L-amino acids, not both, if they were activated on the 3′ ends of tRNA (Figure 4.8). This mechanism specifically channels L amino acids into genetically encoded proteins, generating vast populations of handed catalysts. Note that in such a model, the chirality of the molecule that carries the activated amino acid (functionally, a proto-tRNA) is not important. The chirality at the peptide bond-forming site determines which amino acid enantiomer is channeled into polymers, which then consist of one enantiomer, generating stereospecific molecular surfaces for further reactions.

Cofactors themselves carry a very interesting imprint of early homochirality. One central element of our case for geochemical origins is that there was a progression in the origin of chemically complex systems from inorganic to organic to polymeric catalysts as outlined in Figure 4.9. The progression starts with inorganic catalysts (metal alloys, metal sulfides, metal oxides). These generated organic molecules, including organic catalysts as products. These in turn gave rise to more complex polymers that contained inorganic and organic cofactors as prosthetic groups. In some ways, one can think of early chemical evolution as a process of the evolution of catalysts. One can imagine a sequence in catalytic evolution starting with

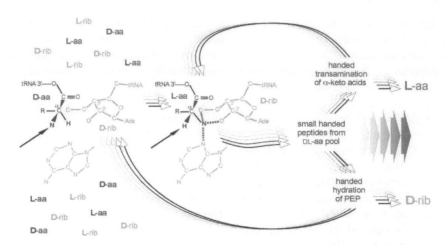

FIGURE 4.8 A food-generated autocatalytic network could introduce chirality into metabolism. The mechanism is a simple stereochemical filter at protein synthesis. R, amino acid side chain; aa, amino acid; rib, ribose; PEP, phosphoenolpyruvate. In organisms that use the acetyl-CoA pathway, the enolase step introduces the first chiral center in the synthesis of sugars via the stereospecific addition of a water molecule to the double bond of PEP (see Figure 4.10). The N atom of the amino group of an activated amino acid (L and D forms) is indicated in red, and its position in the D form is highlighted with an arrow. The hydrogen atoms of the amino group are not drawn. The C atom of the peptide ester bond with tRNA is highlighted yellow and like the D-rib of the tRNA drawn in blue. The figure shows that an activated L-amino acid (L-aa) fits into the modern peptidyl transferase site (center) and is held in place there by interactions with the terminal D-Rib of tRNA, with a nucleobase and with the C atom of the peptide ester bond with tRNA. The amino group of the positioned L-aa attacks the (yellow marked) C-terminal carbonyl carbon of the tRNA-bound growing peptide chain, whereas the D-configuration (left) leaves the amino group in the wrong place for the peptidyl transfer reaction. Hence, during molecular evolution, from a mixture of L and D amino acids, only those of one carbon configuration is incorporated into protein by virtue of the chance stereochemistry of the initial peptidyl transferase reaction catalyzed by a protoribosome. With some enolase activity producing more D sugars (ribose, rib), peptide synthesis at the peptidyl transferase reaction gets more stereoselective because only activated amino acids of one configuration will polymerize with monochiral sugars. Note that the autocatalytic cycle requires a sustained source of new precursors (amino acid 'food') to operate (see Section 3.2.5 on RAFs). The configuration of the ribosome active site is redrawn from Steitz T. A. (2005) On the structural basis of peptide-bond formation and antibiotic resistance from atomic structures of the large ribosomal subunit, *FEBS Lett.* 579: 955–958. Redrawn from Martin W., Russell M. J. (2007) On the origin of biochemistry at an alkaline hydrothermal vent, *Philos. Trans. R. Soc. Lond. B* 362: 1887–1925.

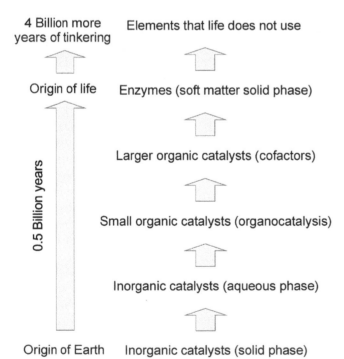

FIGURE 4.9 Schematic depiction of catalyst evolution. Each new catalyst category as it arose, from bottom to top, was not necessarily replaced by the next catalyst type, but expanded and enhanced by it. Solid-state inorganic catalysts were not assimilated by cells when they became free-living. From Wimmer J. L. E., Martin W. F. (2022) Origins as evolution of catalysts, *Bunsen-Magazin* 4: 143–146.

inorganic mineral catalysts in the solid phase first, followed by inorganic catalysts like Fe(II) in solution, followed by the simplest kinds of organocatalysts in solution such as amino acids, with an additional step in complexity coming in the form of cofactors like PLP or NADH and finally peptides capable of interacting with these catalysts coming last. With each step, increases in rate and specificity emerged that could have contributed to the formation of new catalysts and autocatalytic networks.

The process of optimizing biochemical and biological catalysts so as to generate living cells that could autonomously grow and divide took place about 500 million years after the Earth's formation. In the 4 billion years since then, nature has apparently been unable to improve upon the basic principle of the enzyme as the currency of catalytic function (Figure 4.9). Curiously, humans have learned to access the catalytic activity of elements that nature does not use in cells. Elements such as ruthenium, palladium, iridium, and platinum allow chemists to explore catalytic properties that nature never incorporated into metabolism. That does not mean, however, that rare elements such as Pd played no role at origins. Pd is an outstanding catalyst for reactions germane to the origin of life, in particular the otherwise very slow incorporation of ammonium into α-keto acids to yield amino acids, as Kaur H. et al. (2024) have recently shown, but Pd never became part of the repertoire of metals that enzymes use in metabolism. Either it was too rare, or too difficult to coordinate using standard protein and cofactor ligands, or both.

The expansion of catalyst types during evolution leads to autocatalytic relationships. A newly arisen catalyst can positively feed back into the improved synthesis of the preceding type. An important exception is that the original inorganic catalysts required no help from organic compounds to assemble, although to be incorporated into proteins, organic help was required. Many cofactors contain aromatic rings that are synthesized from starting compounds having many chiral centers. During aromatization, chiral centers are lost. It is particularly noteworthy that the active moieties of cofactors do not contain chiral centers (see Figure 3.16). They can catalyze reactions with great stereospecificity, but the stereochemistry of the reaction is governed by (1) the chiral centers of the substrates and (2) the handed environment of the protein active site that brings cofactor and substrate into contact for the reaction. This suggests that early chemical reaction networks could have reached considerable complexity, but as impure stereochemical mixtures, before the advent of genes and proteins. This requires that at least some of the catalytic activities afforded by cofactors could be provided by spontaneously (nonenzymatically) synthesized molecules.

Three more aspects concerning chirality in metabolism should be noted. In the primordial reaction sequences connecting CO_2 to pyruvate through the acetyl-CoA pathway and incomplete rTCA cycle (Figure 3.10), the only chiral intermediate is malate, which still today occurs as both D and L forms in central metabolism across all lineages. This lack of chirality is consistent with the ancient nature of the core backbone of carbon metabolism. In addition, in organisms that use the acetyl-CoA pathway for CO_2 fixation, the homochirality of sugars starts with the stereospecific addition of a water molecule to the double bond in phosphoenol pyruvate, only one conversion away from pyruvate, to yield 2-phospho-D-glycerate (see Figure 4.10). The stereochemistry of that carbon atom never changes throughout central metabolism. A third aspect is that D-amino acids and L-sugars do occur in cells today, but they occur mostly as components of the cell wall, which has only structural rather than catalytic functions. Today, D-amino acids have to be synthesized from their L forms by racemases. Their presence in the cell walls of prokaryotes might reflect an early use of abiotically synthesized, environmentally available D-amino acids. D-amino acids are furthermore very common in antibiotics and are commonly used in nonribosomal protein synthesis, which we will discuss in Section 5.1.

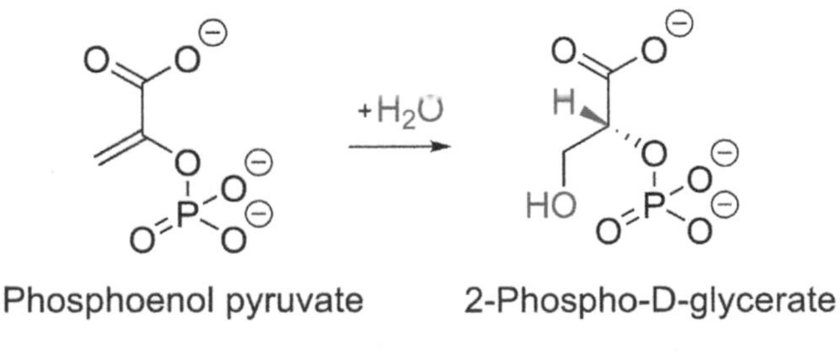

Phosphoenol pyruvate 2-Phospho-D-glycerate

FIGURE 4.10 **The enolase reaction.** In cells that use the acetyl-CoA pathway, the first chiral carbon atom in carbon metabolism is C2 of 2 phospho-D-glycerate. The chirality is introduced in the stereospecific addition of a water molecule to the double bond in phospho*enol* pyruvate in the reaction catalyzed by enolase. In metabolism, handed catalysts introduce chirality into achiral precursors.

The peptidyl transferase activity of the ribosome and the enolase reaction generating the D sugar series from phosphoenolpyruvate, which has no chiral carbons, can be seen as gatekeepers of chirality in modern metabolism. Amino acyl tRNA synthetases also help to ensure the proper chirality of amino acids in proteins. Enolase has an exceptional status among all enzymes of central carbohydrate and sugar phosphate metabolism in that there are alternative enzymes and alternative routes for all other reactions of central carbohydrate metabolism, but there are no alternative enzymes for the reaction catalyzed by enolase, even though the reaction that it catalyzes, the stereochemically specific addition of a water molecule to the double bond in PEP, is mechanistically simple. Similar to the case with the polymerization of only one stereoisomer at the onset of translation, a mixture of D- and L-ribose incorporated into RNA would generate a dysfunctional polymer, because the sugars of an RNA having chiral heterogeneity would point in different directions and with them the nucleobases attached to the sugars. This would give rise to a molecule that could not direct the synthesis of the complimentary strand for lack of base-pairing.

4.2.5 INORGANIC MICROPORES ARE POSSIBLE PRECURSORS OF BIOLOGICAL CELLS

Hydrated pores (rock-bound water, hydroxyl groups) in serpentinizing systems constantly produce iron and nickel metals, alloys, oxides, and sulfides via synthesis or oxidation of molecular hydrogen. These iron-containing compounds can catalyze the hydrogenation of molecular nitrogen and carbon dioxide to ammonia and reduced carbon compounds including α-keto acids. With activated ammonia and hydrogen, the ketoacids can react to amino acids and, in principle, a variety of other N-containing carbon compounds that are derived from amino acids in metabolism, in particular cofactors or nucleobases, which could fuel autocatalytic protometabolic networks. Activated with bicarbonate from dissolved CO_2 under alkaline conditions or with COS, these amino acids can polymerize to peptides via *N*-carboxyanhydrides. At high salt concentrations (low water activity) inside serpentinizing pores, where monomers accumulate by thermophoresis, rates of polymerization to polypeptides and proteins may be able to compete with hydrolysis. In the presence of salts, proteins can fold to generate structures including α-helices and β-sheets, which can further concentrate substrates in their protected (hydrophobic) interior. Of course, this would entail the folding of peptides containing both D- and L-amino acids as well as non-amino acid residues. Proteins made of D-amino acids are the mirror image of the L versions, whereby both can be folded by the same (L) chaperonin (heat shock protein assisting protein folding). In this discussion, when we refer to small peptides with structures, it should be clear that we are referring to very heterogeneous mixtures. Recall however the discussion in the last section about the preference of homochiral peptides.

Layers of peptides folded as β-sheets are especially stable against high temperatures as can be expected near serpentinizing olivine systems and are stable against hydrolysis, but they can only be formed from homochiral peptides. In Section 5.1 we will see that some α-helical peptides can self-replicate without enzymes or RNA. Also, proteins can synthesize small peptides using part of their sequence as a template (*nonribosomal peptide* synthesis in Section 5.1). The micropores in ancient serpentinizing hydrothermal fields contain the necessary physicochemical conditions for such primitive self-sustaining metabolisms and cycles and could be the earliest precursors of biological cells (Figure 4.11).

Biologists have long thought that modern extremophiles living at high temperatures, high salinity, or extreme pH might have preserved relics from the metabolism of the first cells. The first reasoned case for a thermophilic origin of life

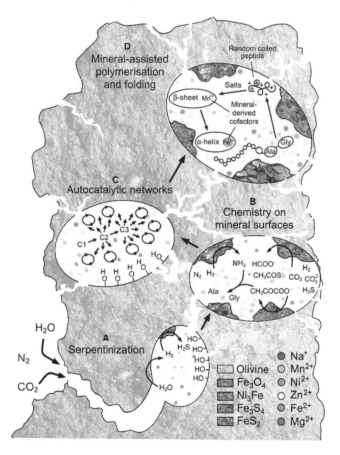

FIGURE 4.11 Serpentinizing systems and chemical reactions at origins. Shown are hydrated pores (rock-bound water, hydroxyl groups) in serpentinizing fields, where minerals like magnetite (Fe_3O_4), iron sulfides (FeS_2, Fe_3S_4), and Ni–Fe alloys (Ni_3Fe) are constantly produced in the presence of sulfide, iron, and nickel ions. With these minerals as catalysts, N_2 is hydrogenated to ammonia and CO_2 reduced to carbon compounds like α-keto acids. With activated ammonia and hydrogen, the α-keto acids can react to amino acids. The low water activity in serpentinizing olivine micropores probably enhanced the polymerization of activated amino acids to polypeptides. Foldable polypeptides derived from prebiotic amino acids are extremely thermostable and resistant to high salt concentrations (halophile). They therefore match the conditions of serpentinizing olivine crystals where serpentinization consumes seawater at elevated temperatures leading to high salt concentration. The polypeptides contain β-sheets and α-helices and can incorporate mineral cofactors. Redrawn from Vieira *et al.* (2020) The ambivalent role of water at the origins of life, *FEBS Letters* 594:2717–2733.

can probably be traced to a 1910 paper by the Russian biologist Konstantin Mereschkowsky. He divided Earth's early history into four phases, Epochs I–IV, as it pertained to origins. In the first epoch, the Earth had a fiery glowing surface; in the second, the fire had subsided, but the surface was still very hot, ca. 100°C, and therefore dry. In the third Epoch, the surface was covered with boiling water (50–100°C); in the fourth, the water had cooled to less than 50°C. Based on comparative physiology and a conviction that the first cells were anaerobic autotrophs and observations of cells that grow at high temperatures, he then concluded that the first forms of life arose in Epoch III, as the Earth was covered in boiling water: The first cells

> could have easily originated within the third period, when the water was still hot, saturated with minerals and devoid of oxygen. The rough conditions under which this plasma [*the cytoplasm of the first cells*] originated would explain its remarkable properties, its unusual tolerance of high temperatures, its tolerance of concentrated solutions of various harmful substances, its ability to live without oxygen and to synthesize its own proteins exclusively from minerals and so on.

With regard to temperature, we recall the insight of Wolfenden that reaction rates are greater at high temperatures and that as temperatures drop, the role of catalysts becomes increasingly important, a strong argument in its own right for thermophilic origins. The idea that the first cells were thermophiles was most recently championed by the German microbiologist Karl Otto Stetter, who isolated many hyperthermophiles, in particular archaea, and brought them into culture for study. Early studies of rRNA in the 1980s suggested that the deepest branches in the tree are occupied by thermophiles. Today there is less confidence in the branching order for the first lineages in rRNA trees, but the idea that the first cells were thermophiles still resonates well with most microbiologists and is consistent with Wolfenden's insights about the kinetics of biochemical reactions.

Concerning salt tolerance, haloarchaea, for example, can endure the high salt concentrations of serpentinized olivine crystals very well and can be preserved in pure salt crystals for long periods of time. In order to counteract the osmotic pressure of the saline environment, they transfer K^+ ions into the cell. Many other halophiles choose a more energy-intensive route to deal with high salt concentrations by synthesizing osmolytes like sugars, glycerol or amino acid derivatives. With these strategies, only organisms with very effective metabolic rates (high rates of ATP synthesis) can survive under extreme saline conditions. Haloarchaea are not primitive lineages though, they are heterotrophs that are evolutionarily derived from methanogens via lateral gene acquisition of genes for enzymes of heterotrophy and oxygen respiration.

Modern phylogenies tend to put methanogens around the base of phylogenetic trees within the archaea suggesting that methanogenesis was the ancestral state for the group. For the bacteria, the phylogenetic consensus is less clear, although the H_2-dependent acetogens seem to have preserved the most ancient genes and traits. Many modern methanogens are thermophiles and furthermore thrive under hydrothermal conditions. They live from H_2, CO_2, and a variety of C1 compounds. They outcompete acetogens at high temperatures. The thermostability of proteins in thermophiles is of fundamental interest. In general, what one observes in globular thermophile proteins (enzymes) is that thermostability increases as the hydrophobic core becomes more hydrophobic via amino acid replacements and the hydrophilic surface becomes more hydrophilic. This increases the stability of the protein toward unfolding. Also, additional ionic residues on the surface and buried within the protein form salt bridges, generating a system of ionic bonds both around the protein and within the protein that confer further stability against unfolding. The buried salt bridges are very stable because the protein must be unfolded to solvate the bridging ions and dissociate the ionic bond. An increased frequency of Cys–S–S–Cys disulfide bonds in thermophile proteins is usually not observed.

4.3 PEPTIDES STABILIZED EARLY AUTOCATALYTIC NETWORKS VIA CATALYSIS

4.3.1 RANDOM PEPTIDES WITH SUITABLE LIGANDS CAN COORDINATE TRANSITION METALS FOR CATALYSIS

In the following we use the electron transfer agent *ferredoxin* as an example of a very ancient polypeptide with an iron-sulfur cofactor to discuss the possible origin of enzymes with covalently bound cofactors. We posit that simple polypeptides with a random sequence of prebiotically acidic and hydrophobic amino acids and a folding determined by geochemical conditions can pick up transition metal ions for catalysis.

Ferredoxins (from Latin: *ferrum* iron, and *redox*) were among the first proteins for which there was abundant protein sequence data from different organisms because they are present in large quantities in the cell (on the order of 80–400 μM in anaerobes), and thus easy to purify for the purpose of protein sequencing. Biochemists could sequence proteins in the 1950s, but not until the late 1970s was it possible to sequence nucleic acids. Ferredoxins can harbor Fe_2S_2 clusters or Fe_4S_4 clusters (Figure 4.12). Their function is the transfer of single electrons as opposed to electron pairs as in NADH. Most bacterial ferredoxins contain cubic Fe_4S_4 clusters as the covalently bound cofactor, also called a prosthetic group (Figure 4.12 B). The iron-sulfur clusters are bound with the thiol side chain of four cysteines to the protein. One iron atom in the cluster changes its oxidation state from Fe^{3+} to Fe^{2+} when it receives an electron and relaxes back to Fe^{3+} when it releases the electron. Each cluster accepts or donates only one electron. The Fe_4S_4 clusters can exist in different oxidation states [$3Fe^{3+}$, $1Fe^{2+}$], [$2Fe^{3+}$, $2Fe^{2+}$], or [$1Fe^{3+}$, $3Fe^{2+}$], so that the average oxidation state of Fe can be on the order of +2.75 to +2.25. These clusters have different redox potentials up to −450 mV for the most energetic electrons.

In early groundbreaking work, Eck and Dayhoff (1966) reconstructed the evolutionary history of ferredoxin from *Clostridium pasteurianum*, a very ancient anaerobic,

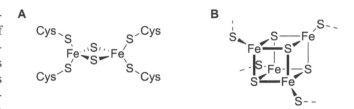

FIGURE 4.12 Structure of iron-sulfur clusters in ferredoxins. (A) Fe_2S_2 cluster in ferredoxins of chloroplasts, mitochondria, and some bacteria. The two iron ions are bridged by inorganic sulfur and bound with the thiol side chain of four cysteines in ferredoxin. (B) Fe_4S_4 cluster in ferredoxins of bacteria. The cubic Fe_4S_4 cluster is bound with four cysteines to the protein. The electrons are delocalized in the symmetric clusters and can be easily detached. Ferredoxins generally have the strongest reducing capability of all soluble proteins but can transfer only one electron at a time. Some ferredoxins have two Fe_4S_4 clusters and can perform two single-electron transfers.

nitrogen-reducing bacterium. They suggested the 'incorporation of its prototype very early during biochemical evolution, even before complex proteins and the complete modern genetic code existed'. According to their analysis, ferredoxin could have evolved from a repeating sequence of only four amino acids ADSG (Alanine A, aspartic acid D, serine S, and glycine G; for hydrothermal synthesis of A, D, and G see Figure 3.12). This short sequence was extended from copies of the genetic material and one mutation to (ADSG)₃(ADDSG)(ADSG)₃. Further mutations might have resulted in the exchange of four amino acids with cysteines which can bind iron-sulfur clusters from the surroundings. The sequence from 29 amino acids doubled to 58 amino acids with two identical moieties. Further mutations and loss of 3 amino acids at one end led to the present 55 amino acid long ferredoxin of *C. pasteurianum*. Margaret Dayhoff did exceptional work on early evolution, her 1966 book *The Atlas of Protein Sequence and Structure* created both the foundation and the first floor of modern bioinformatics, while simultaneously inventing the structure of all modern protein and DNA sequence databases.

A plethora of experiments have shown that peptides can be synthesized on protein templates without RNA (nonribosomal peptides in Section 5.1). Therefore, repeating sequences of peptides can be expected before the genetic code evolved. The simple elongation described above may thus have evolved before RNA and DNA-programmed peptide synthesis developed. Ferredoxins in chloroplasts and human mitochondria have very different amino acid sequences, but similar arrangements of their α-helices, β-sheets, turns, and loops. For example, the four cysteine amino acids binding 4Fe4S clusters in the ferredoxins of *C. pasteurianum* are contained in loop structures at the surface of the protein (Figure 4.13), as in ferredoxins from other sources. A variety of polypeptides containing cysteines can incorporate iron-sulfur clusters into their structure. Even without replication, a collection of simple polypeptides can be expected under primordial conditions that contain iron-sulfur clusters at the active site and pocket structures for substrates.

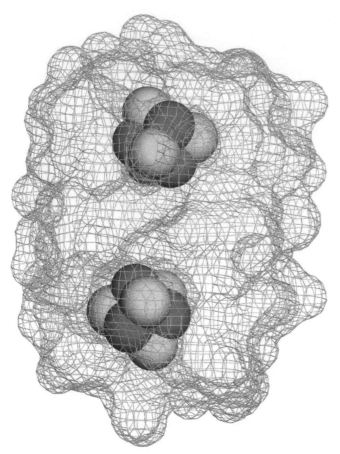

FIGURE 4.13 Three dimensional structure of ferredoxin. Ferredoxin of *C. pasteurianum* from the Protein Data Bank (PDB) entry 1CLF, image generated with PyMol. The iron atoms (brown) of the two 4Fe4S clusters are coordinated by sulfur atoms of cysteine residues in the peptide chain (not shown). In cells, the iron and sulfur atoms of the 4Fe4S clusters are incorporated into the protein structure during the enzymatic process of FeS cluster maturation, but in the presence of cysteine, FeS clusters can form spontaneously (Jordan S. F. *et al.* (2021) Spontaneous assembly of redox-active iron-sulfur clusters at low concentrations of cysteine, *Nature Comms* 12: 5925). In most organisms FeS cluster assembly starts with the addition of sulfur derived from cysteine, but in some methanogens the source of sulfur for FeS clusters is simpler: environmental sulfide (Liu Y. *et al.* (2010) Cysteine is not the sulfur source for iron-sulfur cluster and methionine biosynthesis in the methanogenic archaeon *Methanococcus maripaludis*, *J. Biol. Chem.* 285: 31923–31929).

4.3.2 Biofilms of Tar-Like Polymers Covered Hydrothermal Rock with Hydrophobic Layer

The amino acids glycine (2-amino acetate) and alanine (2-amino propionate) are easily synthesized under abiotic conditions and were likely abundant during early chemical evolution. Peptides synthesized from these amino acids are hydrophobic and might have accumulated at primordial vent settings coating their walls. An important part of the catalytic function of modern enzymes is to exclude water from the active site. Hydrophobic layers of short abiotic peptides together with the 'tar' encountered in many prebiotic chemistry experiments likely fulfilled the same task at origins. Such hydrophobic layers, or films, phase separating

onto the surfaces in rock pores of vent settings could have gradually sequestered abiotically synthesized hydrocarbons and peptides with deposition driven by thermal gradients.

In terms of chemical synthesis, amino acids are much easier to synthesize than nucleotides. Small, linear chemical networks such as those we encountered in Chapter 3—pyruvate synthesis combined with parts of the reverse tricarboxylic acid cycle and reductive amination of ketoacids to amino acids—might have been accelerated by the low water activity in hydrophobic layers. Possibly, organic-coated inorganic cell walls were the precursors of biological cell walls. Such hydrophobic layers might have had a composition similar to 'tholin', a heterogenous tar-like organic material that arises when methane and ammonia react under energy input from UV light or an electric spark. Such compounds are familiar to the origin of life chemists since the first Miller-Urey experiments, though the name tholin (from Greek *tholos*, muddy) stems from a 1979 paper by Carl Sagan. Tholins are reddish-brown disordered organic solids with different functional groups. Such polymer-like material is found on icy bodies and reddish atmospheric aerosols of planets and moons of the outer solar system, like Titan, but not on the modern Earth.

4.4 SUGARS AND RIBOSE ORIGINATED VIA AUTOTROPHIC ROUTES AND CYCLIZATION

4.4.1 Aldehydes or Ketones from Pyruvate Undergo Nucleophilic Attack of Their C=O Group

Chapter 1 outlines the reasoning behind autotrophic origins, an origin of metabolism from CO_2. How does central carbohydrate (sugar and sugar phosphate) metabolism fit into the concept of autotrophic origins? Today sugars are important reservoirs of carbon and energy, in particular as glycogen, a reserve polysaccharide (branched poly α-1,3-D-glucose) found in nearly all prokaryotes. At origins, the main function of sugars was probably the same as in cells today: constituents of cofactors, nucleic acids, cell walls, compatible solutes (compounds required to make cells) and, finally, carbon reserves (compounds required to survive conditions of starvation). At origins, sugars had to be synthesized from CO_2 as needed, there were no sugars lying around for cells to live off.

In today's world, there are abundant sugars. In fact, the vast majority of Earth's biomass consists of glucose because land plants are the greatest contributors to modern biomass and they are made of cellulose, a polymer of glucose (linear poly β-1,4-D-glucose). In modern environments, important catabolic pathways of sugar breakdown are the Entner-Doudoroff pathway and the Embden-Meyerhoff pathway. Both are glycolytic (sugar-splitting) pathways and start with glucose, converting it into pyruvate while storing the released energy from sugar disproportionation in the universal energy currency ATP. The electrons from glucose breakdown are transiently deposited either on the universal redox cofactor NADH or (among archaea) on ferredoxin at the oxidative step of the pathway: glyceraldehyde-3-phosphate oxidation. Although glycolysis is often considered as an ancient pathway, it is not a genuinely primordial pathway, going back to origins. Glycolysis arose from the gluconeogenetic pathway of glucose synthesis from CO_2.

Most prokaryotes, including H_2-dependent autotrophs such as methanogens and acetogens, synthesize glycogen as a reserve carbon compound. This storage compound allows cells to access carbon (C6 backbones), electrons (NADH or Fd^-), and energy (ATP) under conditions in which the environment provides no substrates to run carbon and energy metabolism. Storage and mobilization of glucose as glycogen gave rise to glycolytic pathways, not environmental supplies of glucose. However, ever since the origin of the very first cells, a very common sugar substrate in microbial communities was ribose because cells are about 20% RNA by weight and RNA is 40% ribose by weight, making the cell about 8% pure ribose by weight. Glycolysis is more important for heterotrophs than it is for autotrophs, and it is not an ancient pathway in the context of biochemical origins. By contrast, the enzymatic synthesis of ribose has to be ancient given its essential role in nucleic acid structure. Sugars are not only essential components of nucleic acids, but they are also starting compounds for many amino acid and cofactor biosynthetic pathways (Figure 3.18).

Ribose is part of RNA, ATP, NADH, coenzyme A, and $FADH_2$. D-Ribose 5′-phosphate is typically produced in metabolism from glucose in the pentose phosphate pathway and used for the biochemical synthesis of nucleotides. This would be compatible with a gluconeogenetic origin of sugar metabolism and hence with autotrophic origins. The 'D' in D-ribose refers to the stereochemistry of the chiral carbon atom C4′: D-Ribose has all OH groups on the right-hand side in the Fischer projection (Figure 4.14), L-ribose on the left-hand side. Ribose occurs in five different forms where β-D-ribopyranose is the most frequent configuration in equilibrium whereby ribose in RNA nucleosides is in the β-D-ribofuranose form. The designations 'α' and 'β' distinguish the anomers, which differ by the position of OH at the carbon atom C1 (Figure 4.14).

In older prebiotic chemistry experiments, sugars were obtained from the condensation of formaldehyde (CH_2O) in water by the autocatalytic, base-catalyzed *formose* reaction: $n\ CH_2O \rightarrow n/2\ CH_2OH - HC = O$ (glycolaldehyde) $\rightarrow n/4$ $CH_2OH - CH(OH) - CH(OH) - HC = O$ (*threose*) $\rightarrow\rightarrow$ etc. The reaction is autocatalytic because glycolaldehyde catalyzes the condensation of 2 CH_2O to glycolaldehyde. A complex mixture of C2–C6 aldoses and ketoses is produced, such that isolation of ribose from the formose reaction mixture can be tedious. The formose reaction can also be carried out in the solid state, without the addition of water by using mechanical forces in an oscillatory ball mill for activation. Using that approach, a variety of sugar molecules is formed, however, with fewer side products such as lactate compared to the aqueous phase, as work by Oliver Trapp and colleagues has shown. While the formose reaction may have contributed to the abiotic synthesis of sugars, it is still debated whether this route can readily explain the targeted formation of β-D-ribofuranose for nucleosides. For sugars, with their intensely complex sets of stereochemical isomers, reactions with higher selectivity would appear to be necessary. Somewhat more promising is a route where accumulated glycolaldehyde condenses to threose and its stereoisomer erythrose driven by homochiral dipeptide catalysis. For example, > 80% of the D-enantiomer of erythrose was obtained with L-val-L-val catalysis. However, neither formaldehyde nor glycolaldehyde have been observed as stable products of CO_2 hydrogenation so far (Section 3.1). Also, no current organisms use that route nor is there any indication that they ever did.

Instead, sugar in autotrophs is synthesized by gluconeogenesis which starts from pyruvate which is a stable product of CO_2 hydrogenation (Section 3.1). It was recently reported that the nonenzymatic aldol condensation reaction between $HC=O-CH(OH)-CH_2-P_i$ (D-glyceraldehyde-3-phosphate: G3P) and its isomer $CH_2(OH)-C=O-CH_2-P_i$ (dihydroxyacetone phosphate: DHAP) in ice yields the C6 sugar D-fructose-1,6-biphosphate, an intermediate in gluconeogenesis. That recent report was however predated, as has been known since the 1960s that the uncatalyzed aldol condensation reaction for the non-phosphorylated substrates proceeds at a rate of roughly 1 µM per minute under alkaline conditions (Degani C, and Halmann M., 1967). The reaction is catalyzed in metabolism by aldolase. The aldolase reaction could be accelerated by the amino acids glycine and lysine which may point to early enzyme evolution. G3P is obtained from pyruvate by phosphorylation to phosphoenolpyruvate, isomerization, and reduction; however, these steps of gluconeogenesis have not yet been demonstrated nonenzymatically, though Moran and coworkers have carried out promising experiments in this direction.

There is a further very simple possibility for the synthesis of sugars like ribose. Pyruvate might react with glyoxylate at reducing conditions to different forms of the sugar ribose (Figure 4.15). The hydrides of the methyl group of pyruvates are slightly acidic such that their carbon atom is negatively polarized under alkaline conditions and can attack the positively polarized carbon atom of the carbonyl group of glyoxylate. The nucleophilic attack leads to aldol condensation

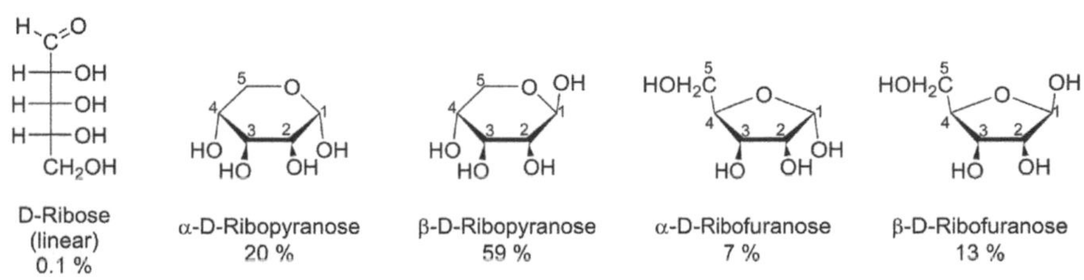

FIGURE 4.14 **Different forms of D-ribose with relative abundances in equilibrium.** The different forms rapidly interconvert, and their equilibrium proportions in water are indicated. The linear (open chain) form is shown in the Fischer projection. Note that cyclic ribose is not a planar molecule but puckered due to steric hindrance.

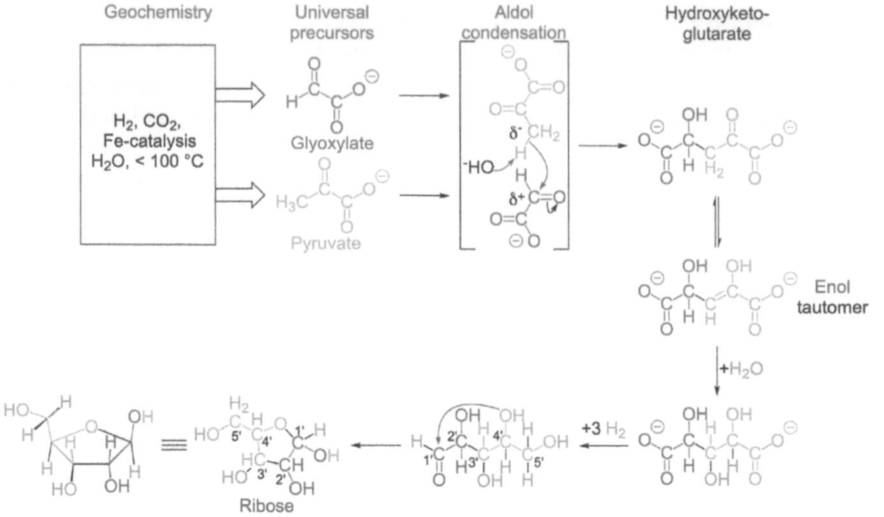

FIGURE 4.15 **Possible pathway of reductive ribose formation from glyoxylate and pyruvate.** The proposed reaction requires a base and transition metals that can catalyze hydrogenation (see text).

(Section 1.1). The condensation product hydroxyglutarate might partially tautomerize to the enol form and add H_2O to the C=C double bond of the enol tautomer. The terminal carboxyl groups are reduced with hydrogen under transition metal catalysis. Depending on the reduction conditions, terminal aldehydes, alcohols, and mixed forms are produced. The aldehyde group is attacked by the C4′ hydroxyl group to produce a furanose form or by the C5′ hydroxyl group to produce a pyranose form (reaction by cyclic hemiacetal formation). The different forms are in equilibrium as described above. Simple reactions of this type, though not yet demonstrated experimentally, deserve further attention.

In autotrophic cells, ribose-5-phosphate is generated by gluconeogenesis and the pentose phosphate pathway. Similar reactions occur nonenzymatically under aqueous conditions if metal ions, particularly Fe^{2+}, are introduced to the reaction mixture. However, the nonenzymatic catabolic pathway to ribose would require anabolic synthesis of glucose first followed by oxidation, whereas glyoxylate might react with pyruvate to ribose directly (Figure 4.15) in a primordial precursor.

4.5 RNA NUCLEOSIDES CAN BE SYNTHESIZED FROM AMINO ACIDS BY HEATING

The synthesis of nucleotides in cells is based on the condensation of amino acids and phosphorylated ribose. Thus, condensation of anhydrous amino acids could lead to interesting heterocyclic products, possibly related to nucleobases or even nucleosides. Indeed, heating of an equimolar mixture of glycine, alanine, and lysine to about 180°C led to several heterocyclic products including pterins with a glycosidic bond to a pyranose ring (Figure 4.14).

4.5.1 IN METABOLISM, NUCLEOTIDES ARE FORMED BY CONDENSATION OF RIBOSE WITH AMINO ACIDS

The synthesis of nucleotides is an important step in early biochemical evolution. It is of interest to consider how the biochemical pathway could have developed. In Section 4.6

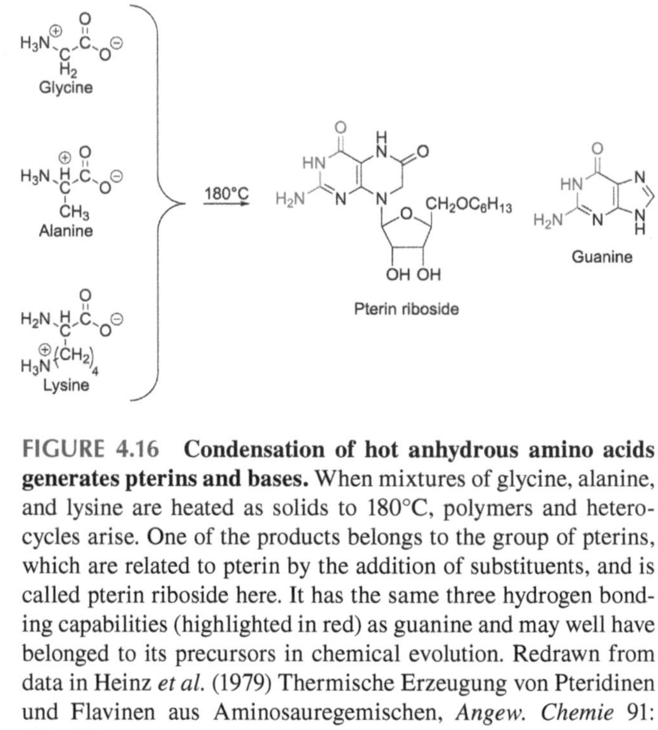

FIGURE 4.16 **Condensation of hot anhydrous amino acids generates pterins and bases.** When mixtures of glycine, alanine, and lysine are heated as solids to 180°C, polymers and heterocycles arise. One of the products belongs to the group of pterins, which are related to pterin by the addition of substituents, and is called pterin riboside here. It has the same three hydrogen bonding capabilities (highlighted in red) as guanine and may well have belonged to its precursors in chemical evolution. Redrawn from data in Heinz *et al.* (1979) Thermische Erzeugung von Pteridinen und Flavinen aus Aminosauregemischen, *Angew. Chemie* 91: 510–511.

we will discuss the organic synthesis of nucleotides by routes that involve cyanide, or cyanide derivatives like thiocyanide, which however lack precursors or intermediates in common with modern biosynthesis. These routes remain of interest but offer no clues as to the origin of the biochemical routes. In fact, the core of central intermediary metabolism across all known life forms (the ones we have to explain; see next paragraph) is highly conserved and very homogeneous across lineages. This conserved chemistry is not only evidence in favor of a single origin of life, it indicates that the core of biosynthetic metabolism (Figure 3.18) was present in LUCA, as was the core of carbon metabolism leading to acetyl-CoA and pyruvate from which central intermediary metabolism emanates. This in turn suggests that there was an appreciable degree of continuity between preexisting spontaneous reactions in geochemical settings and the later, enzymatically catalyzed biological pathways. This principle, that many biochemical reactions go forward under suitable conditions and in the presence of suitable catalysts but in the absence of enzymes, is supported by many new studies. Hence, we focus here on the known biochemical pathways and their possible modifications under primordial conditions.

In the preceding paragraph we mentioned 'known life forms'. This serves as a reminder that the objective of scientific investigations into the origin of life is to explain the origin of the life forms that are around us today. There are also many kinds of reactions that we can imagine or perform in the laboratory that can give rise to important molecules of life, but starting from very different compounds than those used by modern cells or employing reaction conditions that have little to do with the chemical milieu within cells. It is possible to explore such reactions for the sake of their interesting chemistry, but if they do not connect at all with the chemistry of known life forms, then they remain difficult to connect to primordial forms of life in an origins context.

Most nucleobases do not dissolve well in water, suggesting that nucleotides (base plus ribose plus phosphate) must have been incrementally synthesized from water-soluble ingredients like amino acids and phosphate-activated ribose (Figure 4.17). First, we follow the incremental route to pyrimidine nucleotides. Carbamoyl phosphate already contains the C=O and NH group of uracil and cytosine and is a nucleus for pyrimidine cyclization. To get activated carbamoyl, bicarbonate reacts first with ATP or another strong phosphorylating agent (an agent with very exergonic hydrolysis; see Section 1.1) to carboxyphosphate. Then, ammonia (NH_3) displaces the phosphate group to obtain carbamic acid. Further phosphorylation leads to carbamoyl phosphate.

As the next component of incremental synthesis, consider the production of phosphate-activated ribose (Figure 4.17B and discussion above). Ribose might react with a strongly phosphorylating agent to 5-phosphoribosyl-1-pyrophosphate (PRPP), however, the selectivity of phosphorylation is not known in the absence of enzymes and will likely depend on the metal catalyst used.

Aspartate substitutes the phosphate group of carbamoyl phosphate by nucleophilic attack with its amino group reacting to carbamoyl aspartate (Figure 4.17 C). Carbamoyl aspartate dehydrates to dihydroorotate at acidic pH and gets

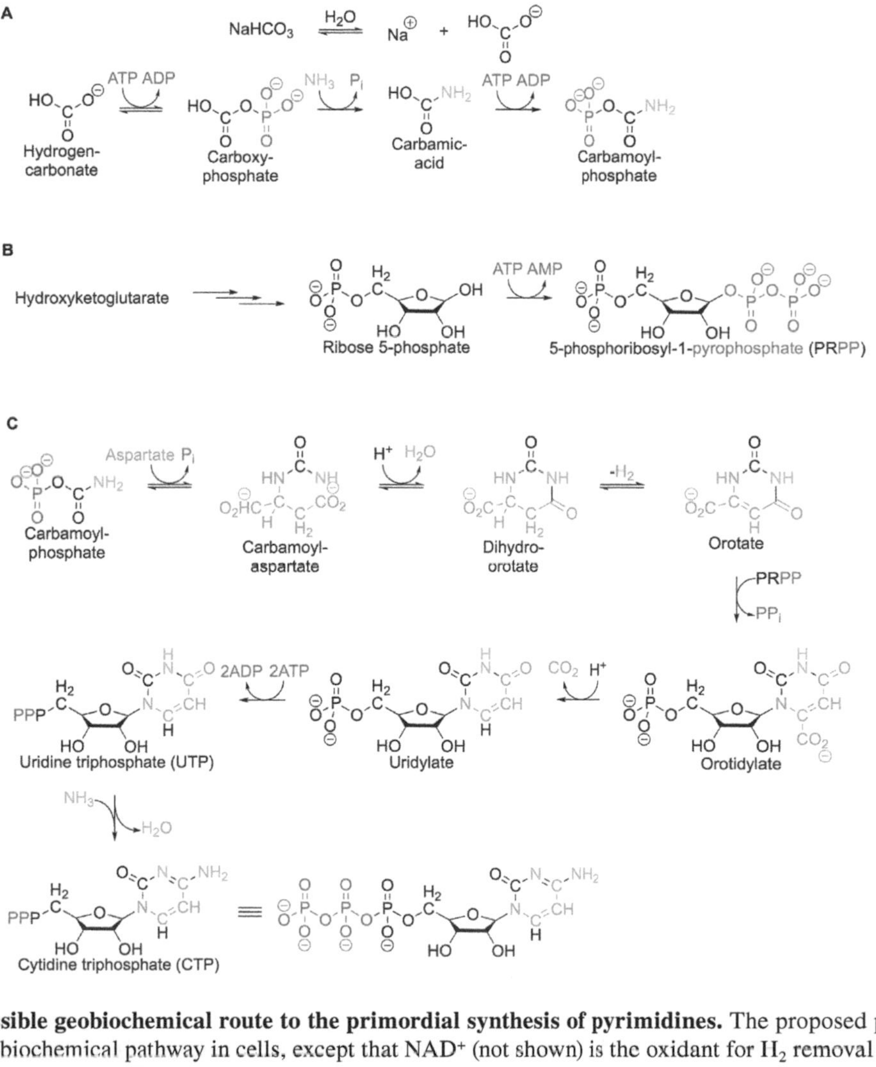

FIGURE 4.17 **A possible geobiochemical route to the primordial synthesis of pyrimidines.** The proposed pathway shown is the conserved, enzymatic biochemical pathway in cells, except that NAD$^+$ (not shown) is the oxidant for H_2 removal in the step leading to orotate. (A) Pathway to carbamoyl phosphate. (B) Pathway to 5-phosphoribosyl-1-pyrophosphate (PRPP) starting from hydroxyketoglutarate (see also Figure 4.14 and the proposal in Figure 4.15). (C) Pathway to uridine triphosphate (UTP) and cytidine triphosphate (CTP). Several of these reactions require activation by phosphorylation. Instead of ATP, agents like metaphosphates or organic phosphates could have been used for abiotic activation. For an explanation of the possible mechanism, see text.

oxidized to orotate in the enzymatic synthesis. Moran and coworkers showed that this reaction sequence runs non-enzymatically, if iron phosphide (FeP) or metaphosphates like P_4O_{10} are used for dehydration (cyclization) and MnO_2 for oxidation (redox potential $E_0 = +1.23$ V at pH 0 for $MnO_2 + 4H^+ + 2e^- \rightleftharpoons Mn^{2+} + 2H_2O$). Carbamoyl phosphate was obtained by a reaction of $NaHCO_3$ or 40 bar CO_2 in H_2O with aqueous ammonia and phosphates.

The nonenzymatic reaction of orotate with PRPP to orotidylate, decarboxylation to uridylate, and phosphorylation to UTP (Figure 4.17 C) has not been investigated yet. Again, strongly phosphorylating and dehydrating agents like metaphosphates will probably be necessary to accomplish the last step. A possible agent is again tetrametaphosphate P_4O_{10} (sometimes misleadingly called phosphorous pentoxide P_2O_5) which is quite stable in water because it forms a protective sticky coating after initial hydrolysis. It is a potent phosphorylating and desiccating (drying) agent with high energy of hydrolysis. The nonenzymatic transformation of UTP to CTP would require activated ammonia (NH_2OH, N_2H_4, or NH_2CONH_2, for example).

4.5.2 Amino Acid Cyclization Reactions to Nucleobases Involves Transition Metal Catalysis

The synthesis of purine (adenine and guanine) nucleotides also requires reactant activation as indicated in Figure 4.18. Most reagents are activated by phosphorylation again. In the first steps, PRPP reacts with activated ammonia to phosphoribosylamine (1), then with glycyl phosphate (Gly) to the amide (2) and (in archaea) with formylphosphate to formylglycineamide (3). Activated ammonia (derived from glutamine in modern organisms) reacts with the carbonyl group of (3) to the amidine (4). The amidine cycles (dehydrates) to form the five-membered imidazole ring of purines (5). Again, in the absence of enzymes, a strongly phosphorylating and dehydrating agent like a metaphosphate or an organic phosphate would have to activate the carbonyl group to be susceptible to nucleophilic attack by the NH group, cyclization, and dehydration. The exocyclic amino group of (5) attacks carboxyphosphate to form the carboxylated imidazole (6) which isomerizes to (7). Again, the carboxylate group is phosphorylated, and the phosphate group is

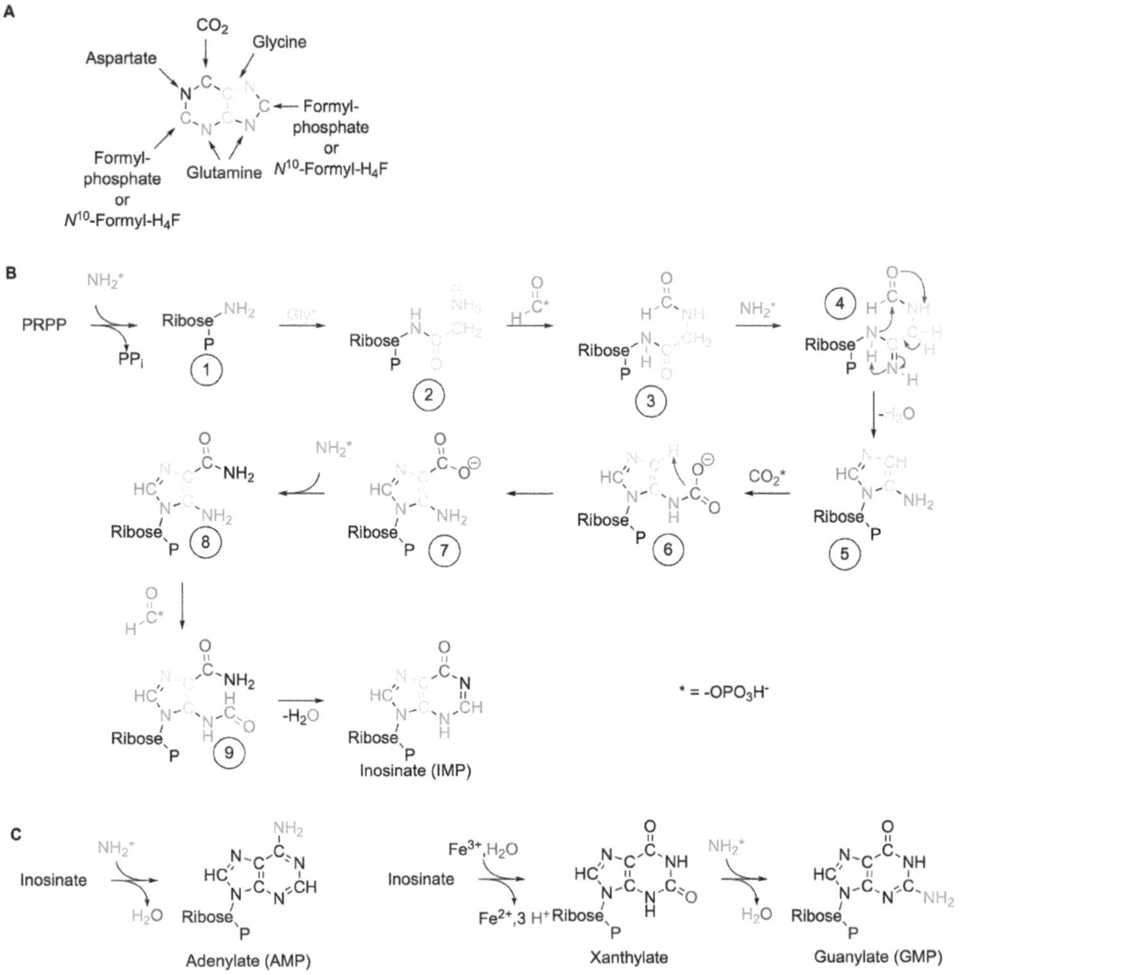

FIGURE 4.18 **A possible geobiochemical route to primordial synthesis of purines.** (A) Origin of atoms in the purine heterocycle. See Figure 3.12. Modified from Lipmann F. (1965) Projecting backward from the present stage of evolution of biosynthesis. In *The Origin of Prebiological Systems and of their Molecular Matrices*, Fox S. W. ed., Academic Press, New York, NY, pp. 259–280. (B) A possible geobiochemical pathway to inosinate. The proposed route corresponds to the enzymatically catalyzed biochemical synthesis of purine nucleotides with simpler components. (C) Pathways to adenylate and guanylate. * indicates activation by phosphorylation or by other means.

displaced by activated ammonia (derived from aspartate in modern organisms) to form amide (8). An activated C1 unit, derived from formylphosphate in archaea or from N^{10}-formyltetrahydrofolate (N^{10}-formyl-H_4F) in bacteria, is added to the exocyclic amine group to form (9). Product (9) cycles with water elimination to form inosinate monophosphate (IMP). Recall that formyltetrahydrofolate is an intermediate of methyl synthesis in the acetyl-CoA pathway.

Inosinate reacts with activated ammonia (from aspartate in modern organisms) to adenylate monophosphate (AMP). Guanylate (GMP) is generated by the addition of water, dehydrogenation (with Fe^{3+} or enzymatically with NAD^+), and reductive amination with Fe^{2+} and NH_2OH, N_2H_4, or NH_2CONH_2 or enzymatically with NH_2 from glutamine). Note that none of the reactions in Figure 4.18 has been non-enzymatically demonstrated so far.

4.6 CYANIDE CONDENSATIONS GENERATE RNA BASES VIA NON-BIOMIMETIC ROUTES

It has been known since the early 1960s that amino acids and nucleobases can also be synthesized more specifically and in high yields with cyanide chemistry. As shown in the following, cyanides and cyanoacetylenes are possible products of a carbon- and nitrogen-rich magma, they were possibly prebiotically available in some geochemical environments. However, we consider the ketoacid pathway to amino acids and the amino acid pathway to nucleosides or some form of amino acid condensation, as in Figure 4.16, to be a more realistic scenario for prebiotic synthesis, because there is no hint that cyanides have played a role in prebiotic synthesis of organics in even the most primitive bacteria and archaea. In fact, cyanides have no role at all as substrates in biosynthetic metabolism, whereas CO_2 is ultimately the starting point for all biosynthetic reactions in metabolism. One should however keep in mind that for the early phases of chemical evolution—before the origin of genes and enzymes—all theories for origins, including autotrophic theories, are strictly dependent upon input of viable biochemical components that were provided by the environment up until the origin of the corresponding biosynthetic pathway. One should therefore keep an open mind about the origin of components before the origin of biochemical pathways. At the same time, even though some CN^- might have been available in some environments, a biosynthetic dependence upon CN^- would have restricted early metabolism to very few and very specific environments, at best. By contrast, CO_2 was available everywhere.

4.6.1 CARBON- AND NITROGEN-RICH MAGMA MIGHT HAVE PRODUCED CYANIDE AND CYANOACETYLENE

Calculations of the C/N/O/H chemistry of shallow and surface hydrothermal vents showed that gases from carbon- and nitrogen-rich magma can react with hydrogen to mM—1 M concentrations of acetylene (C_2H_2), diacetylene (C_4H_2), cyanoacetylene (C_2HCN), hydrogen cyanide (HCN), bisulfite (HSO_3^-), hydrogen sulfide (HS^-), and soluble iron (Fe^{2+}) in hydrothermal water (Figure 4.20A). Cyanoacetylene and

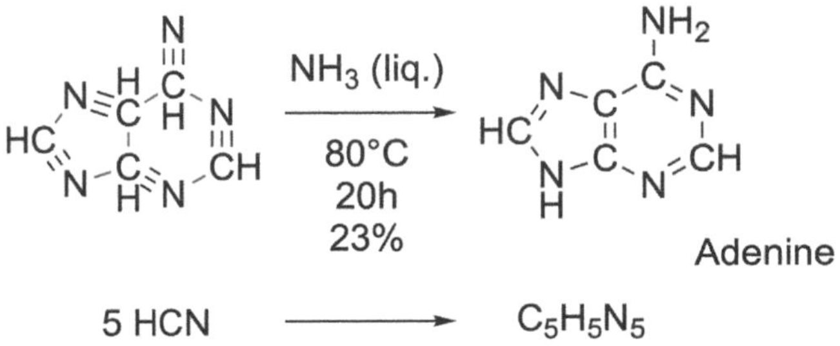

FIGURE 4.19 Formal arrangement of five HCN to adenine. The condensation probably occurs via formamidine NH=CH–NH_2 as an intermediate formed by nucleophilic attack of NH_3 on the carbon atom of C≡N–.

hydrogen cyanide can react to heterocycles at suitable conditions (Figures 4.19 and 4.20B). At high pressures, more cyanoacetylene than hydrogen cyanide is formed, because HCN reacts with C_2H radicals from dissociation of diacetylene to cyanoacetylene: $HCN + C_2H \leftrightarrow C_2HCN + H$. At lower pressure there is more hydrogen and acetylene available, and the equilibrium shifts to hydrogen cyanide.

4.6.2 THE Fe^{2+} CATALYZED REACTION OF C_2HCN WITH NH_3, RIBOSE, AND PHOSPHATE YIELDS NUCLEOTIDES

The nucleobase adenine is a pentamer of HCN and was indeed observed as a product of its cyclization (Figure 4.19). Up to 23% product yield was observed in liquid ammonia, but under realistic prebiotic conditions, the yield is much lower and insoluble black polymers are the main products. Heating of concentrated NH_4CN solutions at 80°C for 16 days gave up to 0.1% adenine yield. Acid hydrolysis of the HCN polymers added a further 0.1% adenine. The solubility of adenine in water is low under any conditions and a combined synthesis of the readily soluble adenine nucleoside offers more opportunities for further reactions. Such synthesis routes are described in the following.

Thomas Carell and coworkers have reported the efficient synthesis of the pyrimidine nucleosides cytidine and uridine from cyanoacetylene, hydroxylamine, urea, and ribose, driven by wet-dry cycles and by reduction with Fe^{2+} (Figure 4.20B).

Nucleophilic attack of hydroxylamine (NH_2OH) on carbon atoms of cyanoacetylene (C_2H-CN) depicted in (1) and hydrogen tautomerization leads to heterocycle (2). The exocyclic amino group of (2) attacks urea (NH_2CONH_2) and displaces ammonia (NH_3) in a Zn^{2+} catalyzed wet-dry cycle (warming to 95°C; 2 days dry-down) to obtain (3). Becker *et al.* suggested that due to flooding or some other mixing process, (3) could come into contact with ribose or another polymerization-suitable sugar unit. After addition of boric acid and heating to 95°C, the ribosylated product (4) was obtained with high yield. Reductive ring opening with small amounts of Fe^{2+} from FeS, FeS_2, or thiols was followed by tautomerization, cyclization, and water elimination to get cytidine nucleoside (6). The addition of the naturally occurring minerals *hydroxyapatite*, *colemanite*, or synthetic *lüneburgite* to the reductive pyrimidine forming mixture, the addition of urea as a catalyst, and evaporation to dryness

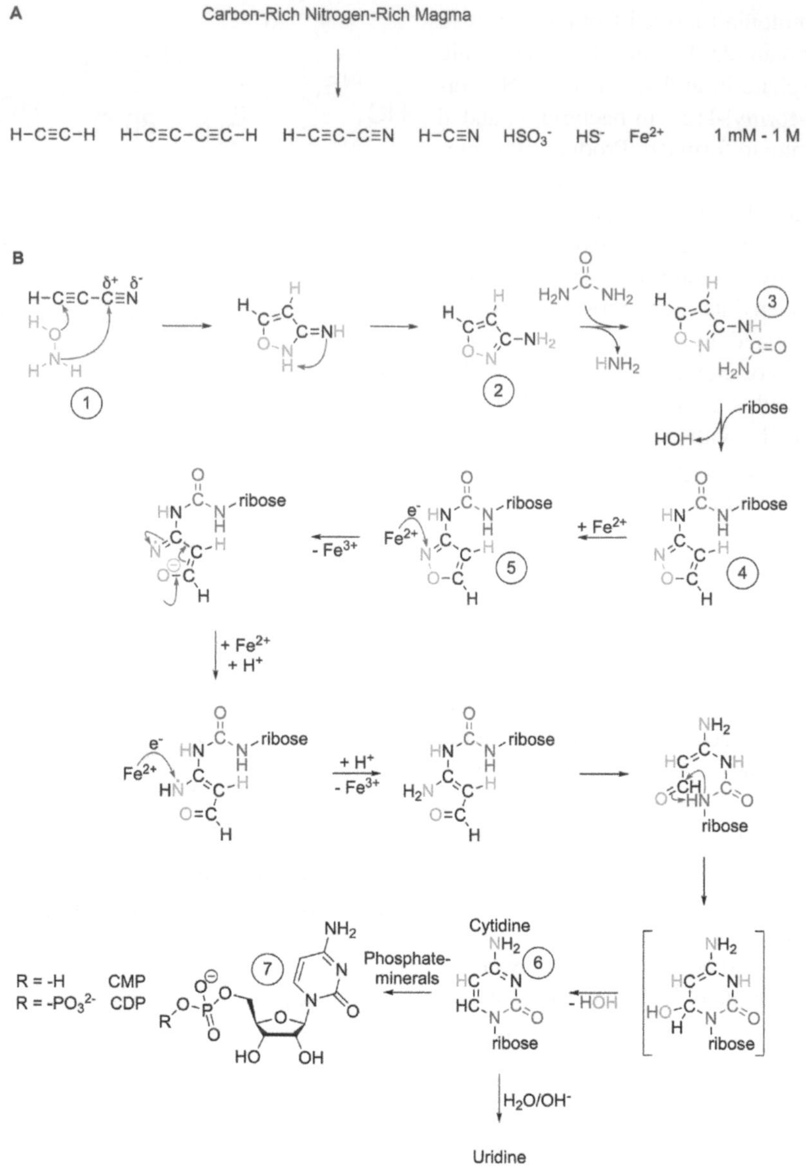

FIGURE 4.20 **Synthesis of nucleotides from cyanoacetylene and hydroxylamine.** (A) Production of cyanides from carbon- and nitrogen-rich magma. (B) Formation of pyrimidine RNA nucleosides from cyanoacetylene and hydroxylamine. CMP = cytidine 5′-monophosphate, CDP= cytidine 5′-diphosphate. Redrawn from Becker S. *et al.* (2019) Unified prebiotically plausible synthesis of pyrimidine and purine RNA ribonucleotides, *Science* 366: 76–82.

at 85°C resulted in (7) as major products. Phosphorylated pyranosides are only minor products. Hydrolysis at alkaline pH led to the synthesis of the uridine-phosphates UMP and UDP (OH⁻ displaces NH₂ and tautomerizes to C=O). Thomas Carell and coworkers showed that the conditions for purine synthesis (not shown here) are compatible with pyrimidine synthesis such that all Watson-Crick bases can be formed concurrently.

Xu, Sutherland, and coworkers demonstrated the cyanide-based generation of pyrimidine and purine nucleosides in a number of organic syntheses. Saladino *et al.* showed that the synthesis of nucleobases is possible by heating formamide at 160°C in the presence of catalysts. In addition, Rafaelle Saladino and colleagues were able to demonstrate the synthesis of a number of cofactors from formamide. Various metal oxides and minerals containing borate, zirconium, iron-sulfur, TiO_2, phosphates, and olivines as well as clays and material from the Murchison

meteorite proved to be effective catalysts of these syntheses. None of these formamide or cyanide-dependent syntheses or variants thereof, however, are observed in modern organisms, even the most primitive ones. Hence nature either never traversed the cyanide or formamide routes or they were dead ends that were replaced by CO_2 and H_2-dependent syntheses. All known organisms and reconstructed early entities like LUCA (Last Universal Common Ancestor) synthesize the building blocks of nucleic acids from amino acids obtained by amination of α-ketocarboxylic acids. These simple acids are obtained by hydrogenation of CO_2 with suitable catalysts under hydrothermal conditions (Section 3.1), they are also the starting point of biosyntheses leading to the cofactors that are required in nucleotide and amino acid synthesis. The prebiotic synthesis of bases probably occurred along the lines of the modern biochemical pathways, rather than cyanide-based chemical reactions.

4.7 ACTIVATED RNA-MONOMERS CONDENSE TO POLYMERS WITH RANDOM SEQUENCE

4.7.1 THERMOPHORESIS IN HYDROTHERMAL PORES ACCUMULATES NUCLEOTIDES AND RNA STRANDS

Only low concentrations of nucleotides, perhaps a few nanomoles per liter of hydrothermal fluid, can be expected from the syntheses discussed in Sections 4.5 and 4.6. If we embrace for a moment the central salient premise of the RNA world concept, namely that very high concentrations of very pure and chemically activated ribonucleotides had to be attained at specific sites to allow efficient polymerization of the phosphate-activated nucleotides, some form of accumulation must take place. This can be either specific binding to surfaces like Montmorillonite as shown in Figure 4.24 or association with molecules adsorbed on surfaces as in affinity chromatography. A third possibility is thermal accumulation on the cold side of the bottom of hydrothermal pores, which has recently been shown to provide an additional means of nucleotide and RNA concentration. A fourth possibility—that an RNA world never existed—would still not preclude the synthetic utility of concentration mechanisms like thermophoresis.

Dieter Braun and coworkers demonstrated the accumulation and polymerization of short DNA strands in a capillary heated with an infrared laser. The degree of reversible DNA hybridization was quantitatively analyzed by Förster Resonance Energy Transfer (FRET) microscopy (Figure 4.21). Only the hybridization of complementary DNA strands could be investigated by this technique, not polymerization (covalent bonding) of nucleotides. FRET for assaying DNA hybridization works as follows. Energy-donating and accepting dyes are coupled to the complimentary + and − strands that can form a DNA molecule. When DNA strands with complementary nucleobases hybridize, the energy-donating and accepting dyes that are attached to the strand ends come close to each other and the donor dye transfers energy to the acceptor dye (dyes marked dark green and yellow in Figure 4.21). Fluorescence is recorded. The energy transfer efficiency E is

$$E = 1 - \frac{I'_D}{I_D}$$

where I'_D and I_D are the donor fluorescence intensities with and without acceptor. Due to quenching by FRET, the donor intensity is reduced $\left(I'_D < I_D\right)$ if the acceptor dye is closeby. Here, $I_D{}'$ is the donor fluorescence intensity of the hybridized strands and I_D is the donor fluorescence intensity of unhybridized strands. The transfer efficiency must be normalized to the efficiency of a 100% hybridized dimer to directly give the degree of 'reversible polymerization' (hybridization). From the known number of nucleobases in an unhybridized strand and the measured degree of polymerization, the mean length of the polymer is obtained: up to 24 monomer strands reversibly polymerize with 120 base pairs/monomer each = 2880 base pairs.

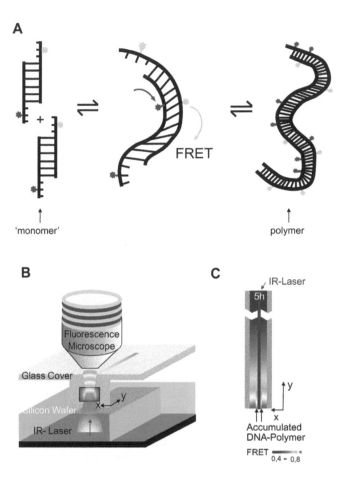

FIGURE 4.21 FRET microscopy for detection of thermal gradient-induced DNA accumulation. FRET stands for fluorescence energy transfer. (A) Two single strands of DNA with 95 base pairs long complementary sequences bind to form double strands (ds monomers) with 25 base pairs long 'sticky' (complementary) single-strand ends. These dsDNA monomers reversibly bind to each other via their sticky ends to form double-strand 'polymers' (hybrids). To measure the grade of polymerization, each sticky end of a 'monomer' is labeled with a complementary fluorescent dye termed donor and acceptor (marked dark green and yellow). In the polymer the two dyes come close to each other (distance r) and the donor dye excited by a light-emitting diode (LED) transfers energy to the acceptor dye through nonradiative dipole-dipole coupling. For this, the donor fluorescence spectrum must overlap the acceptor absorption spectrum. The normalized transfer efficiency is a measure of the degree of polymerization and hence the mean length of the polymer. (B) Set up of the fluorescence microscope. The IR-laser is aligned in the middle of the capillary and scanned along its long y-axis (7 mm long, 100 µm x-width: aspect ratio 7/0.1 = 70) to induce heating there and hence convection and thermodiffusion toward the cool walls and bottom of the capillary. (C) Regions of normalized FRET efficiency (E_{norm}) after 5 hours hybridization time are color marked ($E_{norm} = 0.4 - 0.8 \rightarrow$ blue − red). The thermal gradient induced by the IR-laser leads to convection and thermodiffusion (not shown here) such that the DNA monomers accumulate in the cold zones (regions not heated by the IR-laser) and polymerize there (red zones of high FRET efficiency at the bottom of the capillary). Redrawn from Mast C. B., Braun D. (2010) Thermal trap for DNA replication, *Phys. Rev. Lett.* 104: 188102.

The experimental FRET efficiency E can be used to estimate distances between fluorophores in polymers like DNA or proteins and temporal changes of these distances. According to Förster's theory, the transfer efficiency E depends on the distance r between donor and acceptor

$$E = \frac{1}{1 + \left(\dfrac{r}{R_0}\right)^6}$$

where R_0 is the *Förster distance* at which energy transfer efficiency is 50% (donor fluorescence intensity I'_D has decreased by half). Hence, FRET depends sensitively (r^6!) on small changes in distance in the range of 1–10 nm and FRET efficiency can be used to determine distances between two fluorophores in a protein. Using FRET, not only nucleic acid hybridization but also the structure of folded proteins and changes of enzyme structure with binding of substrates can be elucidated if the fluorophores are bonded to strategical locations in the protein.

The heated capillary used in the experiment typifies a hydrothermal pore with warm and cold sides and a closed bottom (Figure 4.22). The temperature gradient in the hydrothermal pore powers *convection* because water on the hot side expands, becomes lighter, and moves up in the gravitational field. On the cold side, water contracts, becomes heavier, and sinks (circular arrows in Figure 4.22A). Additionally, solved molecules move from the hot to the cold side by *thermodiffusion* (horizontal arrows) counterbalanced by diffusion out of the zone of higher molecular concentration. By a combination of convection and thermodiffusion, molecules are trapped and accumulated at the pore bottom (darker zone in Figure 4.22A), which allows polymerization as described above.

According to Baaske, Braun *et al.* the analytical solution for accumulation in a rectangular cleft geometry is

$$\frac{c_{bottom}}{c_{top}} = \exp\left[0.42 \cdot S_T \cdot \Delta T \cdot r\right]$$

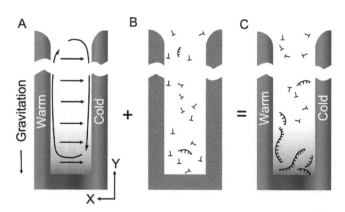

FIGURE 4.22 Accumulation of DNA in the thermal gradient of a water-filled rock pore. (A) Thermal trapping due to convection (circular arrows) and thermodiffusion (horizontal arrows). (B) Slow polymerization and polymer hydrolysis limit polymer length to short DNA strands under dilute primordial conditions. (C) Thermal accumulation increases monomer concentration and polymerization rate such that polymer decay gets counterbalanced. Redrawn from Mast C. B. *et al.* (2013) Escalation of polymerization in a thermal gradient, *Proc. Natl Acad. Sci. USA* 110: 8030–8035.

where c_{bottom} and c_{top} are the molecular concentrations at the bottom and top of the cleft, S_T is the Soret coefficient (see thermodiffusion in Section 3.2), and ΔT is the temperature difference between hot and cold side and $r = $ length/width is the aspect ratio of the cleft. Hence accumulation increases exponentially with S_T, ΔT, and length of the cleft or pore at a given width. For a single nucleotide, $S_T = 0.015$ K^{-1} at 1.7 mM NaCl concentration. Thus, at $\Delta T = 30$K, accumulation is 7 for $r = 10$, and about 110 for $r = 25$ (black curve in Figure 3.17). *S_T increases with the molecule or polymer length such that longer molecules are preferentially accumulated in an exponential manner*, essentially because of slower diffusion. For double-stranded DNA with 100 base pairs, an experimental $S_T = 0.075$ K^{-1} at 1.7 mM NaCl concentration was obtained. Hence accumulation is about 10^4 for $r = 10$ and 10^{10} for $r = 25$ at $\Delta T = 30$K. Consider that a lower temperature difference can be compensated by a longer pore or by concatenated shorter pores. Elongated pores of Lost City's vent mounds are typically $\leq$ 1mm long with $\approx$20–100 μm diameter, hence aspect ratios are between 10 and 50. In the numerical simulations of Baaske, Braun *et al.* accumulation did not change much for a large variety of dented, bent, incised, inclined, and reservoir-connected pores or clefts because regions of reduced accumulation are linked by mass diffusion.

4.7.2 HEATED GAS BUBBLES INSIDE HYDROTHERMAL PORES CAN PHOSPHORYLATE NUCLEOSIDES

Wet-dry cycles with evaporation to dryness as used in the experiments described in Section 4.6 are hard to reconcile with a deep-sea hydrothermal vent scenario. However, similar cycles are possible in hydrothermal pores with gas bubbles and can be used to phosphorylate nucleosides, for example.

Hydrothermal water in porous rocks moves due to thermal and pressure gradients. Gases from magma degassing from depth can infiltrate into the cycling water to form microbubbles. Experiments showed that molecules accumulate on the warm side of the gas-water interface by the continuous *capillary flow* along the capillary walls. At the interface, several prebiotically important processes can occur. It could be shown that accumulated *Hammerhead ribozymes* cleave their RNA substrates more efficiently at the interface. Self-complementary RNA forms hydrogels, DNA is encapsulated in vesicles, RNA nucleosides are phosphorylated and wet-dry cycles induced by evaporation on the warm side of the capillary and condensation on the cold side drive crystallization at the gas-water interface (Figure 4.23). Newly formed crystals act as seeds for new bubbles.

The experiments show that wet–dry cycles at the gas–water interface lead to 20 times more efficient phosphorylation of cytidine nucleosides than in bulk water at similar conditions. There are other possible scenarios of fluctuating water activity in hydrothermal vents, for example, low or controlled water activity from the variation of salt concentration in serpentinizing olivine crystals (Section 2.2), hydrophobic surfaces of biofilms (Section 4.3), swellable clay minerals (see below) and hydrated transition metal ion catalysts (Section 3.1) and catalyst surfaces (Section 2.3). In hydrothermal vents, water activity can vary to a great extent,

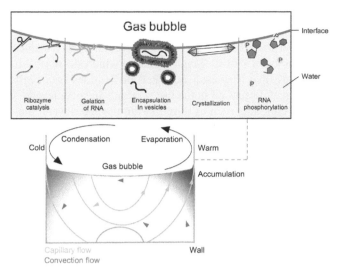

FIGURE 4.23 **Accumulation of biomolecules on the warm side of gas bubbles.** In a rock pore filled with water and a gas bubble, water can flow against gravity to the interface between the pore wall and gas by capillary action. Capillary flow (marked yellow) occurs due to cohesion within the liquid (surface tension) and adhesion of the liquid to the capillary walls. In a narrow tube, the combination of these forces propels the liquid up the walls to an extent which depends on the wall diameter. For example, water rises 70 mm in a 0.2 mm glass tube, but only 0.7 mm in a 2 cm radius tube. Remember also that pull-up of water in porous paper results from capillary forces. In a thermal gradient the combination of convection (drawn dark green) and capillary flow (drawn yellow) leads to the accumulation of molecules at the tip of the meniscus between bubble and wall (drawn red) on the warm side. Focused evaporation of water at the tip enhances molecular flow and accumulation. The evaporation shifts the gas bubble a little and dries molecules in the meniscus at the gas-water interface. The evaporated water condenses on the cold side of the bubble and forms water droplets which shift the gas bubble back to its initial position. Then, the dried molecules dissolve again. Accumulation and wet-dry cycles can trigger important biochemical processes indicated in the inset of the figure and described in the text. Redrawn from Morasch M. *et al.* (2019) Heated gas bubbles enrich, crystallize, dry, phosphorylate and encapsulate prebiotic molecules, *Nat. Chem.* 11: 779–788.

and many different kinds of physical and chemical effects result from that, accordingly.

4.7.3 Ribonucleotide Triphosphates React in Water to Short RNA Strands and Pyrophosphate

Even if high concentrations of nucleotides (nucleotide monophosphates) have accumulated in a cool spot or at a gas bubble, then their polymerization to RNA strands remains difficult, because this process is endergonic. For example, hydrolysis of the phosphate ester bond of AMP in RNA releases about 9 kJ/mol energy such that the reverse polymerization requires at least the same input of energy (activation) to proceed. In solution, hydrolysis of the activated monomers and polymers is the principal reaction pathway and only ~1% yields of dimers and trimers are achieved in equilibrium. In most RNA oligomerization

experiments, imidazole has been used for activation of the nucleotide monomers.

4.7.4 Activated RNA Strands Can Diffuse into Clay Minerals and Condense to Longer Strands

Montmorillonite is a clay mineral formed by the weathering of volcanic ash. Deposits up to 16 m thick have been found in the western United States. In layered minerals like montmorillonite (a magnesium aluminum silicate), RNA oligomers of up to 14 monomer units in length form at the interlayer when the mononucleotides are activated at the 5'-end with a good leaving group like imidazole or pyrophosphate. Montmorillonite consists of two tetrahedral sheets of silica sandwiching a central octahedral sheet of alumina (Figure 4.24). Mg^{2+} and Fe^{2+} often replace Al^{3+} in the octahedral layers. Their lower positive charge is compensated by incorporation of 'exchangeable cations' which are bound to the negatively charged silicate groups and held the layers in a 'deck of cards' structure together. Montmorillonite swells upon addition of water. The amount of expansion is largely due to the type of exchangeable cations contained in the sample. For example, the predominant sodium cation leads to clay swelling to several times its original volume by building up extended hydration shells. Similarly, montmorillonite can adsorb organic compounds and bind them in the interlayers leading to expansion of the clay. Of course, the binding energy of the organics by the interlayer atoms must be larger than the binding energy which holds the layers together.

How does nucleotide polymerization proceed in montmorillonite interlayers? First, activated nucleotides and

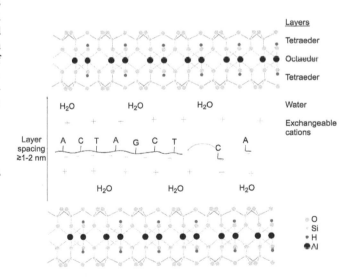

FIGURE 4.24 **Schematic diagram of montmorillonite clay catalyzing RNA oligomerization.** Activated nucleotides add to short primer RNA strands which diffuse from solution in the clay interlayer and get concentrated there. The negatively charged RNA strands interact with the exchangeable cations in the interlayer. Addition of salt and Mg^{2+} ions enhances polymerization. When activated monomers are added continuously to the primer, oligomers with length > 40 arise. The distance between the layers is 1–2 nm and larger. Silicon yellow, oxygen grey, hydrogen purple, aluminum black; activation of nucleotides C and A is indicated with hooks. Redrawn from https://upload.wikimedia.org/wikipedia/commons/0/0c/Montmorillonite-en.svg

short RNA primers from solution diffuse in the interlayer of the clay mineral and get attached there by interaction of the negatively charged phosphates in the RNA backbone with the bound cations. Van der Waals interactions between the purine and pyrimidine rings and the silicate groups strengthen the bonding. Water activity is low at the interlayer—most water is bound to the ions of the clay as hydrated water. Therefore, hydrolysis of RNA oligomers is slow, and polymerization can compete. The addition of salt enhances polymerization. With alkali and alkaline Earth metal ions as exchangeable cations, the catalytically most active clays were obtained. Ferris and coworkers showed that oligomers with length > 40 arise when 5'-phosphor-imidazole activated monomers are added continuously to primer RNA in the catalytically active interlayer. The oligomers contain 2',5'- and 3',5'-links, pyrophosphate linkers, and sometimes cyclic nucleotides.

Polymerization in clay minerals shows some *sequence selectivity*. Initially, 2',5'- and 3',5'- phosphodiester bonded dimers are formed, however, the 3',5'- linked dimers react about five times faster than the 2',5'-linked isomers to give trimers. Therefore, the 3',5'-forms accumulate in the larger oligonucleotides. Purine nucleotides at the 3'-position react about five times faster than pyrimidine nucleotides at that position and therefore also accumulate. Ferris and coworkers found that longer oligo(C)s obtained in montmorillonite initiate the reaction of (imidazole) activated guanine nucleotides to complementary oligo(G)s. Hence, template-directed synthesis of oligonucleotides is possible to some extent, at least for sequences with only one kind of nucleobase. The error rate of replication of mixed sequences has not been investigated yet in montmorillonite. However, it is known that template-directed synthesis is strongly repressed when monomers of opposite handedness are used as templates to synthesize the complementary strand. Indeed, partial inhibition of replication in the form of shorter oligomers was found in montmorillonite-catalyzed polymerization, when racemic mixtures of nucleotide enantiomers (containing D and L ribose) were used. The formation of short RNA oligomers was observed in laboratory analogues of alkaline hydrothermal chimneys based on iron sulfide-silicate precipitates. RNA dimers can be obtained from unactivated adenosine monophosphate (AMP) and oligomers up to tetramers from AMP-activated at the 5'-phosphate by imidazole.

4.7.5 Thermal and pH Gradients Can Drive Cyclic RNA Replication

Notwithstanding the ubiquitous problem of chain-terminating side reactions in the presence of chemicals other than activated ribonucleotides, the experiments described above suggest that prebiotic RNA replication (if it occurred) could have been driven by thermal gradients (Figure 4.25). Complementary strands of DNA and to a minor extent strands of RNA spontaneously aggregate to form double strands. Aggregation is exergonic because the loss of entropy in the orderly aggregates is overcompensated by the hydrogen bond energy released as heat to the environment. At temperatures > 90°C, the double strands dissociate into single strands. Single RNA strands with complementary

nucleobases can internally fold to stem-loop structures, also called 'hairpins'. Depending on the number of hydrogen bonds and their stability, stem-loop structures stay stable at temperatures ≤70°C. Single strands, which do not fold to hairpins, may sequester activated nucleotides (see Figure 4.25 and discussion in the last section). In specific catalytic surroundings such as montmorillonite interlayers, the ranked nucleotides can polymerize to RNA oligomers; however, the error rate may be high.

If we embrace the proposition that concentration-diffusion, capillary-flow, or pressure-gradient induced flow transports newly formed double strands to regions with higher temperatures where they dissociate again, then convection can transport single strands to cooler places where they may form hairpins in case of complementary sequences or become replicated, *provided* that a constant supply of activated monomers is introduced into the system. Otherwise, hydrolysis outpaces polymerization. In such conditions, a thermal loop might be able to replicate RNA strands if activated nucleotides and mineral catalysts are available. However, exponential replication requires a constantly doubling concentration of activated RNA precursors without compounds that induce chain-terminating side reaction, a condition that strains the plausibility of origins scenarios entailing exponential nucleic acid replication. A similar dissociation-replication mechanism can take place in an environment with a strong pH gradient. At high pH >9, guanine deprotonates to an anion (Figure 4.25, right side) and is not able to form a base pair anymore. Due to the loss of guanine's strong hydrogen bonds, double-stranded DNA or RNA destabilizes and dissolves into single strands. At pH < 5, some of the acceptors of hydrogen bonds involved in base pairing are protonated and cannot form hydrogen bonds anymore. Again, double strands dissolve. Hence, nucleic acids can form double strands and hairpins only in a limited pH range of around 7 and at temperatures <90°C. It follows that double-strand dissociation followed by replication may occur in thermal or pH cycles.

We try to avoid the term self-replication, using the more generic term replication instead, because most experiments that involve RNA synthesis from templates and activated monomers (replication) involve some form of catalyst, for example, montmorillonite. Hence it is better to speak of replication, to avoid the impression that the templates require no help from other substances for synthesis of the complementary strand and a second round of error-free synthesis to generate the template (self). Also, we note the strict condition for replication that a constant stream of activated, correctly base pairing and extendable monomers is supplied in such a form that no chain-terminating monomers or chain termination reactions are encountered in a world of prebiotic, nonenzymatic synthesis. That is a more or less unimaginable condition under any real-life geochemical conditions. Even thermophoresis, which effectively concentrates monomers, will concentrate all kinds of monomers without a priori selectivity for their ability to engage in accurate polymerization.

Ligation of small oligomers that hybridize on complimentary templates might be more likely as a mechanism to produce populations of longer polymers than de novo synthesis from monomers, but that mechanism does not lead to 'populations' of homogeneous molecular 'individuals'.

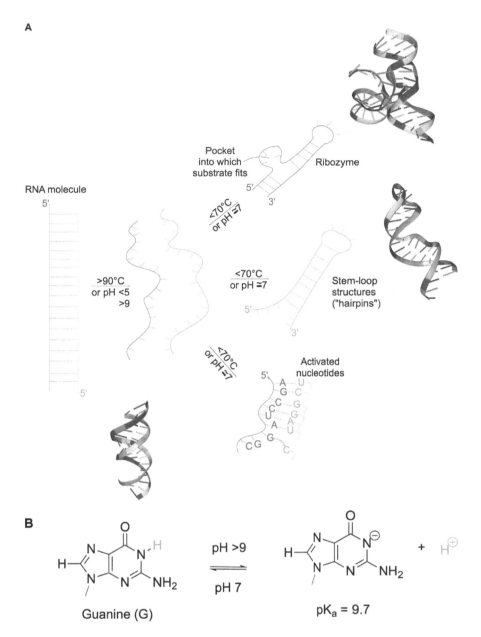

FIGURE 4.25 Schematic depiction of RNA replication in a thermal cycle. (A) Formation of hairpins and replication of RNA in a thermal or pH loop with the help of mineral catalysis. RNA hairpins can form catalytically active loops and pockets into which substrate molecules fit (ribozymes, Section 4.8). (B) At high pH >9, guanine deprotonates to an anion which cannot form base pairs anymore such that double strands dissolve. For further explanation and discussion of the limitations of this scenario (see text). Images from 2MXL and 2OUE in the protein data bank PDB.

We do not expect that RNA phenotypes are formed that can undergo the kinds of replication, natural variation, and natural selection that are germane to traditional versions of the RNA world. In laboratory versions of the RNA world, the accumulation of molecules is not limited by the availability of their monomers but is limited by some form of selective value, whereby synthesis and precursors are taken as a given. In the geochemical context of a real world starting from rocks, water, CO_2, H_2 and a source of N, monomer synthesis and activation is almost certainly going to be the limiting principle governing nucleic acid polymer accumulation. In cells, RNA does not self-replicate, it synthesizes protein. DNA does replicate during proliferation. Nucleic acid replication requires complicated enzymatic machinery (proteins). Nonetheless, there is an extensive literature on RNA molecules that needs to be discussed in the origins context, which we cover in Section 4.8, though with the caveat that there are no replicating or self-replicating RNA molecules in prokaryotic cells.

4.7.6 SINGLE-STRAND RNA FOLDS TO HAIRPINS BY HYDROGEN BONDING OF COMPLEMENTARY BASES

Figure 4.25 displays some RNA stem-loop structures obtained from the protein data bank (www.rcsb.org/). Many RNA structures with internal folding are known, and many have been analyzed by X-ray crystallography and NMR techniques.

Nucleobase tautomers, posttranscriptionally modified nucleobases, non-canonical base pairing, and binding of cofactors enrich the number of possible RNA structures.

Keto-enol and amino-imino tautomerism lead to 20 different tautomers of guanine, for example. The most important guanine, adenine, and cytosine tautomers were identified by isomer and tautomer-specific vibrational laser spectroscopy. RNA and DNA contain nucleobases that have been modified after transcription (a selection is shown in Figure 4.26).

Overall, 93 modified nucleosides have been reported in RNA, 79 of them in transfer-RNA (tRNA), 28 in ribosomal RNA (rRNA), 12 in messenger RNA (mRNA), 11 in small nuclear RNA (snRNA), and 3 in other RNAs.

An estimated 60% of the nucleobases in structural RNA bind in Watson–Crick G···C and A···U base pairs, also called

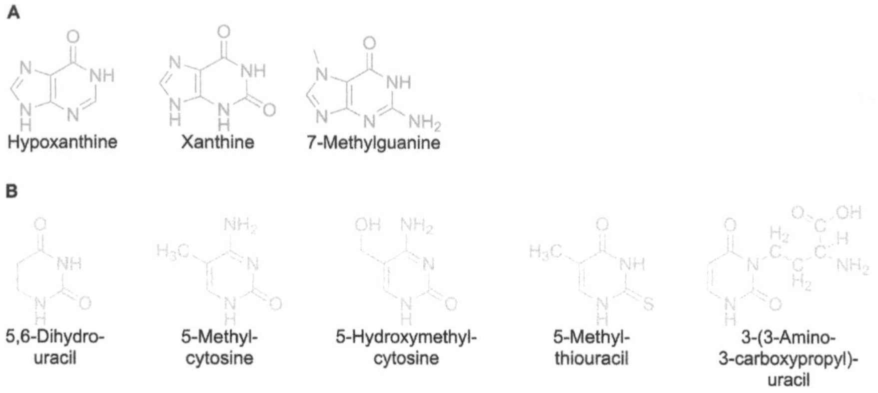

A

Hypoxanthine   Xanthine   7-Methylguanine

B

5,6-Dihydro-   5-Methyl-   5-Hydroxymethyl-   5-Methyl-   3-(3-Amino-
uracil   cytosine   cytosine   thiouracil   3-carboxypropyl)-
                                                                                    uracil

FIGURE 4.26 **Some examples of modified nucleobases.** (A) Purines. Hypoxanthine is produced from adenine and xanthine from guanine through deamination (replacement of the amine group with a carbonyl group). (B) Pyrimidines. Uracil results from cytosine through deamination; 5,6-Dihydrouracil can be obtained from uracil by hydrogenation. Uracil can react with amino acids to 3-(3-amino-3-carboxypropyl)-uracil), for example.

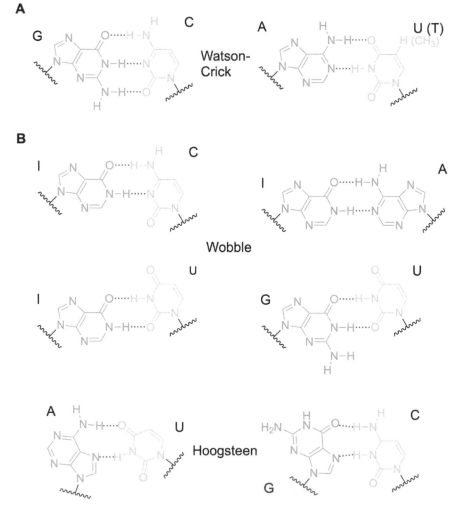

FIGURE 4.27 **Canonical and non-canonical base pairs in RNA.** (A) Canonical Watson-Crick G···C and A···U base pairs. (B) Non-canonical Wobble (I–C: hypoxanthine-cytosine, I–A: hypoxanthine-adenine, I–U: hypoxanthine-uracil, G–U guanine-uracil). I stands for the hypoxanthine nucleoside Inosine. Hoogsteen base pairs (A–U: adenine–uracil; G–C: guanine–cytosine).

canonical base pairs (Figure 4.27A). The most important non-canonical base pairs are *Wobble* and *Hoogsteen* base pairs (Figure 4.27B) which commonly occur in tRNA stems (for example the G···U wobble base pair) and recognition (for example at the third position in the anticodon nucleobase triplets of tRNA). The non-canonical base pairs are typically less stable than the Watson-Crick base pairs and exist in numerous arrangements, including those of the homo pairs G–G, A–A, and C–C. The Hoogsteen base pairs A–U and G–C (Figure 4.27B) exist in equilibrium with the corresponding Watson-Crick base pairs and the physicochemical conditions determine the preferred form.

It is likely that a vast diversity of nucleobases and base pair variations existed during RNA evolution before the modern Watson–Crick base pairs and the modern genetic code developed as prevailing motifs (Section 5.2).

4.8 RNA CAN CATALYZE DIVERSE CHEMICAL TRANSFORMATIONS, BUT NOT REDOX REACTIONS

The 'RNA-world' concept is largely based on the idea that during early evolution, RNA passed information from one generation of molecules to the next (inheritance by replication) and simultaneously catalyzed chemical reactions at origins, which would seem to solve the 'chicken and egg' problem of what came first: protein or DNA. By this means, RNA enzymes (*ribozymes*) could have accelerated early chemical reactions before proteins assumed this task.

Experimentally it could be shown that RNA, selected from a random RNA pool for specific attributes and replicated, can indeed catalyze a number of biochemical reactions. A random RNA pool is a collection of relatively stable RNA strands with an immense number of different shapes, many of them stem-loop structures ('hairpins'). Catalytically active RNA structures (*ribozymes*) can be selected from a random RNA pool by *affinity chromatography* and replicated with the polymerase chain reaction (*PCR*). Hence, affinity chromatography followed by replication of the fittest (or those which are fit enough) is a kind of 'evolution in the laboratory' or in vitro evolution.

We consider ATP affinity chromatography as an example (Figure 4.28). Experiments showed that 17 different RNA sequences were able to bind ATP sufficiently strong to be fixed in the column ('first generation'). They were flushed out with excess ATP, separated, and replicated with the polymerase chain reaction. Replication leads to errors (mutation) of the selected RNA and therefore to a larger pool of partially optimized RNA (*affinity maturation*). This *selection and replication with errors* procedure is repeated several times (say 8 'generations') and the final RNA products are analyzed by NMR or x-ray spectroscopy to obtain their structures. Of the 17 fittest RNA strands, 16 form a loop with the sequence given in Figure 4.28. The loop shapes a deep pocket into which the adenine ring fits. Each of these RNAs binds ATP very efficiently with a dissociation constant of 50 μM (7 mM dissociation products ATP and RNA each for 1M ATP-RNA aggregate, for example).

RNA can be selected with this method for a great number of binding (*aptamer*) and catalysis (*ribozyme*) capabilities.

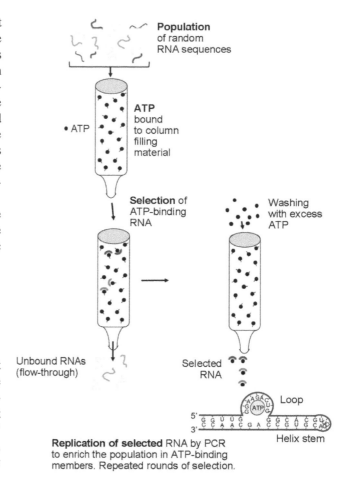

FIGURE 4.28 Selection of RNA by ATP affinity chromatography. A large pool of RNA molecules of random sequences is obtained from combinatorial chemistry; however, the number is generally much smaller than the total possible number of random sequences of an RNA strand of a given length. This limited 'population' is selected for its capability to bind ATP by passing the RNA through a column with ATP bound to the column-filling material. The ATP-binding RNA (drawn blue and red) is released from the column by washing with excess ATP and then replicated with the polymerase chain reaction (PCR). Selection by ATP binding and replication to enrich the population with the fittest members is then repeated several times. The final RNA products are analyzed.

For PCR replication, the selected RNA must be reverse transcribed to DNA first, then the DNA replicated and transcribed back to RNA. For replication, the short single-strand DNA obtained from the selected RNA must be attached to complementary sequences at the ends of a DNA template (*primer*) such that the DNA polymerase enzyme can dock at the template and assemble a daughter DNA strand from free nucleotides. Because both mother (+) and daughter (–) strands produce new strands, the original DNA template is exponentially multiplied (product is doubled at every cycle assuming 100% reaction efficiency).

In principle, by affinity chromatography, a pool of proteins with random structures could likewise be selected for avidity or enzymatic activity. Replication of proteins is more difficult, however, than replication of RNA. Experiments designed to generate sequence variation in proteins are currently only feasible through nucleic acid mutation and subsequent *in vitro* or *in vivo* translation steps using ribosomes. The late Dan Tawfik was a pioneer of such techniques in the study of protein evolution, investigations that shed a great deal of light on the question of how proteins, in particular their catalytic sites, evolve.

4.8.1 IN VITRO SELECTED RIBOZYMES CAN PERFORM MANY ENZYMATIC TASKS

Ribozymes selected by affinity chromatography from a random pool (*artificial ribozymes*) can perform a large variety of enzymatic tasks. The mechanisms of their reactivity are largely unknown and just a few heuristic functions can be listed here. Ribozymes have been shown to catalyze many different kinds of reactions, among them reactions involving RNA monomers and polymers. Examples are illustrated in Figure 4.29. They include:

1. Hydrolysis of cyclic phosphates attached to RNA-ribose at positions 2' and 3'.
2. RNA self-cleavage (Ribozymes can cleave and splice themselves or other RNA).
3. RNA ligation (linear bonding of two RNA strands).

4. RNA branching (branched bonding of two RNA strands leading to 3D networks).
5. RNA lariat formation (condensation to an RNA loop).
6. RNA phosphorylation with thiol-substituted ATP
7. Template-directed RNA polymerization (triphosphate-activated nucleotide binds to the template and to ribose-3'OH of RNA-primer bonded to the template).
8. Pyrimidine nucleotide synthesis (thiouracil binding to diphosphate-activated RNA-ribose).

Ribozymes have also been shown to catalyze reactions with biosynthetic functions. Examples are illustrated in Figure 4.30. They include:

9. Peptide bond formation (a ribozyme simulates peptide (P) and amino acid (A) site of a ribosome to polymerize phenylalanine bonded to the ribozyme via a sulfur bridge with adenylated methionine. Methionine binds to the carboxyl group of biotin, a water-soluble B-vitamin, as an amide; biotin is immobilized by strong bonding to avidin beads).

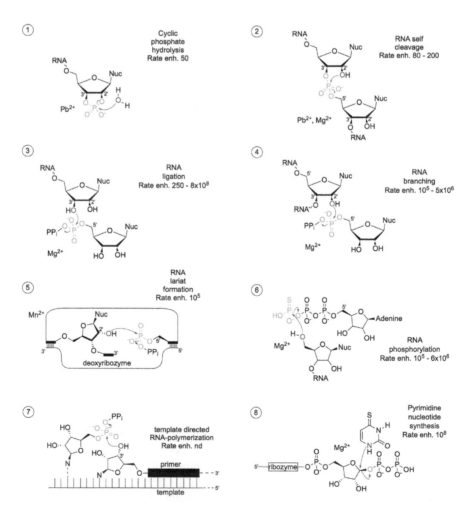

FIGURE 4.29 Artificial ribozymes. Rate enh.: Rate enhancement compared to rate of RNA without catalytic region. Rate enh. nd: Rate enhancement not determined. Pb^{2+}, Mg^{2+}, Mn^{2+}, Ca^{2+} and Zn^{2+} are catalytically active cations. N = Nucleobase, PPi = pyrophosphate, Bio = biotin. For discussion of the ribozyme activities, see text. The numbering in the text corresponds to the numbering in the figure. Redrawn from Silverman S. K. (2007) Artificial functional nucleic acids: Aptamers, ribozymes, and deoxyribozymes identified by in vitro selection. In *Functional Nucleic Acids for Sensing and Other Analytical Applications.* Lu Y., Li Y. eds., Springer, New York, NY.

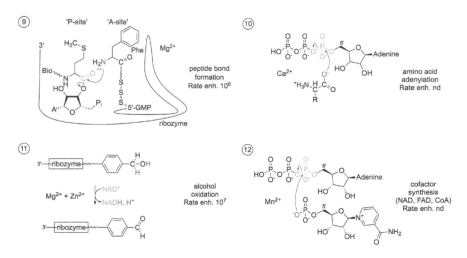

FIGURE 4.30 Artificial ribozymes that catalyze reactions with biosynthetic functions. Abbreviations as in Figure 4.29. For discussion of the ribozyme activities, see text. The numbering in the text corresponds to the numbering in the figure. Redrawn from Silverman S. K. (2007) Artificial functional nucleic acids: Aptamers, ribozymes, and deoxyribozymes identified by in vitro selection. In *Functional Nucleic Acids for Sensing and Other Analytical Applications.* Lu Y., Li Y. eds., Springer, New York, NY.

10. Amino acid adenylation (activation of amino acid as in ribosomal peptide synthesis).
11. Alcohol oxidation (an alcohol covalently attached to a ribozyme is oxidized to aldehyde by NAD+).
12. Cofactor synthesis (for example nicotinamide adenine dinucleotide (NAD+); flavin adenine dinucleotide (FAD), coenzyme A (CoA)). In the figure, NAD+ synthesis is shown.

Ribozymes can also catalyze the aminoacylation of transfer RNA (Figure 4.31). In the first step, phenylalanine derivatives with good leaving groups (LG) are bonded to biotin/avidin beads packed in an affinity column. Then a large number of RNA strands (called an 'RNA library' or 'RNA pool') with 70-nucleotide (nt) long random sequences bonded to the 5'-end of a glutamine transfer RNA (tRNA) is loaded onto the affinity column. Only those RNA strands which bind selectively to the phenyl group (Ph) of the phenylalanine derivative (Phe-LG) (drawn yellow in Figure 4.31) and displace LG with the aligned (drawn green) 3'OH end of their tRNA (drawn red) are fixed to the solid package of the chromatographic column. In the next round, from the random RNA pool ribonuclease-like ribozymes are selected which cleave tRNA-Phe-biotin from the Phe-ribozyme. The fittest Phe-ribozymes are PCR amplified, sequence analyzed and a 'mini-ribozyme' is constructed from their most important sequences. The 'mini-ribozyme' is further selected for Phe recognition. In further experiments, Phe derivatives with hydrophilic leaving groups were used which raised the water-solubility of amino acids with hydrophobic side chains like phenylalanine.

In other RNA affinity chromatography experiments, Michael Yarus and coworkers selected a 29-nt long RNA catalyst which successively forms the aminoacyl ester Phe–RNA, and then peptidyl-RNA (Phe–Phe–RNA) with phenylalanine adenylate (Phe–AMP) as substrate. Hence, artificial ribozymes managed important parts of translation. The authors suggest that uncoded, but RNA-catalyzed peptide synthesis preceded coded peptide synthesis.

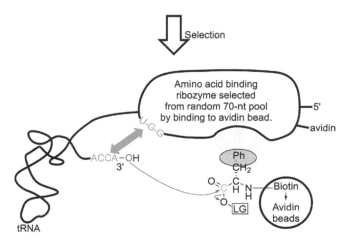

FIGURE 4.31 Ribozyme directed aminoacylation of tRNA. For explanation, see text. Redrawn from Suga H. *et al.* (2011) The RNA origin of transfer RNA aminoacylation and beyond, *Phil. Trans. R. Soc. B* 366: 2959–2964.

4.8.2 SELECTION OF RIBOZYMES FROM A RANDOM RNA POOL REQUIRES VERY SPECIAL CONDITIONS

In the laboratory, ribonucleic acids can both self-replicate and catalyze the reactions discussed above. This suggests an early *RNA-world* preceding transcription and translation of nucleic acids to protein enzymes. Is the scenario of a primordial RNA world realistic? There are several arguments against this tempting hypothesis.

First, the synthesis of the RNA building blocks is complicated and needs several steps at very specific conditions. Especially a one-pot synthesis of the purines under prebiotic conditions is hard to imagine. The synthesis of amino acids is much easier to accomplish if starting from CO_2 and NH_3. Recall that the biochemical synthesis of nucleotides is based on amino acids, suggesting that their synthesis and polymerization to peptides preceded RNA chemistry.

Second, ribozyme selection from a random RNA pool needs complicated PCR replication with sophisticated enzymes, which were not available at the origin of life. In our view, accumulation of amino acids by thermophoresis, convection, and capillary flow, spontaneous polymerization to polypeptides in mineral sheets and structural selection of the polypeptides by geochemical conditions (salt, pH, interaction with mineral surfaces, and biofilms) is a more realistic setting for generation of the first enzymes (though not genetically encoded enzymes) at the origin of life. Further on, biologically active peptides can be synthesized on protein templates without RNA (Section 5.1). Hence, reproducible replication is not unique to nucleic acids, peptides can do it too.

Third, ribozymes are negatively charged and hence have problems binding to negatively charged or hydrophobic substrates. By contrast, amino acids have charged, polar, and nonpolar side chains and peptides can bind all different sorts of substrates.

Therefore, it seems more realistic (to us) that solid-state transition metals, soluble transition metals (ions), small molecules (amino acids and cofactors), and peptides dominated catalysis at the origin of life. Ribozymes probably played an (unavoidable) role in (self-)cleaving of RNA and molecular interactions related to translation in (proto-)ribosome-like structures. However, their role as biosynthetic catalysts in the context of metabolism was possibly, or probably, always as limited as it is today. In modern cells, catalytic activities of RNA are few. Ribozymes are sometimes associated with viruses, for example, hammerhead ribozymes that catalyze RNA cleavage and ligation. In prokaryotes, the RNA component of RnaseP (Ribonuclease P cleaves RNA) is active in tRNA processing. The peptidyl transferase activity of the ribosome is an RNA-dependent activity. Splicing reactions are RNA-dependent activities. RNA activity in modern cells is usually involved with processing RNA.

One could argue that RNA-dependent catalytic activities were more widespread in early evolution and were replaced by proteins and cofactors. But if we look at the catalytically active moieties of cofactors today, there is nothing to suggest that the bases, sugars, or phosphates of RNA could easily catalyze such reactions, especially redox reactions. Perhaps most tellingly, the ability of RNA to tightly coordinate transition metals as reactive site catalysts is so far unknown. And given that so many cofactors are required for the synthesis of the bases themselves (Figure 3.18), caution is needed in the context of extrapolation of what RNA *can* do under laboratory conditions and what it *did* do or might have done in early evolution. At the same time, recent findings from Thomas Carell's group underscore the concept of a *modified* RNA world in which modified bases germane to tRNA in translation today—covalently carrying amino acids and other functional moieties, with properties typically associated with modern peptides—could have vastly expanded the early catalytic repertoire of RNA molecules. This could have also included the involvement of cysteine residues and transition metal coordination.

The chemistry of catalytic RNA and its ability to be replicated and selected has opened vast new fields of exploration for chemists, in much the same way that gene technology or more recently gene editing gave biologists fundamentally new tools to manipulate cells and organisms. The question is how heavily these technologies bear upon life's origin, if at all, as opposed to showing us what is possible and how we can learn more about the chemistry of life. As it stands, RNA in prokaryotes today is mainly involved in processing itself and in catalyzing the peptidyl transferase reaction. It is possible that it has always been that way and that the synthesis of RNA from amino acids, cofactors, and sugars in metabolism recapitulates its origin in much the same way that DNA is derived from RNA in metabolism. Indeed, the DNA-from-RNA biosynthetic polarity via the enzyme ribonucleotide reductase has always been part of the argument that RNA came before DNA in evolution. By the same measure, RNA likely came from small molecule networks in evolution, as opposed to a central tenet of the traditional RNA world theory that RNA generated metabolism. In general, the bases of RNA do not function in redox reactions in metabolism. There are no known enzymes that use RNA bases to coordinate transition metals for catalysis involving CO_2 and N_2 fixation or H_2 activation, yet in metabolism, transition metal catalyzed reactions are essential for RNA synthesis. This suggests that transition metal-based catalysis preceded RNA-based reactions in biochemical evolution.

* * *

4.9   CHAPTER SUMMARY

Section 4.1 At the dawn of life there were no sophisticated enzymes to catalyze metabolic reactions. Transition metal catalysts produced by serpentinization could have enhanced the synthesis of biomolecular building blocks and their polymerization via autotrophic routes. For some reactions, phosphorylating or dehydrating agents were necessary for the activation of the reactants. 2-Oxoacids from the acetyl-CoA pathway and tricarboxylic acid cycle react with activated ammonia compounds and electrons from Fe^0 or with ammonium salts and freshly precipitated FeS_2 or $Fe(OH)_2$ as electron donor and catalyst to diverse biological amino acids.

Section 4.2 For efficient polymerization to peptides, amino acids must be activated by phosphorylation, thiolation, or anhydride formation with COS, HCO_3^- or other dehydrating agents. COS reacts on a FeS_2 catalyst surface selectively with the amino group and not with the carboxyl group or non-amino side chain groups to form an activated cyclic N-carboxy anhydride, which reacts with another amino acid to generate a dipeptide after decarboxylation. Activated peptides are remarkably stable for hydrolysis. Water activity affects peptide structure: in pure water, alternating homochiral leucine–lysine poly-peptides fold as random coils whereas in a salt solution they fold to thermostable and hydrolysis-stable bilayers of β-sheets with the hydrophilic side chains of lysine pointing outwards into the salt solution and the hydrophobic side chains of leucine pointing inwards. Likewise, larger proteins from homochiral acidic and simple hydrophobic amino acids form hydrophobic core/pocket structures in their interior while protecting hydrophilic outer layers in salt solutions. Homochirality of the amino acids is necessary for this selective folding but a food-generated autocatalytic network can introduce chirality into the metabolism of protein synthesis in a very natural way. High salt concentration and low water activity in serpentinizing olivine micropores may have enhanced the polymerization of activated amino acids to thermostable and halophilic proteins with β-sheets and α-helices which can incorporate mineral cofactors for catalysis. The evolution of catalysts probably started with inorganic mineral catalysts in the solid phase, followed by inorganic catalysts like Fe(II) in solution, followed by simple organocatalysts such as amino acids in solution, later more complex cofactors like NADH and finally peptides capable of interacting with these catalysts.

Section 4.3 The electron transfer protein ferredoxin is a very ancient polypeptide with an iron-sulfur cofactor. Ferredoxin is a very ancient protein. Ferredoxin from *C. pasteurianum* contains internal duplications as an example of nature starting with small and simple molecules and then progressing to larger more complex structures. A variety of primordial polypeptides containing cysteines could bind iron-sulfur clusters sustainably. Even without replication, a collection of simple polypeptides with nonrandom sequences can be obtained by the criterion of structure-dependent stability against hydrolysis. An important component of the catalytic function in modern enzymes is to exclude water from the active site. The amino acids glycine and alanine are easily synthesized under abiotic conditions and their hydrophobic peptides might have accumulated at primordial vents coating their walls together with the 'tar' encountered in many prebiotic synthesis experiments. These hydrophobic layers could have gradually sequestered abiotically synthesized hydrocarbons and peptides with deposition driven by thermal gradients. Organic-coated inorganic cell walls may have been the precursors of biological cell walls.

Section 4.4 Sugar in autotrophs is synthesized by gluconeogenesis which starts from pyruvate, a stable product of CO_2 hydrogenation. The nonenzymatic aldol condensation between D-glyceraldehyde-3-phosphate (G3P) and its isomer dihydroxyacetone phosphate (DHAP) yields the C6 sugar D-fructose-1,6-biphosphate, an intermediate in gluconeogenesis. G3P is obtained from pyruvate by phosphorylation, isomerization, and reduction. Pyruvate can, in principle, react nonenzymatically with glyoxylate at reducing conditions to different forms of the sugar ribose but this reaction has not yet been demonstrated.

Section 4.5 The synthesis of nucleotides in cells is based on condensation of amino acids and phosphorylated ribose. Condensation of anhydrous glycine, alanine, and lysine by heating leads to several heterocyclic products including pterin riboside which has the same three hydrogen bonding capabilities as guanine and may well have belonged to its precursors in chemical evolution. Most nucleobases do not dissolve well in water, suggesting that nucleotides must have been incrementally synthesized from water-soluble ingredients like amino acids and phosphate-activated ribose. Only the nonenzymatic cyclization of carbamoyl phosphate with aspartate to orotate as a precursor of uridine- and cytidine triphosphate has been demonstrated experimentally so far.

Section 4.6 Cyanide- and formamide-based generation of nucleobases and nucleosides has been amply demonstrated in the laboratory. None of these syntheses or variants thereof, however, are observed in modern organisms, even the most primitive ones. Hence nature either never traversed the cyanide or formamide routes or they were dead ends that were overrun by CO_2, NH_3 and H_2-dependent syntheses.

Section 4.7 With fluorescence energy transfer (FRET) microscopy it could be shown that thermophoresis accumulates nucleotides and RNA strands in water-filled pores. Also, heated gas bubbles inside hydrothermal pores can enrich and phosphorylate nucleosides. Accumulation and activation by phosphorylation are prerequisites for nucleotide condensation to RNA strands. Short primer RNA strands and mononucleotides can diffuse into swellable clay minerals like montmorillonite (a magnesium aluminum silicate) and concentrate there. They condense to longer strands in the clay interlayers when the mononucleotides are activated at the 5'-end with a good leaving group like imidazole or pyrophosphate. Thermal and pH gradients can in principle drive cyclic RNA replication by dissociation of double strands to single strands at higher temperatures and mineral-catalyzed addition of complementary activated mononucleotides to the single strands at lower temperatures. The single RNA strands can fold into stem-loop structures ('hairpins') which may be enzymatically active. RNA replication however requires a constant stream of activated, correctly base pairing and extendable monomers such that no chain termination reactions are encountered in a world of prebiotic, nonenzymatic synthesis. This scenario is unlikely, therefore. In cells, RNA does not replicate, it synthesizes protein. DNA does only replicate during proliferation which requires complicated enzymatic machinery (proteins). Nucleobase tautomers, post-transcriptionally modified nucleobases, non-canonical base pairing, and binding of cofactors enrich the number of possible RNA structures. Probably, very diverse nucleobases and base pair variations were realized during RNA evolution before the Watson–Crick base pairs and the modern genetic code developed as prevailing motifs.

Section 4.8 Catalytically active RNA structures (ribozymes) can be selected from a random RNA pool by affinity chromatography and replicated with the polymerase chain reaction (PCR). For example, RNA is selected for its capability to bind ATP by passing the RNA mixture through a column with ATP bound to the column-filling material. The ATP-binding RNA is released from the column by washing with excess ATP and then replicated with PCR. Ribozymes selected by this method can perform a large variety of enzymatic tasks. The selection and catalysis observed in the laboratory suggest an early RNA world preceding translation of nucleic acids to protein enzymes. Selection of ribozymes from a random RNA pool requires very special conditions, however. Nucleotides are difficult to synthesize compared to amino acids, PCR replication requires sophisticated enzymes and ribozymes are negatively charged and hence have problems binding to negatively charged or hydrophobic substrates, moreover they cannot catalyze the metabolic redox reactions. RNA activity in modern cells is usually involved with processing RNA. However, modified bases in RNA strands can bind amino acids, which could have expanded the early catalytic repertoire of chemically modified RNA molecules.

* * *

PROBLEMS FOR CHAPTER 4

1. *Synthesis of peptides*. Describe four different reaction paths to condense amino acids to peptides.

2. *Hydrolysis of activated peptides*. Why are activated (phosphorylated) peptides less sensitive to hydrolysis at alkaline conditions than activated (phosphorylated) amino acids?

3. *Peptide structures*. How does water activity in salt solutions affect peptide structure?

4. *Homochirality of peptides*. How might have homochiral peptides originated during the evolution of peptide synthesis?

5. *Catalyst evolution*. Discuss the possible order of catalyst development during molecular evolution.

6. *Inorganic cell walls*. List some arguments that speak in favor of micropores in serpentinizing olivine as possible precursors of biological cells.

7. *Peptide catalysis with inorganic cofactors*. How can random peptides coordinate transition metals for catalysis?

8. *Sugar synthesis*. Specify modern and possible primordial nonenzymatic pathways for sugar synthesis.

9. *Nucleotide synthesis*. What molecules react to form nucleotides (a) during metabolism in biological cells and (b) in the laboratory?

10. *RNA polymerization*. In prebiotic chemistry experiments, where and how do RNA strands polymerize into longer strands without the help of enzymes?

11. *RNA replication*. Is continuous (cyclic) RNA replication a realistic scenario in the prebiotic world?

12. *Ribozyme selection*. How can catalytically active RNA structures (ribozymes) be selected from a random RNA pool in laboratory experiments?

13. *Ribozyme catalysis*. Give some examples of reactions that ribozymes catalyze and some that do not.

5 Template-Directed Synthesis of Polymers

5.1 DARWINIAN-TYPE EVOLUTION REQUIRED REPLICATION OF INFORMATIONAL POLYMERS

5.1.1 BIOLOGICALLY ACTIVE PEPTIDES CAN BE SYNTHESIZED ON PROTEIN TEMPLATES WITHOUT RNA

In Section 4.2, we saw that changes in water activity, also in hydrothermal systems, can lead to the synthesis of activated amino acids and salt-tolerant proteins. The formation of important structural features such as β-sheet or α-helix could have been influenced by geochemical conditions that impact salt concentration and water activity, leading to the accumulation of peptides enriched in stable structures. We also saw that the information needed for proteins to fold is contained within their amino acid sequence and that hydrophobic packing and hydrophilic/electrostatic interactions can generate protein structures without external folding factors. In environments where water activity fluctuates, such as in serpentinizing systems, this can lead to small proteins of nonrandom sequence.

Peptide structures persist, however, only if resynthesis of a specific peptide is feasible or if a peptide can synthesize a copy of itself. There are modern examples from cells that go in this direction but with limitations. A 32-amino acid α-helical peptide, a domain of a yeast transcription factor, can template its own 'synthesis' by autocatalytic ligation of its two constituent fragments, a 15-residue, and a 17-residue fragment, if its thioactivated fragments are added to the peptide template in water at pH 7. The aggregation of the fragments with the template is driven by hydrophobic packing interactions of leucine and methionine and electrostatic interactions of arginine and glutamic acid, forming interhelical recognition surfaces. By this aggregation, the thioester-promoted condensation of the fragments is accelerated and a 'self-replicating' peptide is formed. Product formation followed a sigmoidal growth pattern as expected for autocatalysis (Section 3.2). The reaction rate was accelerated ~500-fold relative to the background rate constant of peptide formation without a template. Is this example relevant to origins? Perhaps. In the context of prebiotic RNA synthesis, there are two ways to get larger molecules: polymerization of activated mononucleotides or ligation of preexisting oligonucleotides. The ligation of small oligonucleotides from larger molecules is an accepted mechanism in RNA chemistry; the yeast transcription factor fragments provide an analogous example for peptides. In this context, hydrophobic interactions could allow β-sheets with hydrophobic side chains to bind small activated peptides of hydrophobic amino acids if the recognition surface is large enough. Similarly, β-sheets with acidic side chains would require

bivalent cations like Mg^{2+} for alignment and polymerization $(X^-\cdots Mg^{2+}\cdots X^-)$.

Peptide synthesis directed by a peptide template was possibly an intermediate in the evolution of protein synthesis before the origin of the ribosome because it provided a means to generate populations of peptides with either (1) specific sequences or (2) similar physicochemical properties, without the need for the existence of a genetic code. The classical example of a protein-directed synthesis of peptides with specific sequences and properties is nonribosomal peptide synthesis, NRPS. NRPS is very widespread among modern prokaryotes, where its biological function is to synthesize peptides as secondary compounds (typically antibiotics). In contrast to protein synthesis on ribosomes, NRPS often incorporates D-amino acids into peptides and operates without RNA. It has long been known that homogenates from the bacterium *Bacillus brevis* synthesize the cyclic decapeptide antibiotic Gramicidin S. The synthesis proceeds without RNA-based translation because it is resistant to treatment with RNAse, which cleaves RNA strands. Instead, the bacterium uses an enzyme complex consisting of several proteins as a template for the decapeptide.

5.1.2 NONRIBOSOMAL PROTEIN SYNTHESIS (NRPS) WAS A POSSIBLE PRECURSOR TO TRANSLATION

As originally proposed by Fritz Lipmann, NRPS proceeds in three steps: activation, thioester formation, and condensation. The first step involves activation by reaction of an amino acid with ATP to form an aminoacyl-adenylate (aminoacyl-AMP). We will later see that the same kind of activation takes place in ribosomal protein synthesis. The activated amino acid binds to an amino acid-specific SH-containing group of the NRPS complex (Figure 5.1). In this way, the amino acids are aligned in sequence according to thioester formation by SH groups in the complex. Peptide bonds are then formed between the amino acids (both D and L can be incorporated) by a carrier protein in the complex with a *phosphopantetheine* prosthetic group: a stretched, 20 Å long thiol. The pantetheine arm with a dipeptide ($A^1 \cdot A^2$) linked to its thiol terminator binds its thioester-activated carboxyl to the free amino (NH_2) group of the next peripherally bound amino acid. After transpeptidation and transthiolation, the pantetheine arm with the newly elongated tripeptide attached swings to the next thiolated amino acid, which is lined up for polymerization to the tetrapeptide. This procedure continues until the terminal amino acid is reached and the synthesis of the cyclic peptide closes by reaction with the N-terminal amino group.

The sequential formation of peptide bonds resembles transpeptidation and translocation in ribosomal protein

DOI: 10.1201/9781003378617-5

A

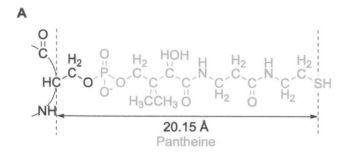

20.15 Å
Pantheine

B

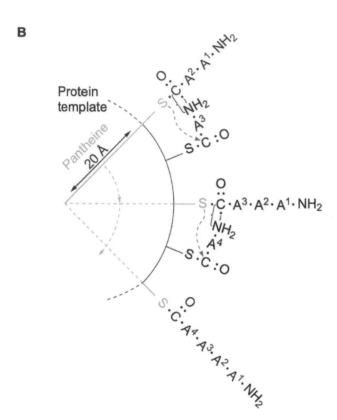

FIGURE 5.1 **Protein-directed peptide synthesis using a pantetheine arm for transpeptidation.** (A) The rotatable pantetheine arm. (B) The amino acids are activated by adenylation with ATP/Mg2+, then noncovalently linked to the enzyme, and finally covalently bound to an enzyme SH as thioester (–S–CO–). The dotted lines and arrows (drawn blue) show the motion of the pantetheine arm collecting the peripherally bound amino acids (arrows drawn red). A1–A4 denotes the respective rest of the amino acid residue without specification. For further explanations, see text. Redrawn from Lipmann F. (1973) Nonribosomal polypeptide synthesis on polyenzyme templates, *Acc. Chem. Res.* 6: 361–367.

synthesis (Figure 5.4), while the sequence of pantetheine-mediated thioester condensation steps resembles bacterial fatty acid synthesis. Although the principle of protein-dependent synthesis of specific peptides very well could reflect a genuinely primordial mechanism of reproducible expression of information into peptide sequences, it should be stressed that no modern NRPS system can be a direct holdover of processes that predate the ribosome, because all modern NRPS enzyme complexes are synthesized on ribosomes using genes and the genetic code. However, if peptides with appreciable NRPS activity were among the nonrandom sequences generated by fluctuating water activities, for example, then they would provide a seed of function that would lead to more highly sequence-specific (nonrandom) peptides with distinct stereochemical and catalytic properties. Moreover, because it is a *catalyst*, the first appreciable NRPS activity generates many products with a very similar sequence and, if they have catalytic activity, a similar catalytic activity. Recalling what we learned in Chapter 1, enzymes just accelerate reactions that tend to occur anyway, such that products of NRPS could readily become incorporated into autocatalytic network elements or form new ones. Modern products of NRPS generally do not have relevant catalytic activities in metabolism; instead, they tend to act as specific inhibitors of non-self proteins. Interactions with metabolism in the cells in which they are made are likely counter-selected during evolution.

Figure 5.2 displays a more modern and detailed representation of NRPS of the decapeptide Tyrocidine by the bacterium *B. brevis*. The biosynthesis of Tyrocidine is analogous to Gramicidin S and is achieved by three synthetase proteins with ten modules in all, one for each amino acid of the decapeptide. Each protein module contains subdomains for adenylation (A), subsequent transfer of the activated amino acid to the phosphopantetheinyl cofactor of the neighboring thiolation domain (T), followed by peptide condensation (C), in specific modules epimerization (E) and finally cyclization (Te).

Figure 5.3 shows the different processes A, T, C, and Te in more detail.

The structural diversity of nonribosomal peptides produced by bacteria is enormous, and many of them are medically relevant. For example, β-*lactam* (cyclic amide) antibiotics include *penicillins* and *cephalosphorins* and account for more than half of all prescribed antibiotics worldwide. These antibiotics are produced by different fungi and bacteria by the formation of the common intermediate *isopenicillin N* from a tripeptide synthesized by NRPS. The β-lactam antibiotics inactivate the transpeptidation reaction of cell wall biosynthesis, leading to cell lysis. Further examples of medically relevant NRP are cyclic lipopeptide antibiotics (*Daptomycins*) for the treatment of skin infection, *Cyclosporin A* with anti-inflammatory and immunosuppressant activity to prevent graft rejection and treat autoimmune diseases, glycopeptide antibiotics such as *vancomycin*, the antitumor and anti-HIV active quinoxalines, *capreomycin* with antibacterial activity against multidrug-resistant strains of *Mycobacterium tuberculosis*, and the anticancer active *Bleomycin* glycopeptides. Numerous pathogenic microorganisms obtain iron (Fe^{3+}) by secretion of iron-chelating molecules called *siderophores* and their readsorption after Fe^{3+} charging. Siderophores are typically nonribosomal peptides and form soluble Fe^{3+} complexes that can be taken up by the cell via active transport.

The structural diversity produced by NRPS is caused by the number of modules used and their order, the amino acids recognized and activated by their A domains, and the addition of modifying domains such as E domains in selected places. Nonribosomal peptides can aggregate

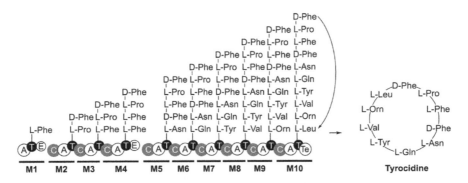

FIGURE 5.2 A more modern representation of NRPS of Tyrocidine from *Bacillus brevis*. Schematic representation of tyrocidine NRPS with ten modules (M) distributed over three peptide synthetase enzymes. Orn is ornithine. Note the presence of D amino acids in the peptide product. The subdomains of the protein synthetases accomplish activation of the amino acids by adenylation A, transfer of the adenylated amino acid to the thiolation domain T with phosphopantetheine as acceptor (see Figure 5.1A), peptide condensation C and depending on the amino acid position, epimerization E (changing one asymmetric center, for example, epimerization of L-Phe to D-Phe), and finally cyclization Te. The thiolation domain T with pantetheine is used repeatedly in every module and not only once as a swiveling arm as in Figure 5.1. For further assignment, see text. Redrawn from Feinagle E. A. *et al.* (2008) Nonribosomal peptide synthetases involved in the production of medically relevant natural products, *Mol. Pharm.* 5: 191–211.

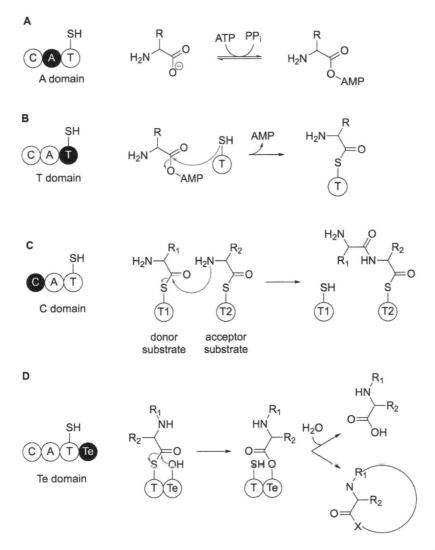

FIGURE 5.3 Reactions catalyzed by the core domains of NRPS modules. (A) Adenylation domain, catalyzing the formation of aminoacyl AMP by nucleophilic substitution; (B) thiolation domain, catalyzing the formation of the aminoacyl thioester by nucleophilic substitution; (C) condensation domain, catalyzing the formation of a peptide bond; T1 and T2 designate T domains from neighboring NRPS modules; (D) thioester domain, catalyzing the aminoacylester formation on the Te domain and then either hydrolysis or cyclization of the peptide. The 'X' depicts either nitrogen or oxygen. The '–S' represents the pantetheine cofactor bonded to each T domain. Redrawn from Feinagle E. A. *et al.* (2008) Nonribosomal peptide synthetases involved in the production of medically relevant natural products, *Mol. Pharm.* 5: 191–211.

with modifying enzymes such as transferases and oxidases which vary the NRP further. Nonproteinogenic amino acids extend structural diversity and all enzymes for their synthesis (NRPS and modification) are coded within the same structural gene or *operon*.

5.1.3 RNA-INDEPENDENT PEPTIDE SYNTHESIS COULD PROMOTE THE ORIGIN OF COMPLEXITY

There are many reasons to think that a form of nonribosomal peptide synthesis served as a precursor to ribosomal protein synthesis before the existence of the genetic code. However, this line of thinking confronts us with the problem that there are no transitional forms of the ribosome known, and all known forms of NRPS are performed by proteins that are synthesized on ribosomes. Yet it is generally true that the evolution of molecules operated forward in time from the simple to the complex. That is, the ribosome is an example of a complex RNA molecule that cannot have arisen overnight in one step. There had to have been transitional intermediates that were smaller and less complex and that led to the origin of rRNA and the ribosome. For example, it is possible that the original 16S and 23S rRNAs that form the core of the prokaryotic small and large ribosomal subunits, respectively, were each initially composed of smaller RNA molecules that aggregated and folded to assemble a functional subunit. This happens today in the mitochondria of the photosynthetic protist *Euglena gracilis*, where both the large and small subunit rRNA genes are split such that each subunit is composed of two smaller rather than one large rRNA molecule.

The *Euglena* example represents a highly derived trait, not an ancient trait. However, it serves to illustrate the principle that combinations of RNA molecules can aggregate to form larger complexes with novel structures and functions. Though the ancestral ribosome did not have to be as big as the modern one, the problem remains that we have no evolutionary intermediates that would help us reconstruct the origin of the ribosome. That is, there is no evolutionary grade of primitive to advanced ribosomes in prokaryotes that could serve as a model for intermediate stages in ribosomal evolution to help us better understand the origin of such a complex structure. In Chapter 6, we will however see that, within modern ribosomes, conserved elements of structure and function can be pinpointed that permit inference of what a proto-ribosome might have looked like and how it might have operated. In this chapter, we will see how components of the genetic code might have emerged from interactions of small molecules.

The problem of 'no intermediate forms' is very general in the field of early evolution. The same problem is encountered at the origin of the ATP synthase and at the origin of cells. There are no proto ATP synthetases of the rotor-stator type; nor are there protocells that have a complexity less than that of a free-living cell. Nonetheless, we know that some simpler intermediates must have existed, even though there are no molecular descendants around.

How does NRPS figure into this context? It provides a conceptual intermediate. NRPS complexes can reproducibly synthesize small peptides with remarkable sequence specificity. They achieve this through the recognition of specific amino acids by SH-containing groups of the complex and the sequential activity of condensation reactions that connect both D- and L-amino acids via peptide bonds. Before the origin of the genetic code, it is possible that one or many spontaneously synthesized peptides (for example, one resistant to hydrolysis; see Section 4.2) possessed such a rudimentary amino acid condensing activity. Such fortuitous activities of enzymes are far more common in modern proteins than one might think. Such side activities of proteins are collectively termed promiscuity, and they can be a property of either the canonical active site of the enzyme or another site of the protein.

The initial amino acid condensing activity of a primordial NRPS catalyst need not accelerate the reaction to the same degree that an enzyme does. Recall that in the case of reflexively autocatalytic food-generated networks (RAFs, Chapter 3), acceleration on the order of 10% above the noncatalyzed reaction rate can be sufficient for a RAF to emerge. If the initial NRPS activity has some degree of sequence specificity in the NRPS product, that is, if the products of the reaction it catalyzes are peptides similar in sequence, then a collection of related sequences and structures can ensue. This all assumes, of course, that activated monomers capable of polymerization are constantly provided as food for the reaction network. There is no heredity for such an activity, but it can organize activated monomers. If one of the products of the first round of NRPS has an NRPS activity, then it can amplify the polymerization process such that more monomers are channeled into peptides. This process, viewed as a sequence of rounds in which peptide synthesis was an activity of an NRPS product, gives rise to an autocatalytic cycle in that products of one step promote the synthesis of products in the next (Figure 5.4).

NRPS can, in principle, lead to molecular amplification of organic catalysts. How does peptide autocatalysis square off with RNA-based concepts of molecular amplification? In Figure 4.29 we presented an example of template-directed RNA polymerization by an artificial ribozyme. A triphosphate-activated nucleotide binds to the template via G···C and A···U and to ribose 3′OH of an RNA primer bonded to the template. One at a time, the primer sequence is elongated. While these results seem promising at first sight, they require—in addition to pure and activated nucleotides—PCR-like amplification and specific polymerization, which were not available at the origin of life. This, in addition to the circumstance that bases stem from amino acids and cofactors in biosynthetic metabolism, is another reason to think that a simple and versatile form of peptide catalysis probably preceded RNA catalysis in chemical evolution. RNA catalysis was furthermore likely limited to a few functions such as (self-) cleavage before ribosomal peptide synthesis evolved, where ribozymes play a decisive role.

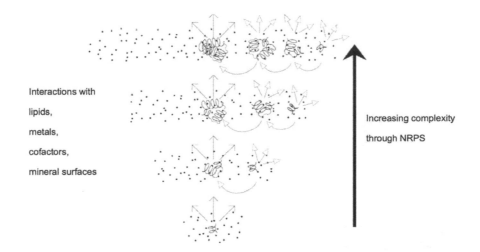

Interactions with
lipids,
metals,
cofactors,
mineral surfaces

Increasing complexity
through NRPS

FIGURE 5.4 **NRPS can generate catalysts and autocatalytic loops.** Given activated monomers, a basic NRPS activity can, in principle, organize amino acids into specific sequences, some of which can also possess or enhance NRPS activity (feedback arrow in the figure). Then, an autocatalytic process results that organizes amino acids into specific structures but without a heritable basis. This autocatalysis will lead to exponential accumulation (not shown in the figure) of the NRPS-enhancing peptides if there are no processes which remove them. At the same time, the peptides can impact small molecule reactions if they catalyze or decelerate them. This kind of activity could have led to locally confined enzyme-like activities that could have promoted the emergence of molecular complexity before the origin of the ribosome. See text.

5.2 THE ORIGIN OF RNA-TEMPLATE-DIRECTED PEPTIDE SYNTHESIS IS STILL UNRESOLVED

5.2.1 TRANSLATION ON RIBOSOMES IS PERHAPS THE MOST COMPLEX PROCESS CHEMISTRY EVER INVENTED

Of all novelties ever recorded in evolution, ribosomal protein synthesis was probably the most important breakthrough invention. This event coupled retrievable and heritable information to evolvable catalytic function. It allowed peptides to be synthesized by sequential addition of amino acids, one at a time, coded by nucleobases in an RNA strand (the precursor of modern *messenger RNA*, abbreviated mRNA). Ribosomal peptide synthesis is shown schematically in Figure 5.5 and explained in the legend.

The origin of ribosomal peptide synthesis is not known, and several pathways have been proposed to explain its evolution. The origin of translation directly involves the origin of the ribosome and the origin of the universal genetic code. Among experts, there is still no consensus of any appreciable depth or breadth concerning the origin of the code. Hundreds of papers have been written about the origin of the code, and each one tends to relate a different narrative. Some references with proposals for the origin of the ribosome and the code can be found in the bibliography of this book; we will consider four models.

The genetic code is organized as triplet codons. The standard universal code has 61 sense codons (ones that specify the 20 universal amino acids) and 3 termination codons. The 61 sense codons in mRNA recognize tRNAs, and the termination codons can recognize *release factors*, proteins that have a similar size and shape as tRNA but terminate the peptide synthesis process. Some organisms incorporate selenocysteine via tRNA as a 21st amino acid; some organisms incorporate pyrrolysine via tRNA as a 22nd amino acid. Most amino acids are specified by more than one synonymous codon, and many of these synonymous codons are recognized by tRNAs that harbor specific modified bases in their anticodon loop and elsewhere in the molecule.

For many decades scientists have pointed out that there are many correlations between the properties of amino acids and the properties of bases that make up the codons. The crux of the code, and arguably its origin, is the physical and chemical mechanism that associates specific amino acids with specific tRNAs. Today that is the function of aminoacyl tRNA synthetases (AARSs), of which there are roughly 20, one for each amino acid. The AARSs activate the amino acids via ATP-dependent adenylation and then ligate the activated amino acid onto the 2′ or the 3′ hydroxyl of the corresponding tRNA. We will discuss AARSs in some detail below; here we simply point out that they are responsible for identifying a specific amino acid and a specific tRNA and ligating the former onto the latter. In that manner, they read the structural information in tRNA and amino acids, process it, and realize the genetic code via the amino acid–tRNA ligation reaction. The subsequent polymerization of correctly charged tRNAs according to the mRNA codon sequence is, by comparison, a mechanical operation at the ribosome in which the information that was stored during the tRNA charging process is translated one bit at a time into a specific structure that emerges from the ribosome as a polypeptide sequence. Clearly, AARSs play a central role in the manifestation of the code in modern cells. One might say that they not only preserve the code, but they generate the code. That said, we consider in the following four proposals how the code might have arisen based on structural and functional molecular interactions.

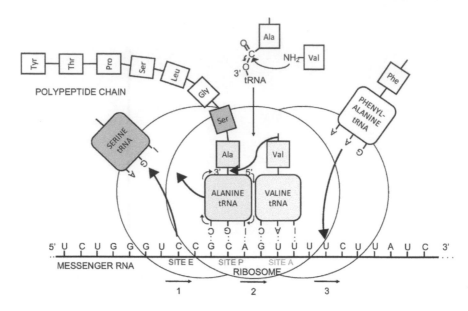

FIGURE 5.5 Simplified scheme of ribosomal protein synthesis (translation). The ribosome is drawn as sliding along on messenger RNA (mRNA). On the left side of the figure, transfer RNA (tRNA), liberated from its serine charge after peptidyl transfer, leaves the ribosome at site E ('empty site'). In the middle, transpeptidation of the peptide chain to valine tRNA leaves alanine tRNA empty and the ribosome moves to the right (translocation). Then, the peptide with valine tRNA is at site P ('peptide site') and the empty alanine tRNA leaves the ribosome at site E. On the right, elongation continues by codon–anticodon binding of phenylalanine tRNA at site A ('amino acid site'). Site P can also be regarded as the donor site and site A as the acceptor site. The 5′→3′ arrows symbolize the antiparallel binding between the anticodon on tRNA and the codon on mRNA. C, U, G, A, and I represent cytidylic, uridylic, guanylic, adenylic, and inosinic nucleotides in mRNA and tRNA. G–C and A–U: Watson–Crick hydrogen bonding; I: special nucleotide appearing in the third position of some anticodons.

5.2.2 Scenario I: Peptides and Predecessors of Nucleic Acids Are Synthesized Synchronously

Of the four possibilities for the origin of the code we discuss here the scenario that is least similar to modern peptide synthesis, namely the coupled synthesis of peptides and β-linked polyesters as evolutionary precursors of nucleic acids. Francis and coworkers proposed malic acid and malamide as monomers of a β-linked poly-malamide template strand for binary-coded peptide synthesis (Figure 5.6). Malic acid (malate) is an intermediate of the tricarboxylic acid cycle and has been obtained nonenzymatically from aldol condensation of glyoxylate and pyruvate obtained from the acetyl-CoA pathway (Section 3.1).

A complementary nascent strand can couple to the template strand (drawn blue in Figure 5.7) with specific hydrogen bond interactions between the short and the long side chains. A polypeptide strand is condensed to the terminal of the nascent poly-malamide strand. A thioester-activated alanyl-malamide monomer (drawn magenta) reacts with the nascent poly-malamide strand (red arrow right) and extends the polypeptide chain by producing a new peptide bond (red arrow left). If alanine is connected specifically to the short side chain of a malamide–thioester monomer, whereas glycine is connected specifically to a monomer with a long side chain (from the reaction of β-alaninamide with malic acid), then a sequence of short and long side chains (binary code 1 and 0) can program a sequence of glycine and alanine in a Gly–Ala polypeptide.

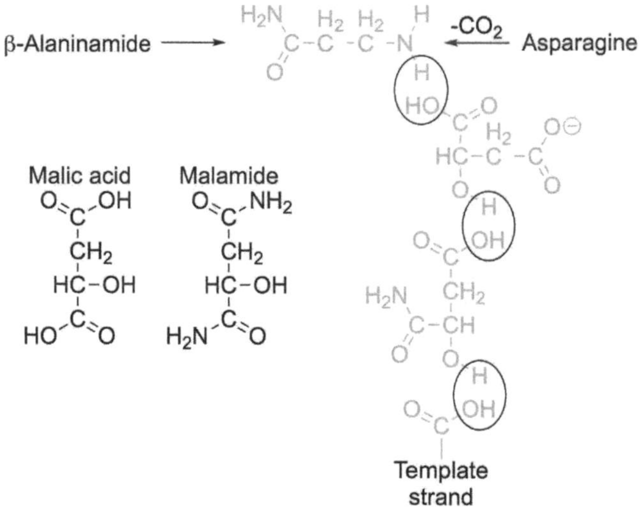

FIGURE 5.6 Template strand from condensation of malic acid, malamide, and β-alaninamide. These condensations can generate short and long side chains for a binary code (information polymer). The carboxyl groups must be activated for condensation. Redrawn from Francis B. R. (2015) The hypothesis that the genetic code originated in coupled synthesis of proteins and the evolutionary predecessors of nucleic acids in primitive cells, *Life* 5: 467–505.

The malamide side chains extended with β-alaninamide can exist in different conformations (rotations drawn with blue arrows in Figure 5.8). The hydrogen-bond donor and acceptor orientations (black arrows) of the four configurations correlate to the hydrogen-bond orientations of the four RNA nucleobases U, C, G, and A. By ring closure

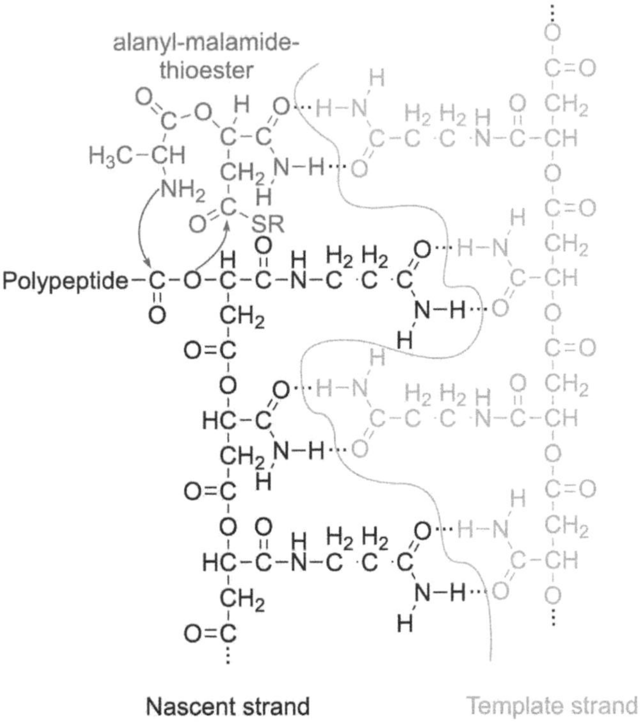

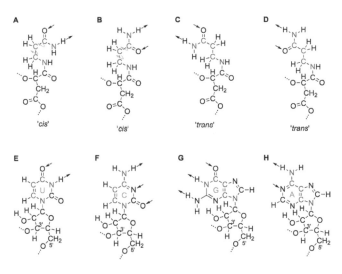

FIGURE 5.7 Coupled synthesis of poly-(β-D-malamide) and polypeptide. For explanation, see text. Redrawn from Francis B. R. (2015) The hypothesis that the genetic code originated in coupled synthesis of proteins and the evolutionary predecessors of nucleic acids in primitive cells, *Life* 5: 467–505.

FIGURE 5.8 Poly-malamide derivation of purines and pyrimidines. The extended side chains can exist in four 'cis' and 'trans'-like configurations as indicated by the blue arrows around the C–C single bonds. Their hydrogen-bond donor and acceptor orientations (black arrows) correlate to the H-bond orientations of the four RNA nucleobases. Redrawn from Francis B. R. (2015) The hypothesis that the genetic code originated in coupled synthesis of proteins and the evolutionary predecessors of nucleic acids in primitive cells, *Life* 5: 467–505.

with formaldehyde and further reaction steps, the four nucleobases and ribose could have evolved from the poly-malamide unit with the long side chain.

Copley and coworkers proposed covalent linking of α-keto acid precursors to the reactive 2′-OH group of RNA ribose, reductive amination of these α-keto acids to amino

acids, and polymerization of these amino acids to peptides. In essence, their proposal was that the proteinogenic amino acids were synthesized on tRNA, as opposed to being formed first and then ligated to it. The dinucleotide that can catalyze the synthesis of a particular amino acid by reductive amination and side chain elaboration is proposed to contain the nucleobases for a doublet genetic code specifying that amino acid. The authors proposed a doublet genetic code that preceded the modern triplet code, a proposal that is germane to many theories about the origin of the code. There are no experiments yet directly the propositions of Francis or of Copley and coworkers.

5.2.3 Scenario II: Amino Acids Covalently Bind to Nucleobases of RNA Strands and Polymerize

In scenario II, we describe a model for aggregation based on the covalent bonding of amino acids to the nucleobases of RNA and polymerization of the aligned amino acids to peptides. The sequence of the amino acids in the peptide is then determined by the sequence of the nucleotides in the RNA. Indeed, transfer RNAs contain many modified bases, and some of them are altered by amino acids (Section 4.7). The most common are adenosine nucleotides with an amino acid (aa) such as threonine or glycine connected with urea to the N⁶-amino group of adenine (Figure 5.9). Recently, Carell and coworkers reported the synthesis of aa6A-nucleosides with several amino acids, among others aspartic acid (Asp), glycine (Gly), threonine (Thr), and serine (Ser), and their incorporation into RNA and DNA.

In essence, this proposal has the code emerging from amino acid–RNA interactions, but involving covalent linkages

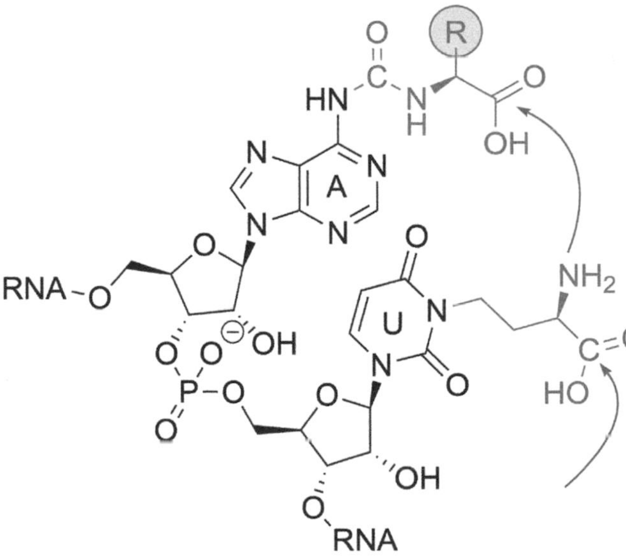

FIGURE 5.9 Amino acids covalently linked to RNA strands polymerize to peptides. A = adenine; U = uracil. The N6-amino group of the upper nucleotide adenine is linked with urea to the upper amino acid (R=−CH(OH)CH₃: threonine). The lower nucleotide uracil is linked with its NH-group to the hydroxyl group of serine. The amino acids must be activated for polymerization by thiolation or phosphorylation. Redrawn from Müller F. *et al.* (2022) A prebiotically plausible scenario of an RNA–peptide world, *Nature* 605: 279–284.

that today would be seen as tRNA modifications. Hydrogen bonding of OH, NH, SH, or COOH-containing side chains of activated amino acids with nucleobases of an RNA strand, followed by polymerization of the amino acids is also conceivable and would be autocatalytic because the reverse hydrogen bonding of the product peptide side chains to activated nucleotides is then feasible too. The concept opens up new ways of thinking about RNA–amino acid interactions. There are no experiments yet, however, that investigate the interaction of selected RNA strands with such amino acids.

5.2.4 Scenario III: The RNA Activators of Amino Acids Couple to RNA-Collector Hairpins

In the early 1970s, Hans and coworkers developed a scenario for coded peptide synthesis that resembles modern biochemical translation but is based on aggregates of simple RNA hairpins and recognition of the activated amino acids without enzymes.

To start, we consider the formation of RNA strands with self-complementary sequences, which are prerequisites for RNA hairpins (Figure 5.10). Assume we have two identical nucleic acid double strands which have been replicated by reactions described in Sections 4.7 and 5.1. Consider now that one of the strands is rotated by 180° before condensation. Then, the condensed double strand R^+: R^- separates due to a change in temperature or pH value, allowing the separated self-complementary strands R^+ and R^- to fold internally to

stem-loop structures with a few unpaired nucleobases at the loop. The most stable folding with parallel stem moieties is achieved with three nucleobases at the top. Loop curvatures with two bases are too open for hydrogen bonding of the stem bases, whereas loops with four bases pointing out are too open for stable hydrogen bonding near the loop (too large loop). Kuhn and coworkers presume that parallel hairpin strands of this type lead to optimal aggregation of the hairpins for interaction with an open RNA strand, called a *collector* (Figure 5.10). Such hairpin aggregates could be better protected against hydrolysis than single hairpins. Bivalent cations such as Ca^{2+} or Mg^{2+} diminish repulsion of the negatively charged phosphates in the hairpins and support aggregation.

The nucleobases at the terminals of the collector strand are not complementary; therefore, the collector is 'open' at the ends and the nucleobases positioned there serve as trapping centers for the protruding bases at the hairpin loops (Figure 5.10). The hairpins diffuse along the collector strands until they find a complementary base triplet. There, they stick. This is a fast process, because diffusion along a collector strand is one dimensional and much faster than three-dimensional diffusion in a free solution. Only hairpins with a matching base triplet and parallel stems 'crystallize' to an aggregate; other hairpins hydrolyze or are lost to diffusion. For example, a collector with helix structure adds only hairpins with helix structure. The lateral attraction between the helices, however, should not be stronger than the hydrogen bonding in the base pairs; otherwise, the

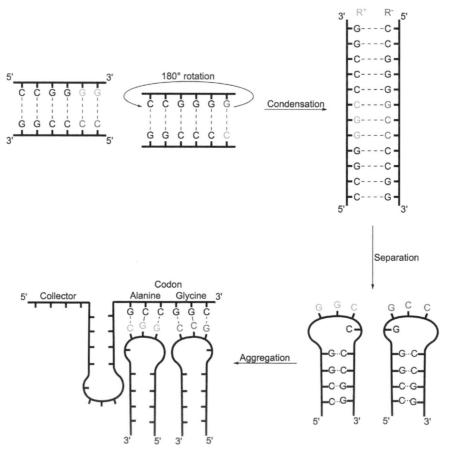

FIGURE 5.10 RNA stem-loop structures ('hairpins') and aggregation with a 'collector' strand. The hairpins are proposed to be precursors of tRNA and the collector precursor of mRNA. For further explanations, see text. Adapted from Kuhn H. (1972) Self-organization of molecular systems and evolution of the genetic apparatus, *Angew. Chem. Int. Ed. Engl.* 11: 798–820.

selective attachment of hairpins to the collector gets lost and a 'clump' with random connectivity develops.

What happens if such an aggregate interacts with an activated amino acid? It is remarkable and quite unusual, that in ribosomal peptide synthesis, the amino acids are activated by transfer of adenosine monophosphate (AMP) from ATP, not simply by transfer of a phosphate group. After this *adenylation*, the remaining pyrophosphate hydrolyzes, which makes the phosphorylation irreversible. Kuhn and coworkers proposed that the nucleotides GMP, CMP, and UMP may have been activating agents as well during the evolution of translation. In that case, the *nucleobases of the activators can hydrogen bond to nucleobases at the hairpin terminal*. In this way, the hairpin–amino acid interaction is specific without coupling by an enzyme, just by Watson–Crick pairing (G . . . C in Figure 5.11). The hairpins are hydrogen bonded to an RNA collector ('messenger') strand via codon–anticodon interaction (not shown in Figure 5.11).

In Kuhn's model, G–C, C–G, A–U, and U–A interactions can specifically couple four amino acids to 'their' cognate RNA hairpins. The reaction rates of the four activators with the amino acids must be different to couple them specifically

to 'their' amino acids. Four activators could couple four different amino acids to transfer hairpins. Four amino acids, for example, glycine, alanine, aspartate, and glutamate, are sufficient to generate quite complex polypeptides with hydrophilic and hydrophobic domains (Section 4.2).

The hairpins loaded with their corresponding amino acids bind to the collector hairpin by codon–anticodon hydrogen bonds (Figure 5.12). Then, the aligned amino acids can react to peptides at neutral or alkaline conditions. The newly synthesized peptide, drawn blue in Figure 5.12, detaches from the transfer hairpins but sticks to the collector–hairpins aggregate and stabilizes it against hydrolysis. Peptides which keep the collector open for aggregation with hairpins will be favored. The error-prone replication of the collector will generate a pool of slightly different collectors. Then, *the collectors with a code (sequence of base triplets) generating peptides with the most effective aggregate stabilization will survive chemical attacks and prevail* (molecular evolution in a pool of collectors driven by preferential stability against hydrolysis as opposed to preferential replication).

Larger peptides will evolve which envelop the whole aggregate of collectors and hairpins. Such large peptides

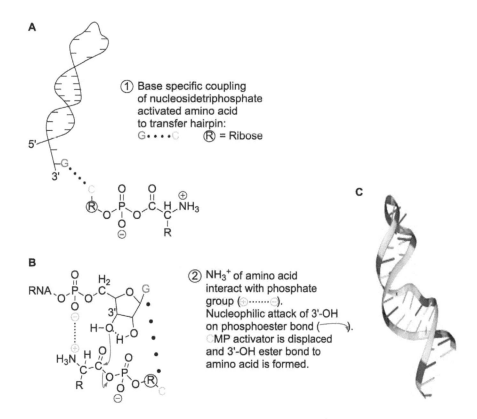

FIGURE 5.11 Coupling amino acid activators to complementary bases in RNA hairpins. (A) Base specific coupling of nucleoside monophosphate-activated amino acid to RNA-transfer hairpin: G···C. (R) = Ribose; R = amino acid side chain. Adapted from Kuhn H. (1972) Self-organization of molecular systems and evolution of the genetic apparatus, *Angew. Chem. Int. Ed. Engl.* 11: 798–820. (B) The terminal RNA phosphate group interacts with the protonated amino group and blocks this reactive group: $-O-\cdots NH_3^+-$. It is important that the amino group is blocked; otherwise, it may react with ATP (phosphorylation of the amino group as in creatine phosphate). Recall that the amino group of the amino acids in an automated peptide synthesizer must also be blocked from cycle to cycle. Guanine at the RNA-terminal binds to cytosine of the activated amino acid: G···C. The two interactions fix the activated amino acid in the correct position for bonding to ribose. Then, the hydroxyl group at the 3'-carbon of the terminal ribose undertakes a nucleophilic attack on the carbonyl group of the phosphoester bond. The activator CMP is displaced, and an ester bond of ribose 3'-OH with the amino acid is formed. (C) Segment of the Influenza A RNA virus containing a 3' splice site to encode a virus-essential protein. Green: guanine; red: adenine; yellow: cytosine; cyan: uracil (PDB Data bank: Hairpin 2MXL). The 39 nucleotide-long segment is taken as an example for the possible structure of a primitive RNA transfer hairpin. See however Krzyzaniak A. *et al.* (1994) The non-enzymatic specific aminoacylation of transfer RNA at high pressure, *Int. J. Biol. Macromol.* 16: 153–158, where Phe was specifically coupled to tRNA(Phe) and Met to tRNA(Met) at 6 kbar pressure without activation of the amino acid and without enzymes.

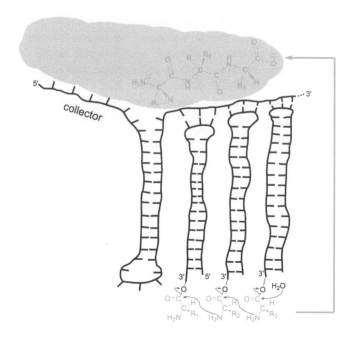

FIGURE 5.12 **Peptide synthesis in RNA hairpin—collector aggregates.** The newly synthesized peptide sticks to the aggregate of the RNA collector and RNA hairpins and stabilizes it against hydrolysis (peptide drawn blue). The collector code is optimized for the most effective peptide stabilizer of the RNA collector–hairpins aggregate by molecular evolution. See text.

can interact with a specific amino acid, its activating agent, and a specific RNA hairpin. By this means, precursors of the enzymes which activate amino acids and couple them to 'their' tRNA (*aminoacyl tRNA synthetases*: AARSs) arise. Such peptides will autocatalytically enhance their own synthesis. Similarly, peptides which bind nucleotides to collector templates and condense them to complementary collectors (*proto-RNA replicases*) enhance their own synthesis. Collectors synthesizing proto-RNA replicases will be favored in the pool of collectors. The role of aminoacyl tRNA synthetases will be discussed in detail in Section 5.2.5.

Recall that the essence of the genetic code is the mechanism that associates specific amino acids with specific tRNA species. Kuhn's proposal posits that the code, the specific charging of hairpin tRNA molecules with specific activated amino acids, emerged from hydrogen bonds and electrostatic interactions of the two partners as they were held in place on an RNA surface. Artificial ribozymes can also couple activated amino acids to tRNA (Figure 4.30), and catalyze their synthesis to peptides (Figure 4.29). A scheme based on aggregation of RNA hairpins as proposed here, however, needs no complicated polymerase chain reaction (PCR) as artificial ribozymes do. Instead, it will autocatalyze its synthesis by producing its own enzymes. The primary driving force for the formation of hairpin aggregates is protection against the attack of water and other chemicals. The enhancing of tRNA–amino acid recognition by RNA strands acting as a 'primer' for aminoacylation cannot be excluded at this stage of molecular evolution (Figure 5.13). In principle, they could develop into AARS ribozymes as precursors of RNA-containing proteins, only to be later replaced by AARSs enzymes. This is possible, but there is no hint of such synthetase ribozymes from present bacteria or phylogenetic reconstructions of primordial first cells.

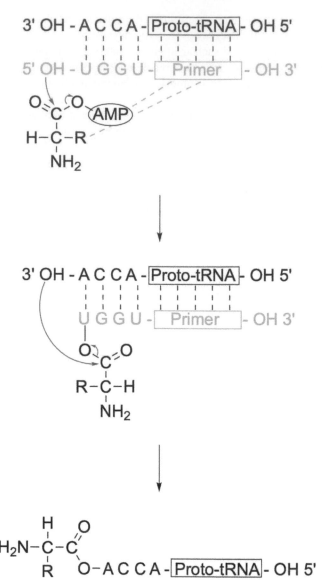

FIGURE 5.13 **An RNA strand acting as a primer for proto-tRNA aminoacylation.** An RNA strand with complementary sequences to the RNA transfer hairpin (black dashed lines) and interaction with ATP/AMP and amino acid (blue dashed lines) could enhance recognition compared to the interaction of tRNA with solely the amino acid activator (see Figure 5.11). An RNA strand acting as a primer for proto-tRNA aminoacylation shows a primer strand and AMP-activated amino acid interacting with it but AMP is not necessary if the primer is triphosphorylated at its end.

Above we have described the coevolution of RNA and peptides based on self-replicating aggregates that contain amino acids and RNA. In ribosomes, the functional roles of encoding RNA (mRNA) and catalytic RNA (rRNA) are supported by proteins as well, and the aggregates described above could be seen as precursors of ribosomes in an RNA–peptide world.

5.2.5 SCENARIO IV: AARS URZYMES ACYLATE TRNA ACCEPTOR STEMS SPECIFICALLY

Scenarios I–III assume that early translation proceeded without the help of proteins. In scenario III, the first peptides were synthesized via recognition of the nucleotide activators of the amino acids by RNA hairpins followed by polymerization of the amino acids. Subsequently, the newly synthesized peptides docked to the RNA collector–hairpin complex, stabilized it against chemical attack, in particular hydrolysis, and then coevolved with the RNA complex. We furthermore assumed that complexes with the most stable noncovalent binding of peptides to RNA survived. These peptides could have developed into the first enzymes, which would then bind optimally to the transfer hairpins and to the nucleotide-activated amino acids. In Section 5.2.5, we look closer at the possible fate of these peptides ('urzymes') by comparison with the properties of their modern correspondents.

Again, the essence of the genetic code is the mechanism that associates specific amino acids with specific tRNA species. In contemporary protein synthesis, this mechanism is provided by enzymes called *aminoacyl tRNA synthetases* (AARSs), which activate amino acids by adenylation and link them to their specific transfer RNA (tRNA). The enzymatic reaction mechanisms involved are well known. First, ATP and amino acid bind at the AARS active site (Figure 5.14A) and react to an AMP-activated amino acid,

releasing pyrophosphate. Subsequent hydrolysis of pyrophosphate makes the activation irreversible. Rarely, other nucleoside triphosphates including GTP or UTP can activate the amino acid (see scenario III and the work of Fujiwara *et al*). In a second step, the 2'- or 3'-OH of the terminal adenosine (3'-OH of A 76 in Figure 5.14) of the tRNA performs nucleophilic attack on the carbon of the acyl group of the aminoacyl-adenylate (Figure 5.14B) and reacts with the amino acid by displacing adenosine monophosphate (AMP).

Figure 5.14 shows that the amino acid reactant, the activating agent ATP, and the terminal adenosine at position 76 of tRNA are recognized by numerous interactions with the AARS. The boxes in Figure 5.14 contain strictly or functionally conserved amino acids (those with similar physicochemical properties) that are especially important for recognition in class II AARS (see below).

At least one AARS exists for each of the 20 proteinogenic amino acids. The 20 AARSs can be divided into two classes I and II (Table 5.1). The two classes of AARS share a common function in that they interpret the code by linking each amino acid with its cognate tRNA. Yet they differ in several remarkable properties that we will consider in the following. Within the two classes, their sequences and structures are related. Across the two classes, the sequences and structures are not related in any conventional manner. The question of whether the two classes of AARS share a common ancestry is a surprisingly complicated issue, as we well see.

Synthetases from the two classes bind to different sides of tRNA. This is possible because the terminal CCA arm of tRNA can adopt two different conformations in complexes with the synthetases (Figure 5.15). In one conformation, the CCA arm follows the direction of the tRNA helix (recognized by class II synthetases; Figure 5.15) whereas it forms a loop in the other helix conformation (recognized by class I synthetases; Figure 5.15). The separate recognition of two faces of tRNA enlarges the number of interactions

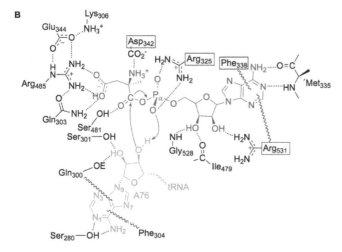

TABLE 5.1
Classification of Aminoacyl-tRNA Synthetases

Class I	Class II
Arg	Ala
Cys	Asn
Gln	Asp
Glu	Gly
Ile	His
Leu	Lys
Met	Phe
Trp	Ser
Tyr	Pro
Val	Thr

Note: The existence of two classes of aminoacyl tRNA synthetases has been known since the early 1990s (see Schimmel P. 1991) Classes of aminoacyl-tRNA synthetases and the establishment of the genetic code, *Trends Biochem. Sci.* 16: 1–3.

Source: The list shown here is from Carter C. W., Wills P. R. (2018) Hierarchical groove discrimination by Class I and II aminoacyl-tRNA synthetases reveals a palimpsest of the operational RNA code in the tRNA acceptor-stem bases, *Nucleic Acids Res.* 46: 9667–9683.

FIGURE 5.14 Amino acid adenylation and aminoacylation reaction. The scheme is derived from the crystal structure of yeast Aspartyl-tRNA synthetase (AspRS). A. Amino acid (drawn blue) activation with formation of aspartyl-adenylate and pyrophosphate (drawn magenta) by nucleophilic substitution (red arrows). B. Amino acid transfer by nucleophilic substitution (red arrows) of AMP by the 3'-OH group of terminal (A76) adenosine ribose (drawn blue) of tRNAAsp. The boxes contain residues strictly or functionally conserved in class II AARS. Redrawn from Caravelli J. *et al.* (1994) The active site of yeast aspartyl-tRNA synthetase: Structural and functional aspects of the aminoacylation reaction, *EMBO J.* 13: 327–337.

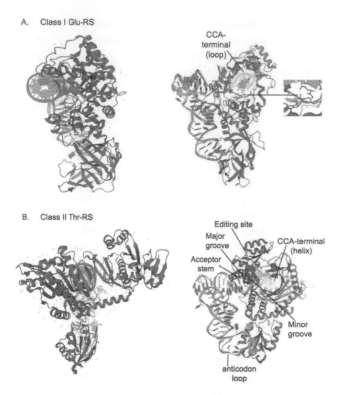

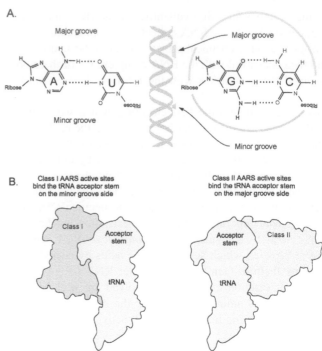

FIGURE 5.15   Class I and class II of aminoacyl tRNA synthe-tases. The terminal CCA arm is in a loop conformation for class I enzymes (glutaminyl-RS; top graphs) and in a helical conforma-tion for class II enzymes (threonyl-RS; bottom graphs). Therefore, class I and class II synthetases (drawn red) recognize different faces of the tRNA molecule (drawn green). In the left graphs, the CCA arm of the tRNA is turned toward the viewer; in the right graphs, the CCA arm is in the plane as in Figure 5.17. In general, class I enzymes bind to the minor groove and class II enzymes to the major groove of the acceptor stem helix of tRNA. The inset shows adenylated glutamine near the 3'OH-terminal of tRNAGln. Images from 1EUY.pdb (Class I) and 1QF6.pdb (Class II). See de Pouplana L. R., Schimmel P. (2001) Two classes of tRNA syn-thetases suggested by sterically compatible dockings on tRNA acceptor stem, *Cell* 104: 191–193.

FIGURE 5.16   The major groove and minor groove in nucleic acids. (A) There are larger and smaller sides of a Watson–Crick base pair because the sugar residues are not diametrically opposed across the central axis of the double helix. (B) The active site of Class I AARSs (indicated in beige) binds to the minor groove, and that of Class II enzymes (indicated in purple) binds to the major groove of the acceptor stem helix of tRNA (indicated in blue). Prepared from information de Pouplana L. R., Schimmel P. (2001) Two classes of tRNA synthetases suggested by sterically compatible dockings on tRNA acceptor stem, *Cell* 104: 191–193 and Carter C. W., Wills P. R. (2018) Hierarchical groove discrimi-nation by Class I and II aminoacyl-tRNA synthetases reveals a palimpsest of the operational RNA code in the tRNA acceptor-stem bases, *Nucleic Acids Res.* 46: 9667–9683.

and may have been necessary to specifically bind the AARS to 20 different tRNAs. Class I enzymes acylate the 2'-OH of the terminal adenosine of tRNA, while class II enzymes acylate 3'-OH (except for Phe-tRNA). Most class I AARSs are monomers, and most class II AARSs are dimers. Some AARSs contain an editing (proofreading) domain: the flex-ible CCA arm of tRNA can move the attached amino acid from the activating to the editing site. If the amino acid fits into the editing site, it is removed by hydrolysis (proofread-ing activity).

Class I enzymes typically bind to the minor groove and class II enzymes to the major groove of the accep-tor stem helix of tRNA. Because the two glycosidic bonds of a base pair are not diametrically opposed (Figure 5.16), every base pair has two sites: a larger one, which defines the major groove and a smaller one on the site of the minor groove.

Note that most AARSs bind to both the acceptor stem at the CCA terminus and the anticodon loop at the head of tRNA, regardless of class. For example, each base of the threonyl-tRNA anticodon sequence 5'-CGU-3' binds with

hydrogen bonds to the synthetase. The interactions of G and U seem to be more important because the anticodons GGU and UGU of tRNA^Thr bind just as efficiently to the threonyl RS. Some synthetases or fragments thereof, how-ever do not bind to the anticodon loop but still recognize 'their' amino acid and tRNA (as discussed in the follow-ing).

Figure 5.15 shows that many sites of tRNA interact with the AARS. The interaction frequencies of single bases have been quantitatively determined by NMR or crystal structure analysis of a large number of AARS–tRNA complexes. The sizes of the circles in Figure 5.17 are proportional to the frequency with which they are used as recognition sites by AARSs. Hydrogen bonding to the acceptor stem, part of the T stem, and the anticodon loop are especially important for recognition.

The 3'-OH linked amino acid and the anticodon trinu-cleotide are situated at opposite regions of the two-domain L-shaped tRNA structure. The distance between the 3' ter-minal and anticodon loop is ≈ 76 Å. All AARSs are com-posed of two domains. The first domain binds to the tRNA acceptor stem and T helix near the 3'-terminal of tRNA. It contains amino acid sequences that are conserved within

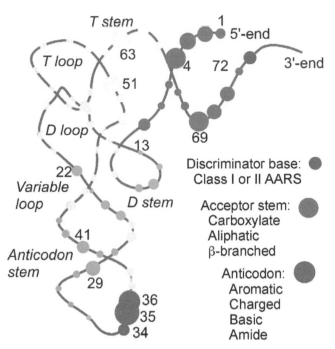

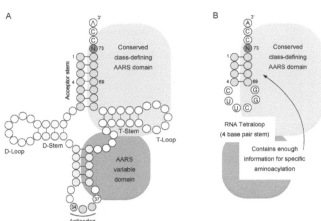

FIGURE 5.17 Function in an idealized tRNA structure. Circles indicate nucleotides; the sizes of the circles are proportional to the documented frequency with which they are used as recognition sites by AARSs. The colors of the circles represent the assignment of the nucleotides to the amino acid acceptor stem (drawn red), T stem (contains T: ribothymidine; drawn yellow), D stem (contains UH2: Dihydrouridine; drawn green), anticodon stem (drawn light blue), and anticodon (drawn grey). The discriminator base 73 (drawn in dark blue) specifies the AARS class. The numbers indicate the position of the nucleotides in the base sequence from the 5′ end to the 3′ end of tRNA. Redrawn from Ibba M., Söll D. (2000) Aminoacyl-tRNA synthesis, *Annu. Rev. Biochem.* 69: 617–650. Functional information about the properties of side chains of amino acids contained in the acceptor stem and anticodon from: Carter Jr. W. C., Wolfenden R. (2015) tRNA acceptor stem and anticodon bases form independent codes related to protein folding, *Proc. Natl Acad. Sci. USA* 112: 7489–7494.

FIGURE 5.18 Cloverleaf structure of tRNA and AARS domain interactions. (A) Schematic cloverleaf structure of tRNA from *E. coli* and simplified representation of tRNA synthetase as a two-domain structure bound to tRNA. The binding region between the conserved domain of AARS and tRNA is indicated in blue, and the binding region of the non-conserved, variable domain is indicated in red. The colors of the bases are as in Figure 5.17. See Schimmel P. *et al.* (1993) An operational RNA code for amino acids and possible relationship to genetic code, *Proc. Natl Acad. Sci. USA* 90: 8763–8768; Carter Jr. W. C., Wolfenden R. (2015) tRNA acceptor stem and anticodon bases form independent codes related to protein folding, *Proc. Natl Acad. Sci. USA* 112: 7489–7494. For *E. coli* tRNA cloverleaf see Crécy-Lagard V., Jairoch M. (2021) Functions of bacterial tRNA modifications: from ubiquity to diversity, *Trends Microbiol.* 29: 41–53. (B) RNA tetraloop with four base pairs, UUCG loop, discriminator base N, and CCA terminal. Even RNA stem-loop structures of this size are recognized and specifically bound by tRNA synthetases (symbolized as a blue oval). Similarly, minihelices consisting of 12 base pairs, microhelices with 7 base pairs, and 'open' (no loops) RNA double strands (duplexes) with 5, 7, 9, and 12 base pairs are specifically aminoacylated by AARS. These structures must merely contain specific base pair sequences of the acceptor stem. See Shi J. P., Martinis S. A., Schimmel P. (1992) RNA tetraloops as minimalist substrates for aminoacylation, *Biochemistry* 31: 4931–4936.

the respective class. The conserved motifs identify AARSs as class I or class II enzymes (conserved class-defining domain, binding region drawn blue in Figure 5.18). The second domain (nonconserved domain, binding region drawn red in Figure 5.18) varies in size with some being large enough to interact with the anticodon, but others not. For example, a fragment of methionine tRNA synthetase specifically aminoacetylated tRNA^fMet despite lacking the anticodon binding domain.

The conserved motif in class I AARS is the $\beta_5\alpha_4$ Rossmann nucleotide binding fold, which consists of five parallel β-strands connected by four α-helices. This motif defines class I domains. The conserved motif in class II AARS is composed of seven antiparallel β-strands flanked by three α-helices and serves as nucleotide cofactor binding domain too. Examples of nonconserved domains in class I enzymes are the N-terminal topology of methionine-RS with predominantly α-helices and glutamine-RS with a β-barrel structure (β-sheets aligned as toroid). Examples of nonconserved domains in class II enzymes are an N-terminal β-barrel domain in aspartyl-RS and an extended antiparallel

coiled-coil domain (several α-helices are coiled together like the strands of a rope) in serine-RS. Obviously, the structural motifs of nonconserved domains in different AARSs of the same class are different.

If acceptor stem binding domains of AARSs suffice for specific binding of tRNA, then short RNA hairpins consisting of only the acceptor stem sequences might bind specifically too. Indeed, experiments showed that short RNA hairpins and duplexes with acceptor stem sequences can be specifically aminoacylated by their respective synthetases. Even RNA tetraloops (acceptor stem consisting of only four base pairs; Figure 5.18B) suffice—the single-stranded N^73 'discriminator' base and certain base pairs within the first four pairs of the acceptor stem (1···72 to 4···69) contain enough information for specific aminoacylation.

Statistical analysis from 1268 tRNA sequences of viruses, bacteria, archaea, eukaryotes, and chloroplasts, but not mitochondria, revealed a common 11 base pairs long double-strand acceptor stem with a significant content of anticodon-codon pairs at nucleoside positions 1, 2,

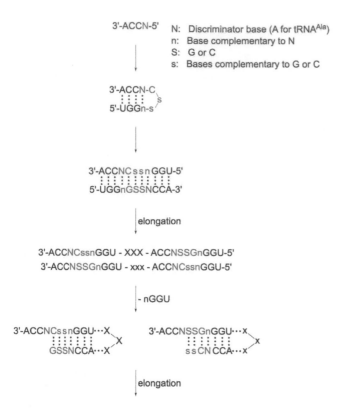

The Anticodon migrates from the acceptor stem inwards to form a protruding anticodon loop for codon recognition

FIGURE 5.19 Hypothetical conversion of a 3′-ACCN-5′ tetramer to a proto-tRNA. Initially the anticodon-like sequence Css (drawn red; s = G or C) was in the 11 base pairs long double-strand acceptor stem and could be recognized by proto-AARS. Subsequently, by self-priming and self-templating, an anticodon sequence XXX and xxx could have migrated from the acceptor stem to an exposed position of the hairpin where it could bind to codons of linear RNA strands. Simultaneously, a second non-conserved enzyme domain added to the AARS which interacted with the anticodon loop and ensured reliable anticodon recognition and RNA template-directed peptide synthesis. Redrawn from Rodin S. *et al.* (1996) The presence of codon-anticodon pairs in the acceptor stem of tRNAs, *Proc. Natl Acad. Sci. USA* 93: 4537–4542 Copyright (1996) National Academy of Sciences USA.

and 3 (see Figure 5.18). Figure 5.19 shows how the original 5′-NCCA-3′ (N: 'Discriminator' base) tetranucleotide self-templates and loops by the addition of a Css(s: C or G) sequence to a hairpin with 11 nucleobases. This structure self-templates to an 11 bp long double strand (third sequence from above in Figure 5.19). Recombination of the two strands leads to two helices with the same flanking complementary repeats and with anticodon-codon-like Css–GSS pairs at the ends. Loss of the nGGU sequence at the 5′ terminal end and further recombination and self-templating shift the anticodon inwards to a protruding position. There, at some distance from the acceptor stem, it forms a loop with protruding nucleobases XXX and xxx, which are able to interact with coding sequences of linear RNA strands (codon–anticodon interaction). Recognition is supported by the addition of a nonconserved domain to the AARS, which interacts with the anticodon loop and ensures reliable anticodon recognition and thus reproducible anticodon-codon binding as the basis of RNA template-directed peptide synthesis.

5.2.6 The Genetic Code Might Contain Evidence for Its Origin

A remarkable aspect of the relation between class I and class II AARSs is that tRNAs with complementary anticodons are recognized by AARSs from complementary genetic coding sequences (Figure 5.20). This prompted the idea that at some early stage of evolution of the code, the single strands of RNA double strands encoded with a part of their sequence tRNAs and with another part of their sequence the corresponding urzymes (proto-AARS).

The different side chains of amino acids impact protein folding. The acceptor stem of tRNA contains information that impacted the folding of the early proteins (Figure 5.17). Furthermore, acceptor stem recognition probably preceded anticodon recognition in the evolution of tRNA structure. Carter and Wolfenden correlated nucleotide sequences both in the acceptor stem and the anticodon triplet of the tRNAs with experimental properties of the coded amino acids. They used the water-to-cyclohexane (w→c) distribution coefficients of amino acid side chain model systems as a measure

A

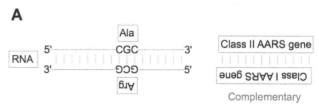

B

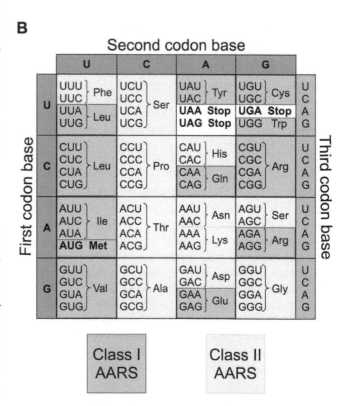

FIGURE 5.20 AARS and tRNA complementary and the genetic code. (A) The coding strands of the conserved domains of the Ala class II and Arg class I AARS show a reverse complementary relationship, that is, their coding sequences trace to the complementary strands of the same double-stranded nucleic acids. tRNAs with complementary anticodons are recognized by synthetases from complementary genes. (B) Structure of the universal genetic code with the classes of AARS that activate tRNA for the codon indicated.

of hydrophobicity, and the vapor-to-cyclohexane (v→c) coefficients to estimate the chain size of the amino acids. From the experimental distribution coefficients, they obtained the equilibrium constants and free enthalpies $\Delta G_{w \to c}$ and $\Delta G_{v \to c}$.

Using a two-bit code for each tRNA base, Carter and Wolfenden expressed these experimental free enthalpies as linear and cross-term combinations of the binary base codes with coefficients which were determined by the standard mathematical method of *multivariate regression*. In that way, they obtained information about the functional coding of acceptor stem and anticodon bases (see Figure 5.17). The coefficients showed that the acceptor stem of tRNA is the seat of the information that specifies the size of the respective amino acid side chain. By contrast, the anticodon information preferentially specifies hydrophobicity. Further analysis showed that the acceptor stem also specifies aliphatic side chain β-branching (Val, Ile, Thr). Both the acceptor stem and anticodon encode carboxylate side chains demonstrating the important role of carboxyl groups at early and later stages of protein evolution. The anticodon triplets, but not the acceptor stem correlate with aromatic, positively charged, basic, and amide side chains, suggesting their later addition to a more primitive code based on acidic and aliphatic side chains. Hence, acidic groups and the surface area determined by side chain size and branching were probably the first encoded properties, whereas hydrophobic, aromatic, positively charged, and amide side chains were encoded later after the anticodon had evolved. The AARS class also contributes to the recognition of amino acid properties determining protein folding. Amino acids of class I AARS such as Leu, Ile, Val, and Met (Figure 5.20B) have aliphatic side chains and are found in hydrophobic cores inside of proteins, whereas amino acids activated by class II AARS such as Asp and Lys remain largely on the surface. The correlations are not strict, however. For example, the aromatic amino acids Tyr and Trp do not fit this hydrophobicity pattern, but these two aromatic amino acids are generally seen as late additions to the code, mainly because of their complicated biosynthetic pathways.

At the outset of this chapter, we found that processes similar to nonribosomal protein synthesis (NRPS) involving aminoacylation of thiols to form thioesters might have preceded translation. We saw that translation involves peptide synthesis on ribosomes with the help of the genetic code involving aminoacylation of the 2′ or 3′ OH of tRNA. We also saw that AARSs actually generate the genetic code because they provide the physical, informational link between anticodons and the corresponding activated amino acids. Are there any traces of connections that might hint at an earlier form of translation? Possibly. Jakubowski and others have shown that several AARSs will readily aminoacylate the free thiol residue of coenzyme A forming aminoacyl thioesters, as in NRPS. It is possible that this activity of AARSs represents a molecular fossil, an enzymatic relic from a time when two kinds of protein enzyme synthesis were used by the first protocells. Recall that the products of modern NRPS are typically inhibitors with antibiotic activity, not catalysts, but the initial function of NRPS-generated proteins could well have been catalytic. It is impossible to say whether thioester formation by AARSs reflects an intermediate state in the evolution of translation because we cannot say for sure that something like NRPS preceded translation, it is an interesting thought nonetheless.

5.3 MODIFIED BASES IN TRNA AND RRNA

5.3.1 THE RIBOSOME AND THE GENETIC CODE REQUIRES MODIFIED BASES TO OPERATE

Many tRNAs require conserved modified bases for functionality (pseudouridine or dihydrouridine, for example) and many tRNA anticodons require modified bases in or near the anticodon to generate an accurate translation of the encoded mRNA sequence. That is, modified bases are as old as the code itself. What is the function of modified nucleosides in tRNA (Figures 5.18A and 5.21)? The phylogenetic reconstruction of genes that trace to the genome of the Last Universal Common Ancestor (LUCA) revealed the presence of genes for many RNA-modifying enzymes, particularly the enzymes that modify nucleosides in tRNA. Several of these enzymes are methyltransferases. There are 28 modified bases, mainly occurring in tRNA, that are shared by bacteria and archaea, and many of them are methylated variants, sometimes thiomethylated variants.

A possible interpretation is that in the very early phases of evolution, before there were enzymes, there was a large pool of diverse modified RNA nucleosides that were synthesized geochemically before highly specific biosynthetic pathways evolved and that these modified bases supported the evolution of translation and the code. When the enzymatically catalyzed pathways arose, they enabled specific synthesis of the four canonical bases G, C, A, and U plus ribose, whereby essential base modifications had to be enzymatically re-introduced. That is, at least some base modifications could be extremely ancient, antedating the code. What role do the modified bases play? If we compare their positions in Figures 5.17 and 5.18, we see that modified bases do not seem to play a significant role in direct recognition. Modified bases are not often used as recognition sites by AARSs, except inosine in the anticodon, which is essential for codon–anticodon interaction to work.

Methylation increases hydrophobicity via the displacement of water molecules by the methyl groups, weakening hydrogen bonds in neighboring base pairs. Hence some base modifications may be kind of a 'lubricant' to form the rather stressed cloverleaf structure of tRNA and enhance tRNA mobility during translation. Indeed, many chemical modifications of the nucleotides of ribosomal RNA involve methylation of ribose, not the base, and are concentrated around the peptidyl transferase site, where they are essential for tRNA ribosome interactions.

Some modifications are however highly conserved at the tRNA anticodon loop and play an essential role in codon anticodon interactions, especially for degenerate codons where 'wobble' in Watson–Crick base pairing is required. In addition to pseudouridine (Ψ) in the T loop and dihydrouridine (UH$_2$) in the D-loop (Figure 5.18), one of the most frequently modified nucleobases in tRNA is the first anticodon position, base 34 (Figure 5.21B). During translation (Figure 5.5), base 34 interacts with the third codon position of mRNA. The third codon position is often degenerate, as shown in Figure 5.20, meaning that the third codon position can be any nucleobase. Modifications of the base at position 34 (the first anticodon position) help it to accommodate this degeneracy by reducing the specificity of its canonical A::U or G:::C hydrogen bonding, permitting it to interact with any third codon position base. This is also known as wobble at the third position. It stems from tRNA base modifications. All rRNA and tRNA base modifications are performed by enzymes on the canonical bases G, A, C, and U during the RNA maturation process in cells. This can entail dozens of enzymes in a prokaryotic cell. The modified bases of tRNA in *E. coli* alone involve 20 tRNA positions and require 60

enzymes for their introduction, this corresponds to roughly 1.5% of all *E. coli* protein-coding genes (Figure 5.21B). The fact that cells have not dispensed with tRNA modifications as part of the genetic code in 4 billion years of evolution attests to the antiquity and essentiality of modifications.

In the modern cell, RNA must fold and function in the presence of a roughly two-fold excess of protein by weight

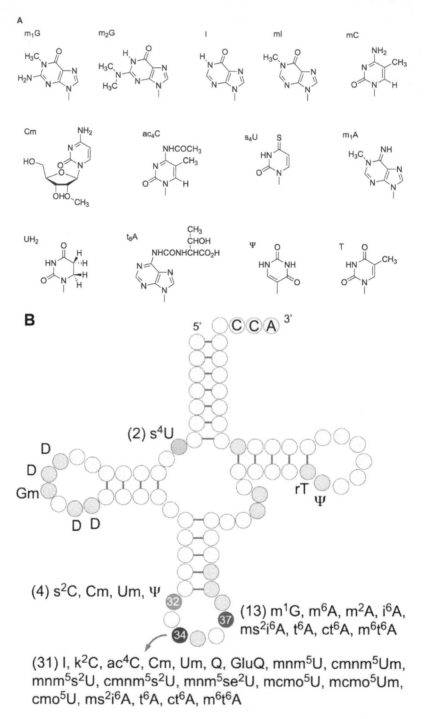

FIGURE 5.21   Modified bases and nucleosides in tRNA. (A) Modified nucleosides are abbreviated as follows: 1-methylguanosine (m1G), dimethylguanosine (m2G), inosine (I), 1-methylinosine (mI), 5-methyl-cytosine (mC), 2′-methoxycytosine (Cm), 4-acetylcytosine (ac4C), 1-methyladenine (m1A), N6-Threonylcarbamoyl adenosine (t6A), dihydrouridine (UH$_2$), 4-thiocytosine (s4U), pseudouridine, (Ψ) and ribothymidine (T). Modified bases are very common in tRNA. Lorenz C. *et al.* (2017) tRNA Modifications: Impact on structure and thermal adaptation. *Biomolecules.* 7: 35. (B) The modified bases of tRNA in *E. coli*. Modifications are shown next to the position. Numbers in parentheses: Number of genes involved in the modifications. Numbers within the circles representing bases: Position number in the standard tRNA numbering scheme. Red shading highlights positions where modifications are common. ac, acetyl; s, thio; se, seleno; mo, methoxy; n, amino; m methyl; t, threonyl; i, isopentenyl; k²C, 2-lysyl cytidine; Q, queuosine; cmo5U, uridine 5-oxyacetic acid; D, dihydrouridine; rT, ribothymidine. Redrawn with permission from: Crécy-Lagard V., Jairoch M. (2021) Functions of bacterial tRNA modifications: from ubiquity to diversity, *Trends Microbiol.* 29: 41–53.

(cells are 50% protein and 20% RNA by dry weight). It is possible that similar mass ratios could have existed at the very early stages of molecular evolution, and that the avoidance of irreversible binding to molecular surfaces of peptides within the milieu where RNA was evolving had a considerable impact upon structure, perhaps as much as intramolecular base pairing has. RNA from biological systems has many moieties that can bind to other molecules, making it a 'sticky' form of molecule. Methylations as in the modified nucleosides serve to reduce non-specific molecular interactions between tRNA and rRNA during translation (see Section 5.2.5 for the origin of translation). Given the chemically reactive environment required for prebiotic synthesis, the synthesis of a highly specific set of pure (unmodified) RNA precursors at origins seems unlikely. Chemically modified bases seem more likely and the kinds of modifications— methyl groups, acetyl groups, sulfur (Figure 5.21)—fit well with the natural chemistry of a hydrothermal environment.

5.3.2 THE CODE REFLECTS ANCIENT PROCESSES

From the foregoing, an admittedly sketchy but structurally and functionally consistent picture of the evolution of the genetic code emerges, one in which the genes for the two classes of AARSs seem likely to have been among the very first genes whose information was translated into protein using a primitive ur-code. They may have been the first translated genes altogether. Initially, short RNA hairpins, amino acids, and nucleotide activators formed aggregates, but aggregation was not particularly specific (see Section 5.2.4). Amino acids and activators reacted in these aggregates, and the activated amino acids esterified 3′-OH of the hairpins and polymerized by transesterification. Within aggregates, polymerization was much more efficient than in free solution due to the proximity of the reactants in the complex and some alignment (diffusion in <3 dimensions). Closely neighbored activated amino acids in RNA–peptide aggregates polymerized more rapidly than those in free solution. The exclusion of water from such complexes added stability to the system by counteracting hydrolysis and prolonging its persistence as a small and imperfect—but autocatalytic—network.

Short peptides having interaction with the acceptor stem of the RNA hairpin and with amino acids and nucleotide triphosphate activators could have helped to increase the efficiency of aminoacylation. The peptides developed into proto-AARS. Hydrolysis and other chemical attacks were outcompeted by fast polymerization and shielding; however, there was no or very little contact of the short RNA hairpin in the proto-AARS complex with coding linear RNA strands. At this primitive stage, peptide synthesis was not yet RNA template directed, just more efficient than in free solution. Later, with the increasing size of the RNA hairpins achieved by self-templating and self-priming, the anticodon sequence migrated from the acceptor stem to exposed positions of the hairpin at some distance from the aminoacylation site where the anticodon could interact with the codon of an RNA template (proto-mRNA). A second, non-conserved protein domain extended the proto-AARS to interaction with the anticodon. From that point on, specific aminoacylation was achieved by recognition of tRNA acceptor stem and anticodon sequences such that codon–anticodon interaction was reliably connected to a specific amino acid. In this manner, modern translation could have gradually developed.

This leads to a situation where the basic structure of the code emerges in such a way as to reinforce itself—not with a forward-looking purpose but simply because it can. This is similar to the proliferation of transposable elements in genomes, they accumulate because they can, autocatalytically producing more of themselves. It is distinctly different from the idea of a 'frozen accident', because it was not accidental; it was autocatalytic. Remember that tRNAs with complementary anticodons are recognized by synthetases from complementary genes (Figure 5.20A). Hence, the genes for tRNA and the corresponding AARS genes are closely connected and may well have coevolved: A changed tRNA gene with a changed or unchanged anticodon sequence could only 'survive' and continue to 'translate' if the corresponding AARS gene also changed. Its protein had to better match the new tRNA structure and vice versa, a change in the AARS gene in a region important for tRNA–AARS interaction might well have led to a change of the tRNA gene with a better-adapted tRNA structure—an autocatalytic process leading to more optimal tRNA–AARS interaction.

The primordial role of AARSs as very early and perhaps the first enzymes coded by genes has appeal in that the first function fulfilled by the code cements it in place. Since the early 1990s, this basic idea that AARSs are as ancient as the ribosome and generated the code has pervaded much, but not all, literature on the origin of the code, and with various degrees of emphasis.

5.3.3 TRANSLATION EMERGED FROM ENVIRONMENTALLY SYNTHESIZED COMPONENTS

Aminoacyl tRNA synthetases are among the most universally distributed proteins among genomes, which makes perfect sense because no organism can express genes without being able to ligate the correct amino acid to the correct tRNA for protein synthesis. AARSs preserve and express the code in that the interaction between structural information in AARSs, tRNA (even without the anticodon), and amino acid are integrated into a single output signal: properly charged tRNAs destined for mechanical processing at the ribosome. AARS genes are universal among all organisms, because without them translation would not be possible.

But, in biology, there are always exceptions (or apparent exceptions). Some bacterial genomes that encode ribosomal RNA and some tRNAs are so highly reduced that they only encode 122 proteins among which no AARSs are to be found. The immediate question is: How do they make the 122 proteins that their genome encodes? The answer is that those highly reduced bacteria themselves have endosymbiotic bacteria within their cytosol, as Carol von Dohlen and her colleagues found. The endosymbiotic bacteria encode about 500 proteins, among them all 20 AARSs needed to properly charge tRNAs for protein synthesis. During growth, one or the other endosymbiont lyses from time to time, releasing its cellular content to the 122-protein-encoding bacterial host. The bacterial host uses 'previously owned' AARSs to charge its tRNAs. But even then, the two genomes together only encode 622 proteins which is not enough to support a free-living lifestyle. Accordingly, the two bacteria are not free living; they live within specialized cells (called bacteriocytes) of a specialized organ (called a bacteriome) within the body of a mealybug (a sap-sucking insect that afflicts many ornamental plants). The bacterial consortium taps into the metabolism of the mealybug in such a way that the function of about 600 genes is no longer needed because it can obtain the missing compounds from the insect cytosol. In fact, that is the whole basis for the bacterium-bacterium-insect symbiosis because plant sap does not have a high enough amino acid content to support mealybug growth. The insect is just like humans in its dietary requirements for amino acids, it can only make the 11 non-essential ones, the 9 essential ones are made by the bacterial consortium and they are exported to the insect bacteriocyte cytosol. The bacterial consortium cannot synthesize the non-essential amino acids, which it must obtain from the insect. There are a number of interdependences going on there, whereby the plant has to energetically finance the whole thing, supplying sugar for the sapsucking mealybugs to respire in their oxygen consuming mitochondria. Sugar is made in the plant's leaves by photosynthesis, but even the plant cannot access nitrogen by itself, it has to obtain nitrate or ammonium from the soil. There, N_2-fixing bacteria have done their job to keep everyone supplied with N for amino acid biosynthesis to create the AARS problem in the first place.

Why, one might ask, should we bring up this highly derived modern example in an origin of life context? Throughout this book, various biological precursors are discussed as being supplied by the environment as essential starting material for a series of subsequent steps from H_2, CO_2, and N_2 toward chemical complexity. We encountered the concept of a last universal common ancestor of all cells that possessed the genetic code and a conserved core of central intermediary metabolism but was not a free-living cell. Instead, it was confined to a set of inorganic compartments within which both it and its molecular constituents arose. All molecules that were required for that series of transitions before the origin of genes, enzymes, and translation had to be supplied by a natural and, by definition, abiotic source. In that sense, the concept of a LUCA that was only half alive, dependent upon a supply of precursors delivered by the external environment, is nothing radical in the slightest; it is completely in line with the way life works today. Without exception, all origin of life theories, including hydrothermal vent theories (which have the advantage of sustained far from equilibrium synthesis), require an abiotic supply of precursors.

In modern environments, primary producers generate fixed carbon, but comparatively few of them can fix nitrogen by themselves. Everything else on the planet lives from that. Most modern primary production is dependent upon the sun, but some of it, deep in the crust, is still independent of the sun, as it always has been, powered by the chemical energy of serpentinization. For life to have taken hold 4 billion years ago, it had to have been able to live from the simple organic substances that the early earth provided in geochemical settings. To escape from vents, the first microbes had to be able to synthesize everything they needed to live (amino acids, cofactors, high-energy chemical bonds) from the compounds that were environmentally available outside hydrothermal vents. Outside the immediate vent environment, the source of carbon was CO_2, the source of electrons was H_2, the source of nitrogen was NH_3 or N_2, and the source of chemical

energy was the natural tendency of H_2 and CO_2 to undergo exergonic reactions in the presence of suitable catalysts. That means that the first genuinely free-living cells would have necessarily been H_2-dependent chemolithoautotrophs.

The genome-based inference of the physiology of LUCA is in line with the H_2-dependent reduction of CO_2 in simulated vent conditions using vent mineral catalysts, and it is in line with the circumstance that acetogens and methanogens, despite being holdovers from the most ancient split in the tree of life, the one separating bacteria and archaea, are H_2 dependent, chemolithoautotrophic lineages. To attain such a sophisticated level of enzymatic and physiological independence, an umbilical cord of organic compounds from geochemical reactions had to be supplied for the first cells to arise. This is an underlying premise germane to all origin of life theories—the assumption that the environment provided a constant supply of biologically relevant but structurally imperfect precursors for further reactions that gave rise to the biological chemistry of cells. In that view, when biochemical pathways arose, they served to replace a geochemical supply of starting compounds.

5.4 ACTIVATED RNA STRANDS CAN READILY RECOMBINE WITH OTHER RNA MOLECULES

5.4.1 RECOMBINATION BETWEEN RNA MOLECULES COULD HAVE ACCELERATED INFORMATION PROCESSING

Ribozymes still play an important role in the ribosomes of modern organisms and in RNA cleavage. We assume that RNA cleavage by RNA and RNA-self-cleavage are very old and fundamental properties of nucleic acids. The discovery that RNA molecules can exhibit catalytic properties came from studies of intron removal from rRNA transcripts in the ciliate *Tetrahymena*, the pure RNA cleaved itself, and from the discovery that the catalytic activity of the tRNA transcript processing enzyme RNaseP from *E. coli* resides in its RNA component. Those insights led Walter Gilbert to write:

> If there are two enzymic activities associated with RNA, there may be more. And if there are activities among these RNA enzymes, or ribozymes, that can catalyze the synthesis of a new RNA molecule from precursors and an RNA template, then there is no need for protein enzymes at the beginning of evolution. One can contemplate an RNA world, containing only RNA molecules that serve to catalyze the synthesis of themselves.

That is where the term "RNA world" entered the literature. It was a great idea. However, it turned out that apart from the peptidyl transferase reaction, the reactions catalyzed by RNA in the cell are limited to splicing and RNA processing. The exciting prospect that 'there may be more' was not fulfilled. Splicing reactions are transesterifications and are fully reversible from the thermodynamic standpoint. Splicing is simply RNA recombination.

The mechanism of reversible self-cleavage of phosphodiester bonds in ribozymes is known from the detailed quantum mechanical/molecular dynamics calculations of Kumar and

Marx and from isotope experiments (Figure 5.22). The self-cleavage is initiated by the polarization of the ribose 2'-OH by the neighbored guanine in position 8 (G8). The hydroxyl proton changes position to P–O such that the 2'-oxo anion can attack the phosphate group. The nucleophilic attack displaces the oxygen of the next ribose (cleavage of the phosphoester bond). The oxo anion is stabilized by the attachment of a proton from adenine in position 38 (A38). After proton tautomerism (red arrows), a cyclic phosphate has formed at the 2'–3' terminal of the cleaved fragment.

RNA self-cleavage is initiated by internal activation (G8 proton in Figure 5.22). For cleaving and inserting into extraneous RNA, the attacking RNA must be activated, and the recipient RNA adsorbed on a mineral surface or fixed otherwise (Figure 5.23). By insertion, an RNA molecule with new information, new activities, or both, is formed.

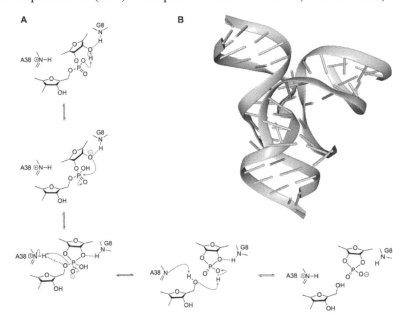

FIGURE 5.22 **Calculated reaction mechanism of self-cleavage in ribozymes.** (A) Self-cleavage of the hairpin ribozyme 2OUE (protein data bank: pdb). For an explanation of the mechanism, see the text. Redrawn with permission from Kumar N., Marx D. (2018) Mechanistic role of nucleobases in self-cleavage catalysis of hairpin ribozyme at ambient versus high-pressure conditions, *Phys. Chem. Chem. Phys.* 20: 20886–20898 (B) Crystal structure of the hairpin-ribozyme 2OUE.pdb.

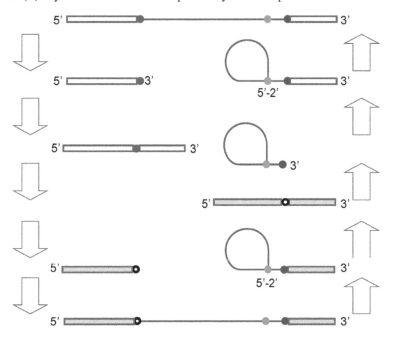

FIGURE 5.23 **One of many possible routes for recombination (RNA sequence transfer).** RNA as genetic material is highly recombinogenic. The inserting RNA (equivalent to a lariat intron) is first excised from a larger RNA molecule, via transesterification reactions, becomes free to diffuse, binds to a second RNA molecule (the one with a black circle between 5' and 3'), and inserts via the reverse of the same transesterification reactions. The reactions are thermodynamically freely reversible (see large arrows). The other 5' and 3' termini can also undergo transesterification reactions. RNA as genetic material is highly recombinogenic. The lariat form of RNA intermediate occurs in eukaryotic pre-mRNA splicing, which involves proteins that stabilize intermediates, although eukaryotic pre-mRNA splicing is derived from self-splicing RNA molecules (called group II introns) of prokaryotes. The colored dots indicate nucleotides that undergo transesterification reactions during the sequence of ligation and cleavage events.

There is only one RNA strand after insertion, but it is longer (see RNA with a black circle between 5′ and 3′) and contains the genetic information of both molecules or different information depending on the reading frame. Hence, if the ribosome has evolved there are more instructions for peptide synthesis. Because transesterification is fully reversible, both mechanistically and thermodynamically, RNA recombination was unavoidable during early molecular evolution. If RNA molecules diffused between different hydrothermal pores, new combinations could arise. Small RNA molecules could become larger by combination.

This opens up many evolutionary opportunities because recombination gives rise to novelties, many of which may have been useful. One question often arises in this context, namely, which functions would have been the first to confer 'advantage' to the contents of the compartment within which they existed? Obviously, molecules that most rapidly organize available components into more of themselves dominate the molecular population, something that Manfred Eigen recognized early in his exponential molecular growth experiments with Qβ replicase. Moreover, early RNA replication could not be of high fidelity, leading to a tradeoff between the rate of replication between the fastest replicating molecules and the preservation of identity in the face of rapidly accumulating variants.

The result of these two processes (fast error-prone replication and fidelity) gave rise to Eigen and Schuster's concept of *quasispecies*, a population of replicating molecules, their plus and minus strands, and their rapidly emerging variants that tend to mutate away from the fastest replicating phenotype. One of the problems with this kind of an RNA world is that, everything else being equal, small RNA molecules replicate fastest, giving rise to molecular parasites that sequester all available monomers (but that can still ligate or recombine). Conceptually, this kind of RNA world leads to something that looks like a polymerase chain reaction of molecular parasites, but it does not lead to anything with similarity to metabolism or a cell, because neither metabolism nor cells exhibit replicating RNA molecules.

Thus, a scenario where selection among molecules for the fastest replicators has problems with regard to the inevitable endpoint trajectory (pure parasites). But the problems at the foundation of such a scenario are far more severe, namely, in order for anything to replicate even once, a constant supply of exactly its constituents in exactly the right molar ratios (assuming low fidelity replication) and in activated (but unreacted) form without contaminating chain terminating moieties has to be supplied—in the absence of enzymes. In theoretical population genetics that might be imaginable. But looking at the far from equilibrium, but utterly primordial organic synthetic reality that a hydrothermal (or any other real world) setting might offer, a constant supply of pure ribonucleoside triphosphates as in Eigen's Qβ experiments from the 1970s would seem completely out of the question. We have already criticized the concept of an RNA world in Section 4.8 and present some additional considerations in the next section.

5.4.2 THE RULES OF STATISTICS DETERMINED EARLY GENETIC INFORMATION PROCESSING

What is the probability of a correct assembly of a nucleotide by the random combination (synthesis without enzymes in free solution) of nucleobase, ribose, and phosphate and condensation to an RNA strand? The probability for installation of the correct ribose isomer is 1/2 (D-ribose or L-ribose), condensation of ribose and not phosphate with the nucleobase is 1/2, binding of nucleobase to the correct OH-group of ribose is 1/4 (4 ribose OH-groups), binding of phosphate to the correct OH-group of ribose is 1/3, and condensation of the next nucleotide to phosphate at position 3′ (for 5′ → 3′ polymerization) is 1/2. Hence, the total probability is $1/2 \cdot 1/2 \cdot 1/4 \cdot 1/3 \cdot 1/2 = 1/96$. This simple calculation assumes equal reaction probability at the different positions. Overall, the assembly of separate units is statistically unfavorable compared to condensation of poly-phosphorylated ribose with amino acids which cycle to nucleobases (Section 4.5). Also, the nucleobases do not dissolve well in water (except uracil) whereas the amino acids and the nucleosides do.

During early evolution, docking of nucleotides to the mother strand was probably not directed by enzymes but determined by the selectivity of G–C and A–U interaction. Eigen guessed that every 100th nucleotide docked incorrectly and assumed $P = 10^{-2}$ as mutation probability. During evolution, enzymes developed which decreased the probability of incorrect docking and corrected for docking errors. Thus, the length N of RNA strands, which could be replicated without errors, increased. This development stretched to a limit, because occasional replication errors are both unavoidable and necessary as new evolutionary starting material for adaptation to a changing environment. The error probability must be 10^{-6} at least and 10^6 (0.1×10^6 to 14×10^6) is indeed the approximate number of base pairs in bacterial DNA. The current measurement of the error rate in *E. coli* (roughly 5 million base pair genome size) is 10^{-3} per generation (one new mutation for every 1000 new cells) or roughly 0.2×10^{-9} per genome per generation, under conditions of a rich medium. This is a very low error rate to adapt to a changing environment, but in nature, the incorporation of new sequences from other species dwarfs the accumulation of new mutations in natural *E. coli* strains and ensures adaptation by other means, namely, lateral gene transfer.

How long does it take until a mutation dominates in a pool of molecular phenotypes? The number of generations n can be roughly estimated by a simple calculation performed by Kuhn and coworkers. The basic idea is that a mutant lives longer than the normal form and therefore replicates more often in a 'copy phase' (phase with favorable environmental conditions) and dies less frequently in a 'death phase' (phase with unfavorable environmental conditions). Overall, the replication frequency is $r > 1$ for the mutant compared to $r = 1$ for the normal form. Let us assume that after a 'death phase' one mutant survives. This mutant increases to r mutants after one generation and to $r \cdot r$ mutants after the next generation. After n generations, r^n mutants survive. Let us further assume one mutation in 10^9 molecules such that a pool of 10^9 normal molecules is needed for the mutant to emerge. Replication frequency $r = 1.1$ means that the mutant replicates 10% more than the normal form. If $r^n > 10^9$, then the mutants are in the majority. After 218 generations ($1.1^{218} = 1.06 \times 10^9$), the mutation prevails. Assuming one copy cycle per day, the mutants prevail after 218 days.

Is the primordial replication accuracy sufficient for mutants to dominate a pool in a few hundred generations or does the mutation change before to a new, possibly failed mutation? The probability of an RNA copy error without the help of enzymes is $P = 10^{-2}$ for one nucleotide according to Eigen's guess and P·N for an RNA strand with N nucleotides. In addition, these erroneous copies add up in every generation. Thus, the total copy error after n generations is P·N·n and must be smaller than one for the survival of the correct single mutant: P·N·n < 1 or P·N < 1/n. Assume that primordial RNA strands were short, say $N = 20$. Then P · N = 10^{-2} · 20 = 1/5 per replication, hence one copy error after five generations. That is much more frequent than the value of 218 generations necessary for a mutation to dominate. In this broad-stroke estimate, base pairing selectivity is not sufficient for selection of the best mutations. Further factors must have improved RNA replication accuracy early on; this could have entailed RNA–peptide aggregates. Recent experimental work of Frenkel–Pinter and coworkers is in line with this view.

5.4.3 The Virtues of a Slow Start

Let's step back for a minute from the idea that there was some kind of fast or exponential replication in some kind of RNA world, similar to exponential phase *E. coli* with a doubling time of 20 minutes given optimal conditions and an optimal set of substrates. Maybe we are looking at the wrong biological example. *E. coli* lives in our intestines and only doubles every 20 minutes in the laboratory anyway (thank goodness, otherwise we would be consumed by our intestinal flora). What if we look at low-energy environments, where cells are just surviving, not growing but also not dying? Such low-energy environments can be found among other places in marine sediment where nutrient flux from the surface is low. In such environments, doubling time is the wrong concept because the cells do not actually double in mass, they get very small, and they just convert available carbon into a 'new' cell without doubling. The appropriate term is 'turnover time', such that the atoms in the cell are replaced by new ones, regardless of whether a cell division has occurred. In low-energy environments, turnover times are estimated to be on the order of hundreds or even thousands of years. That is barely alive. It is a non-living state of potentially animate matter. It is possible that this kind of state might resemble the organization of matter before the first cells started dividing. In this very early state, polymers folded into stable conformations, harnessing what energy there is to maintain their form, battled against hydrolysis, separated the phase of hydrophobics from solutes, and so on. That kind of organic gunk, or peptide-laden ribofilm, would however still have properties of catalysis, so that reactions could take place if substrates or activated substrates came along. This kind of 'slow start' state would be closer to the organizational state of a cell. Cells are assembled in a spontaneous process anyway.

In that view, the chemical reactions catalyzed in hydrothermal vents by their organic reaction products would simply lead to the accumulation of molecules, some or many of which were very similar to those used by present biological systems. If we look at the prebiotic synthesis literature openly, regardless of what one starts with, as long as the reaction is performed under anaerobic conditions and has suitable reactive carbon, nitrogen, hydrogen, and oxygen starting compounds as well as effective catalysts, the products that accumulate tend to be biological molecules or structurally very similar to them. That is just another way of saying that amino acids, bases, even cofactors, and the carbon backbones of central intermediary metabolism tend to have a low energy of combustion, that is, they are stable compounds that will tend to accumulate in thermodynamically controlled reactions.

The concept of a slow start, or a ribofilm, places catalysis by stable compounds and exergonic reactions in the foreground. That basic configuration has a great deal in common with the metabolism and physiology of modern cells. Whereas modern chemolithoautotrophs accumulate roughly one carbon atom into cell mass for every 20 that goes through the cell as an exergonic reaction producing acetate or methane, the corresponding ratio for the first inorganically catalyzed chemical systems was probably closer to one to 20 million, we might guess. Although, the situation might not be that dire, if the right catalysts are provided. If we look once again at the H_2 plus CO_2 experiments of Preiner *et al.* we see that they only observed main products that are integral to metabolism—on average 100 mM formate, 100 µM acetate, and 20 µM pyruvate. The main 'waste' product formed, formate, only exceeded pyruvate by four orders of magnitude.

With the origin of genes and proteins, the process of molecular organization is instantaneously freed from the almost suffocating constraints that geology and physical chemistry pose. We try to avoid the term molecular self-organization because the organic world did not arise in free solution, it arose within an inorganic world where surfaces adsorbed, bound, catalyzed, concentrated, and partitioned organic compounds which were formed from H_2, NH_3, and CO_2. The term molecular self-organization implies independence from the environment. Molecular organization leaves room for contributions from the environment.

5.4.4 The Function of the First Genes

Which functions would have been the first to confer 'advantage' to the contents of the compartment within which they existed? In an RNA world, molecules that most rapidly organize available components into more of themselves dominate the molecular population. But if the first competing molecules are protein catalysts, not RNA, what would be the first kind of function to tip the scale in the direction toward increasing complexity that ends up at life? One could make an argument that the *very* first protein catalysts whose genes would be fixed as irreplaceable in the system would be those that channel H_2 and CO_2 into the basic starting material of metabolism: C1, C2, and C3 units. But if inorganic catalysts such as Ni_3Fe can serve that function, then in natural hydrothermal conditions, the backbone of carbon metabolism into formate, acetate, and pyruvate unfolds not only without enzymatic help, it does so with *specificity*. With H_2, CO_2, and mineral catalysts generating a C1–C3 feedstock for metabolism, the first protein-coding genes would not be essential to direct that flux. They could improve it, but the constant flux of a basic chemical feedstock underlying metabolism exists under primordial vent conditions (more CO_2 than today) without enzymes.

Upon reflection, it becomes clear that the property that would most strongly impact autocatalysis toward molecular complexity *in the presence of genes and proteins* would be proteins that cement the genetic code—any genetic code—in place. That points directly to aminoacyl tRNA synthetases as the very first functionally selected protein-coding genes, as Paul Schimmel, Charlie Carter, Richard Wolfenden, and others have been saying for some time. If we look at the matter openly, AARSs do not interpret the code, they actually generate it. The interactions between AARSs, tRNAs, and amino acids produce the association of information in the acceptor stem and the anticodon of an individual tRNA with the amino acid that is specified by that codon in the genetic code (Section 5.2.5).

The identities of the amino acids ligated onto tRNA are the result of AARS catalytic site specificity for tRNA charging. In that sense, the identity of the 20–22 amino acids that made it into the code would reflect their availability, that is, their local concentrations at the site where the code arose, combined with their acceptance at the AARS catalytic site. The evolution of the two main AARS families, Class I and Class II, together with their differentiation into the ca. 20 specific AARSs that have preserved the code until today would thus appear as the most likely candidate for the first genes. *The AARS family determined the rules for transforming heritable sequence into fitness-impacting structure and thus started Darwinian evolution.* Moreover, AARSs transformed the relationship between sequence and structure into a *constant*, as opposed to a variable, whereby the first output of that constant (AARS genes and proteins) cemented itself in place. In four billion years of subsequent evolution, genes have varied and tRNAs have varied but those variants that did not adhere to AARS substrate specificities were eliminated, as were AARS variants that altered the basic code (see also discussion at the end of Section 5.2).

In that sense, if we look at the genetic code for what it is, namely the result of amino acid and tRNA substrate specificities of AARSs enzymes, the puzzle of its origin would appear to dissolve into thin air. The table of the genetic code (Table 5.2) is nothing more than the result of AARS substrate specificities. There is a long tradition in origins thinking and literature that assumes the origin of the code to have preceded the origin of protein synthesis, such that AARSs came to interpret a code that existed prior. If that premise is false, namely if the origin of the code emerged from virtually no specificity to high specificity among the first gene families—the AARSs themselves—then the problem of the origin of the code shifts status from intractable to a process of working out the details. Retracing the phylogeny of the two AARS families could only solve the problem if molecular evolution were neutral during that early phase (a prerequisite for phylogenies to work), a premise that is not sound under any circumstances because the structures were intensely selected, and the substitutions that led to AARS specificity were not neutral. Given that, the origin of the code cannot be considered as solved, but if we relax the RNA world concept, or let go of it altogether, the nature of the problem with regard to the origin of the code changes fundamentally. It is entirely possible, if not likely, that NRPS preceded AARS-tRNA dependent translation. But NPRS would not require an RNA-dependent code. A key

observation is that the genes for tRNA and the corresponding AARS genes are closely connected (Figure 5.20 A) and probably have coevolved.

In the next chapter, we will turn to the origin of the ribosome and the first steps toward the origin of self-contained, self-sufficient, self-replicating free-living cells.

* * *

5.5 CHAPTER SUMMARY

Section 5.1 Changes in water activity as in serpentinizing olivine minerals can lead to the synthesis of thermostable and salt-tolerant proteins with hydrophobic pockets in the interior. Such peptide structures can persist, if a peptide can synthesize a copy of itself by forming interhelical recognition surfaces or if the synthesis is directed by a peptide template. Sustainable replication of more complicated proteins with highly ordered sequences is possible with RNA-template-directed synthesis. Nonribosomal peptide synthesis (NRPS) is widespread among modern prokaryotes. For example, the bacterium *B. brevis* synthesizes the cyclic decapeptide antibiotic tyrocidine with a protein complex consisting of ten protein modules, one for each amino acid. Each module consists of an adenylation domain, subsequent transfer of the activated amino acid to the phosphopantetheinyl cofactor of the neighboring thiolation domain, a peptide condensation domain, and in specific modules epimerization and finally cyclization. Specificity for the added amino acid is achieved by the pattern of the recognition surface of the adenylation and thiolation domains of the respective module.

Section 5.2 Ribosomal protein synthesis was probably the most important invention in evolution. Ribosomes condense amino acids in the order specified by codons of three nucleotides each of a messenger RNA (mRNA) strand to form a peptide chain. To perform this task, the ribosome contains three neighboring RNA binding sites A, P, and E and slides along on mRNA. At the 'Amino acid site' A the anticodon of a transfer-RNA (tRNA) loaded with its specific amino acid by an AARS binds to the complementary codon of mRNA. At the 'Peptide site' P transpeptidation of an already synthesized peptide chain to the amino acid at site A prolongates the peptide chain. Then, the ribosome moves one codon in the 3' direction (translocation), so that the free tRNA at site P is now at the 'Exit site' E, where it leaves the ribosome. Four possible scenarios are discussed for the origin of this complicated machinery in order of proximity to modern translation.

I. *Polyester template with sidechains for a binary code binds a complementary nascent strand for synchronous condensation of amino acids.* Peptides and a nascent strand from the condensation of malic acid, malamide, and β-alaninamide are synthesized synchronously. The nascent polymalamide strand has short and long side chains for a binary code (information polymer: possible evolutionary precursor of nucleic acids) and can couple to a complementary template strand. A peptide strand is condensed to the terminal of the nascent strand and a thioester-activated amino acid–malamide monomer reacts with the peptide–poly-malamide ester bond and extends the peptide chain by one amino acid and the poly-malamide strand by one monomer. The malamide side chains extended with β-alaninamide can exist in different conformations which correlate to the hydrogen-bond orientations of the four RNA nucleobases and may have been their precursors.

II. *RNA template binds amino acids with its nucleobases covalently and specifically.* Amino acids bind to specific nucleobases of RNA strands and polymerize to peptides. Then, the sequence of the amino acids in the peptide is determined by the sequence of the nucleotides in the RNA. Indeed, transfer RNAs contain many modified bases, and some of them are altered by binding to amino acids. The process can become autocatalytic because the reverse hydrogen bonding of the product peptide side chains to activated nucleotides is feasible too.

III. *RNA template binds nonenzymatically formed RNA–amino acid complexes noncovalently.* The nucleotide activator of an amino acid couples via Watson–Crick bonding (without enzyme) to a specific RNA stem-loop structure ('hairpin'; precursor of tRNA). Then, the activated amino acid binds to the 3′-terminal of 'its' hairpin and the hairpin attaches via an anticodon nucleotide sequence at its loop to the codon of the RNA template ('collector': precursor of mRNA). Further hairpins with covalently bound amino acids attach to the collector. Then, the aligned amino acids react to a peptide which sticks to the aggregate of RNA collector and RNA hairpins and stabilizes it against hydrolysis. The collector code is optimized for the most effective peptide stabilizer by molecular evolution. Larger peptides will evolve which envelop the whole aggregate of collector and hairpins; some of them may interact with a specific RNA hairpin and 'its' amino acid and activating agent. Such peptides may evolve to the precursors of aminoacyl-tRNA synthetase (AARS) enzymes (called 'urzymes'); they will autocatalytically enhance their own synthesis.

IV. *RNA template binds RNA-amino acid complexes formed by 'urzymes' noncovalently.* This scenario approximates modern translation. How do the hairpins and urzymes of this scenario compare to modern tRNA and AARSs enzymes? The early transfer hairpins and urzymes were certainly short oligomers but experiments showed that short peptides with interaction only to a few bases of the acceptor stem of the RNA hairpin and to amino acids and nucleoside triphosphate activators already increase the efficiency of aminoacylation considerably. Possibly, there was no or very little contact of the short RNA hairpin in the proto-AARS complex with coding linear RNA strands. At first, peptide synthesis was not RNA template directed, just more efficient than in free solution. Later, with the increasing size of the RNA hairpins achieved by self-templating and self-priming, the anticodon sequence migrated from the acceptor stem to exposed positions of the hairpin at some distance from the aminoacylation site where the anticodon could interact with the codon of an RNA template (proto-mRNA). A second, non-conserved protein domain extended the proto-AARS to interaction with the anticodon. Notably, tRNAs with complementary anticodons are recognized by AARS from complementary genes, that is, their coding sequences trace to the complementary strands of the same double-stranded nucleic acids. Thus, at some early stage of evolution of the code, it is possible the single strands of RNA double strands (*both* single strands!) encoded tRNAs in part of their sequence and the corresponding urzymes (proto-AARS) with in another part of their sequence. Specific amino-acylation was achieved by both recognition of tRNA acceptor stem and anticodon sequences. In this way, modern translation could have gradually developed.

Section 5.3 Many tRNAs and the ribosomal RNA (rRNA) require conserved modified bases for functionality. Possibly, there was a large pool of diverse modified RNA nucleotides that were synthesized geochemically before highly specific biosynthetic pathways evolved, and some of them supported the evolution of translation and the code. Often the bases are modified by methylation which increases hydrophobicity via displacement of water molecules by the methyl groups, weakening hydrogen bonds in neighboring base pairs. Hence some base modifications may be kind of a 'lubricant' to form the rather stressed cloverleaf structure of tRNA and enhance tRNA mobility and rRNA flexibility around the peptidyl transferase site during translation. Methylations reduce non-specific molecular interactions between tRNA and rRNA during translation. The third codon position can often be any nucleobase (the code is degenerate) and modifications of the nucleobase at the corresponding anticodon position reduce the specificity of its Watson–Crick bonding so that it can bind with any nucleobase in the third codon 'wobble' position.

Section 5.4 Recombination between RNA molecules is very efficient and could have accelerated information processing markedly. The inserting RNA is first excised from a larger RNA molecule by self-cleavage of phosphodiester bonds and transesterification reactions, becomes free to diffuse in the form of a lariat, binds to a second RNA molecule, and inserts via the reverse of the same transesterification reactions to one longer RNA strand containing the genetic information of both molecules or different information depending on the reading frame. If RNA molecules diffused between different hydro-thermal pores, new combinations could arise. Based on base pairing selectivity it is possible to guess the mutation probability during early replication and the time necessary for a mutation to dominate a pool of molecular phenotypes. Calculations show that new mutations occur before a mutation can dominate the pool—base pairing selectivity is not sufficient for selection of the best mutations. RNA–peptide aggregation for example must have improved RNA replication accuracy in early chemical evolution.

* * *

PROBLEMS FOR CHAPTER 5

1. *Nonribosomal peptide synthesis.* How does nonribosomal peptide synthesis (NRPS) work? Describe the main chemical steps.

2. *Translation.* How does ribosomal peptide synthesis (translation) work? Describe the main chemical steps.

3. *Primordial template-directed peptide synthesis.* Summarize four different scenarios of primordial template-directed peptide synthesis with one sentence each.

4. *Urzymes.* Outline a route by which RNA transfer hairpins and urzyme peptides could have developed.

5. *AARS and tRNA genes.* How are the gene sequences of the core domain (urzyme) of the two classes of AARS related? How does tRNA codon complementarity come into play?

6. *Modified nucleobases.* What is the role of methylated nucleobases in tRNA and rRNA?

7. *Degenerate code and wobble pairing.* Calculate the degeneracy of the code and describe how several tRNAs recognize more than one codon.

8. *Transfer RNA.* List some features common to all tRNAs.

9. *AARS–tRNA binding.* Describe some properties of amino acids and tRNAs that AARS enzymes utilize to discriminate between different amino acids.

How do the two classes of AARS differ in terms of the topology of their tRNA binding, that is, which side of the helix do they bind?

10. *DNA reading frame.* What is an open reading frame (ORF) in a nucleotide sequence? What happens if a mutation leads to the insertion or deletion of a nucleotide in the ORF? What kind of mutations are silent in an ORF?

11. *Energy cost of protein synthesis.* How many ATP and GTP molecules are consumed to synthesize a protein from 300 uncharged amino acids (as an example) given a functioning ribosome and required tRNAs?

12. *mRNA codons.* Use the mRNA codons in Figure 5.20 B to obtain the peptide sequence of the following transcribed mRNA: 5'-AUGCUUGUUGCUGGUCGUUAG-3'. What happens if U in position 6 is replaced by C? What happens if U in position 5 is replaced by C?

13. *Peptidyl transferase.* Describe the mechanism of peptide transfer during translation and the location (which molecule) of the active residue that performs the peptidyl transfer reaction.

14. *RNA recombination.* Explain briefly how recombination (RNA sequence transfer) works and whether it requires external energy input.

15. *Mutation probability.* In very early molecular evolution, was base pairing selectivity alone sufficient for the selection of the best mutations?

6 Innovations on the Path to Cellularity

6.1 FIVE MAIN HURDLES NEED TO BE OVERCOME IN THE TRANSITION TO FREE-LIVING CELLS

Chapter 5 outlined the emergence of a molecular system that is capable of primitive translation and protein synthesis using heritable and variable information that is stored in genes. With the help of heritable variation underlying the origin of new catalysts (enzymes and cofactors), the origin of almost any protein-based molecular innovation found in modern chemolithoautotrophic anaerobes became possible. That does not mean, however, that the origin of translation made the transition to the free-living state easy or guaranteed. On the contrary, the process of escape from the vent as a free-living cell might be just as hard as the origin of translation. The chemical reactions that we have discussed so far require solid-phase inorganic catalysts in a hydrothermal setting and continuous chemical disequilibrium afforded by serpentinization. At the organizational level described in Chapter 5, genes and their encoded proteins are young, short, dispersed in space, and not well defined in sequence. Their information resides in chemically labile RNA and is hence not securely stored. Translation exists but ribosomes have yet to take on a discretely physical form.

In free-living cells, all of the information and the catalysts (proteins and cofactors) that are needed to synthesize a fully self-sufficient daughter cell are contained within one small cell that is about 1 µm on a side that contains about 10,000 ribosomes and one DNA molecule. Such a cell can diffuse in the environment as a diaspore (a unit of dispersion) and colonize new environments, as long as the proper nutrients are available. The transition from exergonic chemical reactions at a serpentinizing hydrothermal vent to free-living cells is anything but trivial. The evolutionary transition to escape is not just a matter of 'pack up your bags and go'. The analogy is closer to 'pack up your bags, go to the garage and build a car from whatever stuff is lying around, fill it up with some made-to-order fuel you happen to have in unlimited supply, start the engine, then get those bags that you already packed, then get in, and go'.

This chapter and Chapter 7 will cover chemical evolutionary steps needed to support escape: the transition of a translating system fed by geochemical energy to a full-fledged, *self-replicating* free-living cell. The origin of translating chemical reactions which we encountered in Chapter 5 depends on chemical compounds, physical compartments, and catalysts that the hydrothermal system supplies. Free-living chemolithoautotrophs make new and viable copies of themselves from a few gases and mineral salts. Given inorganic substrates, they double. They are self-sufficient in terms of coding, catalysts, and energetics. Once they escape from their inorganic housing, there is no turning back, and the Earth is irreversibly inhabited. The key innovations required for escape are:

1. The origin of DNA as a stable genetic storage medium
2. The origin of genomes large enough to direct synthesis of a free-living cell
3. Lipids and cell walls to insulate reactions from solid-phase catalysts
4. Tapping the energy of a geochemical ion gradient (the ATP synthetase)
5. Replacing the natural ion gradient with ion pumping to become energetically self-sufficient.

These innovations involve the origin of novel functions, new enzymes encoded by genes. The first three will be the subject of this chapter, while Chapter 7 will look at energetic hurdles that need to be surmounted for escape as free-living cells and early physiological diversification.

6.1.1 AN AUTOTROPHIC ORIGIN WAS PROPOSED IN 1910, DEEP-SEA VENTS WERE DISCOVERED IN 1978

Chapters 4 and 5 were focused on intermolecular interactions. As we turn to the process of cellularization and escape, it is a good time to recall the physical environment and the conceptual context within which the hydrothermal theory operates. The idea that the first cells were anaerobic autotrophs (the theory of autotrophic origins) has a long tradition in microbiology. The basic theory has been around for about 100 years, as documented in Figure 6.1. Based on observations from microbial growth, the Russian biologist Constantin Mereschkowsky proposed in a 1910 paper published in German that the first life forms (1) had a minimal size, inaccessible to the microscope; (2) lacked organization (today we would say prokaryotic organization); (3) had the ability to survive temperatures close to the boiling point (were thermophiles); (4) could live without oxygen (were anaerobes); (5) had the ability to synthesize proteins and carbohydrates (the latter without the help of chlorophyll) from inorganic substances (were chemolithoautotrophs); and (6) exhibited resilience against alkaline solutions, concentrated salt solutions, sulfur compounds, and diverse toxins (translation of Figure 6.1).

Those 1910 thoughts are very close to what proponents of autotrophic origins at hydrothermal vents are saying today. The idea that the mineral-like FeS clusters in ferredoxins are ancient was current in 1966, as the classical paper by Eck and Dayhoff attests. The idea that acetogens and methanogens are ancient was familiar to Decker and colleagues in 1970, well before anyone knew that methanogens are archaea, and well before anyone knew about deep-sea hydrothermal vents. The first deep-sea hydrothermal vents were reported by Corliss and colleagues in 1978. They were of the black smoker type (ca. 400°C hot effluent) (Figure 6.2A) and they were immediately discussed at length by John Baross and colleagues in the context of

DOI: 10.1201/9781003378617-6

origins (Figure 6.2B), although too much early debate centered around the issue of temperature.

Lost City, the first example of an 'off-ridge' submarine hydrothermal vent, with much cooler effluent (ca. 70°C), H₂-producing, and highly alkaline (ca. pH 10) (Figure 6.3), was reported in 2001. There has not been much time to explore the chemistry and catalytic properties of Lost City-type (off-ridge) hydrothermal vents

as they relate to origins. Several properties of off-ridge vents set them apart from black smokers. Black smokers are situated directly above magma chambers at spreading zones; their hydrothermal current comes into close contact with magma and thus emerges as ca. 400°C hot, often acidic fluid, and they have a short life span on the order of decades. Off-ridge vents like Lost City are situated kilometers away from the spreading zone such that

Forderungen, welche unumgänglich an die ersten Organismen gestellt werden müssen.	Eigenschaften der Bakterien, welche diesen Forderungen entsprechen.
1. Minimale Größe, unerreichbar für das Mikroskop.	1. Die bakteriellen Nebel bestehen aus unter dem Mikroskop unsichtbaren bakterienartigen Organismen — den Biokokken [107]).
2. Abwesenheit von Organisation.	2. Bei solch einer geringen Größe, entsprechend dem Gesetze der Abhängigkeit der Organisation von der Größe, können die Biokokken keine Organisation haben.
3. Fähigkeit, hohe Temperaturen nahe am Kochpunkte auszuhalten.	3. Die Bakterien vertragen in vegetativem Zustande eine Temperatur bis 98°, in reproduktivem Zustande bis 150°.
4. Fähigkeit, ohne Sauerstoff leben zu können.	4. Die größte Mehrzahl der Bakterien kann ohne Sauerstoff leben.
5. Fähigkeit, Eiweiße und Kohlehydrate (letzteres ohne Vermittlung des Chlorophylls) aus unorganischen Stoffen zu bilden.	5. Die Bakterien sind fähig, Eiweiß und Kohlehydrate (letzteres ohne Vermittlung des Chlorophylls) aus unorganischen Stoffen zu bilden.
6. Widerstandsfähigkeit in bezug auf Lauge, starke Salzlösungen, Schwefelverbindungen und verschiedene Giftstoffe.	6. Bakterien vertragen Lauge, stark konzentrierte Salze, Schwefelwasserstoff, große Dosen verschiedener Giftstoffe.

FIGURE 6.1   Early ideas about autotrophic origins. See text. The table is from Mereschkowsky C. (1910). Theorie der zwei Plasmaarten als Grundlage der Symbiogenesis, einer neuen Lehre von der Entstehung der Organismen. *Biol. Centralbl.* 30: 278–288, 289–303, 353–367; Kowallik K. V., Martin W. F. (2021) The origin of symbiogenesis: An annotated English translation of Mereschkowsky's 1910 paper on the theory of two plasma lineages, *Biosystems* 199: 104281.

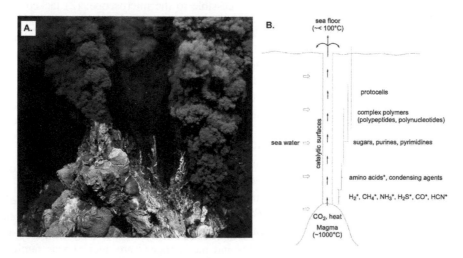

FIGURE 6.2   Original ideas about microbial origins at vents. (A) A black smoker in 2980 meters depth on the Mid Atlantic Ridge. Image from MARUM—Center for Marine Environmental Sciences, University of Bremen (CC-BY 4.0) www.marum.de/en/Discover/Deep-Sea.html. (B) This figure is redrawn from Corliss J. B., Baross J. A., Hoffmann S. E. (1981) An hypothesis concerning the relationship between submarine hot springs and the origin of life on Earth, *Oceanol. Acta* 4: 59–69. See also Baross J. A., Hoffman S. E. (1985) Submarine hydrothermal vents and associated gradient environments as sites for the origin and evolution of life, *Orig. Life* 15: 327–345. The asterisks indicate compounds that as of 1981 had been reported in hydrothermal systems, although later work found that many of the early reports of amino acids in hydrothermal systems stemmed from decaying cells. See however abiotic glycine in Nobu M. K. et al. (2023) Unique H₂-utilizing lithotrophy in serpentinite-hosted systems, ISME J. 17: 95–104.

FIGURE 6.3 Spires at Lost City Hydrothermal Field. Photograph from www.flickr.com/photos/51647007@N08/5277 251233, with creative commons license CCBY https://creative-commons.org/licenses/by/2.0/. See Kelley D. S. *et al.* (2005) A serpentinite-hosted ecosystem: The Lost City hydrothermal field, *Science* 307: 1428–1434.; see also Früh-Green G. L. *et al.* (2003) 30,000 years of hydrothermal activity at the lost city vent field, *Science* 301: 495–498.

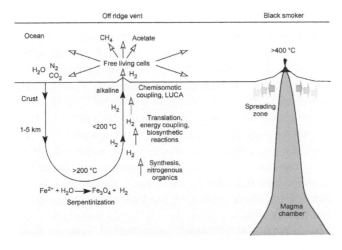

FIGURE 6.4 Serpentinization at an off-ridge vent (left) and chemical processes at origins. Right, a black smoker system at a spreading zone with effluent too hot for life. Modified from Preiner M. *et al.* (2018) Serpentinization: Connecting geochemistry, ancient metabolism and industrial hydrogenation, *Life* 8: 1–22. Regarding the source of reduced nitrogen, see Shang X. *et al.* (2023) Formation of ammonia through serpentinization in the Hadean Eon, *Sci. Bull.* 68: 1109–1112.

their circulating water does not come into contact with magma, having an exit temperature typically below 90°C, and they have a much longer life span on the order of tens of thousands of years. Although H_2 production through serpentinization can take place at both types of vents, the synthesis of catalysts such as Fe_3O_4 and Ni_3Fe, the abiotic synthesis of organics (formate, methane), their long life span, alkaline pH and milder temperatures make the off-ridge vents particularly interesting in modern day models of early Earth environments.

Serpentinizing systems share many similarities with chemolithoautotrophic life: a continuous source of chemical disequilibrium, redox chemistry, reactants, reductants (H_2 generated from rock water interactions), catalysts, and compartments to concentrate products. In the coming sections, we will look at the concepts in the left-hand panel of

Figure 6.4 in greater detail to provide a framework for the daunting molecular transition from simple chemical reactions to free-living cells.

6.1.2 SPONTANEOUSLY FORMING SYSTEMS OF INORGANIC COMPARTMENTS WERE THE PRECURSORS OF CELLS

Schleiden and Schwann independently recognized in 1838 that plants and animals are made up of cells. In the nearly 200 years since, no exceptions to that rule have been found. All life is organized as cells. Cells are biologically formed compartments that separate the cell's content—the cytosol and substances dissolved therein—from the environment. The cell membrane and cell wall permit the cell to retain the compounds and catalysts that are required for its chemical reaction to go forward. The first life forms must also have been organized as cells. In cells, the boundary that immediately surrounds the cytosol is a semipermeable membrane consisting of a lipid bilayer and proteins. In microbes, the membrane is surrounded by a turgor-resistant cell wall that faces the environment.

At origins, what came first: the rigid wall or the semipermeable membrane? Many traditional theories for origins start with vesicles that spontaneously form in solution from abiotically formed lipids (vesicles first, Figure 6.5A). Three problems with vesicles first theories concern the questions of (1) how the molecular constituents of the future cytosol could evolve without diffusing out into the environment before their encapsulation in vesicles, (2) how the constituents of cytosol would come to be encapsulated into preformed vesicles, and (3) how external molecular components ('growth' substrates) would traverse the bilayer so as to become part of the cytosol. Hydrothermal theories (Figure 6.5B) start with naturally and continuously forming inorganic compartments that provide a natural concentrating mechanism and consist of important catalysts that synthesize more complex organic contents. The inorganic precursors of cell walls serve as systems of naturally forming compartments, catalytically walled territories within which molecular contents could accumulate and react. Gases, which are the starting substrates for autotrophic theories, can diffuse freely across membranes; they require no importers.

Modern vents of serpentinizing hydrothermal systems form elaborate networks of inorganic microcompartments. An example from the Lost City hydrothermal field is shown in Figure 6.6. The chimneys at Lost City can reach 60 m in height. They are formed of calcium- and magnesium carbonates that precipitate at the vent ocean interface. Lost City has been active for more than 30,000 years. As we saw in Chapter 1, such serpentinizing systems have been in existence since the Hadean, when the first liquid water rained out onto the early Earth. There are even indications that serpentinization-like H_2-producing reactions are taking place today on Enceladus, an icy moon of Saturn. The general setting of a serpentinizing system is not unique to Earth. This and the simplicity of metal-catalyzed reaction of H_2 with CO_2 suggest that the fundamental chemical reactions of life as we know them need not be specific to the Earth.

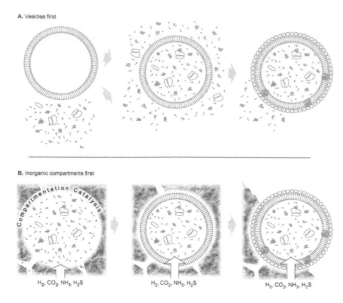

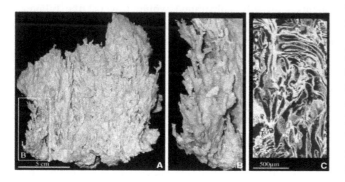

FIGURE 6.5 Vesicles first *versus* compartments first at the origin of cells. (A) In vesicle theories, the components of cytosol are formed by some means, surrounded by lipid vesicles and the turgor-resistant wall arises late. (B) In hydrothermal theories, naturally forming inorganic microcompartments containing transition metals serve as catalysts and as a concentrating mechanism by providing a physical barrier to diffusion. Initially, the catalytic activity of transition metals in the solid-phase converts gases into organic solutes. As enzymes evolve the catalytic activity is gradually transferred to the cytosol. The white arrow in (B) indicates the free diffusion of gases. Part (B) is modified from Martin W., Russell M. J. (2003) On the origins of cells: A hypothesis for the evolutionary transitions from abiotic geochemistry to chemoautotrophic prokaryotes, and from prokaryotes to nucleated cells, *Philos. Trans. R. Soc. B Biol. Sci.* 358: 59–83.

FIGURE 6.6 Naturally forming microcompartments at Lost City Hydrothermal Field. Note the scale of the micrograph in C. Modern cells are about 1 μm in diameter. Reproduced with permission From Kelley D. S. *et al.* (2005) A serpentinite-hosted ecosystem: The Lost City hydrothermal field, *Science* 307: 1428–1434.

In Chapters 2–5 we discussed transitions in the level of molecular complexity in a far-from-equilibrium chemical system that could reach the point of protein synthesis and a genetic code (Chapter 5). At that point, Darwinian evolution can set it: generation of molecular progeny having natural variation and natural selection upon variation in those units. Darwinian selection cannot operate in a homogeneous system in a free solution. Selection requires units that can be selected relative to other units.

In the hydrothermal vent model, naturally forming inorganic microcompartments themselves are not the units of selection. Instead, selection acts upon the organic contents of those compartments (Figure 6.7). Organic contents (proteins, nucleic acids, catalysts, metabolites) can diffuse among compartments, leading to new combinations of molecular contents in different compartments that, by virtue of their molecules, have different properties (natural variation). Figure 6.7 outlines the general process involved, with emphasis on the progression of catalysts from inorganic (bottom) to more complex organic states (top) with intermediate steps labeled on the left. The large grey arrow indicates a flux of hydrothermal water that constantly brings H_2 to the system to react with CO_2 carried into the vent from seawater through mixing.

The effluent from modern serpentinizing systems has typically very low, in some cases even no, CO_2. This is the result of the low solubility of calcium and magnesium carbonates under alkaline conditions ensuing from H_2 production during serpentinization. At origins, reactions of H_2 with CO_2 sequestered in the crust are possible, as are reactions taking place in the downdraft, as CO_2-rich water enters the crust as hydrothermal current to react with nascent H_2 as it arises from serpentinization (or mixing of effluent with marine CO_2 at the vent). Paolo Sossi and colleagues estimate between 10 and 100 bars of CO_2 in the Earth's primordial atmosphere (~10^5 to 10^6 times more than today's) after cooling, generating an N_2–CO_2 atmosphere very similar to that of modern Venus. The vast amount of atmospheric CO_2 was eventually transferred to the mantle as carbonate via solution in the ocean, followed by precipitation as carbonates and subduction (Chapter 1). In the early oceans, there were orders of magnitude more CO_2 available than are available in modern oceans.

In modern serpentinizing hydrothermal systems that have little or no dissolved CO_2 in their effluent, levels of dissolved formate are often quite high, in the 30–150 μM range, indicating that precipitated carbonates are leeched out of the host rock involving their reduction to formate. Formate is not only an intermediate of the acetyl-CoA pathway, it is a growth substrate for many acetogens and methanogens. The main message of Figure 6.7 concerns temporal order of process, not their specific physical locality.

6.1.3 PROTOMICROBES ARE ORGANIZED CHEMICAL SYSTEMS CAPABLE OF PRIMITIVE TRANSLATION

In Chapters 3–5 we discussed possible intermediate states in the organizational grade from CO_2 reduction to autocatalytic networks to a system exhibiting a primitive form of translation. The intermediate result of the microbial origin process is a molecular entity that can perform translation. We designated as a protomicrobe an organized chemical system capable of performing a primitive and heritable kind of translation on the basis of exergonic reactions of H_2, CO_2, and NH_3, but that is not yet a free-living cell (Figure 6.7, top, for example). One might ask whether a protomicrobe is alive or whether it constitutes a form of life. In our view, a protomicrobe is not alive because it is

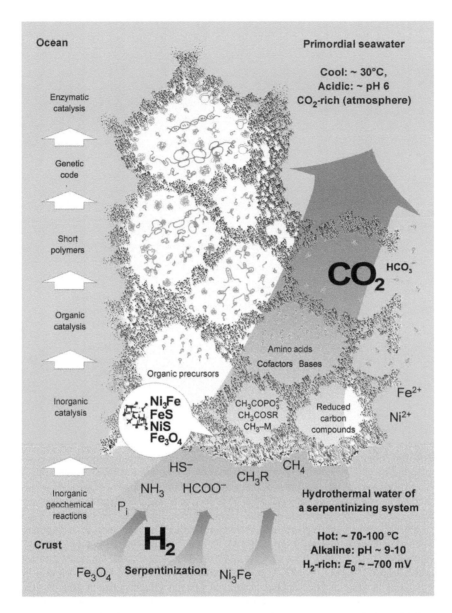

FIGURE 6.7 **The origin of a protomicrobe.** A transition from energy-releasing reactions of H_2 and CO_2 to a confined last universal ancestor capable of translation and possessing the genetic code, a protomicrobe. Modified from the original figure by one of us in Martin W., Russell M. J. (2003) On the origins of cells: A hypothesis for the evolutionary transitions from abiotic geochemistry to chemoautotrophic prokaryotes, and from prokaryotes to nucleated cells, *Philos. Trans. R. Soc. B Biol. Sci.* 358: 59–83.

not free living. It can proliferate within inorganic territories and evolve by having genetic material and translation; it can make catalysts that synthesize more of its own kind, but it is not free living in the way that modern microbes are free living. Protomicrobes are instead confined to the catalytically walled compartments within which they arose. In the present theory, the last universal common ancestor, LUCA, is not a free-living cell but a protomicrobe confined to the inorganic compartments where its components are synthesized and restrained from diffusion. By virtue of genomic reconstructions (Figure 6.8), it had a ribosome, the genetic code, a rotor-stator ATP synthetase, an H^+/Na^+ antiporter, and abundant transition metal catalysts and was dependent upon geochemical processes which supplied essential precursors and the proton gradient generated by serpentinization that could be harnessed by an ATP synthetase.

Because the microcompartments of hydrothermal systems form continuously at the surface of a growing precipitate, the 'growth' and evolution of protomicrobes within an expanding system of territories takes place at those expanding surfaces, where contact with the ocean water maintains an ion gradient. For protomicrobes to make the transition to the free-living state that can disperse as units of growth and selection that can colonize new habitats, some critical innovations are needed. An overview of the innovations required in the transition from a series of energy-releasing reactions between H_2, CO_2, and NH_3 to a free-living state as H_2/CO_2-dependent microbes are outlined in Figure 6.9. Three innovations crucial for escape: Information (a DNA genome and a ribosome that can synthesize proteins), self-enclosure (lipids and cell walls), and bioenergetics (utilization of pre-existing ion gradients and the coupling of exergonic reactions to the formation of an ion gradient).

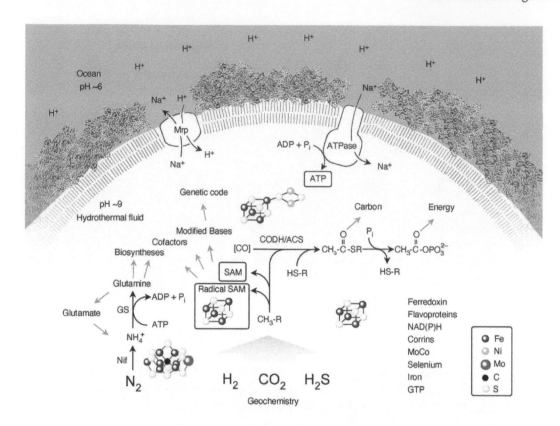

FIGURE 6.8 **LUCA, a protomicrobe.** LUCA was capable of translation and possessed the genetic code as well as nucleic acid genes capable of mutation and replicated by catalysts, hence it was able to evolve. As the microcompartments of hydrothermal systems continuously form at the surface of the precipitate, the 'growth' and evolution of protomicrobes take place at those expanding surfaces, where contact with the ocean water maintains an ion gradient. The lipid-like molecules forming a semipermeable barrier constitute generic hydrophobic compounds and are not genetically encoded lipids. Radical SAM enzymes use a [4Fe–4S] cluster to reductively cleave S-adenosyl-L-methionine (SAM) usually to generate an S-adenosyl radical for activation of C–H bonds, cofactor biosynthesis, peptide and RNA modifications etc. Figure modified from Weiss M. C. *et al.* (2016) The physiology and habitat of the last universal common ancestor, *Nature Microbiol.* 1: 16116; and Weiss M. C. *et al.* (2018) The last universal common ancestor between ancient Earth chemistry and the onset of genetics, *PLoS Genet.* 14: e1007518. The lack of enzymes for the methyl synthesis branch of the acetyl CoA pathway in LUCA suggests that LUCA was dependent upon geochemically supplied methyl groups (CH_3-R). The findings of Preiner M. *et al.* (2020) A hydrogen-dependent geochemical analogue of primordial carbon and energy metabolism, *Nat. Ecol. Evol.* 4: 534–542, and many reports of abiotic methane synthesis (Etiope G, Schoell M (2014) Abiotic gas: Atypical, but not rare. *Elements* 10:291–296) indicate that geochemical methyl synthesis is not only facile but probably unavoidable under serpentinizing conditions in the presence of CO_2.

6.1.4 THE PROTORIBOSOME WAS A SMALL, STRUCTURALLY ULTRACONSERVED UNIT OF FUNCTION

Serpentinization provides a variety of organic compounds for free, without cost to the protometabolic network. Does this truly represent a case of autotrophic origins? The term *autotrophic* comes from microbial physiology. It means that the growing cell can satisfy its carbon needs from CO_2 alone. In practice, many autotrophs have vitamin requirements (auxotrophies) such that autotrophy is expanded to include organisms that satisfy >50% of their carbon needs from CO_2, as Rudolf Thauer has explained. The state of free-living cells is attained when cells arise that (1) can harness environmentally available CO_2 and energy for growth, (2) can compartmentalize themselves and their progeny from the environment by making their own semipermeable membrane and turgor-resistant cell wall, and (3) that can synthesize all of the roughly 1000 catalysts (proteins) required for

that growth from information that they carry in their own nucleic acids. Those criteria set a very high bar for the level of organization that has to be met before a cell can make the transition to the free-living state (Figure 6.9). In particular, the possession and expression of genes for roughly 1000 proteins and autotrophy bring us to the origin of genomic organization from small coding molecules. Proteins are made by ribosomes. How did the ribosome arise?

In Chapter 5 we discussed the origin of translation and the code from simple molecular interactions. For protein evolution to occur, as indicated in Figures 6.7–6.9, a ribosome must exist. An excellent overview of the early evolution of the ribosome in terms of function and ribosomal protein accumulation is given by Fox. A typical prokaryotic ribosome is 50% protein by weight and binds ~55 proteins in bacteria and ~63 proteins in archaea. However, archaeal and bacteria ribosomes only share 33 ribosomal proteins in common; these are universal and thus were present in LUCA. The common ancestor of the bacterial and archaeal ribosome, the

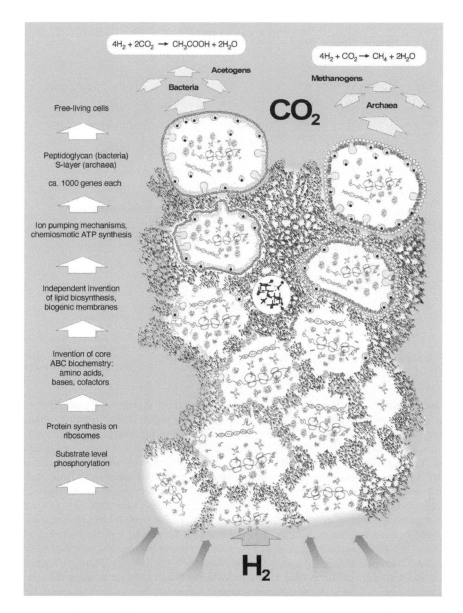

$$4H_2 + 2CO_2 \rightarrow CH_3COOH + 2H_2O$$

$$4H_2 + CO_2 \rightarrow CH_4 + 2H_2O$$

Acetogens

Bacteria

Methanogens

Archaea

CO_2

Free-living cells

Peptidoglycan (bacteria)
S-layer (archaea)

ca. 1000 genes each

Ion pumping mechanisms,
chemiosmotic ATP synthesis

Independent invention
of lipid biosynthesis,
biogenic membranes

Invention of core
ABC biochemstry:
amino acids,
bases, cofactors

Protein synthesis on
ribosomes

Substrate level
phosphorylation

H_2

FIGURE 6.9 Emergence of cells. Transition from a confined last universal ancestor capable of translation and possessing the genetic code to free-living microbes. A sequence of major innovations is listed in the left panel. Modified from Martin W., Russell M. J. (2003) On the origins of cells: A hypothesis for the evolutionary transitions from abiotic geochemistry to chemoautotrophic prokaryotes, and from prokaryotes to nucleated cells, *Philos. Trans. R. Soc. B Biol. Sci.* 358: 59–83. See text.

ribosome of LUCA, was simpler in terms of its ribosomal protein content than either bacterial or archaeal ribosomal (Figure 6.10, top panel). Twenty-one additional bacterial-specific ribosomal proteins evolved on the way to the last common ancestor of bacteria after divergence from LUCA, while archaeal ribosomes possess 33 additional archaeal-specific proteins in the methanogen lineage, and up to 38 additional ribosomal proteins in other archaeal lineages. The bacterial- and archaeal-specific ribosomal proteins are well conserved within the groups, these proteins were added post-LUCA during evolution toward the last bacterial common ancestor, LBCA, and the last archaeal common ancestor, LACA, before diversification into the first bacterial and archaeal lineages (Figure 6.10, top panel). Looking deeper, how, and from what, did the ribosome of LUCA arise?

Possible intermediates in the origin of translation could have included nonribosomal peptide synthesis and structurally simple RNA interactions. While those could have served

as functional precursors in the origin of protein synthesis, the origin of the ribosome itself was a key event in the molecular lineage leading to LUCA. A very useful concept, based on the extreme structural conservation of the peptidyl transferase center (PTC) observed by Yonath and coworkers (Bose *et al.*, 2021), is that of the *protoribosome* (Figure 6.10, bottom panel). The large ribosomal RNA subunit provides the peptidyl transferase center. The catalytic activity of the peptidyl transferase reaction is provided by rRNA, not by a ribosomal protein. The PTC is the site where the growing peptide chain of peptidyl tRNA (at the P site) is transferred to the next aminoacyl tRNA at the A (acceptor) site. The RNA structure around the PTC exhibits detectable symmetry, this region of the structure is called the protoribosome, and it is the site where peptidyl transferase reaction takes place.

Importantly, synthetic RNAs that capture the salient structural elements of the PTC dimerize *in vitro* and furthermore exhibit peptidyl transferase activity *in vitro*, suggesting that

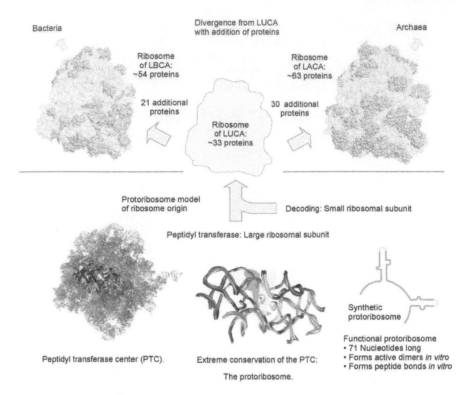

FIGURE 6.10 **A model for the early evolution of the ribosome.** The lower panel sketches the protoribosome model for ribosomal origin as proposed by Ada Yonath and her team. Lower left: The protoribosome region in the structure of the *E. coli* ribosome large subunit. Reproduced with permission from Agmon I. C. (2016) Could a proto-ribosome emerge spontaneously in the prebiotic world? *Molecules* 21: 1701. Lower center: Closeup of the protoribosome showing the axis of semisymmetry (red point and arrow), suggesting an ancestral dimeric nature. The structures of ribosomes for bacteria, archaea, eukaryotes, and mitochondria are superimposed (light and dark-colored lines), and the structural conservation is extreme. Modified with permission from Bose *et al.* (2021) Origin of life: Chiral short RNA chains capable of non-enzymatic peptide bond formation. *Isr. J. Chem.* 61: 863–872; Agmon I. *et al.* (2005) A symmetry at the active site of the ribosome: Structural and functional implications, *Biol. Chem.* 386: 833–844. Lower right: A synthetic PTC. Several different but structurally related synthetic PTCs dimerize and exhibit peptidyl transferase activity [Bose *et al.* (2022) Origin of life: Protoribosome forms peptide bonds and links RNA and protein dominated worlds, *Nucleic Acids Res.* 50: 1815–1828]. The position of the CCA ends of the tRNAs at the P and A sites of the synthetic protoribosome are indicated by color shading. The upper panel of the figure outlines the differential addition of proteins to the ribosome in the lineages leading to bacteria (*E. coli*) and archaea (*T. kodakarensis*). Bacterial and archaeal structures were kindly provided by Natalia Mrnjavac using the structures in PDB. LBCA: Last bacterial common ancestor: LACA: Last archaeal common ancestor. Ribosomal protein numbers from Fox G. (2010) Origin and evolution of the ribosome, *Cold Spring Harb. Perspect. Biol.* 2: a003483. and Londei P., Ferreira-Cerca S. (2021) Ribosome biogenesis in Archaea, *Front. Microbiol.* 12: 686977; and from the PDB structures (N. Mrnjavac).

this essential core function—peptide synthesis from aminoacyl tRNAs—could have been exhibited by a small, stably folding RNA molecule that seeded the origin of the ribosome (Figure 6.10, bottom panel). The heart of the protoribosome model is this: The functional viability of a small structural element of a large molecule encompassing its active site can provide enough selectable function to allow it to act as the seed for its further evolution, provided that the molecule can be faithfully resynthesized in a recurrent manner. We will encounter this functional seed concept again in Section 6.4 in the context of aminoacyl tRNA synthetase origin.

The large ribosomal subunit provides the site for the peptidyl transferase reaction and performs the elongation of nascent proteins by channeling them into the exit tunnel. The small ribosomal subunit participates in translation initiation, aids in selecting the correct reading frame for translation, and helps to ensure the fidelity of codon-anticodon interactions. An accretion model proposed by Petrova *et al.* for the origin of the ancestral small ribosomal subunit starts from a small conserved core and recursively adds RNA helices and domains that generate the functions ultimately conserved

across all ribosomal small subunits. Within the bacterial and archaeal domains, little variation is observed in ribosomal proteins, barring losses in highly reduced lineages such as endosymbionts. Many of the 33 ribosomal proteins shared by bacteria and archaea are encoded in ancient operons with conserved gene order (see Figure 6.13).

This brief account for the origin of the ribosome should not obscure the circumstance that there are dozens of different theories in the literature that address the origin of the ribosome that we have not discussed, whereby most of them operate by the same basic principle: start with something small that performs a beneficial function and modify that initial molecule as it grows in size and function over evolutionary time. The starting point of Yonath's protoribosome model rests upon a particularly robust foundation because the initial function is the same as the modern function and the initial molecule, only a fraction the size of the ribosome itself, has remained catalytically active and structurally unchanged within the ribosome, despite ~4 billion years of lineage-specific opportunity for variation after divergence from LUCA.

6.1.5 DNA as Genetic Material Is Stable; Its Monomers Are Synthesized by Enzymes

If the first cells had 1000 or more genes, they had to have DNA genomes, because of the inherent instability of RNA. RNA is vulnerable to hydrolytic cleavage by OH⁻ ions and has an innate tendency to undergo recombination, as a few examples can illustrate. Self-excision and self-splicing are intrinsic properties of RNA, in part due to the fact that the 2′ and 3′ OH-groups of ribose in RNA are equally reactive. Recall that the aminoacyl tRNA synthetases can activate amino acids by the formation of aminoacyl bonds with the 2′ or the 3′ OH of the terminal tRNA adenosyl residue. In many reactions of RNA, the 2′ OH-group of ribose moieties in RNA exerts nucleophilic attack on the phosphodiester bond of RNA molecules (see Chapter 5). This leads to an RNA strand break (RNA self-cleavage) or in case of several cuts, to excision of a piece of RNA. The 2′ OH-group of ribose in RNA can also be attacked by 5′ phosphates in RNA, leading to lariat structures. Terminal 3′ OH-groups on ribose can also attack phosphodiester bonds in RNA molecules. This leads to thermodynamically reversible transesterification reactions, causing RNA strands to branch or recombine with other RNA strands. Nucleophilic attack by the 3′ OH of ribose on the 5′ triphosphate terminus of another strand concomitant with displacement of pyrophosphate leads to ligation and longer RNA molecules (see Figure 5.22). Similarly, RNA can be inserted into another RNA strand (see Figure 5.23), a reaction called reverse splicing that takes place in modern cells. RNA is sensitive to alkaline hydrolysis, which leads to ribonucleosides having 2′ or 3′ phosphate bonds.

The foregoing examples underscore the point that RNA not only folds stably (the structural basis of function in tRNA, rRNA, and ribozymes), it is highly reactive in terms of forming new phosphoester bonds. Those are desirable properties for a molecule that will end up serving as (1) the carrier of information to make proteins (mRNA), (2) the carrier of amino acids reading that information (tRNA), and (3) the catalyst for peptide formation (rRNA, carrier of the peptidyl transferase activity) at translation. But they are very disadvantageous properties indeed if the molecule is to serve as a stable repository of information that can be inherited to progeny and maintained with fidelity. For genomes to become large and stable, so that incremental advances in catalytic efficiency could become fixed, rather than undergoing recombinational or hydrolytic meltdown, DNA was required. DNA is stable against both nonenzymatic recombination and against hydrolysis because of the lacking 2′ OH-group. Protomicrobes that did not invent a stable alternative to RNA as an information repository would have been unable to reach the level of ~1000 genes and the free-living state. The information in an RNA genome, whether united in one molecule or dispersed across many, would have been unstable because RNA is chemically too reactive, too prone to recombination and cleavage events, which would disrupt coding sequences. As the first carrier of genetic information, RNA's tendency to undergo recombination was beneficial because it enabled the first molecular carriers of coding information to explore sequence space rapidly through recombination, thus promoting the formation of useful catalysts. Yet to reach the level of information required

to express 1000 genes—or 1500 genes as in the smallest genome sizes of H_2-dependent chemolithoautotrophs—a repository of information chemically more stable than RNA was essential. The reason is evident: If sequence variants with positive impact on molecular fitness cannot be saved or stably inherited, they cannot contribute to stable evolutionary advances, regardless of how beneficial their effects might be. Stable increases in molecular fitness require stable storage of retrievable information, that is the function of DNA.

In the terms of standard organismal evolution, fitness refers to the contribution of individuals to progeny. Before the origin of free-living cells, we have neither individuals nor organisms. In the terms of replicating RNA molecules initially outlined by Eigen, fitness refers to the ability of individual molecules to replicate rapidly and with high enough fidelity to retain their identity. But the system at this stage must be able to harness energy and perform translation, it has to be far more complex than an RNA molecule. In the terms of exergonic chemical reactions capable of primitive translation confined to inorganic compartments, molecular fitness refers to the ability of a compartment's organic molecular contents to sustain its own exergonic reactions so as to permit continued protein synthesis. This requires a stable form of molecular memory—DNA—that can save the information underpinning beneficial innovations. The chemical origin of DNA comprises two stages: monomer synthesis and monomer polymerization.

First, we consider the synthesis of DNA monomers, 2′-deoxyribonucleosides. A nonenzymatic pathway has been reported by Trapp and colleagues involving reactions of acetaldehyde and formaldehyde with purines and pyrimidines, the 2′deoxyribose is synthesized on the preexisting base with high specificity and with the correct N-glycosidic bond. This kind of reaction could provide a prebiotic source of DNA precursors. It is however also possible that widespread use of DNA and 2′-deoxyribonucleotides emerged after the origin of translation. In that case, monomer synthesis could have proceeded as in metabolism, starting from ribonucleotides and involving protein-catalyzed reactions. In cells there is only one enzyme known that generates DNA monomers: ribonucleotide reductase (RNR). There are three related forms of ribonucleotide reductase. They all use a thiyl radical (–S·) to abstract a proton to initiate the reaction mechanism, but they differ by the radical they use to remove a hydrogen atom (H·) from the –SH group to generate the –S· radical. The evolutionarily derived class I RNR uses O_2 and a diiron center (Fe–O–Fe) to generate a tyrosyl radical (Tyr–O·). The class II RNR enzyme performs homolytic cleavage of the C–Co bond in adenosylcobalamine (B_{12}) to generate an adenosyl radical (Ado·, adenosine 5′–CH_2·).

The class III RNR enzyme performs homolytic cleavage of a C–S bond in S-adenosyl methionine to generate an adenosyl radical, which then abstracts H· from a main chain glycine to produce a glycyl radical that generates the thiyl radical in the reaction mechanism. At the structural level, all three RNRs are related and furthermore related to pyruvate: formate lyase, an ancient enzyme that cleaves pyruvate with CoASH to generate formate, and acetyl-CoA, also using a radical mechanism. One RNR enzyme generates all four dNDPs (deoxynucleoside diphosphates)

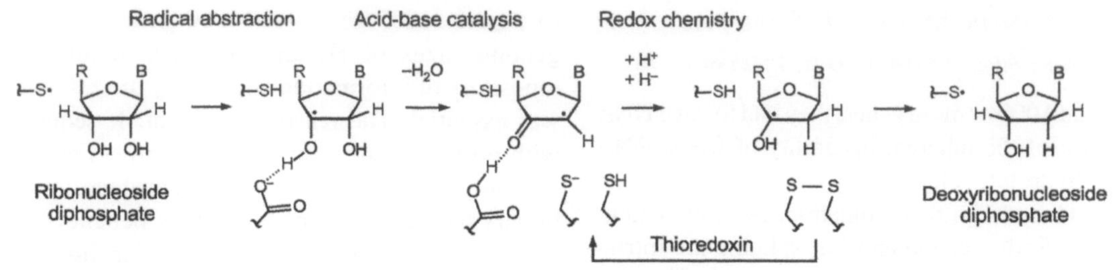

FIGURE 6.11 **DNA monomers are generated by ribonucleotide reductase.** The mechanism involves radical abstraction and migration, acid–base catalysis for water elimination, and thiol-based redox chemistry. B: base; R: $CH_2OPO_3PO_3^{3-}$. RNR reduces all four NDPs to dNDPs at the same active site. The thiyl radical that starts the reaction resides on a main chain cysteine of RNR. The reductant (H– donor) is a cysteine sulfhydryl (–SH), which upon hydride donation forms a disulfide (–S–S–) that is re-reduced by thioredoxin for the next catalytic cycle. Modified from: Lundin D. *et al.* (2015) The origin and evolution of ribonucleotide reduction, *Life* 5: 604–636.

from NDP precursors. The enzymatic reaction mechanism (Figure 6.11) involves radical abstraction and migration, acid–base catalysis for water elimination, and thiol-based redox chemistry. These three reaction types are typical for proteins but not at all typical for RNA. DNA thus probably arose not only after RNA but also after proteins. With a catalytic means of generating dNTPs, DNA synthesis on RNA templates was possible.

6.1.6 THE RNA-TO-DNA TRANSITION WAS CATALYZED BY ENZYMES

How might the transition from RNA to DNA as an information storage molecule have occurred? Figure 6.12 provides a terse proposal entailing six stages. The process of

translation does not fundamentally require DNA; hence, it is possible, if not probable, that DNA monomers and polymers arose post-translation and with the help of proteins. In stage 1, we start from the premise that enzymatic RNA-dependent RNA polymerase (RdR Pol) activity was present at a very early stage in the evolution of LUCA. Such RdR Pol activity persists to the present in viruses with positive-strand RNA genomes. Before the advent of DNA, RdR Pol activity would allow RNAs associated with translation to accumulate, albeit as both a functional + and (assumedly) nonfunctional template—strand, being replicated with a modest fidelity and with the disadvantages that RNA reactivity and instability carry for long term information storage.

Before DNA can become an information storage molecule, dNTP monomers are required (stage 2), the product

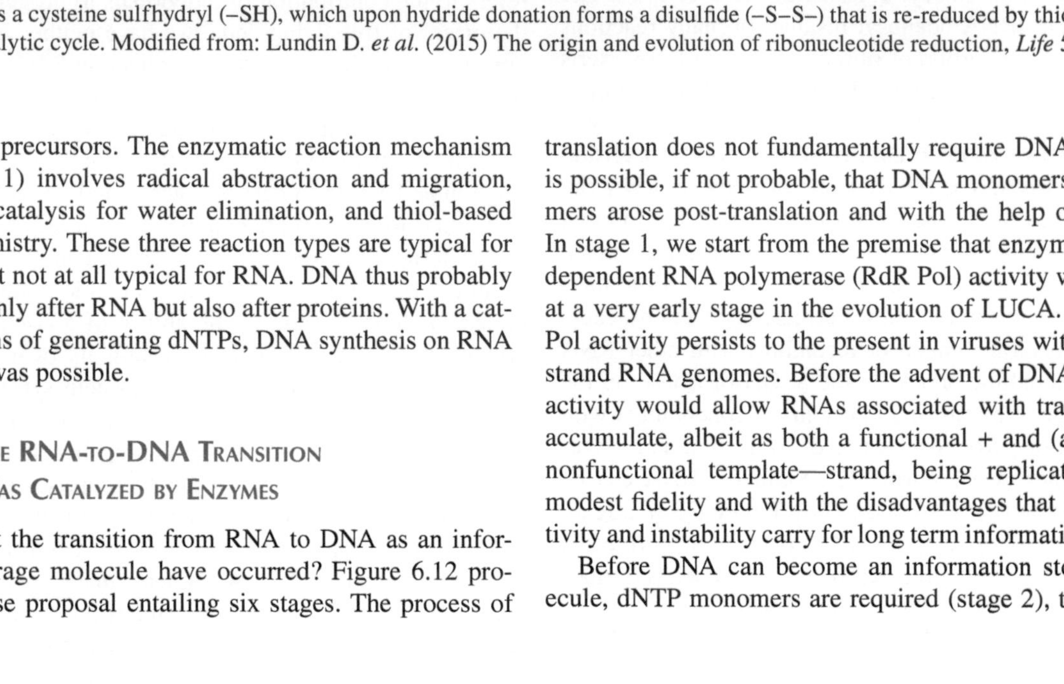

Stage	Enzymatic activity	Process enabled	Molecular consequence	Evolutionary consequence	Virus-encoded?
6.	DNA ligases, recombinases	Linkage of functions	Codiffusion and coexpression of gene combinations		Yes
5.	DNA dependent DNA polymerase (DNA Pol)	Replication of DNA without an RNA intermediate	High fidelity, high stability transmission of information transcribed by RNA Pol		Yes
4.	DNA dependent RNA polymerase (RNA Pol)	Transcription of a specific strand of a DNA template into RNA	Many functional RNA copies generated from a single DNA molecule		Yes
3.	RNA dependent DNA polymerase; Reverse transcriptase (RT)	Deposition of information from RNA into single- and double-stranded DNA	Low fidelity, high stability transmission of information transcribed by RdR Pol		Yes
2.	Ribonucleotide reductase (RNR)	Synthesis of DNA monomers	Substrate synthesis enabling dNTP incorporation		Yes
1.	RNA dependent RNA polymerase (RdR Pol)	Replication of RNAs required in translation	Low fidelity, low stability proliferation of a translating system, transcription		Yes

FIGURE 6.12 **Stages in the transition from RNA to DNA.** All activities are still found among viruses today. The main DNA replication machineries of bacteria and archaea are not closely related, suggesting a very late (independent) origin of DNA-based replication on the way from LUCA to free-living archaea and bacteria [Koonin E. V., Martin W. (2005) On the origin of genomes and cells within inorganic compartments, *Trends Genet.* 21: 647–654.]. All of the activities listed are also found in viruses and phages [Koonin E. V. *et al.* (2020) The replication machinery of LUCA: Common origin of DNA replication and transcription, *BMC Biol.* 18: 61, 1–8], including ribonucleotide reductase [Harrison A. O., Moore R. M., Polson S. W., Eric Wommack K. (2019) Reannotation of the ribonucleotide reductase in a cyanophage reveals life history strategies within the virioplankton, *Front. Microbiol.* 10: 1–16].

of RNR, possibly a class III type RNR, as cobalamin is also required in the acetyl-CoA pathway. Reverse transcriptase activity (RT, stage 3), another activity that is today virus-associated, enabled DNA synthesis on an RNA template. RT first synthesizes a complementary DNA copy (cDNA) strand on the RNA but it can also synthesize the second DNA strand (DNA-dependent DNA polymerase activity), displacing the RNA from the RNA:DNA hybrid, so as to generate double-stranded DNA molecule from a single-stranded RNA template. RT allows the synthesis of DNA as a stable information repository. Even if the DNA, whether linear or circular, cannot be replicated (RT lacks the ability to displace a DNA strand in a DNA:DNA hybrid), its encoded information containing RNA molecules can be recopied back into DNA, as in some viral replication cycles. DNA as an information repository allows useful innovations to remain useful, stable, and diffusible across compartments (Figure 6.9). Stage 4 corresponds to the onset of DNA-dependent RNA polymerases that are specialized for transcription, generating many RNA copies per DNA template, amplifying the information stably stored in DNA.

DNA-dependent DNA replication, a function endogenous to all cells but also encoded by many viruses, would allow genes made of DNA to multiply (stage 5), generating new possibilities of combinations of chemically stable and expressible genes across compartments via diffusion. DNA ligases and recombination activities (stage 6), essential for viral integration into DNA genomes, permit the synthesis of larger DNA molecules encoding several genes (or dozens of genes or more), that provides an opportunity for codiffusion and coexpression of functions across compartments. This permits physical linkage between genes and a first opportunity for the inheritance of functions (pathways, for example) that require the expression of several genes. Such a stepwise assembly process from plasmids carrying one gene to large plasmids and ultimately genome-sized collections of genes was only possible with stable and largely inert DNA, not with its more reactive RNA predecessor. Today, many prokaryotes carry a genome consisting of large plasmids >100 kb in size or megaplasmids (plasmids larger than ~350 kb in size), rather than one contiguous DNA molecule. Plasmids almost certainly played an important role as organizational intermediates in the origin of microbial genomes. Compartments that accumulated the 1000 to 1500 functions required

for the free-living state via sequestration of many different DNA molecules had the ability to generate free-living cells (Figure 6.9). In contrast to a number of views in the literature, we do not interpret the significance of viruses as the inventors of any of these functions. Viruses and phages merely utilize these activities, in the same way that mobile genetic elements do.

Stages 1–6 outlined in Figure 6.12 entail many assumptions, foremost a continuous supply of energy and NTP monomers for polymer synthesis from core biosynthetic metabolism across all stages. In addition, it is clear that only a few new gene variants or combinations of genes will be useful. Most genetic novelties will lead to dead ends, their expression will simply consume resources without feeding back into the synthesis of more of the essential components—amino acids, nucleotides, and cofactors. That will impose homeostasis, limiting the proliferation of deleterious gene combinations but favoring combinations that increase molecular fitness. Gene combinations that increase the flux of carbon, nitrogen, and energy into amino acid and nucleic acid monomers will naturally increase the copy number of the underlying and physically proximal genes. As long as energy and nutrients are in constant supply, positive feedback loops that organize available nutrients into genes and proteins are possible. Yet all positive effects on molecular fitness hinge upon functional translation. As a consequence, it is possible if not likely that the first multigene assembly for protein-coding genes likely directed the synthesis of the ribosome itself.

Ribosomal protein operons are indeed conserved across many bacterial and most archaeal genomes and are even conserved to some extent across the archaea-bacteria divide (Figure 6.13). They encode proteins of the ribosome, of translation (elongation factors), of transcription (RNA polymerase, terminators), of the SRP-dependent secretory (sec) protein cotranslational insertion pathway (Figure 6.17), a few tRNAs and other ancient functions. This ancient operon is likely a relict from the very first association of coding functions. While biosynthetic functions (amino acids, cofactors, bases) are beneficial, without translation genes for such functions they are useless. In addition, built into all models for origins is the notion that the environment initially supplied the basic building blocks for proteins, nucleic acids and cofactors.

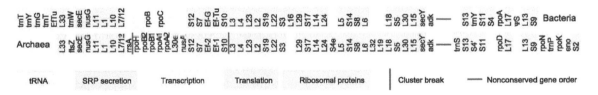

FIGURE 6.13 Conserved gene order for ribosomal protein operons. Bacteria: last common ancestor of bacteria; archaea: last common ancestor of archaea. Functional categories are color coded. Genes outside the color-coded regions include rrsA (alanyl AARS), adk (adenylate kinase), ftsZ (division), and eno (enolase). Note that the conservation is based on the genomes of a small sample of bacteria and archaea and is not strict, as the operon has undergone many fragmentations during evolution. Yet the probability that the observed gene order is the result of convergence is basically nil. Ef: translation elongation factor; nus: transcription terminator; rpo: RNA polymerase subunits. Data compiled in references [Wächtershäuser G. (1998) Towards a reconstruction of ancestral genomes by gene cluster alignment, *Syst. Appl. Microbiol.* 21: 473–477; Stoebe B., Kowallik K. V. (1999) Gene-cluster analysis in chloroplast genomics, *Trends Genet.* 15: 344–347; Wang J. *et al.* (2009) Many nonuniversal archaeal ribosomal proteins are found in conserved gene clusters, *Archaea* 2: 241–251]

Accordingly, one can envisage intermediate evolutionary stages and micro-environments where the inorganic geochemical surroundings still helped to supply monomers for a purely organic translational system. Figures 6.7–6.9 explicitly show how such intermediate states were important in the evolution of increasingly complex, specific, and efficient catalysts during the process of cell origins. Recall that all theories for the origin of life, not just the hydrothermal theory, require an input of chemical constituents from the environment. This in turn requires the existence of catalysts in the environment where the first cells arose. Although the process of catalyst evolution was already discussed in Chapter 4, Figure 6.14 provides a reminder that the evolutionary significance of the origin of ribosomes and translation was to provide enzymes—organic catalysts—that are heritably encoded by genes. That was the final major step in the evolution of catalysts and a prerequisite for the origin of microbial cells. The origin of DNA was essential in that process because it provided a molecular memory bank that allowed the diversification of enzymes into the 1000 or so different catalysts that the first cells required for life. RNA can store information but not stably in the reactive surroundings of a hydrothermal environment. DNA provided a stable repository for the information that specified both the synthesis of those catalysts and the ribosome that was required for their synthesis. The first protocells were of course unaware of how powerful the tool was that they had discovered: evolvable catalysts synthesized on ribosomes.

Although we saw in Chapter 5 how nonribosomal protein synthesis could give rise to reproducible catalysts without

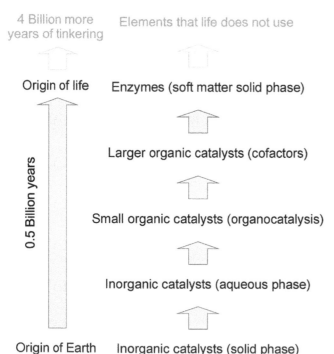

FIGURE 6.14 Schematic depiction of catalyst evolution. Note that each new catalyst category, from bottom to top, was not necessarily replaced by the next catalyst type but expanded and enhanced by it. All catalyst types, except solid-state inorganic catalysts, were eventually assimilated by the cells when they became free living. From Wimmer J. L. E., Martin W. F. (2022) Origins as evolution of catalysts. *Bunsen-Magazin* 4: 143–146.

the genetic code, in the process that led to the origin of free-living cells, there is no inorganic or organic substitute imaginable for the function of mRNA-based translation on the ribosome. The ribosomal operon was clearly the keystone genetic unit. Even today, prokaryotic cells are roughly 40% ribosomes by weight. That is, almost half the cell is dedicated to synthesizing protein and the cell is 50% protein by weight. Today, translation on ribosomes is the sole supply of all the catalysts that cells require for life. In contrast to translation, DNA replication proteins are not conserved across the bacterial-archaeal divide, indicating a very late emergence of DNA *replication* (as opposed to early DNA *synthesis*; see Figure 6.12) in the process leading to the emergence of free-living bacteria and archaea.

6.1.7 Catalysis Is More Important Than Replication

In contrast to large segments of the literature on origins and early evolution, RNA-based replication plays a comparatively minor role in our model. The concept of an RNA world rests on the assumption that there was a time in early molecular evolution when populations of self-replicating RNA molecules existed that underwent selection leading to RNA molecules with various catalytic activities forming a simpler ribonucleic precursor to life as it operates today. As we discussed in Chapter 5, the concept of evolution among replicating RNA molecules traces to work by Eigen and Spiegelmann in the 1970s using a viral protein called Qβ replicase, a phage-encoded RNA-dependent RNA polymerase that synthesizes RNA from RNA templates *in vivo* and *in vitro*. In the presence of an RNA template and NTPs, Qβ replicase synthesizes + and—strands rapidly but with a high error rate, naturally leading to the accumulation of such +/– template pairs that replicate most rapidly, outcompeting other polymers for monomer substrates.

This kind of greedy replication sequesters all NTPs, inevitably giving rise to what are called molecular parasites. Molecular parasites invariably stifle the evolutionary process, leading only to the accumulation of the most efficient RNA replicators (parasites), which rapidly consume available resources to generate more of their own kind as an end in itself, a molecular *Selbstzweck* (end in itself). Parasites generate one kind of molecule, themselves, and consume all the substrates in the process. Mutations during parasite replication do generate diversity, but that diversity represents variations on a theme of the fast-replicating phenotype— the fastest replicator. Recombination between non-replicating RNA molecules or even new *de novo* RNA synthesis, though slow, can generate much more sequence diversity than errors during fast replication, especially if activated nucleotides are limiting.

Life processes require a great diversity of molecules and catalysts, particularly at origins. For an early molecular system to increase in mass or complexity, efficient catalysts are clearly more important than efficient replicators, because catalysts can replenish the supply of organics underpinning the flow of carbon and nitrogen to the NTP supply whereas replicators tend to generate one kind of catalyst—parasitic ones that make more of themselves. Good catalysts can, in

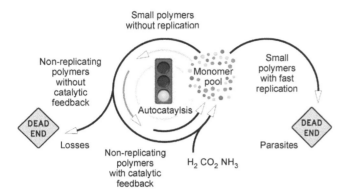

FIGURE 6.15 Autocatalytic sets, dead ends, and networks.
Autocatalysis can lead to small polymers that lack catalytic activity, to product accumulation, or to small polymers that replicate quickly (parasites), or both. Non-replicating polymers with catalytic activities that accelerate reactions of the cycle are more important for early metabolic evolution than fast-replicating (parasitic) or noncatalytic products, which siphon off resources without promoting reaction rates. The problem of molecular parasites was recognized by Eigen M. (1971) Self organization of matter and the evolution of biological macromolecules. *Naturwissenschaften* 58: 465–523. One set of solutions to the parasite problem involves compartmentation (Bresch G. *et al.* (1980) Hypercycles, parasites and packages, *J. Theor. Biol.* 85: 399–405; Branciamore S. *et al.* (2009) The origin of life: Chemical evolution of a metabolic system in a mineral honeycomb? *J. Mol. Evol.* 69: 458–469). Our current model focuses on catalysis rather than replication, thereby solving the parasite problem in a different way, although compartments are essential to the geochemical/hydrothermal origins model.

principle, give rise to autocatalytic cycles (Chapter 3) and increase molecular fitness without the need to undergo replication themselves (Figure 6.15).

Both fast-replicating parasites and non-replicating products without catalytic activity that feed back into monomer synthesis, siphon off monomers down unproductive paths, leading to dead ends (Figure 6.15). Non-replicating products that possess even a small amount of catalytic activity leading to the synthesis of more monomers (the defining property of metabolic networks in cells) can channel starting compounds (H_2, CO_2, NH_3, P_i) into the accumulation of more monomers. For that they need not be suited to replication. In fact, self-replication invariably leads to molecular parasitism and a dead end. However, if there is no replication at all then the catalytic functions of polymers are restricted to the lifetime of the molecule, that is, its first hydrolysis or cleavage event. Clearly, at some point, replication needs to come into play, but from the standpoint of catalysis, because of the unavoidable parasitism issue, the later in chemical evolution that RNA starts to replicate, the better.

A closer look reveals that RNA replication does not even belong to the set of functions that prokaryotes need for life. In living cells, RNA serves in gene expression: rRNA, tRNA, mRNA, 7SRNA (part of the signal recognition particle required for insertion of proteins into membranes), RNAse P (tRNA splicing), and intron splicing. RNA replication is restricted to RNA viruses, RNA phage, and viroids, circular RNA molecules <100 nucleotides long that have no capsid and encode no protein. RNA viruses and phage are replicated by RNA-dependent RNA polymerases that they encode in their genomes, like Qß. Viroids on the other hand encode no proteins at all, they are replicated by side activities of host proteins, either DNA ligase or DNA-dependent RNA polymerase. It is possible, if not likely, that early evolution never hinged upon a phase in which (self-) replication of RNA molecules was a dominant process. Autocatalytic sets can proliferate by generating more of their own components, but none of the elements of the set needs to be capable of self-replication. Starting from an environmentally supplied food set, the elements of autocatalytic sets simply need to exert catalysis on reactions that feedback into the set, and the amount of catalysis required for an autocatalytic set to arise is surprisingly small. As we saw in Chapter 3, a rate increase of about 10% per catalyzed reaction will suffice to generate an autocatalytic set.

In Figure 6.15, there might seem to be a paradox in that replicating molecules are disadvantageous while non-replicating molecules, if they provide catalysis that feeds back into the system, are advantageous. How can such a system increase its mass beyond a collection of single molecules if we do not assume a role in replicating informational RNA from the start? There is, in principle, a solution.

We saw in Chapter 4 that peptides that are stable against hydrolysis can preferentially accumulate in an environment of changing water activities, an environmental parameter that hydrothermal systems provide (Figure 4.9). This provides a means of accumulating proteins (macromolecules) that have distinct properties or catalytic ability. It turns out that double-stranded RNA is much more stable against hydrolysis than single-stranded RNA. Zhang *et al.* recently showed that under pH-neutral conditions, double-stranded RNA had such slow rates of decay in water that it appeared to be stable for up to a decade, far longer than single-stranded RNA. The functional RNA molecules in cells are always double stranded or have extended double-stranded components in their structure, rRNA and tRNA in particular. Even the 'urzyme' model for the origin of AARS enzymes and the genetic code, which we will encounter in the next section, involves reverse complementary strands, which can be generated either by two complementary RNA strands meeting or a single RNA molecule folding to make a hairpin structure. It is possible that RNA stability played an underappreciated role in the accumulation of functionally (catalytically) important RNA molecules in early chemical evolution. Stability and replication both serve the same purpose: increasing the frequency of a particular sequence variant.

6.2 THE FIRST GENOMES ENCODED PERHAPS 1000 PROTEINS

Coming back to the problem of genome size, in the context of assembling a genome large enough to support a free-living chemolithoautotrophic lifestyle, we mentioned that perhaps ~1000 genes might have been required, based on modern cells. For example, *Methanopyrus kandlerii*, a thermophilic methanogen that grows on H_2 and CO_2, has a genome size of 1.69 million bases and encodes 1692 proteins. But the genomes of the first cells need not have been that large. Three factors bear upon the issue: (1) the active sites of enzymes can be encoded by small peptides, (2) peptides can exhibit functional promiscuity, and (3) some reactions might not have strictly required proteins.

6.2.1 The First Proteins Were Small

How big might the first active enzymes have been? Modern proteins are, on average, about 330 amino acids long, requiring roughly 1000 bp per gene. However, the first protein-coding reading frames were probably shorter than today's, an idea that goes back to Margaret Dayhoff. There are reasons to think that the first encoded enzymatic functions started out from small peptides with low catalytic rates and low substrate specificity. Furthermore, they might have tended to consist of regions or domains (units of stable folding) that mainly comprise the active site followed by extension, domain duplication, or both. Two examples underscore this view.

The first is the Urzyme model of AARS origin. Modern aminoacyl tRNA synthetases are large proteins on the order of 500 to 900 amino acids long. Carter and colleagues have shown that tryptophanyl tRNA synthetase (TrpRS) can be pruned down considerably in size, with a loss in activity, however. Fragments of TrpRS protein that include the active site but have less than 20% of the mass of the active TrpRS dimer can still accurately charge $tRNA^{Trp}$. The specific activities of the fragment are roughly 10^4- to 10^5-fold lower than that of the native enzyme. Although such loss in specific activity is dramatic, the core fragment, which they called the minimal catalytic domain (MCD), is still a highly effective catalyst in that it catalyzes the TrpRS reaction at a rate that is 10^9 times higher than the rate of the spontaneous, uncatalyzed reaction. The MCD represents the conserved core of structural motifs shared by the 10 class I AARS enzymes, possessing both intact ATP-binding and amino acid-binding sites. The more of the full-size enzyme structure that is retained, the greater the substrate specificity (discrimination) of the charging reaction becomes.

Was amino acyl tRNA synthesis only possible once encoded proteins, translation and the genetic code existed? Yarus and coworkers reported short RNA molecules that were able to activate amino acids by catalyzing the formation of aminoacyl adenylate (aa-AMP) and PP_i from amino acids and ATP, the first reaction of AARS. Other RNA molecules they characterized were able to synthesize aminoacyl tRNA from tRNA and aa-AMP, the second reaction of AARS. The smallest ribozyme they found is the sequence GUGGC-3′, which aminoacylates GCCU-3′ with Phe-AMP such that they observed the Phe dimer and larger oligopeptides of phenylalanine. Hence, even a function analogous to peptidyl transfer (Figure 6.10) can be performed by this small ribozyme.

Although the core functions of translation (including the peptidyl transferase reaction itself performed by the ribosome in modern cells) can be catalyzed by RNA, note that none of those three activities belong to self-replicating RNA molecules. The RNA molecules involved in translation have to be replicated to increase their absolute numbers, even though they do not have to be replicated in order to exert their activity, nor do they have to be replicated in order to increase their relative numbers—resistance to hydrolysis

can do that (Section 6.5.2). It is possible that the role of RNA in early evolution was mainly one of stable folding (for example, as double-stranded RNA as outlined at the end of Section 6.4) and catalytic activity to perform translation and modify RNA, rather than rapid replication. While RNA can catalyze essential functions related to RNA, almost all other catalytic functions in cells are provided by amino acid side chains (and cofactors) in enzymes. The example of the minimal catalytic domain of TrpRS suggests a general path for enzyme evolution: an enzyme that starts small and slow with modest specificity for an important reaction can readily become subject to selection for greater rate and substrate specificity at the expense—translation consumes 75% of the ATP budget in modern cells—of requiring a longer polypeptide chain.

Another example of small peptides exerting beneficial functions on activity comes from the study of a ribozyme called RNA polymerase ribozyme (RpR). The RpR RNA requires ~0.2 M Mg^{2+} for polymerase activity, but small peptides 15–20 amino acids long derived from ribosomal proteins permit RpR activity at 1 mM Mg^{2+}. To function, the peptides need to be rich in positive charges (lysine and arginine) and can even be replaced by small homopolymers of orinithine or diaminobutyric acid. Ribosomal proteins and RNA binding proteins in general tend to be rich in positive charges that interact with backbone phosphates. As in the case of TrpRS above, functions that start out with very short peptides having low specificity can be selected for greater specificity of interaction and, in the case of enzymes, increased rate, while proteins increase in length and folding stability over evolutionary time. This all requires, of course, that there are genes that can replicate and that transmit their variation into functional catalysis via the genetic code and translation at the ribosome.

6.2.2 Some Proteins Can Perform Multiple Functions

Another factor that might have allowed the very first proteins and genomes to be smaller than today's is that a single protein can be multifunctional, performing several functions at the same time. There are several ways this can occur. The first is promiscuity, which designates the utilization of the active site for more than one enzymatic function. This can encompass the traditional observation of broad substrate specificity for structurally related substrates at the active site leading to the evolution of more specific gene families. The 2-oxoacid oxidoreductases (the 2-OAOR family) are a well-studied example, encompassing enzymes that catalyze the reversible oxidative decarboxylation of oxalate, pyruvate, 2-oxoglutarate, 2-ketoisovalerate and indole-pyruvate for the generation of low potential (ca. −500 mV) reduced ferredoxins. The common enzymatic mechanism and sequence conservation of 2-OAORs indicates evolution from a common ancestor with broad substrate specificity. As another example, some detoxifying enzymes such as glutathione S-transferases have inherently broad substrate specificity. This kind of functional promiscuity involves interaction of the active site with structurally unrelated substrates, but it can also promote an evolutionary transition in which a protein that folds in one conformation can undergo a transition to fold in a different conformation, giving rise to fundamentally new functional constraints on the protein and a fundamentally new evolutionary trajectory.

Multifunctionality can also involve 'moonlighting', which typically involves highly specific binding sites (and activities) present on the protein, but distinct from the canonical active site. Moonlighting originally designates a person's second job, often a night job. In physiology it is used when one protein performs more than one function. Examples are the glycolytic enzyme enolase, that also functions in the RNA degradosome in *E. coli* or tRNA transport to mitochondria in yeast. The glycolytic enzyme glyceraldehyde 3-phosphate dehydrogenase (GAPDH) can function in iron metabolism. The citric acid cycle enzyme aconitase can function in iron metabolism, like GAPDH, or in the maintenance of mitochondrial DNA. Many examples of moonlighting are known and some proteins have dozens of different moonlighting functions. Thioredoxin, which reduces ribonucleotide reductase as we saw in Figure 6.11, moonlights as a structural subunit of phage T7 RNA polymerase. The list of examples for moonlighting activities is long.

6.2.3 How Much Function Was Needed?

Yet another factor bearing on the genome size of the very first microbes concerns uncatalyzed reactions. For example, S-adenosyl methionine is an essential cofactor involved in many methylation reactions, and its methyl transfer reactions are generally exergonic. The formation of SAM from the reaction of ATP and methionine is also exergonic, releasing both P_i and PP_i. Yet the reaction occurs spontaneously without enzymes, as does the formation of SAM from methionine and adenosine, albeit at slower rates. There are a very large number of reactions in modern metabolism that, though catalyzed by enzymes today, go forward spontaneously without the help of enzymes. This principle is essential for understanding how the origin of metabolism could have been possible in the first place: Enzymes do not perform feats of magic, they just accelerate reactions that have a natural tendency to occur anyway.

There is a hefty caveat to that axiom though, in that enzymes can increase the spontaneous reaction rate by up to 10^{18} fold. That is, the uncatalyzed spontaneous rates of some biological reactions are so slow that they have half-lives ($t_{1/2}$) of greater than a billion years, meaning they will effectively not take place at all in an early evolution context. Does that mean that enzymes had to invent such reactions? No, cofactor themselves can provide substantial acceleration. One of the most instructive examples for that is the pyridoxal phosphate (PLP) dependent decarboxylation of orotidine 5-phosphate (OMP) catalyzed by OMP decarboxylase. Wolfenden showed that the reaction is massively accelerated by a factor of ~10^{17} by the PLP dependent enzyme. Zabinsky and Toney showed that at pH 8 the total rate enhancement is 10^{18}, whereby the enzyme only contributed roughly 10^8 fold acceleration—PLP alone in solution without the enzyme accelerated the reaction by a factor of 10^{10}. The coenzyme alone provided more rate enhancement than the protein. But even in the 1950s, Metzler and Snell showed that PLP efficiently catalyzed transamination and deamination reactions without enzymes. Thus, the cofactors

themselves, indicated by 'organic catalysts' in Figure 6.7 or 6.14, can supply a required function before enzymes arose and in many cases metal ions alone can provide substantial catalysis to biological reactions, as recent work by Moran and colleagues has shown.

By how much do enzymes accelerate reactions on average? Wolfenden and colleagues have tabulated a number of values for the rate of enhancement ranging from 10^7 to 10^{19}, whereby the limiting factor for such estimates is obtaining the rate of the uncatalyzed reaction, particularly for very slow reactions. The average rate of a reaction catalyzed *in vitro* by an enzyme is about 10 reactions per second (10 s^{-1}), with reaction rates in carbon and energy metabolism (the backbone of metabolism) being faster, ~80 s^{-1}, amino acid and nucleotides being higher than average, ~18 s^{-1}, and maintenance functions being slower, ~5 s^{-1}. It is not essential that biological reactions in the cell generally take place at great speed, but it is essential that they all occur at roughly the same rate.

Two other important insights by Wolfenden about catalysis should be mentioned. First, enzymatic reactions are not under selection to be particularly fast, as we just mentioned; rather they are under selection to take place at roughly the same rate so that the overall reaction of the cell runs smoothly, meaning that different reactions require amounts of rate enhancement that differ by many orders of magnitude. It is fair to assume that reaction rates of enzymatic reactions in the first cells had not been synchronized and that there was still some room for further optimization of rates. A second insight is the effect of temperature. It is well known that biological processes generally tend to increase in rate by about a factor of 2 for every 10 degrees increase in temperature, this is known as $Q_{10} = 2$ in older literature. Q is not constant across biochemical reactions. Wolfenden noticed that the rate constants of the slowest uncatalyzed reactions are most sensitive to temperature; that is, the slowest reactions in cells are more heavily accelerated by higher temperatures than the fastest reactions are. This means that early chemical evolution would have taken place much more rapidly at high temperatures (on the order of 100°C) than at temperatures near freezing. Related to that, he noticed that the rate enhancements produced by enzymes tend to increase with decreasing temperature, meaning that enzymatic rate enhancement would have increased automatically in transitions to cooler temperatures. Both observations argue strongly in favor of the origin of metabolic reactions at temperatures near 100°C, as biologists suspected over 100 years ago (Figure 6.1).

That brings us back to the question at hand: How many genes did the first microbes absolutely require and how big were those genomes? We know that the synthesis of amino acids, bases, and cofactors from H_2, CO_2, and NH_3 requires 400 reactions that are catalyzed by enzymes today and probably in LUCA. A few of those reactions, but not many, are catalyzed by promiscuous enzymes, for example in branched-chain amino acid synthesis. Many of the enzymes involve multiple subunits, but the first polypeptides were probably shorter so perhaps 400 kb were required to encode the basic biosynthetic set. Add roughly 40 enzymes for lipid synthesis and cell wall synthesis, 40 universal ribosomal proteins, 40 more proteins for ribosome biogenesis and translation, 100 proteins for rRNA and tRNA processing and RNA base modifications, 40 more for ATP synthetase subunits, RNA and DNA polymerization, 40 more for membrane proteins and the signal recognition particle (cotranslational insertion), and 100 more for diverse cellular functions (proteolysis and folding, for example), and we end up at another 400 genes or roughly 400 kb. That makes ~800 kb a very approximate lower bound, assuming almost no noncoding or parasitic (mobile) DNA. As the upper bound we might take a modern methanogen like *Methanopyrus*, mentioned above, giving a range of about 0.8 to 1.6 million bases for the size of the first microbial genomes.

We did not mention gene regulation, DNA binding proteins, or enzymatic regulation anywhere in this chapter. We can safely assume that the first microbes were highly specialized to one main nutrient supply (H_2 and CO_2) and hence did not have any complicated decisions to make that would require gene regulation. As Don Bryant aptly described it concerning gene regulation in *Chlorobium tepidum*, a complete specialist for low light, CO_2, and H_2S, 'There are no decisions to make, the state of the genome is 'ON'. If *Chlorobium* finds the growth substrates it needs, promoter affinities produce exactly what is required in exactly the right amounts'. (A promoter is the region of a gene that binds the RNA polymerase, strong binding means high transcript levels, hence high protein levels, without the help of additional transcriptional regulators.)

In Chapter 7 we will see that PP_i-producing reactions vectorialize metabolism. Reactions that produce PP_i are irreversible under physiological conditions because of ubiquitous pyrophosphatases; they ratchet reactions forward in the direction of growth. This keeps the reaction network of life from running backward so that the cell does not dissolve back into the nutrients from which it arose. In Chapter 7 we will see that when the supply of H_2 and CO_2 is terminated, chemolithoautotrophs not only stop growing, but the decomposition of cell mass into H_2 and CO_2 becomes thermodynamically favorable, which gave rise to the first fermentations.

6.2.4   DNA Molecules as Genetic Material

To recapitulate, RNA is too reactive to have served as a reliable genetic material for the complex reaction networks encoded by modern genomes. Stable heredity required DNA, which is synthesized by proteins and is not required for translation at all, bringing stable memory and permanency into chemical reaction networks. Furthermore, prior to the origin of translation, any form of replication was slow, if it occurred at all. All of the replication-associated functions of cells are catalyzed by proteins. It is possible that it has always been that way, suggesting that early replication was probably less important than early catalysis. If there was no selection among RNA molecules, how then, one might ask, could there have been any form of advance in molecular organization at the start? At the onset of molecular evolution, the spontaneous synthesis, folding, and stable conformation of catalytically active organic molecules that could channel environmentally available precursors into the building blocks of catalysts would serve as the origin of greater molecular complexity more than RNA replicators with their unavoidable tendency to evolve toward greedy, parasitic elements.

With the advent of translation, molecular evolution was dramatically accelerated via the availability of new and sustainable catalysts. However molecular evolution did not become truly Darwinian until the products of translation were encoded by reproducibly inherited nucleic acids. DNA afforded stable heritability of genetic innovations, the association of genes to multigene modules of function, the most important and possibly oldest of which is that encoding the ribosome and associated functions itself. The transition to the free-living state required numerous inventions and the origin of enzymes and cofactors underpinning core metabolism processes. The genome size of the first microbes was probably small but how small is very uncertain. Based on inferences of LUCA and observations from modern chemolithoautotrophs, between 800 and 1600 genes (encoded functions) were likely needed to support the first free-living cells. Based on modern examples, the first enzymes were probably smaller than today's, they were possibly less specific in their catalysis and possibly slower in terms of rate. The first enzymes might have been promiscuous in function due to broad substrate specificity at the active site or multifunctional due to functions not localized in the active site.

Because a number of reactions in modern metabolism can take place without enzymes, there is uncertainty about the number of functions that needed to be encoded by the first genomes. Uncatalyzed reactions proceed much faster at higher temperatures, providing a strong case for the thermophilic origin of metabolism (and microbes). We recall that the first microbes did not need to have a doubling time of 20 minutes like *E. coli*. Even if the first cells could get by with some reactions being 10^6 times slower than in *E. coli*, such that their doubling time was 40 years instead of 20 minutes, after a mere 144 doublings and given unlimited substrate, the resulting culture would have still outweighed the Earth in less than 6000 years. At the origin of microbes, we do not have to think in terms of fast replication, because there is plenty of time. We have to think in terms of catalysis because the main stockpile of protein functions underpinning the origin of free-living microbial cells did not exist and hence needs to be evolved.

6.3   LIPIDS AND A CELL WALL PROVIDE COMPARTMENTATION FROM THE ENVIRONMENT

The transition to the free-living habit requires that microbes can compartmentalize themselves from the environment without the aid of external structures. The first walls housing organic metabolism-like reactions were likely inorganic and composed of catalytic transition metals (Figure 6.9). We already pointed out that the interior of serpentinizing olivine microcrystals filled with hydrogen, carbon dioxide, transition metal ions and catalyzing submicron particles of magnetite, greigite, awaruite, brucite, and the like, were suitable environments for reactions of the acetyl-CoA pathway coupled to parts of the reverse tricarboxylic acid cycle. Part of the organic products of these pathways, such as carboxylic acids, amino acids, their polyesters, and hydrophobic polypeptides probably coated the inorganic cell walls with a 'tar' like partly hydrophobic, biofilm-like chemical layer that phase separated onto the walls from the inner aqueous phase ('tholin' in Chapter 4.3). While these heterogenous hydrophobic coatings protected the cell interior from the hydrolytic reaction, they also insulated reactants in the aqueous phase of the compartments from the catalytic activity of the walls.

Such insulation by hydrophobics could, in principle, lead to a different kind of dead end that we encountered with genetic parasites, namely by terminating catalysis. This is because hydrophobic insulation separates solid-phase catalysts from soluble substrates (Figure 6.16), bringing solid-state catalysis to a halt. A hydrophobic insulating layer also offers opportunity for innovations, however, by providing a chemical niche in which (1) soluble catalysts such as cofactors could have increasingly participated in catalysis and (2) organic compounds that were more ordered, with more homogeneous thickness and permeability came to constitute the hydrophobic layers. The natural tendency of fatty acids to organize into double layers of fatty acids with the polar carboxylic ends pointing outward and the hydrophobic tails pointing inwards is well documented. Such bilayers naturally undergo phase separation at the mineral surface. Much the same occurs in modern cells at the cell wall. This insulating layer (the plasma membrane) protects molecules in the aqueous phase from chemical attack at catalytic surfaces while simultaneously restricting the interaction between the aqueous phase and the catalytic compartment walls. As catalytic inorganic walls become coated with hydrophobics, reactions of soluble compounds become increasingly dependent upon soluble organic catalysts. This confers benefit upon compartments with higher activities of soluble organic catalysts: cofactors, small peptides, and encoded proteins. Because the basic starting compounds for the reaction network are gases, which freely diffuse across lipid bilayers, specific import mechanisms for the gas phase starting materials are not required.

This transition from inorganic catalysts in compartment walls (solid-phase, heterogeneous catalysis) to organic catalysts (aqueous phase, homogeneous catalysis) would confer strong selective benefit, in terms of product accumulation, to the contents of compartments that are able to synthesize the most catalytically active cofactors and proteins as substitutes for inorganic catalysts. It would also select mechanisms that promote the insertion of nascent peptides into the lipid bilayer, peptides that could provide access to catalytic metals within the walls, which were the ancestrally active catalytic sites. In modern cells, the insertion of proteins into membranes is the function of the universally conserved signal recognition particle (SRP), a small complex consisting of a 7S RNA and protein (FtsY) that binds the hydrophobic N-terminal leader peptide of membrane proteins as they emerge on the ribosome, allowing their cotranslational insertion into the universally conserved SecY channel of the plasma membranes (Figure 6.17). This is the main and most conserved mechanism that cells use to secrete proteins outside the cell and to insert them into membranes.

Because the SRP and Sec pathway are as universally distributed and highly conserved as the ribosome, they were present in LUCA. In many traditional origins theories, the formation of a membrane is a very early step in biogenesis. In hydrothermal theories, membrane formation is a late step because abiotic lipid synthesis (1) requires catalytic activities provided by inorganic compartment walls in the first place and because (2) bilayer formation insulates the chemical system in the cytosol from catalysis provided by compartment walls, such insulation halts chemical reactions that are dependent upon solid-state catalysts, such as methyl- and acetyl-synthesis in the acetyl-CoA pathway. This places a high molecular fitness value both on a transition to cofactor-based catalysis (transition from inorganic to organic catalysts) and on (3) a mechanism to generate contact between proteins and catalysts present in the inorganic wall. The universally conserved SRP-dependent sec pathway of cotranslational

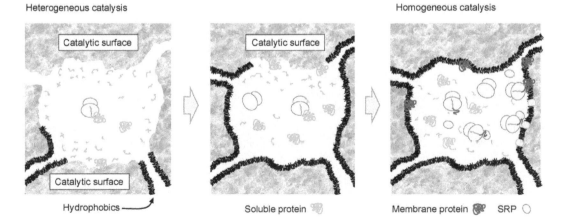

FIGURE 6.16 Accumulating hydrophobics insulate catalysts. The figure depicts an inorganic compartment in a hydrothermal system. At left, hydrophobic substances (black) synthesized in deeper regions of the vent are carried into the pore by hydrothermal fluids. Phase separation leads to the hydrophobic covering of pore surfaces, which have problems now harboring inorganic (solid-state) catalysts that perform important catalytic functions. This insulation renders genes that encode for soluble catalysts that can replace the function of solid-state catalysts beneficially (benefit: product formation). The signal recognition particle (SRP; see Figure 6.17) targets some proteins to the hydrophobic layer, restoring contact between the cytosol and the catalytic activity of the solid-state catalyst. Such interactions could have led to the incorporation of Fe and Ni into FeNiS clusters in enzymes, effectively mobilizing the inorganic catalysts into proteins. A transition between inorganic catalysts and protein-based catalysts is the result.

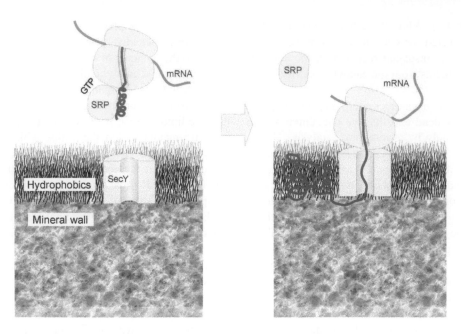

FIGURE 6.17   The conserved signal recognition particle (SRP) and secretory (SecY) channel. Highly schematic. The SRP has a protein (FtsY) and a small (7S) RNA component. Some SRP components are encoded in the ribosomal protein operon (Figure 6.13). In a primordial subsurface context, the hydrophobic layer is opposed to the solid phase rather than to a periplasmic space or cell wall, though primitive membrane–external cell wall material may have evolved quite early after the origin of translation: Subedi B. P. *et al.* (2021) Archaeal pseudomurein and bacterial murein cell wall biosynthesis share a common evolutionary ancestry, *FEMS Microb.*, 2: 1–20.

membrane insertion is such a mechanism. The emergence of the SRP may have mediated the evolutionary transition from catalysis provided by metals in the minerals of inorganic walls to cysteine-coordinated metals in FeS and FeNiS clusters in proteins.

The existence of a primitive SRP presupposes the existence of a primitive ribosome and basic translational capability (see Section 6.4). If molecular organization reaches the level of a system that can store and retrieve information in such a way that allows it to synthesize catalysts (proteins) that have the opportunity to feed back into molecular fitness of the inorganic compartments where they are made, and if the DNA genes specifying the function of those catalysts can be heritably altered to generate variation in catalysts with improved or modified function, then the system has transitioned into a world of very standard Darwinian evolution. At that point, steps toward the origin of microbes become increasingly easy from the conceptual standpoint, almost trivial in comparison to the origin of a system capable of translation from rocks, water, and gases, because standard Darwinian principles can (natural variation and natural selection) be invoked.

6.3.1   Archaea and Bacteria Synthesize Their Lipids Using Unrelated Pathways

To escape, free-living cells will require the genetically encoded synthesis of membrane lipids and turgor-resistant cell walls. In the presence of genes and an existing metabolic core, this does not present a major hurdle. In fact, this hurdle may have been taken twice independently by the ancestors of bacteria and archaea. Modern cells are about 10% lipids by weight, but the membrane lipids of bacteria and archaea differ in chemistry and biosynthetic

origin (Figure 6.18). The membrane lipids of bacteria are fatty acid esters of glycerol-3-phosphate. The fatty acids are made from acetyl CoA by cycles of Claisen condensations and reductions. Wächtershäuser and coworkers obtained unsaturated $C_{3,5,7,9}$-monocarboxylic acids from a one-pot reaction of acetylene (C_2H_2) and carbon monoxide (CO) with nickel sulfide (NiS) as catalyst and CO as reductant. More recent studies found synthesis of >C18 hydrocarbons from H_2 and CO_2 in water over iron and cobalt. The membrane lipids of archaea are isoprene ethers of glycerol-1-phosphate. That is, the stereochemistry of glycerol phosphate in archaeal (C1) and bacterial (C3) lipids is different. The isoprenes of archaeal lipids are synthesized from condensation of four C5 units (isopentenyl diphosphate) which are in turn derived from acetyl CoA by three Claisen condensations, a decarboxylation, reductions, and ATP hydrolysis. Membrane lipids spontaneously form membranes and vesicles, all that is required for membrane formation is their synthesis and an aqueous phase. Prior to the origin of genetically specified lipid synthesis, hydrophobic compounds synthesized by geochemical and non-dedicated biochemical processes could have served the function of the lipid bilayer membrane on the path to free-living cells (Figure 6.9).

Not only are the lipid pathways in archaea and bacteria different, but the enzymes involved are unrelated across the bacterial archaeal divide. This indicates that the corresponding enzymes and pathways arose not only late in the process leading to free-living cells but independently in the lineages leading to bacteria and archaea, that is, after divergence from LUCA (Figure 6.9). The cell wall biosynthesis pathways of bacteria and archaea do share some rudimentary similarities, better seen at the protein structural level than at the sequence similarity level. Bacteria almost always

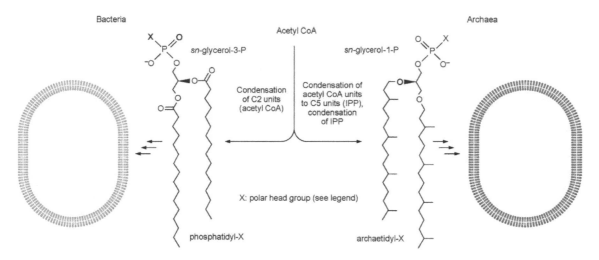

FIGURE 6.18 **Bacterial and archaeal phospholipids.** X: Polar head group. This can be any number of compounds including glycerol, serine, inosine, ethanolamine, myo-inositol, or sugar residues (glycolipids). The sugar in glycolipids can vary, for example, glucose, mannose, galactose, gulose, or *N*-acetylglucosamine. Bacterial lipids are indicated in blue, and archaeal lipids are indicated in red. IPP: isopentenyl diphosphate. The abbreviation *sn* indicates *stereochemical numbering* for the position of phosphate in glycerol phosphate (the precursor dihydroxyacetone phosphate is achiral).

possess murein as their main cell wall component, a polymer of L and D amino acids, sugars, and amines. Archaeal cell walls typically contain pseudomurein, also a polymer of L- and D-amino acids, sugars, and amines but differing in the specific components and their linkages. Cell walls can comprise up to 10% of the dry weight of a cell. It is noteworthy that cell walls contain a relatively high proportion of D amino acids.

Today, D amino acids have to be made from L-amino acids by racemases. At origins, the D amino acids were probably available from nonenzymatic syntheses. That D amino acids are sequestered in the cell wall, where they cannot interfere with protein synthesis or catalysis in enzymes, might be a relict from primordial times. That is, the sequestration of D amino acids into cell wall material might reflect a kind of primordial 'secondary metabolism', a metabolic diversion of interfering amino acid stereoisomers into an inert structural metabolite that is deposited outside the cell membrane. The incorporation of D amino acids in nonribosomal protein synthesis might also be a relict from ancient times when early metabolic systems confined to inorganic compartments were still confronted with abiotically synthesized and non-proteinogenic amino acids. This notion would be compatible with the suggestion by Lipmann, discussed in Chapter 5, that nonribosomal peptide synthesis was an evolutionary precursor to ribosomal protein synthesis.

Inorganic compartments provide a preexisting structure for lipid encapsulation and a natural sequence in which genetically encoded lipids replace abiotic hydrophobics. The proposal that lipid synthesis arose in independent events in the two lineages leading to bacteria (fatty acid esters of glycerol 3-phosphate) and archaea (isoprene ethers of glycerol 1-phosphate), as shown in Figure 6.9, can readily account for the unrelated nature of the pathways: unrelated enzymes, intermediates, and products (Figure 6.18). The alternative hypotheses, that bacteria evolved from free-living archaea or that archaea

evolved from free-living bacteria and changed every aspect of lipid synthesis, including glycerolphosphate chirality, do not seem likely. Another alternative has been proposed that LUCA possessed both archaeal and bacterial lipid synthesis pathways and they underwent differential loss in the emergence of bacteria and archaea. This is also unlikely because it presumes two coexisting and fully functional pathways in a primitive lineage that needs one of the two at most, and furthermore has the general problem that it is just too convenient as an explanatory principle.

That is, we could easily explain all differences between archaea and bacteria (there are many) by the dual presence of traits in LUCA followed by differential loss. That would not only put too many functions in LUCA but would also fail to explain why LUCA needed to evolve the second pathway if it already had one that was functional. Bacteria also independently invented a pathway of isoprene synthesis, the C5 units comprising archaeal lipids (Figure 6.18). The bacterial isoprene pathway starts from pyruvate and glyceraldehyde-3-phosphate and entails 1-deoxy-D-xylulose-5-phosphate (DOXP) as an intermediate; it arose independently of the archaeal isoprene pathway, which starts from acetyl-CoA and entails mevanolate (MVA) as an intermediate. The independent origin of the bacterial and archaeal isoprene synthesis pathways shows that lipid biosynthetic pathways can arise independently during evolution. It has also been shown that archaeal lipids and bacteria lipids can be engineered to coexist in the same cell, just like the bacterial type of lipids and cholesterol coexist in human cell membranes. The problem at origin is one of evolving functional pathways *de novo*, not of losing them. In the same way that the ancestors of bacteria and archaea independently evolved more than 20 different ribosomal proteins each (Figure 6.10) to complete the process of ribosomal evolution, bacteria and archaea evolved their lipid biosynthetic pathways independently after divergence from LUCA.

6.3.2 Was There a Physical Precursor to Cell Division?

To generate units of dispersion that would allow the first microbes to establish new colonies in new physical environments, mechanisms of cell division were required. In addition, within systems of inorganic compartments, the ability to seal off a portion of cytosol containing successful combinations of genes within compartments would exert positive impact on molecular fitness. Prokaryotic cells divide by fission, constriction of an elongated cell into two smaller cells. There are two main cell division routes known among prokaryotes. One is conserved across all bacteria and many archaea, and among archaea it is present in methanogens, the most ancient archaeal lineage. It involves a prokaryotic precursor to tubulin called FtsZ, a structural protein that spontaneously aggregates to form long linear filaments that, under GTP hydrolysis, depolymerize thereby shortening the filaments and converting GTP hydrolysis into physical force (Figure 6.19). The physical force function of FtsZ is very similar to that of its eukaryotic homologue tubulin, except that FtsZ forms linear chains of polymers rather than the 12-membered tubular polymers (microtubules) of tubulin. FtsZ forms a ring along the inside of the cell membrane at the division plane. Upon depolymerization (and with the assistance of additional proteins), the filaments shorten and the ring narrows, thereby constricting the cell in two. A second kind of division system that utilizes cell division (CDV) proteins related to components of eukaryotic vesicle formation systems occurs in some lineages of archaea. FtsZ even tends to occur in the conserved ribosomal operon (Figure 6.12), its widespread presence in archaea and bacteria reflects presence in LUCA.

Leaver *et al.* observed that bacteria without a cell wall and without FtsZ-dependent cell division machinery are able to proliferate using a protrusion-resolution mechanism. Pseudopodium-like protrusions in wall-less *Bacillus subtilis* variants resolved into spherical cells which were at least partly viable and able to further divide into micro-colonies. The colonies showed DNA segregation into the detached cells. What is the mechanism of this form of cell proliferation? Leaver *et al.* proposed that an active force from actin homologues of *B. subtilis* simply pushed extrusions of nucleoids followed by collapse and resealing of the membrane and that this mechanism may have been a form of replication before LUCA developed.

Despite the antiquity of FtsZ, it is possible that simpler physical forces acted within inorganic microcompartments to promote division-like processes. We have already seen that in the laboratory, nucleic acids can be encapsulated into aggregates of lipid vesicles driven by thermal gradients near gas bubbles in hydrothermal pores (Section 4.7). More generally, thermodiffusion and convection can accumulate lipid molecules to form aggregates and vesicles. Under special hydrothermal conditions, shear forces develop in the flow that gently fission these vesicles to smaller vesicles. The daughter vesicles can 'grow' by the accumulation of further lipids. Prebiotic division processes and accretion in protocell cycles could have occurred by a similar mechanism. Braun and coworkers showed experimentally that a parabolic flow and shear forces in a steep temperature gradient reproducibly divide lipid vesicles (Figure 6.20) by a membrane phase transition.

If microbes were to escape their inorganic confines but lack genetically encoded mechanisms for division, they would produce long, continuous, filamentous cells at best. Indeed, the name of Fts genes comes from mutations that give *E. coli* a filamentous phenotype under certain conditions (filamentous temperature sensitive). FtsZ turned out to be a protein that exerts the physical force for division. Division was required by the first microbes to allow them to generate small cells that could diffuse into the environment and disperse. But there is one more crucial barrier to be overcome on the way to the free-living state: Harnessing chemical energy at hydrothermal vents. That will be covered in Chapter 7.

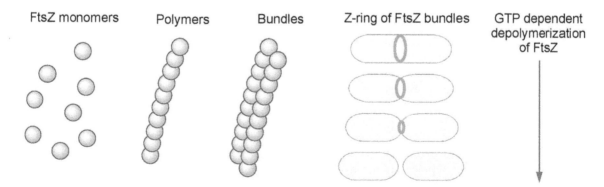

FIGURE 6.19 **Schematic depiction of microbial cell division using FtsZ.** Bacterial and many archaeal cells form a ring of FtsZ bundles at the cell division plane. FtsZ filaments spontaneously assemble and are slightly curved. GTP-dependent depolymerization at one end of the filaments leads to constriction of the Z ring and cell division but requires a number of additional proteins that properly position the ring and anchor it into the cell membrane and wall.

A.

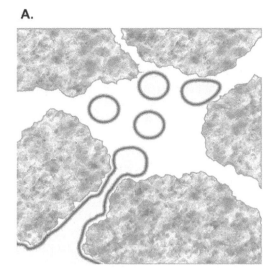

B.

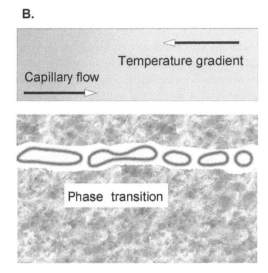

FIGURE 6.20 **Fission of lipid vesicles by membrane phase transition and shear forces.** (A) 'Soap bubble' mechanism of vesicle abscission. (B) Capillary flow of lipid vesicles in a thermal gradient. At temperatures below the phase transition temperature, the spherical vesicles drift with the surrounding medium without being deformed. At temperatures above phase transition, the vesicle membrane becomes deformable and transforms into raindrop-like shapes with additional surface area. Then, shear forces from the fluid flow profile are strong enough to pull the vesicles apart and form tether domains. In the fission phase, the vesicles are divided into smaller particles which become spherical again for minimization of surface area and energy. Vesicles near the capillary wall move the slowest and experience more phase transitions than vesicles in the center of the convective flow. (B) and (C) Vesicle division in narrow cracks and compartments of submerged porous, volcanic rocks. Redrawn Kudella P. W. *et al.* (2019) Fission of lipid-vesicles by membrane phase transitions in thermal convection, *Sci. Rep.* 9: 18808 (CCBY license).

6.4 CHAPTER SUMMARY

Section 6.1 The process of escape from a geochemical housing like a hydrothermal vent required a number of fundamental innovations to attain the state of a free-living cell. Five main (bio-) chemical innovations are necessary to evolve from a translating system dependent on geochemical energy to a fully-fledged, self-replicating free-living cell. The key innovations required for escape were

1. The development of DNA as a stable genetic storage medium
2. The origin of genomes large enough to direct synthesis of a free-living cell
3. The evolution of lipids and cell walls which insulate reactions from solid-phase catalysts and the fixation of these catalysts with proteins inside the cell
4. The conversion of energy from a geochemical ion gradient to chemical energy (the ATP synthase) and
5. The substitution of the natural ion gradient with ion pumping to become energetically self-sufficient

The RNA-to-DNA transition was catalyzed by enzyme activities that are still found among viruses today. Translation changed the nature of chemical evolution by introducing selection upon heritable variation, which rapidly accelerated protein evolution. The ribosome acts as the essential protein-synthesizing robot of the cell; it manufactures all tools that cells require to make a living in their environment. A protoribosome was isolated experimentally and shown to be a small, structurally ultraconserved unit of function consisting of only 70 nucleotides which, however, catalyze the peptidyl transferase reaction and form peptide bonds between activated amino acids. The efficiency and specificity of the ribosome were improved in evolution by adding additional nucleic acid domains and proteins to the complex.

Section 6.2 The assortment of genes into genome-sized collections likely involved combinations of genes encoded on plasmid-sized molecules that later came to be assembled into larger operons and then mini-chromosomal-sized units of selection that allowed co-inheritance of selectable traits. The first such coinherited gene collection likely harbored the RNA and proteins required for the operation and expression of the nascent genetic code. The key players were the protoribosome, tRNA minihelices, and the two classes of aminoacyl tRNA synthetases. This collection of functions made it possible for all proteins that we observe today to evolve; hence, it was likely the starting point of protein evolution. Remnants of this gene organization are still preserved as the ribosomal superoperon in modern microbial genomes. The first proteins were smaller than today's and likely catalyzed more than one reaction each. They were encoded in DNA, meaning that the site of origins must have provided means to assemble up to 1000 DNA-encoded genes in one region of time and space. Inorganic compartments in porous rock could fulfill such a role.

Section 6.3 To escape from the inorganic confines of an energy- and electron-supplying hydrothermal vent, cells needed to synthesize lipid membranes and cell walls that could compartmentalize the contents of the cell against the environment. Archaea and bacteria synthesize their lipids using unrelated pathways. Membranes insulated cytosolic contents from the environmentally supplied inorganic solid-phase catalysts that drove the synthesis of organic compounds from H_2, CO_2, NH_3, and H_2S to begin with. This generated selective advantage to enzymatic functions that could, piece by piece, replace the functions provided by solid-state transition metal catalysts with the help of the ribosome and genetic code, both of which are inventions of (bio-)organic chemistry that have no discernable inorganic forebears. This state of organization is highly advanced but still dependent upon exergonic chemical reactions provided by the vent itself. Energetic self-sufficiency is the final step in the process of escape and requires ion pumping to achieve an ion gradient for driving ATP synthase independent of a geochemical ion gradient. For spreading cells must increase their numbers and divide. Bacterial and many archaeal cells form a ring of FtsZ bundles at the cell division plane. GTP-dependent depolymerization leads to constriction of the Z ring and cell division. Fission of lipid vesicles by membrane phase transition and shear forces induced by temperature gradients in hydrothermal channels is possible in principle but would produce filamentous cells at best and is not sustainable because it is not genetically encoded.

* * *

PROBLEMS FOR CHAPTER 6

1. *Free-living cells.* Describe five key innovations required for the escape of cells in the wild.
2. *Hydrothermal vents.* Explain the temperature difference between Black Smokers and a hydrothermal field like Lost City.
3. *Protoribosome.* What functions can small synthetic protoribosomes perform?
4. *DNA monomers.* How are DNA monomers generated by ribonucleotide reductase?
5. *RNA-DNA transition.* Describe the most probable different stages of the transition from RNA to DNA during evolution and give the names of the corresponding modern enzymes.
6. *Signal recognition particle.* What function does the signal recognition particle (SRP) possess today, and what might its ancestral function have been?
7. *Microbial cell division.* Summarize the process of microbial cell division.
8. *Physical cell division.* Describe possible physical (proto-)cell division forces in early evolution.
9. *Size of early genomes.* How many and what genes did the first microbes absolutely require?
10. *The protoribosome.* Extract the essential information from the following paper and summarize it critically: Tanaya Bose *et al.* (2021) Origin of life: Chiral short RNA chains capable of non-enzymatic peptide bond formation. *Isr. J. Chem.* 61: 863–872.

7 Harnessing Energy for Escape as Free-Living Cells

7.1 ENERGETIC HURDLES NEED TO BE OVERCOME IN THE TRANSITION TO FREE-LIVING CELLS

In 2013 John F. Allen wrote: "Life is the harnessing of chemical energy in such a way that the energy harnessing device makes a copy of itself." Chapter 6 brought us to protocells that can perform all the basic functions required for the synthesis of a cell. The decisive hurdle in the transition to the free-living state is, however, the encapsulation of an energy-harnessing system that can provide a continuous supply of chemical energy needed to sustain maintenance and growth. This chapter will cover the biochemical evolutionary steps needed to convert chemical energy available in the environment into chemical energy that will support maintenance and growth in the wild as a free-living cell. We will encounter the physiology of the first free-living microbes, the nature of the habitats they could colonize, what habitats were available, their transitions from H_2-dependent chemolithoautotrophy to primitive fermentative metabolism, and from there to primitive respiratory ATP synthesis. We will consider the key chemical and molecular innovations that supported these early physiological transitions before finally turning to the evolution of molecular novelties that enabled free-living microbes in the wild to be able to escape the constraints imposed by ecosystems rooted in H_2-dependent primary production—photosynthesis. We saw in earlier chapters that the basic energy supply for the evolution of biochemical networks and translation was substrate-level phosphorylation. In modern cells, however, the main supply of chemical energy comes from the ATP synthase at the plasma membrane. The origin of the ATP synthase marks a crucial transition in early evolution on the path to fully fledged free-living cells that can diffuse in the environment as diaspores (units of dispersion) and colonize new environments. With regard to energy harnessing, the key innovations required for escape are:

1. Tapping the energy of a geochemical ion gradient (the ATP synthase)
2. Ion pumping from the reaction of H_2 with CO_2 to replace the geochemical ion gradient
3. Generating reduced ferredoxin from H_2 in a stoichiometrically balanced metabolism.

We learned in Chapter 1 that serpentinization generates alkaline effluent. This alkalinity generates a proton gradient relative to seawater. That geochemical ion gradient was likely the precursor to ion gradients that modern cells harness to synthesize ATP.

7.1.1 Harnessing Geochemical Ion Gradients

In order for cells to escape they not only need genes and proteins, they need energy, and a lot of it. On a gram-for-gram basis, the biochemical energy budget of the very first cells was probably not that much different from modern cells. If anything, the ATP demand of the first cells was higher than that of modern cells because the first cells or protomicrobes were less energy efficient than modern microbes. The ATP requirement per cell depends on how big the cell is. For a given species, how big the cell is depends on how fast it is growing. In general, slow-growing or starved cells are smaller than fast-growing cells, this has been known since work from the 1950s by Moselio Schaechter.

An *E. coli* cell with a 20-minute doubling time on glucose is typically $1 \times 1 \times 2$ μm in size (2 μm³) with an approximate density of 1.1 g/ml which rounds to about 2×10^{-12} g (2 pg) per cell. Slow-growing *E. coli* grown on glycerol has a volume of roughly 0.5 μm³ per cell. In the wild, very slow-growing cells at the surface of marine sediment have a 10-fold smaller volume (0.05 μm³) and starving—but still living—cells buried under 60 m of marine sediment, where doubling times are on the order of years rather than hours, are yet again 10 fold smaller (0.005 μm³ per cell). We can assume that the first cells were less perfect at growth than modern *E. coli*. They probably started out much smaller, slower, and less well-optimized in comparison to a fast-growing, finely tuned modern cell.

The first cells were probably very small and very, very slowly growing, with doubling times possibly on the order of years. Such inefficient machines are a more realistic model for the nature of the first free-living microbial cells that emerged from the subsurface 4 billion years ago than highly tuned *E. coli* cells that divide every 20 minutes on glucose. But for matters of energy budget, *E. coli* is the traditional model, so we can use *E. coli* as a model for energetic costs. When *E. coli* cells are grown on rich medium in the presence of glucose and O_2, they shift to acetate fermentation as the main source of ATP, rather than O_2 respiration, with the consequence that in the presence of glucose and O_2 about 75% of their ATP comes from the O_2-independent process of acetate producing fermentation rather than the energetically more efficient process of O_2 respiration. The reasons behind this phenomenon, called overflow metabolism (or the Crabtree effect) are still discussed, but we recall that in the wild, *E. coli* never has the luxury of rich medium. In nature, most cells grow much more slowly than their maximum growth rates in the laboratory. The slower cells grow, the smaller they are and the more they depend on ion gradients for ATP synthesis.

7.1.2 A Cell's Energy Budget

Today the universal energy currency of cells is ATP. Stouthamer calculated that most of the cell's biosynthetic energy expense goes for protein synthesis, roughly 70% of the biosynthetic ATP budget. An overview of the ATP

DOI: 10.1201/9781003378617-7

costs for an exponentially growing *E. coli* cell is given in Table 7.1. ATP demands differ for different cells and under different growth conditions, but most microbial cells are 50–60% protein by dry weight, and the ATP-consuming steps of translation are universally conserved, as are almost all of the chemical reactions underlying precursor (amino acids and nucleotide) biosynthesis, such that the values in Table 7.1 should be generally applicable across lineages for the cost of biosynthetic processes.

Protein synthesis (including RNA synthesis, which is necessary for protein synthesis) comprises 75% of the biosynthetic ATP expense, 90% if we include ammonium and phosphate import, which was not an expense for protomicrobes until they became fully compartmentalized from the environment. A large *E. coli* cell contains on the order of 5 million protein molecules with an average length of 300 amino acids, or 1.5 billion peptide bonds. At 4 ATP per peptide bond, that means 6 billion ATP for protein synthesis per new cell. Peptide bond formation entails 55% of the ATP expense

FIGURE 7.1   The ribosome and the ATP synthase of *E. coli*. The structures are drawn to the same scale. The ribosome is from *E. coli* (PDB ID 4YBB) as is the ATP synthase (PDB ID 5T4O). The images were kindly prepared by Natalia Mrnjavac with VMD: Humphrey W. *et al.* (1996) VMD—Visual Molecular Dynamics, *J. Molec. Graph.* 14: 33–38). During translation, mRNA is threaded through the 'donut hole' in the ribosome. Structure data from Noeske J. *et al.* (2015) High-resolution structure of the *Escherichia coli* ribosome, *Nat. Struct. Mol. Biol.* 22: 336–341 and Sobti M. *et al.* (2016) Cryo-EM structures of the autoinhibited *E. coli* ATP synthase in three rotational states, *Elife* 5: e21598. Figure modified from: Mrnjavac N., Martin W.F. (2024) GTP before ATP: The energy currency at the origin of genes. arXiv 2403.08744.

TABLE 7.1

ATP Costs per Cell in *E. coli* Grown on Glucose and Ammonium

Polymer	Gram per Gram of Cells	ATP Required per Gram of Cells [mol · 10⁴]	Proportion of ATP Cost per Cell [%]
Protein	0.52		59.1
Amino acid synthesis		14	
Polymerization		191	
RNA	0.16		16.4
NMP synthesis		34	
Polymerization		9	
mRNA turnover		14	
Import of salts			14.9
Ammonium		42	
Phosphate		8	
Potassium		2	
DNA	0.03		3.2
dNMP synthesis		9	
Polymerization		2	
Lipid	0.09	1	0.3
Polysaccharide[a]	0.17	21	6.1
Solutes	0.04	-	-
	101	347	100

Note: The high ATP cost of protein synthesis comes from 4 ATP expended per amino acid polymerized at the ribosome: the PP_i-producing step at aminoacyl tRNA synthesis (2 ATP) and the two GTP-consuming steps at translation. Note that at these values, *E. coli* would produce about 30 grams of cells per mol of ATP, but real growth yields from cultured cells are about 10 grams per mol of ATP. The difference is attributed to maintenance energy and to overflow metabolism; see Szenk M *et al.* (2017), Why do fast-growing bacteria enter overflow metabolism? *Cell Syst.* 5: 95–104. [a] glycogen content can vary widely across culture conditions. Modified with permission from: Harold F. M. (1986) *The Vital Force: A Study of Bioenergetics*, W. H. Freeman, New York, NY. and from Schönheit P. *et al.* (2016) On the origin of heterotrophy, *Trends Microbiol.* 24: 12–25.

(191/347, Table 7.1), meaning that roughly 11 billion ATP are required per cell division, based on biosynthetic costs. In other words, 55% of the cell's ATP is consumed specifically at the ribosome, and 90% is consumed for the purpose of making protein: protein, RNA, and NH_4^+ plus P_i import (312/347) from Table 7.1. In modern *E. coli* living in nature, most of the ATP is synthesized at the plasma membrane by the rotor-stator ATP synthase. In acetogens and methanogens growing on H_2 and CO_2, 100% of net ATP synthesis stems from the ATP synthase at the plasma membrane. Using the structures from *E. coli*, the ribosome, the main ATP consumer in the cell, and the ATP synthase, the main ATP producer in the cell, are shown to scale head to head in Figure 7.1.

7.1.3   RIBOSOME AND ATP SYNTHASE: THE MAIN ENERGY CONSUMERS AND PRODUCERS

The ribosome and ATP synthase are roughly the same size but differ in mass. The ribosome has a mass of about 2,300,000 D while the ATP synthase, which stands just as tall, has a mass of only 520,000 D. They are both, by weight, substantial components of the cell. In *E. coli* with a doubling time of roughly 40 minutes (a large cell), there are about 26000 ribosomes comprising about 40% of the cell's dry weight (*cf.* Table 7.1). In a cell with a 100-minute doubling time the number is about 6800 ribosomes per cell. A cell is about 20% RNA (almost all of that rRNA) by dry weight, and the ribosome is 50% protein by weight. For comparison, exponentially growing *E. coli* has about 3200 ATP synthases per cell. The roughly 8:1 ratio of ribosomes to ATP synthases in a cell reflects the ratio of their catalytic rates:

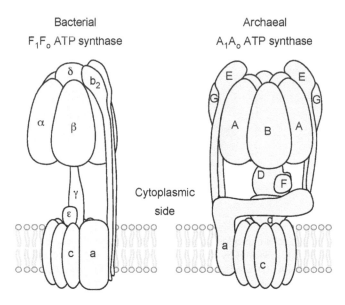

Bacterial
F_1F_o ATP synthase

Archaeal
A_1A_o ATP synthase

Cytoplasmic
side

FIGURE 7.2 Bacterial and archaeal ATP synthases. The enzymes clearly share a common ancestral structure and function using the same rotor-stator principle but exhibit differences across the bacteria archaea divide. The terms F_1F_o stem from early studies of the protein and refer to the cytoplasmic head (F_1) and the membrane (F_o) component of the enzyme. The 'o' in F_o stands for inhibited by oligomycin. Lipids indicate the plasma membrane. F_1F_o schematic structure redrawn with permission from: Mayer F., Müller V. (2014) Adaptations of anaerobic archaea to life under extreme energy limitation, *FEMS Microbiol. Rev.* 38: 449–472. The A_1A_o nomenclature refers to the archaeal enzyme, the structure of which was, however, determined from the *Thermus thermophilus* enzyme, which is an archaeal A_1A_o ATP synthase that was laterally transferred into the bacterial Thermus lineage. Schematic structure and nomenclature for the A_1A_o enzyme with permission from Zhou L., Sazanov L. A. (2019) Structure and conformational plasticity of the intact *Thermus thermophilus* V/A-type ATPase, *Science* 365: eaaw9144.

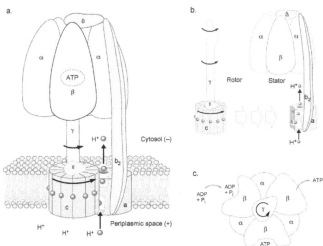

FIGURE 7.3 The ATP synthase is a rotor-stator enzyme. The protein uses proton (pH) gradients to phosphorylate ADP. a. The schematic structure of the E. coli enzyme, which consists of a rotor (c-ring, γ and ε subunits) and a stator (a, b, α, β, and δ subunits). The c-ring is embedded in the plasma membrane. Electron transport during substrate breakdown respiration in E. coli generates a proton gradient, depleting the inside of the cell (cytoplasm) of protons relative to the outside of the cell (periplasmic space). This generates a proton (pH) gradient and charge separation across the membrane, an electrochemical gradient (originally termed a chemiosmotic gradient by Mitchell). The catalytic cycle is shown in c), viewed from the top. Redrawn with permission from Müller V., Grüber G. (2003) ATP synthases: Structure, function and evolution of unique energy converters. *Cell. Mol. Life Sci.* 60: 474–494; and Neupane P. *et al.* (2019) ATP synthase: Structure, function and inhibition, *BioMol Concepts* 10: 1–10.

a ribosome can forge about 10 peptide bonds per second at 4 ATP each, an ATP synthase can generate about 100 ATP per second so that under optimal growth conditions ATP can be synthesized at about the same rate as it is consumed (a prerequisite for growth), whereby the ATP supply limits the speed of translation. Third, from an evolutionary standpoint, both the ribosome and the ATP synthase are among the most strictly conserved and universally distributed molecules across all cells. Like the ribosome, the structure of the ATP synthase differs slightly across the bacterial-archaeal divide (Figure 7.2), but they are clearly homologous and were present in LUCA. The ATP synthase is as old and universal as the ribosome and the code. Why that is, we will discuss shortly.

7.1.4 ATP SYNTHASE FUNCTION

The enzymatic mechanism of the ATP synthase, schematically sketched in Figure 7.3, is exceptional in every regard and very important for early microbial evolution. The base of ATP synthase resides in the cell membrane (Figure 7.3a). The enzyme converts the energy released by the dissipation of a proton gradient into physical rotation of the rotor (c-ring, ε and γ subunits) against the stator (Figure 7.3b). The rotational force is generated by protons passing through the *a* subunit into the c-ring, which rotates so as to release

one proton into the cytosol for every proton taken up from the periplasmic space. Each proton released leads to conformational change that turns the rotor by one c-ring subunit (30° or 1/12th of a turn in *E. coli*, because it usually has 12 *c* subunits in the c-ring). This rotation extends to the knob in the top of the γ subunit, causing it to rotate against the inside of the $\alpha_3\beta_3$ head of the enzyme, thereby inducing conformational changes in β subunits of the stator (Figure 7.3c). These conformational changes affect changes in the active site (dotted oval in Figure 7.3a), causing ADP and P_i in the cytosol to bind, to form a covalent bond as ATP, and to release ATP to the cytosol during the catalytic cycle. A full turn of the γ subunit requires the traversal of 12 H^+ through the enzyme, converting 3 ADP+P_i into 3 ATP. This mechanism of ATP synthesis, and the ATP synthase itself, is as universal and as conserved as the ribosome and the genetic code itself, because the ATP synthase is just as ubiquitous among genomes as the ribosome. Only about 30 proteins are universally present among all microbial genomes by the measure of sequence comparisons, they mainly encode proteins of the ribosome, tRNA charging, and the ATP synthase.

Again, most of the ATP that is produced and consumed in a modern microbial cell flows from the ATP synthase to the ribosome (Figure 7.4). In *E. coli* grown on glucose in O_2, only about 25% of the total ATP production comes from the ATP synthase, the rest comes from acetate-producing fermentations by substrate-level phosphorylation. In acetogens and methanogens, the ATP synthase generates 100% of the

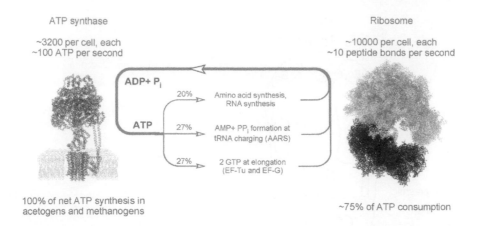

FIGURE 7.4 An ancient energetic relationship. Most of the ATP in modern cells is produced by the ATP synthase and most of it is consumed by protein synthesis. The tRNA charging reaction generates PPi that is cleaved by ubiquitous pyrophosphatases. Adenylate kinase (adk) is often encoded in the ribosomal operon (Figure 6.13). It converts AMP to ADP. GTP is consumed by elongation factors during peptidyl transfer and ribosome movement at translation. ATP is consumed at the PP$_i$ producing a reaction of AARS in tRNA charging with activated amino acids.

ATP except in highly engineered strains. Even in specialized amino acid fermenting bacteria, the ATP synthase provides 50% or more of the cell's ATP production, because fermenters couple fermentations to ion gradient formation and harnessing via flavin-based electron bifurcation (see later sections of this chapter). At the same time, about 75% of the biosynthetic energy budget goes to translation, much of that being consumed through GTP hydrolysis by the elongation factors EF-Tu and EF-G, which induce physical movement of the ribosome relative to the mRNA during peptide elongation in translation.

Powered by a chemical gradient—a pH gradient, the source of which we will turn to shortly—the physical rotation of the ATP synthase is converted into diffusible chemical energy in the form of ATP that, in turn, induces physical movement of the mRNA relative to the ribosome, synthesizing protein in the process. The first microbes cannot have been energetically more efficient than today's. They were likely slow growing, and the slower a microbe grows, the more energy it has to divert to maintenance as opposed to growth.

7.1.5 HOW AND WHY DID ATP BECOME THE ANCESTRAL ENERGY CURRENCY?

How many ATP synthases did ancestral cells need? The maximum catalytic rate of the ATP synthase is about 100 per second in *E. coli*. If *E. coli* needs 10 billion ATP per division, and the ancestral ATP synthase was 10 times slower than *E. coli*, one single ATP synthase molecule, barring a molecular dysfunction or 'injury', could provide the 10 billion ATP needed for *E. coli* replication in 32 years (32 years is 1 billion seconds, a convenient conversion number). If ancestral protomicrobes were 100-fold smaller than *E. coli*, a single ATP synthase operating at 10% of the maximum rate could supply the ATP required for the doubling of the protocell's mass in under 4 months; at 100% of the catalytic rate one single ATP synthase could supply the ATP needed for the doubling of a protocell in 12 days. With 12 ATP functional synthase molecules, the protocell would have enough ATP to divide in a day. By comparison, an *E. coli* cell growing at a modest rate contains about 3200

ATP synthase molecules that supply ATP for about 10000 ribosomes. Modern cells in low-energy environments have a doubling time on the order of months to years. As a recurrent theme, the power of efficient individual catalytic molecules emphasizes the importance of efficient catalysis over fast replication in early evolution.

If we ask which came first, the ribosome or the ATP synthase, the answer is evident: the ribosome. The ATP synthase is a pure protein with no RNA, cofactor, or transition metal prosthetic groups. It is 100% made by the ribosome using information in genes, hence there is no chicken-or-egg dilemma: the ribosome came first. Still, because of its universal distribution and presence in LUCA, the ATP synthase is almost as ancient as the ribosome. A critic might interject that such an exquisite form and mechanism of ion gradient harnessing can hardly be traits of an ancient enzyme. Is that a valid critique? Is the ATPase really so complex? No one has ever criticized the notion that the ribosome and the genetic code are ancient. Figure 7.1 shows that the ribosome is far more complex than the ATP synthase. It is 50% RNA by weight, has four times the mass of an ATP synthase and four times as many proteins, interacts with three tRNAs and mRNA at once, and requires 20 tRNAs, an amino acid pool, and 20 amino acyl tRNA synthases to operate. The ATP synthase only requires a hydrophobic layer (they can form spontaneously), an ion gradient (they naturally exist at alkaline vents; see below), ADP and P$_i$ to operate. Even though the ATP synthase might be the most intricate and complicated enzyme complex in the cell, the evolutionary hurdles to the origin of a functional ribosome are orders of magnitude higher than the hurdles leading to a functional ATP synthase. Given the complexity of the ribosome, the origin of any other known protein, including the ATP synthase, a very complicated molecular machine, is trivial by comparison. Put another way: Any system that can evolve a ribosome *de novo* can—given genes with heritable variation—use that ribosome to evolve any kind of protein, including an ATP synthase.

Notwithstanding the use of GTP in translation (about 1/4th of the energy budget, Figure 7.4), ATP is the main energetic currency of modern cells, indicating that it was the energy currency in LUCA as well. For example, in the 400 reactions of core biosynthetic metabolism that generate amino acids, bases, and cofactors there are 77 reactions that hydrolyze ATP, three that hydrolyze GTP, one that hydrolyzes CTP, while none hydrolyze UTP or TTP. What caused ATP to become the predominant energy currency in metabolic processes?

The prevalence of ATP in metabolism is possibly a frozen accident. If the ancestral rotor-stator enzyme had a substrate specificity for ADP over other nucleotide diphosphates, then once it arose it could constantly generate ATP from ADP and P_i as long as there is an ion gradient and a membrane. Those are not trivial conditions, though, more on that in the next paragraph. A steady ATP supply not only puts a high energy charge on the cell, meaning ATP >> ADP. In *E. coli* the ATP:ADP ratio is roughly 10:1 reaching as much as 30:1 under some conditions. An active ATP synthase could provide a constant energy supply and thus energetically finance the evolution of new genes and proteins. The ATP synthase operates independently of both SLP and carbon metabolism. It continuously supplies ATP in the mere presence of an ion gradient and hydrophobic layer. This supply presents a massive selective opportunity for enzymes that catalyze reactions involving ATP as an energy source or phosphorylation agent. This selects for enzymes that can use ATP rather than acyl phosphates or other energy (or phosphorylation) currency. It also selects for the incorporation of ATP binding domains, one of the most common domains known, into both new and existing proteins during early protein evolution. The substrate specificity of the ancestral rotor-stator ATP synthase imposed ATP upon the enzymes of biosynthetic metabolism as they were arising. It is all the more noteworthy that GTP was not displaced as the energy currency of translation (see **Section 7.1.8**).

Of course, an ATP synthase in a free solution cannot synthesize ATP. It needs an ion gradient and a hydrophobic membrane. What kind of ion gradients powered the first ATP synthases? Modern cells couple exergonic reactions to the pumping of ions from the inside of the cell to the outside, rendering the cytosol more alkaline during the pumping process. To appreciate the scale, if the volume of water in an *E. coli* cytosol is taken as 1 μm³ (=10^{-15} l) and the pH is 7 (10^{-7} mol/l H_3O^+) there are about 60 H_3O^+ and about 60 OH^- ions in the cell at any given time. At pH 8 the number of H_3O^+ ions is about 6, at pH 9 it is less than one H_3O^+ ion per cell. In contrast to the universally conserved ribosome and the ATP synthase, there are no universally conserved proteins across prokaryotes that couple exergonic reactions to ion pumping. In H_2-dependent acetogens and methanogens, the organisms at the focus of hydrothermal theory, the mechanisms of pumping are fundamentally different (see **Section 7.2**) involving different chemical reactions and unrelated proteins.

7.1.6 PROTEIN-BASED ION GRADIENT HARNESSING CAME BEFORE PROTEIN-BASED PUMPING

The ion-gradient harnessing ATP synthase is universal but the proteins that perform ion-gradient formation (pumping) are not. Was there some biochemical pumping machine that existed before the ATP synthase arose? That is unlikely, as its function would involve pumping (at energetic expense) for no molecular benefit other than harnessing at some later date when the ATP synthase evolves. That would be a form of evolution with foresight, a (rightly) forbidden concept in evolutionary biology. We have to assume that evolution does not operate with foresight. In other words, we cannot assume that proteins or genes 'know' in which nonrandom direction they 'need' to evolve to supply a novel function. If protocells had spent their valuable energy pumping protons in the direction of an ion gradient that already existed, without an ATP synthase to harness that gradient, they would have effectively been trying to 'acidify the ocean' at infinite expense from within compartments already possessing an alkaline interior to begin with. Clearly, gradients must have come before pumping, but how?

The simplest solution to this conundrum is that the first primitive ATP synthases, arisen by trial and error, harnessed naturally formed ion gradients that result from the interface of alkaline hydrothermal effluent from serpentinization (pH 9–11) with ocean water. Today's ocean water is pH 8, but early oceans had substantial amounts of atmospheric CO_2 dissolved as carbonic acid (H_2CO_3), a weak acid ($pK_a = 4.5 \times 10^{-7}$), rendering Hadean oceans slightly acidic with a pH near 6.5. The ion gradient for the first ATP synthases could have easily stemmed from the interface of pH 6.5 ocean water with the hydrothermal effluent, which has a pH on the order of 9–11 from serpentinization, as discussed in Chapter 2. One might interject that hydrophobic layers or membranes are required for the function of the ATP synthase: that is true regardless of when one assumes the origin of ion gradient harnessing by the ATP synthase. In the porous structures of the vent substructure or chimney superstructure (see Figure 6.6 for the example of Lost City superstructure), the aqueous phases can meet either directly at apertures, or they can interface across the mineral surfaces of microcompartments (Figure 7.5). Hydrophobics that have accumulated from organic syntheses will spontaneously undergo phase separation (the vinegar-and-oil effect) at the mineral surface, generating a hydrophobic organic barrier to aqueous diffusion, very similar to that generated by fatty acid vesicles or biological lipid bilayers. The hydrophobic layer need not have been as well-ordered as a modern biological membrane, and it could also have harbored hydrophobic peptides or other compounds soluble in the organic phase.

The thickness of the hydrophobic layer needs to be on the order of 4–10 nm for the ATP synthase to function. A thinner hydrophobic layer will not be sufficiently impermeable to protons: the gradient could dissipate without energy conservation (the function of uncouplers in modern cells). If the layer is too thick, then protons will be impaired during entry into the c-ring, or during exit, and the rotary mechanism will halt. If the hydrophobic layer is heterogeneous in thickness, as sketched in Figure 7.5, the ATP synthase can only be functional in regions with a thickness approaching of 4–10 nm. The ATP synthase does not require encapsulated, three-dimensional vesicles to function, it just needs an ion or electrochemical gradient: the enzyme has been experimentally shown to be active in planar (two-dimensional) membranes. Protons that traverse the layer are neutralized by a large

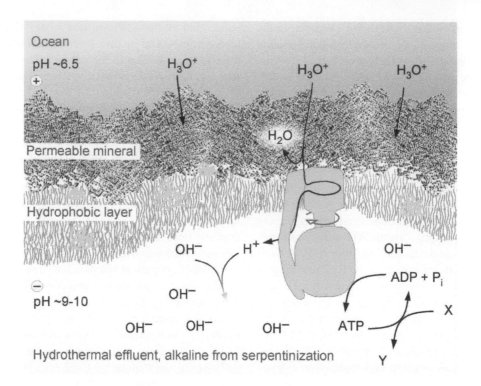

FIGURE 7.5 Harnessing a geochemical ion gradient by an ATP synthase. The + and—signs at the left designate the positively and negatively charged sides of the hydrophobic layer, a bioenergetic convention and reminder that the pH gradient carries an electrical potential, typically 60 mV per pH unit at 25°C according to the Nernst equation for ions with a unit charge of 1. Permeable mineral indicates a portion of precipitate formed at the vent (see Figure 6.6 for a modern example at Lost City). Hydrophobic substances from organic synthesis collected by phase separation at the vent walls form a semipermeable insulator similar to that of a biological lipid bilayer. For the ATP synthase to operate, the insulator must be impermeable to protons. A rotor-stator ATP synthase, the common ancestor of the archaeal and bacterial enzymes, is indicated. See also Martin W., Russell M. J. (2007) On the origin of biochemistry at an alkaline hydrothermal vent, *Phil. Trans. R. Soc. B* 362: 1887–1925.

volume of alkaline effluent (Figure 7.5) and diffusion.

From the standpoint of thermodynamics, one needs to ask whether the natural pH gradient at the chimney of a hydrothermal vent is sufficiently strong to drive an ATP synthase. Some comparisons of values from biologically formed ion gradients in *E. coli* and from calculations are instructive. In growing *E. coli*, respiration generates a proton gradient, ΔpH, of about 0.6 pH units with a cytosolic pH of 7.8 in a medium of pH 7.2. Values of ΔpH can range from ~0.3 to ~1 pH units and tend to increase at higher growth rates. The movement of charged particles across the membrane also generates an electrical membrane potential, ΔV typically meausured in mV, that lies in the range of −80 to −220 mV depending on the growth rate, always with the negative charge on the cytosolic side of the membrane.

For the purpose of comparison, the energy of the proton gradient at a hydrophobic layer in a Hadean serpentinizing system can be estimated from a simple calculation. The Gibbs free energy change for an ion at concentration c_1 on one side of the membrane to move to the other side at concentration c_2 is given by **Equation (7.1)**:

$$\Delta G = RT \cdot \ln\left(c_2/c_1\right) + zF\Delta V$$

where $R = 8.314$ J/mol/K (the gas constant). If we take $T = 363$ K (90°C) as rough estimate for hydrothermal fluid, $c_1 = 10^{-6.5}$ M (pH 6.5) as a value for the pH of the prebiotic

seawater (depends on CO_2 concentration in the seawater after the moon impact) and $c_2 = 10^{-9}$ M (pH 9) as the pH of effluent at an alkaline hydrothermal vent like Lost City, we can calculate the free energy released per proton, ΔG_{H+}. If $c_2 < c_1$, energy is released during proton movement. The charge of protons is $z = 1$, the Faraday constant is $F = 96485.3$ J/mol/V and the membrane potential can vary depending on growth conditions, faster growth yielding higher potentials in *E. coli*.

The free energy released per proton, ΔG_{H+} [kJ/mol], that traverses a biological lipid bilayer membrane from the acidic (+) to alkaline (−) side under various conditions according to the foregoing equation is given in Table 7.2. The membrane potentials of −86, −130, and −220 mV correspond to different studies measured under different growth conditions for *E. coli*. Potentials across sections of chimney at modern submarine hydrothermal vents have been measured by the team of Ken Takai, they can reach up to −500 mV from the effluent side to the ocean side, but across distances of centimeters; the potentials measured in *E. coli* exist across the thickness of a lipid bilayer. Note that the effect of temperature in Table 7.2 is small, in contrast to the strong effects that temperature exerts on the kinetics of chemical reactions. The electrical membrane potential introduced above adds additional energy to the movement of a charged particle (here H^+) compared to the free enthalpy change ΔG for the same reaction without transmembrane potential (see Section 2.4).

From Table 7.2 it is seen that an *E. coli* cell growing with a membrane potential of −130 mV at 37°C and a pH

TABLE 7.2
ΔG per Proton [kJ/mol] across a Lipid Bilayer

Temp	pH$_{in}$	pH$_{out}$	ΔpH	ΔV [mV]			
				−86	−130	−220	−400*
					ΔG/proton [kJ/mol^{-1}]		
20°C	9	6	3	−25.1	−29.4	−38.1	−55.4
	9	6.5	2.5	−22.3	−26.6	−35.2	−52.6
	8	6.5	1.5	−16.7	−21.0	−29.6	−47.0
	8	7	1	−13.9	−18.2	−26.8	−44.2
	7.8	7.2	0.6	−11.7	−15.9	−24.6	−42.0
50°C	9	6	3	−26.9	−31.2	−39.8	−57.2
	9	6.5	2.5	−23.8	−28.1	−36.7	−54.1
	8	6.5	1.5	−17.6	−21.8	−30.5	−47.9
	8	7	1	−14.5	−18.7	−27.4	−44.8
	7.8	7.2	0.6	−12.0	−16.3	−24.9	−42.3
90°C	9	6	3	−29.1	−33.4	−42.1	−59.4
	9	6.5	2.5	−25.7	−29.9	−38.6	−56.0
	8	6.5	1.5	−18.7	−23.0	−31.7	−49.0
	8	7	1	−15.2	−19.5	−28.2	−45.5
	7.8	7.2	0.6	−12.5	−16.7	−25.4	−42.8

Note: The ATP synthase requires between 3 to 3.3 protons per ATP in E. coli, synthesis of one ATP under physiological conditions requires about 47 kJ/mol such that protons need a free energy of movement across the lipid bilayer of roughly −16 kJ/mol in *E. coli*. The amount of energy needed to pump a proton across a biological membrane under physiological conditions is about 20 kJ/mol, a value called the biological energy quantum or BEQ. The membrane potentials of −86 mV, −130 mV, and −220 mV have been measured for E. coli. The pH difference of 0.6 units has also been measured for E. coli. An asterisk (*) next to the value of −400 emphasizes that a membrane potential of that magnitude could arise at the chimney of a hydrothermal vent (see text). Data from: Silverstein T. P. (2014) An exploration of how the thermodynamic efficiency of bioenergetic membrane systems varies with c-subunit stoichiometry of F$_1$F$_0$ ATP synthases, *J. Bioenerg. Biomembr.* 46: 229–241 and Tran Q. H., Unden G. (1998) Changes in the proton potential and the cellular energetics of *Escherichia coli* during growth by aerobic and anaerobic respiration or by fermentation, *Eur. J. Biochem.* 251: 538–543.

gradient of about 0.6 units ($c_1 = 10^{-7.2}$, $c_2 = 10^{-7.8}$) can obtain about −16 kJ/mol from each proton passing through the ATP synthase. For 3 protons that corresponds to roughly −48 kJ/mol. The energy provided per proton passing through the ATP synthase at −130 mV was reported as −3.95 kcal/mol (= −16.5 kJ/mol by Silverstein, who used the value of 3.3 protons required per ATP formed (the number of protons per ATP can vary depending upon the number of c subunits; see Figure 7.3), which yields 54 kJ/mol of energy to synthesize 1 ATP. Rotation of the rotor by 360° yields −198 kJ/mol for 12 protons, hence 66 kJ/mol per ATP for synthesis of 3 ATP.

Tran and Unden calculated the free energy required to synthesize ATP under physiological conditions in an *E. coli* cell, using a measured cytosolic ATP:ADP ratio of 10:1 and a measured cytosolic P$_i$ concentration of 10 mM, as 47.5 kJ/mol per ATP under aerobic or anaerobic growth conditions, suggesting a truly astounding efficiency of energy conservation by the ATP synthase of 88% (47.5/54). Silverstein found that the efficiency of the *E. coli* enzyme indeed approaches 100% under conditions of small membrane potential (−86 mV) and drops to roughly 60% at large potentials of −220 mV. Such energetic efficiency helps to explain the extreme structural and functional conservation of the ATP synthase: much like the code, the ATP synthase is conserved from LUCA, it has not been improved upon in 4 billion years. For cells living in nature, efficient energy

conservation is essential because resources are scarce. The amount of energy that is required to pump an ion across a membrane becomes a critical value for understanding how energy is conserved in cells.

The amount of energy per proton in *E. coli* of about −16 kJ/mol under a pH gradient of 0.6 pH units is vastly exceeded by pH gradients of up to 3 pH units between the alkaline effluent of a serpentinizing system (pH 9–10) and primordial ocean water with a pH of roughly 6.5 (Table 7.2). Even if the ocean had today's pH 8, alkaline effluent with a pH of 9 would still generate a difference of one pH unit, harboring more than enough energy to run an ATP synthase under even modest membrane potentials. In modern vents, alkaline effluent mixes with ocean water in varying ratios within chimneys, such that even diluted effluent would have a pH difference to ocean water.

For growing *E. coli*, the electric membrane potential ΔV contributes far more to the total electrochemical potential across the membrane than ΔpH does (Table 7.2). In the chloroplasts of plants, the membrane potential is small, on the order of −10 to −50 mV, but this weak ΔV is compensated by the large ΔpH across the thylakoid membranes, almost 4 pH units at −10 mV. As seen in **Equation (7.1)**, both ΔpH and ΔV contribute to ΔG per proton. In Table 7.2 we did not take the effects of H$_2$ in serpentinizing systems on potentials into account, except in the last column, where the free energy per proton movement is given for a membrane potential of −400 mV, which is not readily realized in prokaryotic cells but can probably be realized in hydrothermal systems.

In terms of gradient orientation (flow of protons from outside to inside), and the magnitude of the pH, the strength of the natural electrochemical gradient at a Hadean hydrothermal vent and the free energy per proton, it is thermodynamically possible that a primordial ATP synthase could have operated within inorganic compartments utilizing the electrochemical membrane potential generated by the pH difference between the Hadean ocean and alkaline hydrothermal effluent. The molecular prerequisites were in place: a pH gradient, ADP and P_i, protein synthesis, and mechanisms to insert proteins into membranes with the signal recognition particle (SRP).

7.1.7 ASPECTS OF EARLY ENERGY HARNESSING

It has been suggested that simpler gradient conserving enzymes, in particular reversible proton pumping pyrophosphatases, which consist of a single polypeptide chain, might have preceded the ATP synthase. Although that is possible, pyrophosphate is not an energy currency in modern cells, and probably never has been: among the 400 reactions that synthesize amino acids, bases and cofactors in LUCA (see Chapter 3) there are no pyrophosphate utilizing reactions. Might chemiosmosis-driven ATP synthases with a simpler structure and mechanism than the rotor-stator ATP synthase have served ATP synthesis before the rotor-stator enzyme did? Though clearly possible, there is no evidence that such was the case. That is, there are no smaller ion gradient harnessing ATP synthases that are universally conserved across all cells. The rotor-stator ATP synthase is universally conserved and traces to LUCA (Figure 7.2). Just like the ribosome, the ATP synthase is ancient, complex, universally conserved and unimproved upon in 4 billion years. LUCA had ribosomes, it had an ATP synthase as well. Yet the question nags: Was there *really* no simpler version of the ATP synthase? Sir John Walker identified two ATP binding motifs in the ATP synthase primary structure, the Walker A and the Walker B motif. The Walker A motif is also called the P loop because it binds the β-phosphate in a wide spectrum of ATP and GTP binding proteins. ATP and GTP binding domains are very common among proteins. It is likely that the ATP binding domain of the ATP synthase was recruited from a preexisting ATP-utilizing (or GTP-utilizing; see Section 7.1.8) protein at the origin of the ATP synthase. The rotor-stator components of the ATP synthase are not widespread as subunits of other proteins, they are likely an invention at the origin of the ATP synthase.

Similar to the protoribosome, there are no 'living relics' of a proto-ATP synthase, and therefore one can only speculate as to the possible presence or nature of a simpler gradient harnessing precursor, if it existed. The mechanism of the ATP synthase itself might harbor important clues. In the modern ATP synthase enzyme (Figure 7.2), the synthesis of ATP from ADP and P_i occurs in the ATP binding pocket of the active site in the stator β-subunit. When ADP and P_i are bound, ATP formation can take place spontaneously, even without an ion gradient, but the ATP remains bound to the active site, blocking any subsequent rounds of

reaction by the enzyme. This is because the ATP formed is very tightly bound in the active site. The energy input of the reaction is incurred at the step that involves expulsion of the tightly bound ATP molecule to the cytosol. In the modern ATP synthase, this energy is supplied by the conformational change generated by the mechanical motion of the rotor axis γ-subunit that is in turn powered by the movement of protons across the membrane through the c-ring.

Why was ATP not provided by simpler proteins, consisting of only a single membrane spanning polypeptide chain that underwent an ATP synthesizing conformational change upon passage of a single proton. For that matter, why does that not occur today? Recall that the ATP synthase generates one ATP per 3–4 protons. In a single polypeptide, a proton with ca. −12 to −24 kJ/mol as in modern cells with a ΔpH of roughly 1 pH unit, one proton does not present enough energy to synthesize ATP: against a 10:1 ATP to ADP ratio, roughly −47 kJ/mol are needed. Schöcke and Schink showed that the membrane integral pyrophosphatase of prokaryotic cells *does* synthesize one molecule of pyrophosphate (PP_i) per proton, but the free energy of hydrolysis of PP_i ($\Delta G^{0'} = -21$ kJ/mol) is much lower than that of ATP ($\Delta G^{0'} = -32.5$ kJ/mol). A single-proton ATP synthase could, in principle, be imaginable but it would need as fuel protons that carry much more energy each, ΔG_H^+, than is supplied by biogenic ion gradients in modern cells. This would require either a stronger pH gradient, a stronger membrane potential, or both. Is that possible?

The rightmost column in Table 7.2 shows that for the case of a biologically unrealistic, but geochemically possible membrane potential of −400 mV, a variety of conditions yield values of ΔG_H^+ that exceed −50 kJ/mol, which would, in theory, be sufficient to power a single-proton ATP synthase. It is thus thermodynamically possible that proteins existed in early evolution that could harness large membrane potentials so as to generate one ATP per proton. If such proteins existed, they were either lost, or evolved other functions when cells had to generate their own ion gradients, which do not extend membrane potentials sufficiently. It is possible that such proteins never existed and that GTP derived from SLP was the source of phosphorylation potential before the origin of the ATP synthase (see **Section 7.6.1**). Yet it is also possible that protomicrobes possessed proteins that were able to synthesize one ATP per proton in the presence of a chemiosmotic gradient with sufficient thermodynamic strength. ATP hydrolyzing ABC transporters operating on small cations, but in the reverse direction so as to synthesize ATP rather than hydrolyze ATP are one possibility, as are the family of P-type ATPases. These ATPases pump one proton per ATP but do not run backwards as one ATP-per-proton ATP synthases in cells, because cells cannot generate the membrane potential required to synthesize one ATP per translocated proton. Geochemical ion gradients, however, may harbor enough energy to fuel single-proton ATP synthesis (Table 7.2), such that single-proton ATP synthases could have existed in early evolution but were later replaced by the rotor-stator ATPase, which could operate at the membrane potentials that cells can generate by generating about 0.3 ATP per proton. Regardless, known ATP synthases require 3–4 protons per ATP such that the decisive observation is this: by distributing the energetic work of ATP synthesis across three modestly energy rich-protons (instead of one very energy-rich proton) via the rotor-stator ATP synthase, protomicrobes gained the opportunity to harness membrane potentials of a magnitude that they could generate by themselves.

Theories are not to be put forth without need. The proposal that LUCA harnessed naturally formed geochemical proton gradients with an ATP synthase (made by ribosomes) prior to the origin of free living cells is a hypothesis, a component of a theory. What would it explain? (1) It would account for the observed antiquity of the ATP synthase, which operates in all cells. (2) It would account for the lack of comparably universal or conserved ion *pumping* mechanisms even though the *harnessing* mechanism is conserved. (3) It would account for the universality of ATP as the energy currency of cells: with ATP in constant supply (the enzyme lacks a switch to turn it off) via geochemical gradients, enzymes in LUCA evolved to accommodate the existing energy currency (an exception is the ribosome, which is older than the ATP synthase and GTP dependent; see Section 7.1.8). (4) It would account for the universal polarity of ion gradient formation and utilization in cells: alkaline on the inside, with protons streaming into the cytosol during ATP synthesis).

Furthermore, (5) it would account for the otherwise dumbfounding observation that H_2-dependent acetogens and methanogens—lineages that are likely the most ancient physiological types because of the similarity of their core carbon metabolism to geochemical reactions—obtain 100% of their net ATP from ion gradient formation and harnessing, because under physiological conditions, there is not enough Gibbs free energy in the H_2-CO_2 couple to *simultaneously* support net carbon metabolism *and* net ATP synthesis via substrate level phosphorylation. That is, pumping protons allows cells to conserve energy in increments of about −20 kJ/mol (the biological energy quantum, BEQ, the amount of free energy required to pump a proton across a lipid bilayer membrane), whereas SLP operates in increments closer to −70 kJ/mol, more than twice the value of −32 kJ/mol for ATP formation because energy efficiency of soluble processes in anaerobes is about 40%. Protomicrobes solved this problem by coupling H_2-dependent CO_2 reduction to the biochemical generation of ion gradients for ATP synthesis, which required the origin of pumping mechanisms and the origin of a process called flavin-based electron bifurcation, as we will see in **Section 7.2**.

The bioenergetic consequence is that protoacetogens and protomethanogens could have thrived as autotrophic protomicrobes within inorganic compartments of a hydrothermal vent, but they were dependent on geochemical processes. For energy they were dependent upon geochemically supplied ion gradient for ATP and could not escape until they invented biochemical machineries that would enable them to couple exergonic reactions to the pumping of protons from the inside of the cell to the outside. By 'learning' to pump they replaced the geochemical ion gradient with a biogenic ion gradient that they could generate autogenously using catalysts encoded by genes. For carbon metabolism they were dependent upon geochemical mechanisms that would allow them to reduce CO_2 with electrons from H_2 to useful intermediates such as acetyl-CoA and pyruvate. How did protomicrobes solve the problem of H_2-dependent CO_2 reduction? In the absence of chlorophyll-based photosynthesis, cells do not generate midpoint potentials more

negative than −700 mV. The most negative enzymatic reaction potentials generated by cells in the dark are on the order of −680 mV (for one reaction involving anaerobic degradation of aromatics) and require both electron bifurcation and low potential reduced ferredoxin, as work by Matthias Boll and colleagues (Huwiler et al.) has shown.

What is the energetic solution? H_2 rich serpentinizing vents have a midpoint potential on the order of −800 mV (Table 2.1), making H_2-dependent CO_2 reduction energetically facile in the presence of suitable solid-state catalysts such as Ni_3Fe and Fe_3O_4, as Preiner showed. Under such conditions, pyruvate synthesis from H_2 and CO_2 is exergonic and does not require enzymes. But during the process of escape, cells did not have the option of incorporating solid state catalysts into proteins, nor did they have the option of generating cytosolic midpoint potentials of −800 mV. Electron bifurcation (**Section 7.2**) allowed protomicrobes to reduce CO_2 with electrons from H_2, ion pumping allowed cells to generate their own ion gradients for chemiosmotic harnessing from H_2-dependent CO_2 reduction. Those enzymatic mechanisms came to replace geochemical ion gradients and extremely negative geochemical H_2 midpoint potentials required for CO_2 assimilation. Modern acetogens and methanogens can have arisen at hydrothermal vents, and they can generate reduced carbon and synthesize ATP from H_2 and CO_2, but in order to exist as free living cells, they had to invent protein-based mechanisms to replace the naturally preexisting extremely negative midpoint potentials of hydrothermal vents and the geochemical ion gradient of hydrothermal vents. The geochemical mechanism involved solid state catalysts and serpentinization, neither of which can be generated by genes and proteins in cells, and are hence immobile, they cannot be packaged into cells to make them portable. Both required replacement by the invention of biochemical mechanisms.

7.1.8 GTP WAS THE UNIVERSAL ENERGY CURRENCY AT THE ORIGIN OF TRANSLATION

Although ATP is the universal energy currency today, there is a seldom noted curiosity in the energy budget of cells in that GTP is the main energy currency of ribosome biogenesis and function. This suggests that GTP was the energy currency in the environment where the ribosome arose. The ATP synthase is a complicated protein that is made by the ribosome. Hence the ribosome had to exist before the ATP synthase came into existence. Where did the energy come from to power the origin of the ribosome and the origin of the first protein coding genes? In **Chapter 3** we saw that the reactions of H_2 and CO_2 to form acetate and pyruvate can energetically support substrate level phosphorylation involving acyl phosphates. Acetyl phosphate concentrations in modern *E. coli* can reach 3 mM (ATP is roughly 10 mM). Although acyl phosphates were probably crucial at the onset of metabolic reactions, a conspicuous number of ancient biochemical functions are GTP dependent: the succinyl CoA synthase step in the TCA cycle and the phosphoenolpyruvate carboxykinase reaction. In bacteria, guanosine phosphates regulate ribosome biogenesis as ppGpp or pppGpp (guanosine tetraphosphate and pentaphosphate respectively) by governing the stringent response—a genome wide shutdown of transcription, especially rRNA and tRNA transcription, in response to amino acid limitation. Synthesis of the iron-guanylylpyridinol (FeGP) cofactor of methanogen Fe-hydrogenase and pterin synthesis: folate and methanopterin used in the acetyl CoA pathway of acetogens and methanogens, FAD, F_{420} (an archaeal homologue of FAD) and molybdopterin are all biosynthetically derived from GTP.

Those GTP-dependent metabolic reactions are, however, dwarfed in terms of total GTP expense by the role of GTP in translation. The ancient role of GTP is most strikingly conserved in ribosomal biogenesis and function. The GTP dependence of conserved ribosome associated functions (Figure 7.6) strongly suggests that GTP was more widely used in ancient metabolism before ATP became the universal energy currency. GTP is not only the energy currency for the translation step itself (elongation factors), it is the energy currency for initiation and for protein secretion, processes that involve large GTPases. The families of large and small GTPases both trace to LUCA. The small GTPases constitute one of the largest, most diverse and most universal family of proteins. Small GTPases consist mainly of a GTP binding domain and are very widely distributed among bacteria and archaea where they typically function by inducing conformational changes in target proteins upon GTP hydrolysis, modulating diverse functions such as ribosome binding, tRNA binding, Fe^{2+} transport and Ni^{2+} transport, ribosome

biogenesis and tRNA modification. It is very conspicuous that functions associated with the ribosome are primarily GTP dependent, both in bacteria and in archaea. This suggests that GTP was the energy currency of translation both in LUCA and at the origin of the ribosome.

The strictly conserved GTP dependence of ribosomal biogenesis and function suggests that this GTP dependence is a biochemical relict of a very ancient phase of protomicrobe evolution when translation existed and during which the ribosome was taking on its ancestral state—but at a time when GTP was the main energy currency. In that case, this GTP dependence would reflect a well conserved phase of GTP-dependent primordial evolution before ATP had emerged as the universal energy currency. An alternative explanation is that ribosomal function was ancestrally ATP dependent but that ATP became displaced by GTP in translation for some reason that one would have to conjure.

The advent of ATP as a third major energy currency after acyl phosphates and GTP might simply reflect the substrate specificity of the ancestral ATP synthase, an enzyme that generated an ancient energy currency dichotomy (Figure 7.7). From the perspective of modern cells, neither the ribosome (GTP) nor the ATP synthase (ATP) have altered their triphosphate substrate specificity during evolution. Such complete conservation is best explained as preservation of the ancestral state and the operation of very strict functional constraints over billions of years across all lineages. The source of GTP for the ribosome was certainly not the rotor-stator ATP synthase, because it is a protein.

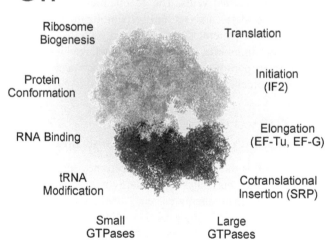

FIGURE 7.6 **GTP is the energy currency of ribosome biogenesis and translation.** Information from: Gibbs M., Fredrick K. (2018) Roles of elusive translational GTPases come to light and inform on the process of ribosome biogenesis in bacteria, *Mol Microbiol* 107: 445–454; Clementi N., Polacek N. (2010) Ribosome-associated GTPases: The role of RNA for GTPase activation, *RNA Biol.* 7: 521–527; Karbstein K. (2007) Role of GTPases in ribosome assembly, *Biopolymers* 87: 1–11; Connolly K., Culver G. (2009) Deconstructing ribosome construction. *Trends Biochem. Sci.* 34: 256–263; Britton R. A. (2009) Role of GTPases in bacterial ribosome assembly, *Annu. Rev. Microbiol.* 63: 155–176. Figure from: Mrnjavac N., Martin W.F. (2024) GTP before ATP: The energy currency at the origin of genes. arXiv 2403.08744.

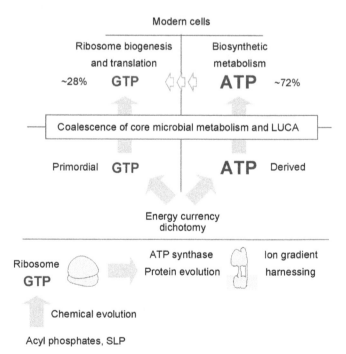

FIGURE 7.7 **GTP before ATP.** The tendency of the ribosome to arise on GTP during ontogeny and to function during translation with GTP suggests that GTP was the energy currency of translation before ATP came to be the universal energy currency. The origin of ATP as a second major energy currency (the energy currency dichotomy) is likely the result of the substrate specificity of the ancestral ATP synthase. The small arrows pointing from ATP to GTP indicate conversion of ATP to GTP for translation in modern metabolism. Figure from: Mrnjavac N., Martin W.F. (2024) GTP before ATP: The energy currency at the origin of genes. arXiv 2403.08744.

Proteins are made by the ribosome. We can thus be sure that the ribosome came before any form of genetically encoded ATP synthase (Figure 7.7). Accordingly, the source of GTP for translation was most likely SLP (or, as pure speculation, single proton nucleoside triphosphate synthases having a ΔG_{H+} per proton on the order of −50 kJ/mol for GTP synthesis; see **Section 7.1.5**). The conspicuous role of GTP in translation and ribosome biogenesis suggests that, as the ribosome arose, the earliest evolution of genes and proteins, in particular those for the synthesis of the ribosome itself, probably took place using GTP as the energy currency.

If one uses the values for *E. coli* as a proxy, the imprint of GTP use in the energy budget of modern microbes is substantial. About 28% of the ATP devoted to biosynthesis (monomer and polymer synthesis) in *E. coli* is converted to GTP (Figure 7.8 a, green shading) by nucleotide kinases for elongation. An additional observation hints at the use of GTP during the origin of translation: the predominance of G in ribosomal RNA. If the ribosome and ancestral rRNAs arose at a time when ATP was the main energy currency, then one would expect A to be the most common base in rRNA, because it was the most common triphosphate available in the system. But G, not A, predominates in rRNA, both in 16S and in 23S rRNA sequences, both in bacteria and in archaea

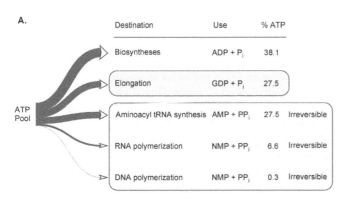

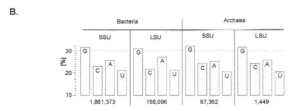

FIGURE 7.8 GTP in metabolism and in the ribosome. A) Biosynthetic energy budget of E. coli. About 28% of the biosynthetic ATP in E. coli is converted into GTP for the elongation step of translation. About 35% of the ATP is dedicated to irreversible processes. GTP use is highlighted in green, irreversible processes are highlighted in yellow. Data from: Wimmer J. L. E. *et al.* (2021) Pyrophosphate and irreversibility in evolution, or why PPi is not an energy currency and why nature chose triphosphates, *Front. Microbiol.* 12: 759359. B) Frequency of the four nucleotides (based on gene sequences, not including modifications) in small subunit (SSU) and large subunit (LSU) of rRNA in bacteria and archaea. Number of rRNA sequences. Data from the SILVA database: Quast Ch. *et al.* (2013) The SILVA ribosomal RNA gene database project: improved data processing and web-based tools, *Nucleic Acids Res.* 41: D590–D596. The excess of G in the distribution of base frequencies is highly nonrandom (p < 10^-300, Smirnov-Komolgorov test). Figure from: Mrnjavac N., Martin W.F. (2024) GTP before ATP: The energy currency at the origin of genes. arXiv 2403.08744.

(Figure 7.8b). This is not just a G+C effect (thermostability of GC base pairing in folding) because G predominates over C, nor is it a purine effect, because G also predominates over A. The excess of G is highly significant. Seen in light of GTP-dependent ribosome function, it appears to be a conserved trait that traces to the ancestral ribosome.

As discussed above in **Section 7.1.3** the predominance of ATP as the energy currency of modern cells is best understood as a property imposed by, and emergent from, the (chance) substrate specificity of the ancestral rotor-stator ATP synthase. A constant supply of ATP (it could have been any other triphosphate) provided by the ATP synthase and the geochemical ion gradient imposed ATP as the energy currency upon the enzymes of biosynthetic metabolism as they were arising, whereby GTP was not displaced as the energy currency of the ribosome.

7.1.9 TRIPHOSPHATES ALLOW EITHER THERMODYNAMIC OR KINETIC CONTROL, DIPHOSPHATES DO NOT

A classical paper by Westheimer addressed the question of why nature chose phosphates in nucleic acids and in energetics. The answer is twofold. As the backbone of nucleic acids, phosphate can form two metastable ester bonds and still retain a charge so that the acids stay in solution. Sulfate diesters are uncharged and insoluble, arsenate can also form two ester bonds and remain charged, but its bonds are very unstable and hydrolyze within about 30 minutes. Phosphate's utility as an energy currency resides in the metastability of its bonds, which can be formed and hydrolyzed under the conditions of the cell. That does not, however, answer the question of why triphosphates such as ATP and GTP are used as energy currencies instead of diphosphates like ADP and GDP, which provide roughly the same energy per phosphoanhydride bond hydrolysis. The synthesis of nucleoside triphosphates might have become essential during the origin of the genetic code and translation because of the irreversibility of the amino acyl tRNA synthesis reaction which is due to irreversible hydrolysis of the PP_i product as explained in more detail below. Recall that that protein synthesis requires 4 ATP per peptide bond, 2 as GTP and one PP_i forming step during amino acyl tRNA synthesis by aminoacyl tRNA synthases (AARS).

We discussed GTP above. Now we look more closely at PP_i. If we consider the 400 universal reactions of central metabolism that generate amino acids and bases, we see that 77 are ATP utilizing, whereby none actually *require* a triphosphate bond from the chemical and mechanistic perspective. Energetically, ADP hydrolysis to AMP and P_i would suffice at the ATP hydrolyzing reactions. Even RNA synthesis and DNA synthesis can operate with nucleoside diphosphates. There is no *a priori* thermodynamic requirement for triphosphates in primordial biosynthetic metabolism. Why triphosphates? The answer is as surprising as it is simple: kinetics. Triphosphates have two high energy bonds. Cleavage of the one between the β and γ phosphates yields ADP and P_i, cleavage of the other yields AMP and PP_i. Pyrophosphate (PP_i) is in turn an agent of irreversibility in biochemical evolution because of the ubiquitous activity of inorganic pyrophosphatases (PPases) that rapidly catalyze the reaction $PP_i + H_2O \rightarrow 2P_i$ which is exergonic with $\Delta G^{0'} = -21$ kJ/mol. PPases remove a substrate for the back reaction of all PP_i forming reactions, making them irreversible. It is a kinetic effect, the rate of the reverse reaction approaches zero because a substrate (PP_i) is lacking. In 1965, Lipmann proposed that PP_i might be an ancient energy currency, but no reactions in core biosynthetic metabolism use PP_i as an energy currency. Instead, PP_i is an agent of irreversibility, as Kornberg proposed in 1962.

Arguably, the first reaction that actually depended upon the use and availability of triphosphates was probably the aminoacyl tRNA synthase reaction itself, the one that links information in RNA to information in protein by charging the right tRNA with the right amino acid. That reaction, which we visited in detail in Chapter 5, is ATP dependent, forming AMP and PP_i. The subsequent hydrolysis of PP_i makes the tRNA charging reaction irreversible, channeling amino acids in the direction of protein synthesis. Pyrophosphatases are ubiquitous and even Mg^{2+} itself exhibits pyrophosphatase activity. Protein synthesis is crucial. About 28% of the purely biosynthetic ATP hydrolysis reactions of a modern cell are consumed by the AARS reaction. If tRNA charging at the AARS reaction were reversible, no charged tRNAs would accumulate for protein synthesis, either in modern cells or at origins. PP_i production at the AARS reaction makes tRNA charging, hence translation, irreversible.

Why is irreversibility at translation so important? The ATP synthase catalyzes a freely reversible reaction. Under some conditions, cells use that reversibility to generate ion gradients at ATP expense; some photosynthesizers do this to generate low potential ferredoxin during nitrogen fixation in the dark, for example; the process is called reverse electron transport. Were there no irreversible reactions in cells, then under an excess of metabolic end products and a dearth of substrates, the microbial growth process could, in principle, run backwards. But that does not happen: When they run out of growth substrate, cells do not grow backwards. Even in extreme energy limited environments, such as deep marine sediment, cells persist as stable structures rather than having their proteins run backwards through the ribosome to synthesize charged tRNAs. What keeps the reaction of life from running backwards? Pyrophosphate. Reactions that cleave ATP into AMP and PP_i are irreversible because ubiquitous pyrophosphatases

remove a substrate for the reverse reaction. Under the conditions of the cell, the reaction can only run in one direction. The PP_i forming step in the 20 AARS reactions that charge tRNAs ensures that once a tRNA is aminoacylated, it stays aminoacylated until it reaches the ribosome. The reaction that sends charged tRNA to become protein is irreversible. This is critically important for cells that live under extreme energy limitation: it keeps them from chemically unravelling. It makes ribosomal protein synthesis irreversible. Danchin and colleagues showed that in growing *E. coli* cells, pyrophosphatase activities are sufficient to drive PP_i concentrations effectively to zero in less than a minute. PP_i is therefore unavailable for the reverse reaction of PP_i producing steps, making them irreversible for kinetic reasons (the reverse reaction rate becomes effectively zero). Because PP_i undergoes immediate hydrolysis, it acts like a ratchet's pawl, or like wheel blocks to keep a car from rolling backwards down a hill.

To summarize, the imprint of GTP in ribosome biogenesis, function and rRNA structure suggests that GTP was the first currency of ribosomal protein synthesis. That process eventually gave rise to the ATP synthase, which in turn likely transformed ATP into the main energy currency in most biosynthetic pathways, the enzymes for which evolved after the origin of the ribosome. However, in core biosynthesis of amino acids, bases and cofactors, only 77/400 (19%) of the reactions in modern metabolism entail ATP hydrolysis as a component of the reaction. The use of GTP at the ribosome has remained conserved across all lineages, with GTP synthesis stemming from ATP and GDP through universal nucleoside diphosphate kinases. The mechanism of the ATP synthase is such that it constantly phosphorylates ADP to ATP in the presence of a pH gradient. This generated an unlimited supply of ATP in the presence of a geochemically generated pH gradient. This constant supply of ATP, independent of carbon metabolism, explains why ATP is the universal energy currency: it was the currency constantly in circulation during the phase of evolution as proteins, which are synthesized on ribosomes, were arising and evolving. An exception is the ribosome itself, where GTP, comprising about 1/4th of the biosynthetic energy budget, remained the universal currency of translation. The structure of the ancestral ribosome and its elongation factors remained dependent upon the ancestral energy currency GTP, with GTP being supplied from ATP.

The use of nucleoside triphosphates instead of diphosphates as universal energy currency in cells reflects a unique property of triphosphates: they allow enzymes to either place a reaction under thermodynamic control or kinetic control, depending upon how the currency is used. No other biological energy currency has the same property. PP_i forming reactions are irreversible under physiological conditions, imparting directionality upon the process of microbial growth.

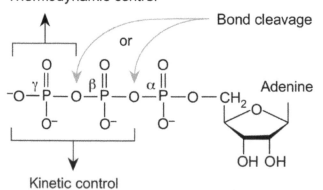

FIGURE 7.9 The unique utility of triphosphates. The options of thermodynamic versus kinetic control posed by triphosphates. Figure from Wimmer J. L. E. *et al.* (2021) Pyrophosphate and irreversibility in evolution, or why PPi is not an energy currency and why nature chose triphosphates, *Front. Microbiol.* 12: 759359 (CCBY license).

7.2 ION PUMPING, BIFURCATION, AND COUPLING

A crucial innovation required for escape is the origin of mechanisms that allow the first microbes to synthesize ATP using the energy that is available in their environment—the reaction of H_2 with CO_2—while simultaneously providing reduced carbon for biosynthesis. It will require coupling the proton-gradient utilizing rotor-stator ATP synthase to a machinery that can generate a proton gradient. This involves the origin of proteins that pump (also called coupling sites in physiology), proteins that couple an exergonic reaction to the pumping of a proton from the inside of the cell to the outside, depleting the cytosol of protons, generating both a chemical (pH) gradient across the membrane and an electrical gradient across the membrane, because a charged particle is actively displaced across a semipermeable barrier. Many acetogens and methanogens use either Na^+ or H^+ in their ATP synthase, and many modern ATP synthases are promiscuous for both H^+ and Na^+, that is, they can use either ion in the same channels. Accordingly, acetogens and methanogens have mechanisms that can pump Na^+, the simplest mechanism involving an antiporter that converts a H^+ gradient 1:1 into a Na^+ gradient.

The utility of Na^+ energetics resides in subtle but important property. Proton gradients are easily dissipated by small acids like acetate, which in the protonated state can readily traverse membranes. This means that acetate, CH_3COO^-, is a decoupler for proton gradients. It wastes the valuable work of proteins that pump protons because it can readily protonate to form acetic acid, CH_3COOH, which is uncharged and can thus freely diffuse across membranes. Acetate cannot transport Na^+ across membranes, however, such that Na^+ gradients are stable also in the presence of decouplers like acetate. Because of this effect, it is likely that proteins capable of converting proton gradients into Na^+ gradients evolved very early in evolution and were of great value to protomicrobes that possessed them, because it allowed them to harness energy in a more sustainable manner. A protein that converts H^+ gradients into Na^+ gradients, a H^+/Na^+ antiporter, was identified in genetic reconstruction of LUCA (Figure 6.8).

Converting a pre-existing proton gradient, for example a geochemical H^+ gradient at a serpentinizing hence alkaline hydrothermal vent, into a more stable Na^+ gradient is energetically free of cost, provided that a H^+/Na^+ antiporter exists. That contrasts sharply to the innovation of generating a proton gradient from chemical reactions that are contained within the cytosol and catalyzed by proteins. The origin of genes for pumping is crucial for the transition from 'life' as a protomicrobe within the confines of a hydrothermal vent to the free-living state, to life in the wild as a self-contained cell that is independent of that geochemical proton gradient. It appears that the coupling problem—the problem of inventing a way to pump ions from the inside of the cell to the outside—was solved independently in the protomicrobial lineages that led to acetogens (bacteria) and methanogens (archaea), in analogy to their independent invention of genetically encoded lipids.

7.2.1 Coupling Pumping to Acetate Synthesis in Bacteria

Physiologically primitive acetogens lack cytochromes and quinones. They pump with the help of interesting proteins called Rnf (for *Rhodobacter* nitrogen fixation), or alternatively with Ech (for *energy-conserving hydrogenase*). We will focus on Rnf here. Rnf is a membrane protein that transfers electrons energetically far downhill, from reduced ferredoxin (Fd^-) to NAD^+. There is enough energy in that redox reaction to power a conformational change in Rnf that pumps a proton from the inside of the cell to the outside. (Ech pumps either Na^+ or H^+ during the Fd-dependent reduction of protons to H_2.) In order

to make Fd^- for the Rnf energy-conserving reaction, acetogens have to send electrons energetically uphill, because under physiological conditions, Fd^- is the reductant for CO_2 reduction reactions, but Fd^- (E_o' ca. -450 to -500 mV) has a more negative redox potential than H_2 under physiological conditions (E_o' ca. -414 mV), meaning that protomicrobes had to be in possession of a trick that modern anaerobes use: flavin-based electron bifurcation (FBEB) or electron bifurcation for short. In electron bifurcation, the electron pair in H_2 is split, one electron goes to a very positive acceptor such that the downhill pull can finance the other electron going energetically uphill to Fd^-. If the average of the midpoint potentials of the two electron acceptors is more positive than that of the donor, the reaction is exergonic (Figure 7.10). Electron bifurcation is ubiquitous among H_2-dependent anaerobes, hence the very first cells used it, too.

How is the pumping reaction at Rnf wired into acetogenesis? As electrons flow from H_2 to CO_2 in the process

Flavin based electron bifurcation

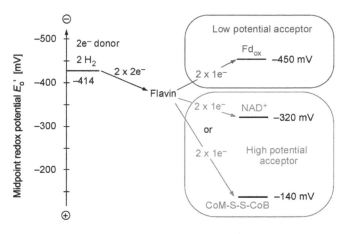

FIGURE 7.10 Flavin-based electron bifurcation. Almost all strict anaerobes and *all* microbes that use H_2 as a reductant in energy metabolism use flavin-based electron bifurcation in some manner. Acetogens and methanogens use flavin-based electron bifurcation to generate reduced ferredoxin from H_2. The electron pair from H_2 is extracted via a hydrogenase and transferred to a flavoprotein (2×2 e^- from 2 H_2 in the figure). From the flavin, one electron is transferred energetically uphill to ferredoxin (Fd; the low potential acceptor), the other is transferred energetically downhill to a high potential acceptor NAD^+ (in acetogens) or the heterodisulfide CoM-S-S-CoB (in methanogens), so that the reaction to Fd^- can go forward. In reactions using H_2 as donor and CO_2 as acceptor with Ni_3Fe as catalyst, electron bifurcation is not needed for the reaction to go forward if the pH of the solution is greater than ~8 for temperatures >25°C because E_o for the dissociation of H_2 to $2H^+$ and $2e^-$ far exceeds -500 mV. Many different compounds can be used as donors and high potential acceptors in flavin-based electron bifurcation, whereby ferredoxin is almost always the low potential acceptor. Redrawn with permission from Buckel W., Thauer R. K. (2013) Energy conservation via electron bifurcating ferredoxin reduction and proton/Na^+ translocating ferredoxin oxidation, *Biochim. Biophys. Acta, Bioenerg.* 1827: 94–113; Müller V. *et al.* (2018) Electron bifurcation: A long-hidden energy-coupling mechanism, *Annu. Rev. Microbiol.* 72: 331–353 and Schut G. J. *et al.* (2022) An abundant and diverse new family of electron bifurcating enzymes with a non-canonical catalytic mechanism, *Front. Microbiol.* 13: 946711 (CCBY license). H_2 can reduce ferredoxin over iron at pH 8.5: Brabender M. et al. (2024) Ferredoxin reduction by hydrogen with iron functions as an evolutionary precursor of flavin-based electron bifurcation. Proc. Natl. Acad. Sci. 121, e2318969121.

of making acetate, they reduce ferredoxin to Fd⁻. Ferredoxin can then transfer electrons to NAD⁺ at Rnf, an energy-releasing reaction that is coupled to proton pumping: it results in a conformation change in Rnf that pumps a sodium ion, Na⁺ (or H⁺), from the inside of the cell to the outside. There are two structures available for Rnf, but the details of the conformational change that results in pumping are not yet fully resolved. The NADH that is formed at Rnf is re-oxidized to NAD⁺ in the series of reactions that convert CO_2 to the methyl group in acetate. If that circuit becomes wired, then the protomicrobe is well on its way to being able to catalyze a balanced stoichiometric reaction of the type $4 H_2 + 2 CO_2 \rightarrow CH_3COOH + 2 H_2O$, which

has a $\Delta G^{0\prime}$ of roughly −95 kJ/mol, sufficient to generate a proton gradient. That ion gradient can be tapped for ATP synthesis, roughly 0.3 ATP per acetate under the conditions in modern anaerobic environments where acetogens grow. The process of pumping converts the chemical energy of a redox reaction into a gradient that can be harnessed for ATP synthesis in a non-stoichiometric manner. This gradient formation allows acetogens to live from a reaction that does not harbor enough energy to make one ATP per acetate (the stoichiometric amount required for SLP) and fix net carbon at the same time.

Note that 1 ATP is consumed to generate formyl-H_4F in Figure 7.11a and that 1 ATP is generated from acetyl

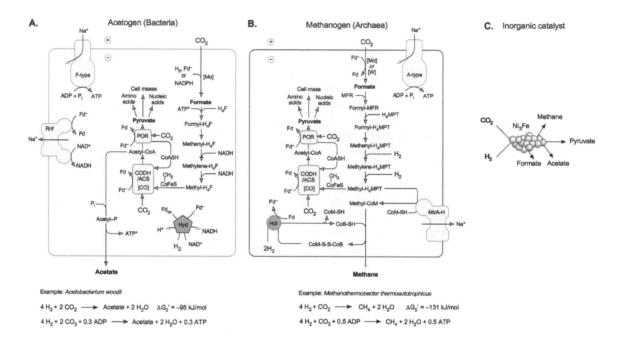

FIGURE 7.11 **Acetogenesis and methanogenesis.** The figure summarizes energy conservation from CO_2 reduction with H_2 in acetogens (bacteria) and methanogens (archaea) that lack cytochromes, very likely resembling the physiology of the first free living cells. The backbone of both pathways consists of pterin-bound methyl synthesis from H_2 and CO_2. The pathways differ with respect to the nature of the pterin—tetrahydrofolate (H_4F) in the case of acetogens and tetrahydromethanopterin (H_4MPT) in the case of methanogens—and with respect to the fate of the methyl group. (A) In acetogens without cytochromes (here using the example of *Acetobacterium woodii*), the methyl group is transferred from H_4F to the cobalt atom of the corrinoid FeS protein (CoFeS) and then to a nickel atom in the active site of CODH/ACS. There it is condensed with nickel-bound CO to form an acetyl group that is removed from the enzyme via thiolysis with CoASH. Acetyl CoA is used for SLP-generating acetate, but without net gain because of ATP-dependent formyl H_4F synthesis, indicated with an asterisk. ATP is obtained via pumping with Rnf (see text). Reduced ferredoxin is obtained with the help of electron bifurcation at the hydrogenase step (Hdr, red circle). The acetogen pathway and coupling is redrawn from (Schuchmann K., Müller V. (2014) Autotrophy at the thermodynamic limit of life: A model for energy conservation in acetogenic bacteria, *Nat. Rev. Microbiol.* 12: 809–821). (B) In methanogens, the methyl group is transferred from a nitrogen atom in H_4MPT to a sulfur atom in coenzyme M, with pumping (see text), and is subsequently reduced to methane by methyl CoM reductase. Reduced ferredoxin is obtained with the help of electron bifurcation at the heterodisulfide reductase step (Hrd, red circle). Roughly 1/20th of the methyl groups from H_4MPT are transferred to the cobalt atom of the corrinoid FeS protein (CoFeS) and then to a nickel atom in the active site of CODH/ACS to generate acetyl CoA for carbon metabolism (not shown here). The step catalyzed by CoFeS (metal to metal methyl transfer) is the reason that both acetogens and methanogens require and synthesize B_{12} (corrins). Methanogens require an additional corrin (F_{430}) at the methyl CoM reductase reaction. The methanogen pathway and coupling is redrawn from Thauer R. K. *et al.* (2008) Methanogenic archaea: Ecologically relevant differences in energy conservation, *Nat. Rev. Microbiol.* 6: 579–591). B_{12} synthesis involves over 20 enzymes. Reduced ferredoxin is obtained with the help of electron bifurcation at the hydrogenase (Hdr, red circle) step. Although the methanogen reaction has a ΔG^0 of −131 kJ/mol, enough to synthesize more than one ATP per methane, growing cells only obtain about 0.5 ATP per methane. In natural environments, the reaction is less exergonic. For 10 Pa H_2, the value of ΔG for the methanogen reaction is −40 kJ/mol (Buckel W., Thauer R. K. (2013) Energy conservation via electron bifurcating ferredoxin reduction and proton/Na⁺ translocating ferredoxin oxidation, *Biochim. Biophys. Acta, Bioenerg.* 1827: 94–113), sufficient to synthesize 0.5 ATP; similar applies to acetogens, which at low H_2 partial pressures in their environments typically obtain only about ~20 kJ/mol from the reaction. (C) A reminder that the overall energy-releasing reaction from H_2 and CO_2 to formate acetate pyruvate and methane is the same in acetogens and methanogens as in the Ni_3Fe catalyzed reactions of Preiner *et al.* (2020), whereby the function of the enzymes and the cofactors (formyl, acetyl, pyruvate and methane synthesis) performed by Ni_3Fe in alkaline hydrothermal vent conditions replaces the function performed by 127 enzymes in modern metabolism (Mrnjavac N. et al. (2023) The Moon-forming impact and the autotrophic origin of life. ChemPlusChem: e202300270).

phosphate, hence there is no net generation of ATP by SLP in acetogens. The yield of only 0.3 ATP per acetate explains in turn, why it was essential to evolve mechanisms that generate an ion gradient from the energy in the H_2/CO_2 redox couple in order to escape into the wild as free-living cells. The conditions of the vent provide, in energetic terms, a free lunch to start (free pyruvate synthesis, free ion gradient), but life outside the vent requires energetic self-sufficiency. In the reaction of H_2 with CO_2, as good as it is for origins at hydrothermal vent conditions, the only way to make that reaction work for simultaneous ATP synthesis and CO_2 fixation in a free living cell is to have electron bifurcation available for Fd^- synthesis and pumping coupled to ATP synthesis in a substoichiometric ratio, 0.3 ATP per acetate in the acetogen example shown in Figure 7.11.

The stoichiometrically balanced map of acetogen energy metabolism in Figure 7.11 represents a state of redox balance: the same number of electrons that enter the cell as substrates leave the cell as end products. Redox balance is a universal principle in energy metabolism. If a cell is not in redox balance, then energy metabolism comes to a halt.

7.2.2 Coupling Pumping to Methane Synthesis in Archaea

How are bifurcation and pumping connected in methanogens? The evolutionary primitive methanogens are H_2-dependent and lack cytochromes and quinones. They also reduce CO_2 with Fd^- that they generate from H_2 so they also have to use electron bifurcation to make Fd^-, but their wiring is slightly different from that in acetogens. The downhill reaction in their electron bifurcation reaction is the reduction of a disulfide, CoB-S-S-CoM, to the thiols, CoB-SH and CoM-SH (see Figure 7.11). The flavin FAD is involved in all electron bifurcation reactions of soluble enzymes known so far, because bifurcation entails a transition from two electron transfers to one electron transfers, a reaction that flavins perform naturally because of the stability of the semiquinone. Pumping in methanogens that lack cytochromes occurs at a membrane-integral methyltransferase named MtrA-H. In methanogenesis, the main energy-releasing reaction is CO_2 reduction to methane. In that process, CO_2 is reduced to the level of a methyl group sequestered on the C1 carrier tetrahydromethanopterin (Methyl–H_4MPT), in which the methyl group is covalently bonded to a nitrogen atom.

In the MtrA-H reaction, the nitrogen-bound methyl group is transferred to CoM in a two-step process involving the corrinoid cofactor. The first step is the transfer of the methyl group from methyl-H_4MPT to Co(I) ($\Delta G^{0'} = -15$ kJ/mol) and then from methyl-Co(III) to CoM-SH ($\Delta G^{0'} = -15$ kJ/mol) to yield CoM–S–CH_3 (methyl-CoM). The transfer of the methyl group from methyl-Co(III) is Na^+ dependent, and is thus implicated in the Na^+ translocation process. The overall reaction involves a sufficiently large change in $\Delta G^{0'}$, ca. −30 kJ/mol to power a conformational change in MtrA-H that pumps two Na^+ ions from the inside of the cell to the outside. Although the methyl transferase reaction does not involve net change in the redox state of the methyl group, the reaction is embedded in the reduction of CO_2 to CH_4, which

is pure redox chemistry. Electron bifurcation and pumping are connected 1:1 by CoM–S–CH_3, the substrate for the methane generating reaction at methyl CoM reductase, the reaction mechanism of which entails a radical intermediate. With the MtrA-H reaction in place, the stoichiometric reaction $4H_2 + CO_2 \rightarrow CH_4 + 2H_2O$, with a $\Delta G^{0'}$ of −131 kJ/mol, is *coupled* to the pumping of ions to form an ion gradient that is harnessed by the ATP synthase to generate about 0.5 ATP per molecule of methane produced. As in the case of acetogenesis, there is not enough energy in the reaction of H_2 and CO_2 to make ATP via substrate level phosphorylation and to fix carbon at the same time.

In protoacetogens, pumping at Rnf replaced the geochemical ion gradient. In protomethanogens, pumping at MtrA-H replaced the geochemical ion gradient. In both cases a pumping reaction had to be coupled to ATP synthesis from CO_2 reduction with H_2 in order for the protomicrobes to energetically master the transition to the free-living state. Both protomicrobial lineages were energetically self-sufficient *within the vent*, which provided the ion gradient for free. Replacing the naturally existing geochemical ion gradient required the invention of pumping, energetically financed at the expense of an exergonic redox reaction. In both lineages the overall exergonic reaction was H_2-dependent CO_2 reduction to acetate or methane as end products. The invention of their energy-conserving pumping reactions occurred independently in the protomicrobial lineages leading to methanogens (archaea) and acetogens (bacteria), because the primordial coupling sites (MtrA-H and Rnf) are fundamentally different, both with regard to the nature of the enzymes that pump as well as in the nature of the specific energy-releasing chemical reactions they harness for pumping (Figure 7.11).

7.2.3 Why Electron Bifurcation If H_2 Can Reduce CO_2 on Metals?

If metabolism evolved from exergonic, nonenzymatic reactions of H_2 with CO_2 on metals as sketched in Figure 7.11c, why do acetogens and methanogens require the complicated enzymatic process of electron bifurcation to reduce CO_2 with electrons from H_2? The answer is not that metals such as Ni_3Fe, shown in Figure 7.11c, somehow change the midpoint potential or the $\Delta G^{0'}$ of the reaction. The metals act as catalysts, and catalysts do not change the equilibrium of a reaction, they just accelerate the rate at which equilibrium concentrations of reactants and products are reached. The reason why electron bifurcation (Figure 7.11a,b) is not required in natural serpentinizing systems that reduce CO_2 to formate and methane and the reason why electron bifurcation is not required for H_2-dependent CO_2 reduction to formate, acetate and pyruvate in laboratory experiments with only metals as catalysts (Figure 7.11c) is explained by the simple but powerful effect of pH on the reaction $H_2 \rightarrow 2e^- + 2H^+$. In water, the reaction is $H_2 + 2H_2O \rightarrow 2e^- + 2H_3O^+$. In alkaline solution, OH^- ions consume the protons produced by H_2 oxidation, $H_2 + 2OH^- \rightarrow 2e^- + 2H_2O$, and pull the reaction to the right by removing a product. Recall that a product of H_2 oxidation are electrons. Alkaline pH shifts the midpoint potential of H_2 oxidation dramatically.

TABLE 7.3

Midpoint Potential of Aqueous Solutions at pH$_2$ vs. pH

Temp	pH	H$_2$ Partial Pressure [atm] (H$_2$ Concentration [mM]*)			
		0.1 (0.078)	1.0 (0.78)	10 (7.8)	100 (78)
		Midpoint Potential vs. SHE, E_o [mV]			
25°C	6	−325	−355	−385	−414
	7	−385	−414	−444	−473
	8	−444	−473	−503	−533
	9	−503	−533	−562	−592
	10	−562	−592	−621	−651
50°C	6	−353	−385	−417	−449
	7	−417	−449	−481	−513
	8	−481	−513	−545	−577
	9	−545	−577	−609	−641
	10	−609	−641	−673	−705
100°C	6	−407	−444	−481	−518
	7	−481	−518	−555	−592
	8	−555	−592	−629	−666
	9	−629	−666	−703	−740
	10	−703	−741	−778	−815

Note: *The H$_2$ concentrations assume a gas phase in equilibrium with the aqueous phase and solubility of H$_2$ according to Henry's law. SHE standard hydrogen electrode. The midpoint potential drops more rapidly with increasing pH than with increasing pH$_2$ because OH$^-$ ions remove H$^+$ according to the reaction H$_2$ → 2H$^+$ + 2e$^-$. Reproduced from supplemental material in Wimmer J. L. E. *et al.* (2021) Energy at origins: Favorable thermodynamics of biosynthetic reactions in the last universal common ancestor (LUCA), *Front. Microbiol.* 12: 793664 (CCBY license).

The magnitude of this effect is summarized in Table 7.3, where the calculated midpoint potential of aqueous solutions of H$_2$ across a range of pH and temperature values that correspond to the pH and H$_2$ concentrations found in the effluent of modern serpentinizing hydrothermal vents are given. The H$_2$-dependent CO$_2$ fixation experiments of Preiner that generated formate, acetate and pyruvate using the hydrothermal catalyst Ni$_3$Fe were performed at 100°C, 10 bar H$_2$, and pH 9—conditions that correspond to those in serpentinizing systems. The midpoint potential a solution at 100°C, under 10 bar H$_2$, and with pH 9 is on the order of −700 mV, as summarized by Eric Boyd and colleagues, values of −800 mV have been measured in some modern serpentinizing hydrothermal systems.

It is clear from the table that the effect of pH is stronger than the effect of H$_2$ partial pressure and stronger than the effect of temperature in the range given. The extremely negative midpoint potentials of the H$_2$ oxidation reaction at pH9 to pH10, values that exist as a result of serpentinization reactions in nature, are the reason that H$_2$-dependent CO$_2$ reduction is facile under serpentinizing conditions. For example, at pH 9, 50°C and 10 mM H$_2$, conditions that exist at Lost City, the midpoint potential of the reaction H$_2$ → 2e$^-$ + 2H$^+$ is not −414 mV (standard physiological conditions), but −609 mV, more than sufficient for efficient CO$_2$ reduction, or even reduction of low potential ferredoxin. Brabender and colleagues recently showed that hydrogen can efficiently reduce a low-potential ferredoxin with H$_2$ at pH 8.5 using native iron as the catalyst. No proteins (other than Fd itself) or cofactors were required. That solves the

energetic problem of how protomicrobes obtained reduced ferredoxin before the origin of electron bifurcation. The alkaline pH of serpentinizing systems shifts the midpoint potential of the H$_2$ oxidation reaction to very negative values, but solid state metal catalysts were required for ferredoxin reduction. Under physiological conditions, electron bifurcation is essential. Under geochemical conditions it is not. Part of the escape process involved the invention of means to perform the energy-releasing and ion-pumping reaction (H$_2$-dependent CO$_2$ reduction) under the physiological conditions of the cytosol, where the cell can regulate homeostasis. The solution involved flavin-dependent electron bifurcation, but at different enzymes (Hdr and Hyd; Figure 7.11) in the ancestors of the methanogenic and acetogenic lineages, respectively, compatible with the findings of Brabender and colleagues that iron and H2 preceeded enzymatic electon bifurcation as a primordial source of reduced ferredoxin.

7.2.4 CHEMICAL ENERGY POWERS ALL REACTIONS OF LIFE

Energy conservation is the most important thing that a cell does, because without a source of biochemical energy, no other processes in the cell can take place. We saw that acyl phosphates could have been an ancestral energy currency, that GTP was likely the energy currency at the origin of ribosomal protein synthesis and that ATP came to be the main energy currency of biosynthesis. The substrate specificity of the ancestral ATPase for ADP supplied ATP as a new currency of biochemical energy generated by a geochemical ion gradient during the early evolution of proteins. This resulted in a permanently high ATP/ADP ratio in protomicrobes, a disequilibrium favoring ATP hydrolysis in a cytosolic environment in which selection would immediately favor enzymes with the ability to use ATP. At the outset of biochemical evolution, $H_2 + CO_2$-dependent methyl and acetyl synthesis supported SLP-dependent energy metabolism. The new energy supply, ion gradient driven ATP synthesis, removed the burden from $H_2 + CO_2$-dependent methyl and acetyl synthesis to supply energy. This allowed fixed carbon to be channeled much more specifically into biosynthetic pathways, via pyruvate in particular, than prior to the existence of a rotor-stator ATP synthase. With the advent of the ATP synthase, H_2 and CO_2 no longer had to supply energy via SLP. The ATP synthase not only afforded protomicrobes a greater energy supply, it allowed them to circumvent SLP and thus devote more carbon flux to the synthesis of organic compounds.

The advent of a rotor-stator ATP synthase marked the decoupling of primordial carbon and energy metabolism, allowing both to go forward at full speed in environments where there was both H_2–CO_2 interface and contact between the alkaline milieu of the vent inside with the neutral milieu of surrounding ocean water. This could have been either near the vent-ocean interface or deeper in the vent, aided by mixing processes. That 'goldilocks' phase of biochemical evolution would not last forever, though, because in order to escape as free-living cells, carbon and energy metabolism would once again have to become coupled, so that the ion gradient was generated out of the $H_2 + CO_2$ reaction. That points to the existence of three phases in early bioenergetic evolution:

1. A start from H_2 and CO_2 reactions as a source of carbon, but also of energy via SLP up to the origin of proteins, acyl phosphates underpinning the origin of GTP-dependent translation at the ribosome.
2. The origin of a protein complex—the rotor-stator ATP synthase—that could harness the natural pH gradient at the vent ocean interface surface, a decoupling of energy metabolism from CO_2 reduction, allowing H_2-dependent CO_2 reduction to devote more carbon to biosynthetic processes without having to support energy conservation as well.
3. The origin of ion pumping, energetically financed by the main energy-releasing reaction of the cell (H_2-dependent CO_2 reduction) so that carbon and energy metabolism were *once again* coupled, but now by the integration of coupling sites (Rnf in bacteria, **Section 7.2.1**, and MtrA-H **Section 7.2.2** in archaea) into the CO_2 reduction reaction sequence. The position of Rnf and MtrA-H in the metabolism of acetogens and methanogens is shown in Figure 7.11. Both membrane integral enzyme complexes tie primordial carbon metabolism—the exergonic redox reaction with H_2 as the electron donor and CO_2 as the acceptor—to ion pumping for ATP synthesis. This configuration

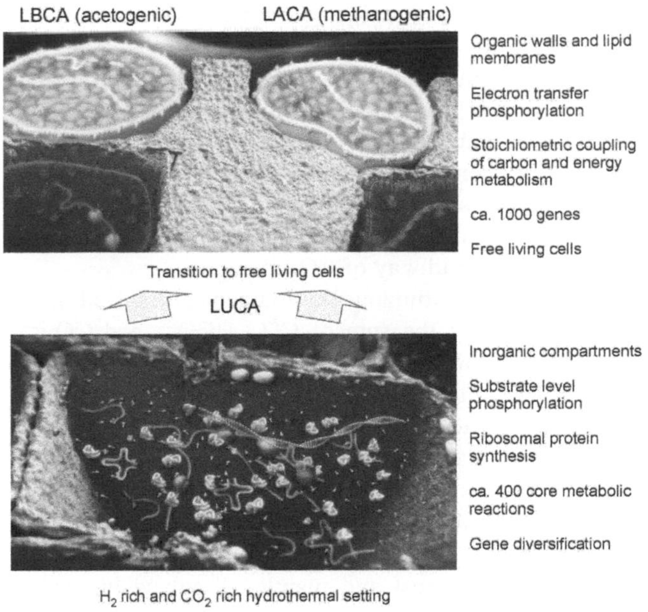

LBCA (acetogenic) LACA (methanogenic)

Organic walls and lipid membranes

Electron transfer phosphorylation

Stoichiometric coupling of carbon and energy metabolism

ca. 1000 genes

Free living cells

Transition to free living cells

LUCA

Inorganic compartments

Substrate level phosphorylation

Ribosomal protein synthesis

ca. 400 core metabolic reactions

Gene diversification

H_2 rich and CO_2 rich hydrothermal setting

FIGURE 7.12 Cellularization and escape. If translation arose from chemical reactions that took place in the environment with the help of solid state catalysts in a hydrothermal setting, then the last universal common ancestor LUCA (bottom panel) was not free living, but confined to the compartments within which it arose and the transition to free living cells (escape, top panel) entailed the origin of the last bacterial common ancestor (LBCA) and the last archaeal common ancestor (LACA). If the exergonic reaction of H_2 and CO_2 drove organic synthetic reactions and the synthesis of building blocks, then LBCA and the first bacteria were likely acetogens while LACA and the first archaea were likely methanogens. Steps on the way to free living cells are outlined in this chapter and in Chapter 6. Images from a short film on the origin of microbes available from www.molevol.hhu.de/en/movies.

was sufficient to support stoichiometric redox balance and the first free living lifestyle. Later in the evolution of free living cells, after their escape from their inorganic confines (Figure 7.12), the general principle of linking redox reactions to ion gradient formation would eventually diversify into a myriad of different donors and acceptors.

7.3 ESCAPE AND FIRST STEPS IN THE WILD

7.3.1 The First Microbes Inhabited Vents in the Crust, Modern Systems Provide Models

The thermodynamics of all reactions and processes that we have discussed so far have been driven by the exergonic reaction of H_2 with CO_2, plus ion gradients after the origin of the ATP synthase. Continuity of that principle leads to the first cells being acetogens and methanogens, which are the only forms of bacteria and archaea, respectively, that can satisfy their carbon and energy needs from H_2 and CO_2 alone. They needed both H_2 and CO_2 for survival and growth, as the overall equations for ATP synthesis in Figure 7.11 show. Because of the initial high CO_2 partial pressure in the atmosphere and bicarbonate dissolved in the oceans, CO_2 was in supply on a global scale. By contrast, H_2 was produced on a local scale via serpentinization in hydrothermal systems. This would tend to restrict growth to sites of H_2 synthesis, or hydrothermal vents. This in turn means that the first free living cells were likely confined to geochemical habitats where H_2 was being made. This is in line with the observations from modern hydrothermal vents, where acetogens and methanogens are always among the most common microbes, and where the acetyl CoA pathway is always the predominant pathway of CO_2 fixation.

In modern serpentinizing environments, dissolved inorganic levels (DIC: the sum of CO_2, HCO_3^-, and CO_3^{2-}) levels are very low, in many cases below detection levels. This is because high pH favors the formation of the CO_3^{2-} ion, which leads to the precipitation of Ca^{2+} and Mg^{2+} carbonates within the vent system. This general reaction of carbonate precipitation is the mechanism that brought the high CO_2 levels in the Earth's early atmosphere down to levels that approach today's. Before the sequestration of carbonates in the crust and mantle, dissolved inorganic carbon levels in Hadean hydrothermal systems—at least in the incoming seawater that entered the hydrothermal circulation current into the crust—were certainly much higher than today's owing to high atmospheric CO_2 concentration. That leads to the question, though, of whether there was a carbon shortage in the effluent of Hadean hydrothermal systems and if so, how primordial life dealt with it. Modern hydrothermal systems and the organisms that live there provide important clues.

Serpentinization requires host rock with a high Mg^{2+} and Fe^{2+} content and low silicate content (mafic or ultramafic rocks). This is typically found in submarine crust, which eventually returns to the mantle during the process of subduction—the Wilson cycle. But sometimes submarine crust is not subducted at subduction zones, it is obducted, that is, a section of crust is pushed on top of the continental margin, leaving it at the surface. The resulting section of submarine crust that has become thrust upon the surface is called an *ophiolite*. Water drawn into ophiolites (even rainwater or melted snow, which are confusingly called meteoritic water in the literature) can start the process of serpentinization, with alkaline and H_2 rich water emerging at the surface. Ophiolites can host active and sometimes hyperalkaline (pH 11–12) serpentinizing hydrothermal systems. Such sites have the advantage that geological and microbial sampling is much simpler than at the bottom of the ocean.

The microbes inhabiting three such hyperalkaline serpentinizing sites have been characterized using environmental genomics: The Cedars in California, Hakuba Happo on the West Coast of Japan, and the Samail ophiolite on the eastern coast of Oman. The pH in those systems reaches values of 11–12 with very high H_2 concentrations (in the 100 μM to 10 mM range) with almost zero CO_2. Despite the lack of CO_2, these systems nonetheless are rich in acetogens, methanogens and other microbes that use the acetyl CoA pathway. What is their carbon source? Their carbon source is formate. The vent effluent in these systems is rich in formate, 6.8 μM in The Cedars, 8 μM in Hakuba Happo, and 1–2 μM formate in the Samail ophiolite. For comparison, Lost City effluent contains 34 to 140 μM formate. The formate in serpentinizing systems stems from abiotic reactions of H_2 with solid phase carbonates. Formate can serve

as a source of carbon and electrons in many acetogens and methanogens. Some acetogens have a specialized enzyme, hydrogen-dependent CO_2 reductase, that catalyzes the near-equilibrium interconversion of formate to CO_2 and H_2. Acetogens and methanogens are known that can utilize formate as the sole carbon and energy source during growth. Furthermore, formate is an intermediate of the acetyl CoA pathway in both acetogens and methanogens such that it can enter the acetyl CoA pathway directly. Yet for autotrophic growth on formate they require formate oxidizing enzymes such as formate dehydrogenase or H_2-dependent CO_2 reductase that can generate CO_2 for the CO-producing CODH reaction, which is essential to the acetyl CoA pathway.

Although modern serpentinizing systems clearly support microbial life, the sources of nitrogen and phosphate required for growth in such systems are not well resolved. It is nonetheless clear that these essential nutrients are available, otherwise no microbial growth would take place in modern vents. In high pressure laboratory reactors that simulate conditions several kilometers deep within the crust (>300°C and >1000 atm), N_2 is readily reduced to NH_3 by a process involving iron oxide catalysts similar to the Haber Bosch reaction. As outlined in earlier sections, the source of phosphate in early evolution is still discussed. A problem with phosphate is its low solubility as apatite (calcium phosphate), although newer findings reveal that in the presence of Fe^{2+} and at pH <7.5, seawater phosphate could have easily been in the millimolar range, whereas modern cells require micromolar phosphate concentrations in growth media and modern seawater contains roughly 1–100 μM phosphate. A role for phosphite (HPO_3^{2-}) in early evolution remains a distinct possibility. For the reduction of phosphate to phosphite, which is much more soluble than phosphate and furthermore stable in seawater, the midpoint potential $E_0' = -690$ mV is well within the range that serpentinizing systems can generate. However, that does not equate to phosphate reduction because at such low redox potentials, CO_2 would be reduced to formate ($E_0' = -430$ mV) before phosphate would be reduced to phosphite, because CO_2 is the stronger oxidant. There are substantial amounts of phosphite in some modern environments, but its source is still unresolved. Environmental phosphite is generally thought to stem from biological phosphonates generated from phosphoenolpyruvate conversion to phosphonopyruvate, which contains a P–C bond, via intramolecular rearrangement. But biological phosphonate breakdown yields phosphates, not phosphite.

Although modern serpentinizing systems provide carbon, energy and nutrients, their alkalinity poses peculiar problems to the principle of ATP synthesis using proton pumping and the ATP synthase, because the environmental pH is much higher than the pH inside the cell. If the cell is pumping protons from the inside of the cell to the outside, the return of those protons into the cytosol to generate ATP is against the proton gradient. Nonetheless, alkaliphilic microbes can grow at pH 11–12 and alkaliphilic microbes have been known for decades. The exact details of how alkaliphiles pump are still not resolved in full detail. It was once thought that they utilize Na^+ gradients, but this does not seem to be the case because investigations have shown

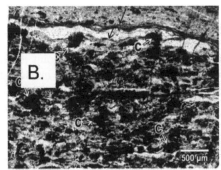

FIGURE 7.13 **Ancient hydrothermal environments.** (A) A 3.3 billion year old hydrothermal vent from the Barberton greenstone belt, South Africa. The vent contains organic matter (B) (clots, labelled 'c') that 'suggest thermal fluids as a principal energy source, providing inorganic (for example, H_2, CO_2) or organic compounds (for example, fatty acids, alcohols, ketones) to fuel chemolitho- and chemo-organotrophic metabolisms'. Reprinted with permission from Westall F. *et al.* (2015). Archean (3.33 Ga) microbe-sediment systems were diverse and flourished in a hydrothermal context. *Geology* 43:615–618.

that the ATP synthases of alkaliphiles studied so far use H^+ not Na^+. The possibility most widely discussed at present is that alkaliphiles pump protons, but that the protons do not enter into equilibrium with the surrounding medium before reentry into the cell via the ATP synthase. This could be possible if pumping complexes and the ATP synthase were in close physical proximity in the membrane, or if pumped protons were restrained from diffusion into the surrounding medium, for example by being channeled along negatively charged residues on the outside of the plasma membrane towards the reentry site on the ATP synthase membrane subunit. Both physical proximity and proton channeling might be operative. Although the exact molecular mechanisms that alkaliphiles use to harness ion gradients for ATP synthesis are still not fully resolved, they grow in hyperalkaline environments and they do use proton gradients for ATP synthesis during growth.

Is there evidence for microbial life in vents on the early Earth? There is no reliable information about specific nutrient availability in hydrothermal vents of the Hadean. There is, however, microfossil and isotope evidence for microbial life in ancient fossilized vents. Some of the oldest evidence for microbial life stems from fossilized hydrothermal vents in 3.3 to 3.8 billion year old formations found in Canada, Australia and South Africa. Although the rocks involved exhibit differing degrees of preservation, the available microfossil and isotope evidence indicates that serpentinizing hydrothermal vents were abundant on the early Earth and that they were inhabited very early in the Archaean.

One might wonder whether the actual site of microbial origin has been preserved somewhere on the ocean floor? This can be excluded, because the oldest submarine crust is only about 250 My old, perhaps slightly older as newer evidence from the Mediterranean seafloor samples suggests. The young age of submarine crust attests to the rapid rate, in terms of geological time, at which new crust emerges at spreading zones and old crust is returned to the mantle via subduction (the Wilson Cycle). The organisms that inhabit submarine crust today colonized that habitat within the last 250 million years, because submarine crust is made from cooling lava, and no organic matter can survive 1200°C magma.

7.3.2 HIGH HYDROSTATIC PRESSURE PROBABLY HAS LITTLE EFFECT ON MOST BIOCHEMICAL PROCESSES

The question often arises concerning the possible effect(s) of pressure on chemical reactions at deep sea hydrothermal systems: Does high pressure favor or hinder particular chemical reactions? Hydrostatic pressure increases with depth in seawater according to

$$\Delta p = \rho \cdot g \cdot h = 1025 \text{ kg/m}^3 \cdot 9.8 \text{ m/s}^2 \cdot 1000 \text{ m}$$
$$= 10.045 \text{ MPa} = 100.45 \text{ bar}$$

for a density $\rho = 1025$ kg/m^3 of seawater with 3.5% salt, depth h = 1000 m and earth's gravitational acceleration g = 9.8 m/s^2. Hence pressure increase in water is roughly 100 bar/km. Sediments and rocks are denser than water and their density varies between 1600 kg/m^3 for sediments and 3500 kg/m^3 for gabbro (deep rock). Assuming an average density of 2550 kg/m^3 we obtain $\Delta p = 250$ bar/km. In the lithosphere, the pressure increases by average 226 bar/km. Cell volume decreases by only ~1% at 200 bar hydrostatic pressure assuming a compressibility coefficient $\kappa = 4.6 \times 10^{-10}$ m^2/N for water. At 1000 bar (10 km water depth; 4 km sediment depth) the volume decrease is ~ 4%. Gases however are compressed very strongly. Similarly, the Gibbs energy of liquid water changes only very little with depth $\Delta G \sim v_{H20,liq} \cdot \Delta p = 0.36$ kJ/mol at 200 bar (2 km water depth), because its molar volume $v_{H20,liq} = 18.02$ ml/mol is small (see Section 2.4). The Gibbs energy of gases however changes considerably with hydrostatic pressure, for example it increases by 13.1 kJ/mol after compression of an ideal gas from $p_0 = 1$ bar to p = 200 bar at 298 K (from thermodynamics: $\Delta G = RT\ln(p/p_0)$).

While there is only a small volume change of the cell liquid with pressure, other parameters like a lower fluidity of membranes due to lipid compression and changed activity of compressed macromolecules might have some influence on cell metabolism, but the effects are probably slight at best. The archaeon *Thermococcus piezophilus* was isolated from a hydrothermal vent more than 5000 m deep, it can grow happily in the laboratory bench at 1 bar of pressure or at more than 1000 bar of pressure. This suggests that very high pressure exerts only minor effects upon its biochemical reactions. A striking example illustrating this point is given by the snailfish (*Pseudoliparis swire*) living in the Marianas Trench below 7000 m hence at more than 700 bar pressure. This vertebrate can reach a size of up to 30 cm, it eats and swims at 7000 m depth like a bottom-feeding fish in a shallow pond. If a fish can run its metabolism, its muscles, its nervous system and behavior at a pressure of 700 bar, microbes can certainly manage their physiological reactions accordingly. Because water and the contents of cells are effectively non-compressible, there is probably not much effect of pressures up to 1000 bar, although gas energies can obviously be much higher at such pressures than at ambient pressure.

7.3.3 INOCULATING THE CRUST

It is estimated that roughly 10^{30} microbes inhabit the crust today. How do microbes come to inhabit the crust? They would have to be brought into the crust by water currents, and the downdraft of hydrothermal circulation would seem the most likely mechanism (Figure 7.14). By inference, the same mechanisms were operating 4 billion years ago, such that once a vent was releasing fully fledged free-living cells into the ocean, the colonization of virtually all habitable H_2 producing regions of the crust went quickly. At that time, hydrothermal currents inoculated new environments where ocean water was bringing CO_2 and serpentinization processes were generating H_2. Newer evidence indicates that in modern

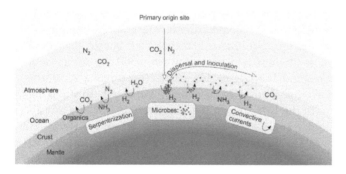

FIGURE 7.14 Inoculating the crust. Schematic spread of microbes in the primordial crust though hydrothermal currents starting from one single site of origins ca. 4 billion years ago. The main nutrients required for growth were CO_2 supplied by the moon-forming impact plus H_2 and NH_3 supplied by serpentinization. Cells spread by the convective currents generated by serpentinization and occupied habitable niches. The process of colonizing the crust could have gone very fast. Although the chemistry of serpentinization might have permitted multiple sites of origin, all life that we know has a single origin because of the universality of the genetic code and central metabolism. The first free living cells would rapidly overgrow all chemically habitable niches accessible via the convective currents of serpentinizing systems. This would inoculate the entire suboceanic crust. Heberling C *et al.* (2010, Extent of the microbial biosphere in the oceanic crust. *Geochem Geophys Geosyst* 11: 1–15) estimate that the total biomass in the modern suboceanic crust is on the order of roughly 10^{30} cells and that this value has probably not changed by more than a factor of about 2 over the last 4 billion years.

environments, H_2 is also emitted via seepage within the earth at low levels at a variety of sites, not only from hydrothermal vents, although the processes underlying such H_2 seepage are not yet known. The crust was almost certainly the first habitat to be colonized. At the surface there was light, yes, but in particular ultraviolet light, which is very damaging. Furthermore, the surface offered very low availability of the energy and electron source required by the first microbes: H_2.

How long did the initial colonization of the Earth's crust take? That is (almost) pure guesswork, but it likely went fast. One can put some bounds on the estimate with the following calculation. An *E. coli* cell doubles in about 20 minutes at maximum growth rate. Given enough substrate, the culture will outweigh the earth in less than 48 hours because 2^{144} is an extremely large number, about 2×10^{43}, such that even if a cell only weighs about 10^{-18} kg, the culture after 144 doublings has a mass of about 2×10^{25} kg, which is more than the mass of the Earth.

So even if our first microbes do not have a doubling time of 20 minutes, but 200 years instead, then it would take them only 30,000 years to outweigh the Earth, and far less to colonize the niches available to H_2-dependent chemolithoautotrophs. If their doubling time was 2000 years then they would colonize all niches in less than 300,000 years. In geological terms, that is more or less instantaneous. But the first niche colonizing microbes would not strive to outweigh the Earth, obviously, they would merely grow, generating a biofilm across habitable environments. In principle, this could have occurred within a few thousand years. Of course, substrates always limit microbial growth, but the calculation makes the point that as soon as microbes reached a stage where free living cells could freely diffuse and double, the colonization process is literally out of control and effectively instantaneous on a geological time scale. Convection and inoculation make the planet irreversibly alive, unless another moon forming impact comes along that melts the entire planet again so as to sterilize it.

Furthermore, the planet is alive in the crust with a curious consequence: any other subsurface sites that might have been on their way to bringing forth life by a similar series of events as we have discussed here are already inoculated with a superior competitor that will be able to organize the available resources into likenesses of the first free living (freely dispersible) cells. On a young planet replete with inorganic nutrients, CO_2, H_2 and water, microbial growth changes everything. Just like today, H_2-dependent chemolithoautotrophs would be able to colonize ancient vent environments.

How extensive was the primordial cell mass that coated the habitable niches of the crust? It was big. Exactly how big is hard to say, even for today's environment. But there are estimates. In 1998 Barney Whitman and colleagues published estimates for the numbers of prokaryotic cells on Earth. They estimated an upper bound of about 6×10^{30} cells (~500×10^{15} grams of biomass, estimating ~10^{-16} kg/cell) total in the combined marine (~3.5×10^{30} cells) and continental (~2.5×10^{30} cells) subsurface, although most of the biomass in the modern marine subsurface is in sediment, which is ultimately fueled today by photosynthesizers and the sun. Ron Milo and colleagues obtained slighter estimates for the total subsurface biomass weight, 70×10^{15} grams of biomass with an uncertainty of a factor 10 ($7-700 \times 10^{15}$ g), corresponding to 0.08 to 8×10^{30} cells, again most of that in sediment, not in crust. To estimate the size of Earth's primordial colony, we would want a value for biomass in the crust. Heberling and colleagues provided such estimates, also looking back in time. They estimated the amount of biomass in the igneous (rocks made from cooled magma rather than sediment) crust below the ocean floor sediment as roughly 200×10^{15} grams of biomass in the modern submarine crust or very roughly 2×10^{30} cells. In terms of absolute numbers, their estimate might easily be too high by about a factor of 10 compared to more recent estimates compiled by Flemming and Wuertz, which is unimportant.

The important aspect is that Heberling *et al.* calculate that the volume of habitable crust hence the amount of biomass (the number of cells) in the submarine crust at any given time has not varied by more than about a factor 2 over the last 4 billion years. This is because the amount of submarine crust, its porosity for water (habitable niches for cells), and its general chemistry stay relatively constant over time. The crust was the first habitat to be colonized and it has remained colonized as a niche even though the rocks themselves that make up the crust are constantly returned to the mantle via subduction, reappearing as magma at spreading zones during Wilson cycles. The more recent estimates of Flemming and Wuertz suggest that the total modern subsurface biomass comprises on the order of 0.7×10^{30} cells. They furthermore estimate that 20–80% of all cells in the subsurface exist as biofilms—cells attached to inorganic surfaces and to other microbes growing on those surfaces. All modern estimates for the distribution of biomass on Earth indicate that about 90% of all cells on Earth inhabit the subsurface. It has probably always been that way. If life were to arise by serpentinzation on other planets or moons, the same would probably apply.

7.3.4 Fermentations Followed Autotrophy

Hydrothermal vents have finite life spans. A vent that was colonized for 100,000 years (the estimated age of Lost City) will ultimately run dry, with crust processes sealing off veins or water flow such that serpentinization and H_2 production cease. In such a case primary production (accumulation of new cell mass) comes to a halt for lack of H_2. This leads to a thermodynamic reversal: In the absence of H_2, amino acid fermentations, which typically produce H_2 as a waste product, become exergonic. That applies for amino acid fermentations of both bacteria and archaea. The H_2 producing fermentations are only exergonic in the absence of H_2 and are particularly exergonic in the presence of syntrophic associations with methanogens that can avidly scavenge any H_2 produced by fermentations. Cells are about 50–60% protein by dry weight, hence there is vast opportunity for fermenters to live from the protein of dead and dying cells at a vent run dry.

Almost no biochemical innovations are needed for a chemolithoautotrophic bacterium or archaeon to become an amino acid fermenter. The amino acid synthesis pathways are in place, they are reversible, they can run backwards if thermodynamics permit. Peptidases, present in autotrophs for housecleaning purposes (removal of misfolded proteins for example), can become coopted into the general secretory (Sec) pathway, which exports proteins through the plasma membrane via the signal recognition particle (SRP) which recognizes and targets these proteins to the plasma membrane. Thus, peptidases can be secreted, exit the cell and digest proteins in the environment. Released amino acids (or small peptides) can then be imported and serve as a source of carbon, nitrogen and energy.

Energy conservation in typical amino acid fermentation involves breakdown of the amino acid down into the corresponding 2-oxoacid via deamination. The 2-oxoacid is oxidatively decarboxylated in a CoA-dependent reaction, generating a thioester that serves as a substrate for acyl phosphate synthesis and ATP synthesis via SLP (Figure 7.15). These are mechanistically very simple enzymatic reactions involving substrate conversions (activities) that

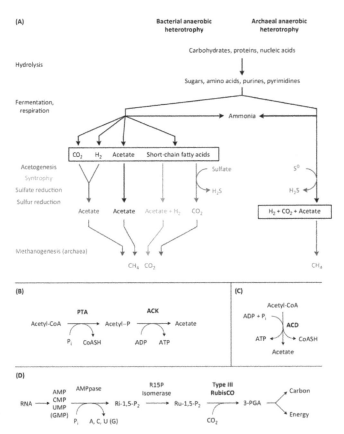

FIGURE 7.15 **Ancient fermentations.** (A) End products of the bacterial fermentations (left) and archaeal fermentations (right) are boxed. The conversion of fermentation end products via acetogenesis, syntrophic associations, and sulfate reduction, as well as the terminal conversion of those products to methane are indicated. Note that some use elemental sulfur as the terminal electron acceptor in facilitated fermentations. (B) The typical bacterial pathway of ATP formation from acetyl-CoA via substrate-level phosphorylation (SLP). (C) The typical archaeal pathway of ATP formation from acetyl-CoA via SLP. Abbreviations: PTA, phosphotransacetylase; ACK, acetate kinase; CoASH, coenzyme A (free thiol); ACD, acetyl-CoA synthase (ADP forming). (D) The short pathway of 3-phosphoglycerate (3PGA) formation from the ribose ring in AMP (redrawn from Sato T. *et al.* (2007) Archaeal type III RuBisCOs function in a pathway for AMP metabolism, *Science* 315: 1003–1006). Abbreviations: AMPpase, AMP phosphorylase; R15P isomerase, ribose-1,5-bisphosphate isomerase. AMPpase will utilize all four NMPs as substrates, but GMP with lower activity. Abbreviations: Ri, ribose; Ru, ribulose; 3-PGA, 3-phosphoglycerate. Reproduced with permission from: Schönheit P. *et al.* (2016) On the origin of heterotrophy, *Trends Microbiol.* 24: 12–25.

were already present in H_2-dependent chemolithoauto-trophs, hence requiring little in terms of catalytic finesse and nothing in terms of evolutionary invention. If fueled by pep-tidases, the terminal reactions of amino acid biosynthetic pathways can simply run in reverse, channeling 2-oxoacids into acetyl-CoA synthesis and ATP synthesis via SLP. Such fermentations yield roughly 1 ATP per amino acid. *Clostridium tetanomorphum*, for example, obtains 0.95 ATP per glutamate that it ferments to acetate, butyrate, CO_2 and H_2, as Buckel has shown.

In addition to protein, cells also consist of ~20% RNA. RNA is, in turn, ~40% ribose by weight, such that cells are ~8% ribose, which, if imported, can be channeled into *de novo* RNA synthesis or utilized for ribose fermentation path-ways. In modern marine sediment, an unexpected observa-tion is the very high frequency of genes for an enzyme called ribulose-1,5-bisphosphate carboxylase (Rubisco), which is well-known as the CO_2-fixing enzyme in the Calvin cycle, the least ancient and energetically most expensive pathway of CO_2 fixation known (see Table 3.1). The presence of such a 'wasteful' enzyme in energy limited environments was ini-tially puzzling. However, Atomi and colleagues have shown that this unexpected Rubisco is actually involved in RNA fermentation. In heterotrophic and fermentative habitats, the bases from nucleoside monophosphates (AMP etc.) stem-ming from RNA breakdown are removed via phosphoroly-sis. This yields ribose-1,5-bisphosphate, which is converted by a simple isomerase into ribulose-1,5-bisphosphate, which is then converted by Rubisco into the glycolytic/gluconeoge-netic intermediate 3-phosphoglycerate for fermentative car-bon supply and energy metabolism via SLP (Figure 7.15D). This is likely the ancestral function of Rubisco.

7.3.5 PUMPING DURING FERMENTATIONS

In an environment such as an extinguished vent, where geochemical H_2 synthesis—hence primary production by chemolithoautotrophs—has stopped, fermentations could proceed until all fermentable substrates have been consumed. Fermenters that can pump ions during the fermentation pro-cess can extract slightly more energy out of the fermentation reaction. Pumping can be coupled to H_2 production, using H^+ as the terminal acceptor. Both acetogens and methanogens are known that possess a very ancient protein called Ech, for energy converting hydrogenase, which was originally characterized in methanogens. Ech is a membrane-integral, reversible ion pumping hydrogenase. It can transfer electrons from ferredoxin to protons and use that exergonic reaction to generate an ion gradient (either Na^+ or H^+) in the process, or it can operate in reverse, harnessing an ion gradient to gen-erate reduced ferredoxin from H_2, an important anaplerotic (intermediate-generating) function that it fulfills in metha-nogens. In H_2-producing fermentations, Ech can oxidize fer-redoxin to generate H_2 and an ion gradient for chemiosmotic ATP synthesis. Ech is a very useful tool linking ion gradients and H_2 metabolism hence it is very widespread among bacte-rial and archaeal anaerobes, including fermenters.

Like most membrane proteins involved in ion gradient formation, Ech is a multisubunit protein. The subunit struc-ture of Ech from an acetogen and a methanogen is shown in Figure 7.16, where it is compared to a protein with a very similar function in the archaeal fermenter *Pyrococcus furio-sus*, namely Mbh (for *membrane-bound hydrogenase*). In the fermenter *P. furiosus*, Mbh couples H_2 production to ion gra-dient formation for ATP synthesis: $Fd^{2-} + 2H^+ \rightarrow Fd + H_2$. The comparison of Mbh and Ech in Figure 7.16 illustrates

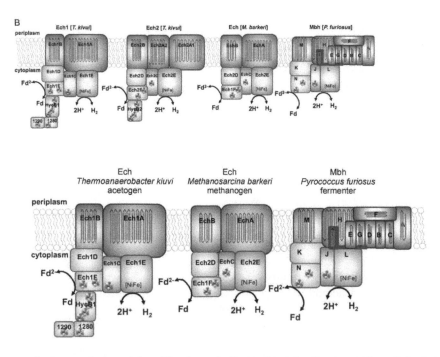

FIGURE 7.16 Some ancient pumping proteins. The figure indicates the subunit composition of the ion pumping hydrogenases Ech and Mbh. The proteins pump H^+ or Na^+ in the hydrogen producing reaction, or they can utilize ion gradients to generate reduced ferredoxin from H_2 (ion transport not indicated for simplicity). The complexes do not contain heme. Redrawn with permission from: Schoelmerich M. C., Müller V. (2019) Energy conservation by a hydrogenase-dependent chemiosmotic mechanism in an ancient meta-bolic pathway, *Proc. Natl Acad. Sci. USA* 116: 6329–6334 and Yu H. *et al.* (2018) Structure of an ancient respiratory system, *Cell* 173: 1636–1649.

a principle that is central to all of protein evolution: modularity and combination of functional domains. The ferredoxin reducing and H_2 oxidizing domains of Ech and Mbh are drawn using similar colors to indicate structural similarity. The Ech1E and MbhL subunits are related to each other and both are in turn related to the catalytic domain of FeNi hydrogenase. Yet the ion pumping membrane domain of Mbh (subunits A-I) is unrelated to the functional homologues in Ech (EchA and EchB). Instead Mbh subunits A-I are related to the ion pumping domain of NADH dehydrogenase (a very common enzyme also known as Complex I in the human mitochondrial respiratory chain). This in turn has as its core ion pumping subunits a protein related to the Na^+/H^+ antiporter discussed above. This antiporter is found in many acetogens, methanogens and furthermore traces to LUCA (Figure 6.8). Modularity and novel combinations of functional domains are a central theme in the evolution of proteins involved in membrane bioenergetics, and patterns of relatedness among protein complexes can become very complicated as we will see later in **Section 7.3.8.**

The yield from amino acid fermentations places limitations on the amount of fermenter biomass that can be generated from primary producers. This is because fermentations permit the synthesis of about one ATP per amino acid taken up as substrate, but protein synthesis requires 4 ATP per peptide bond plus about 1 ATP per amino acid imported. Thus 5 amino acids from a decomposing cell must be consumed to generate one new peptide bond in a growing cell. Cells are mostly protein by weight, therefore one can readily calculate that about 5 grams of dead (fermented) cells can generate at most one gram of fermenting cells in the first generation. In the second generation the same ratio applies such that the sequential mass conversion ratio from generation to generation is, starting at 1, at best $1 \rightarrow 1/5 \rightarrow 1/25 \rightarrow 1/125$ and so forth. Accordingly, there are energetic limits imposed by the cost of protein synthesis as to how much fermenter biomass can be generated from dead cells, and this is why in modern marine sediment, there is always a vector in the direction of net death of cells, not net growth.

When fermentations have run their course in an extinguished vent, what remains are a few viable cells and a high concentration of unfermentable substrates: compounds such as lipids, fatty acids and fermentation products that cannot be disproportionated because they are too reduced. This generates a situation in which anaerobic respirations would enter into modern ecosystems, because respiration allows the cell to use external (environmental) electron acceptors with a midpoint potential more positive than that of H_2. But in a primitive crust ecosystem, oxidants would have been rare.

In modern ecosystems, elemental sulfur (S^0, typically in the form of S_8 rings) can accumulate due to the presence of O_2 which oxidizes reduced sulfur compounds in the environment to S^0. Elemental S has a midpoint potential of $E_o' = -260$ mV and can serve as an electron acceptor, yielding H_2S, in the same manner that H^+ can serve as an electron acceptor yielding H_2 ($E_o' = -414$ mV). The archaeal fermenter *Pyrococcus furious* has an isoform of Mbh, called Mbs (for membrane-bound sulfur reduction), that is almost identical in subunit structure to Mbh, but it uses S^0 as the acceptor for electrons from fermentations via ferredoxin,

generating small reduced sulfane species ($S_3{}^{2-}$) that disproportionate into H_2S and S^0. This reaction pumps ions in the same manner as Mbh, generating an ion gradient for ATP synthesis that augments fermentative SLP. This kind of supporting respiration is also called a facilitated fermentation. In the reducing environments of primitive ecosystems, S^0 was however likely to be rare at best, if it existed at all.

In order to extract energy from unfermentable substrates, modern microbes use two basic strategies: syntrophic interactions and anaerobic respiration. Syntrophic interactions typically involve two microbial partners, one that engages in a very close to equilibrium (or slightly exergonic) H_2 producing reaction (for example propionate or butyrate oxidation) and a second partner that scavenges the H_2 as a pulling reaction to improve the thermodynamics of the first. Without an H_2 scavenging partner, some fermentations are not exergonic (Table 7.4).

In modern environments, the H_2 scavenging partner is typically a methanogen or a sulfate reducer. In our inference at this point, methanogens abound, yet sulfate reducers would not have yet evolved because they require kinds of electron carriers and redox-active proteins that primitive acetogens and methanogens do not possess and that we have not encountered so far. Those electron carriers are heme and quinones.

The redox active proteins are called cytochromes, the collective designation for proteins that bind heme, irrespective of their evolutionary affinity. Pumping mechanisms that depend on Rnf, Ech, and Mbh (as well as Mbs) are 'standalone' complexes that, in contrast to most modern bioenergetic systems (see Section **7.3.8**), pump without the help of cytochromes or quinones. With methanogens present, a portion of the unfermentable substrates can be degraded, but not completely, because methanogens stop growing at $\sim 10^{-5}$ bar H_2. That leaves anaerobic respiration as an opportunity for microbes to survive in a vent environment that has a large supply of unfermentable substrates. Fermentations and respirations are both redox reactions. Fermentations use an internally generated electron acceptor

---

TABLE 7.4

Effects of Syntropy on Changes of Free Energy, ΔG [kJ/mol]

Reaction	ΔG (1 bar H_2)	ΔG (10^{-5} bar H_2)
Fermenters		
Propionate$^-$ + 2 H_2O → acetate$^-$ + CO_2 + 3 H_2	72	−21
Butyrate$^-$ + 2 H_2O → 2 acetate$^-$ + H^+ + 2 H_2	48	−22
Methanogens		
4 H_2 + CO_2 → CH_4 + 2 H_2O	−131	−15

Note: Gibbs free energy calculated at 25°C for CO_2 and CH_4 at 0.1 bar, for H_2 as indicated, and for other compounds at 10 mM. Reproduced with permission from Stams A. J. M., Plugge C. M. (2009) Electron transfer in syntrophic communities of anaerobic bacteria and archaea, *Nat. Rev. Microbiol.* 7: 568–577.

that stem from metabolism. In respirations, external electron acceptors are used. The starving fermenters are under intense selection for the origin of mechanisms that could allow them to respire.

There are electron acceptors available in extinguished hydrothermal vents that hosted serpentinization: Fe^{3+} in magnetite (Fe_3O_4) that was formed in the serpentinization process that generated H_2 and allowed the vent to become colonized in the first place. The use of magnetite as an electron acceptor for microbial growth is a comparatively recent discovery. In 1995, Kostka and Nealson found that *Shewanella oneidensis* (then named *Shewanella putrefaciens*) could grow anaerobically on lactate as the electron donor using Fe_3O_4 as the electron acceptor, probably using the reaction

$$2\,Fe_3O_4 + lactate + 11\,H^+$$
$$\rightarrow 6\,Fe^{2+} + acetate + 6\,H_2O + HCO_3^-$$

with a calculated $\Delta G = -46$ kJ/mol for 10 mM lactate, 15 mM HCO_3^-, 1 mM Fe^{2+} and 1 μM H^+, but the energetics are not simple because the redox chemistry of Fe^{3+} mineral reduction is highly dependent upon environmental conditions such as pH and the form of the mineral, including particle size. At very acidic pH the reduction of Fe^{3+} to Fe^{2+} in solution has a midpoint potential of +770 mV, but at pH 7 the reduction of Fe(III) citrate to Fe(II) citrate is less positive, $E_o' = +380$ mV, while the reduction of $Fe(OH)_3$ to $FeCO_3$ has an E_o' in the range of +100 to +200 mV (Table 7.4). The reported value of $\Delta G^{0'}$ for the Fe_3O_4/Fe^{2+} couple is −314 mV and too negative to support lactate oxidation, but as we note in Table 7.5, the thermodynamics of many reactions involving iron reduction are complicated by the conditions used and the nature of the reduced product. The fact that the bacteria reproducibly grew and dissolved the magnetite used as substrates in the experiments to Fe^{2+} is nonetheless proof that the energetics of the reaction are ultimately favorable and that an iron species in the magnetite used for growth was readily reduced with electrons from lactate.

Shewanella growing by oxidizing lactate (a nonfermentable substrate and very common fermentation product) while transferring the electrons to Fe_3O_4 suggests a solution to how our starving fermenters could survive if they could find a way to transfer electrons from the breakdown of acetate or lactate derived from fermentations to solid phase Fe_3O_4. The solution that modern cells employ for such tasks is possibly the same solution that the very first microbes discovered and employed: they synthesize a series of multiheme cytochromes (proteins containing from 4 to 10 covalently bound heme molecules), which can form protein-based nanowires that generate a conduit for electrons from the cell to solid state Fe_3O_4 in the environment. The nanowires can either contain, or be entirely made of, multiheme cytochromes. As we will see, multiheme cytochromes could have enabled survival in an environment that harbors unfermentable substrates (fermentation end products) in abundance and an insoluble electron acceptor (Fe_3O_4). Utilizing that electron acceptor required however the origin of cytochromes, and that, in turn, required the origin of heme, the topic of the next section.

7.3.6 TETRAPYRROLES, HEME, CYTOCHROMES, AND QUINONES

Most, but not all, prokaryotes contain heme and cytochromes. Many methanogens and acetogens lack heme and cytochromes, they likely arose before heme was invented. Heme is a tetrapyrrole. Its central coordination site is occupied by an Fe atom that acts as a single electron transfer cofactor in electron transport chains. The ancestral route to tetrapyrrole biosynthesis starts from glutamyl-tRNA. The glutamyl moiety is reduced to glutamate-1-semialdehyde and then isomerized to 5-aminolevulinic acid. There is a second, evolutionarily derived route to 5-aminolevulinic acid via succinyl-CoA and glycine that arose later in bacteria. Two molecules of 5-aminolevulinic acid are condensed to porphobilinogen, four molecules of which are condensed via two reactions to the universal precursor of all tetrapyrroles, uroporphyrinogen III (UroIII) (Figure 7.17).

The only tetrapyrrole required in the acetyl-CoA pathway is the corronoid cofactor cobalt-β-aqua-(5,6-dimethyl-benzimidazolyl cobamide), which is structurally very similar to cobalamin, except that the sixth coordination site of Co is free, rather than being bound by an adenosyl or protein side chain residue, as Dobbeck and colleagues have shown in the structure of the CoFeS protein. The corrin biosynthetic pathway is the longest tetrapyrrole synthetic pathway (20 enzymatic steps from UroIII); like the acetyl-CoA pathway, it occurs in bacteria and archaea and the enzymes are homologous. It is the most ancient tetrapyrrole pathway. This might seem counterintuitive at first sight, but recall that a single metal catalyst, Ni_3Fe or Fe_3O_4, can replace all of the enzymes and cofactors of the acetyl-CoA pathway to pyruvate, which amounts to 127 individual enzymes total, if we consider the 10 enzymes of the acetyl CoA pathway plus the enzymatic pathways to synthesize the cofactors required (Figure 7.11).

The corrin cofactor dimethyl-benzimidazolyl cobamide is at the heart of carbon bond formation in the acetyl-CoA pathway because it transfers the pterin-bound methyl group to a nickel atom in CODH/ACS that then undergoes carbonyl insertion of CO to form the Ni-bound acetyl intermediate (see Figure 7.11). That methyl transferase reaction involves the rare enzymatic transfer of a methyl group from one metal atom—Co in the corrinoid cofactor (Figure 7.17)—to another metal atom (Ni in an FeNiS center in CODH/ACS). This methyl transfer reaction is catalyzed by the corrinoid FeS protein CoFeS. The enzyme can be seen as the biological replacement for the metal-to-metal methyl transfer reaction in the abiotic Ni_3Fe catalyzed pathway (Figure 7.11c). In 4 billion years, microbes have not discovered a metal-free alternative to the cobamide-dependent methyl transfer reaction catalyzed by CoFeS, suggesting that there is no metal-independent alternative to the reaction that can be readily evolved. This is similar to the case with nitrogenase discussed in **Chapter 2**. The 20-enzyme biosynthetic effort required to synthesize the corrinoid cofactor (and cobalamin) underscores how essential it was for protein-synthesizing protomicrobes to evolve an enzymatic alternative to the far simpler, but biologically immobile, Ni_3Fe catalyzed H_2-dependent reaction. The cobamide-dependent methyl transferase reaction was the

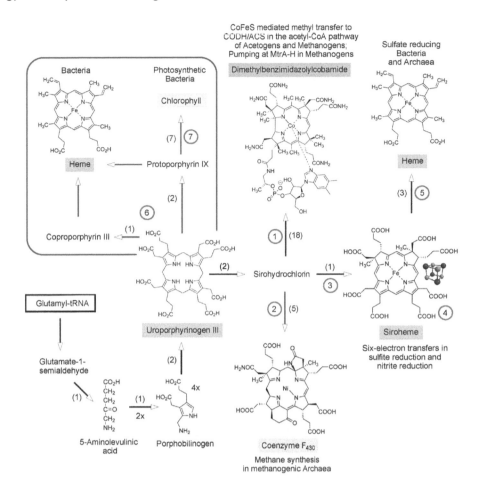

FIGURE 7.17 Schematic summary of tetrapyrrole biosynthesis and evolution. Numbers in parentheses next to arrows indicate the number of enzymatic steps involved. Circled numbers next to arrows (1–6) indicate the probable order in which the pathways arose, cobalamin being the most ancient (see text and Decker K *et al.* (1970) Energy production in anaerobic organisms. *Angew Chem Int Ed* 9: 138–158). There are evolutionarily derived, O_2-dependent routes to chlorophyll and to cobalamin, only the ancestral O_2-independent routes are indicated here. Blue arrows and the blue line at upper left indicate bacterial specific pathways, the red arrow is an archaeal specific pathway. Redrawn with permission from information in Bryant D. A. *et al.* (2020) Biosynthesis of the modified tetrapyrroles—the pigments of life, *J. Biol. Chem.* 295: 6888–6925, and information in Dailey H. A. *et al.* (2017) Prokaryotic heme biosynthesis: Multiple pathways to a common essential product, *Microbiol. Mol. Biol. Rev.* 81: 1–62. The review by Bryant *et al.* (2020) is insightful in terms of biosynthesis but overlooks the most ancient function of corrins among the Co coordinating tetrapyrroles (corrins), namely methyl transfer in acetogenesis and methanogenesis (cf. Decker *et al.*, 1970). The structure of the corrinoid cofactor dimethylbenzimidazolylcobamide is from the structure of the CoFeS protein in Svetlitchnaia T. *et al.* (2006) Structural insights into methyltransfer reactions of a corrinoid iron-sulfur protein involved in acetyl-CoA synthesis, *Proc. Natl Acad. Sci. USA* 103: 14331–14336. Dimethylbenzimidazolylcobamide is a form of cobalamin (vitamin B_{12}). It is used for methyl transfer by CoFeS in the acetyl-CoA pathway of acetogens and methanogens, but a slightly modified version of the cofactor, 5-hydroxybenzimidazolyl cobamide, is essential in the ion pumping methyl transferase reaction catalyzed by MtrA-H in methanogens (see Figure 7.11) (see Shima S *et al*). Structural basis of hydrogenotrophic methanogenesis. *Annu Rev Microbiol.* 74:713–733.

solution to replace the Ni_3Fe catalyzed H_2-dependent reaction, a prerequisite to the transition to the free living state. It is perhaps significant that the pumping reaction of hydrogenotrophic methanogens catalyzed by the methyl transferase MtrA-H (see Figure 7.11) also requires the corrin cofactor. This places corrin-dependent methyl transfer reactions at two key steps of methanogen biology: carbon metabolism (CoFeS) and energy metabolism (MtrA-H). This in turn, reflects the antiquity of Co-bound methyl groups in these reactions and, in the simplest interpretation, an origin of carbon and energy metabolism in a geochemical environment rich in geochemically furnished, metal-bound methyl groups (see Figure 6.8).

As the only archaeal-specific tetrapyrrole biosynthetic pathway, methanogens require the nickel containing tetrapyrrole F_{430} in the last step of methane synthesis, catalyzed

by methyl-CoM reductase (Mcr): CH_3-S-CoM + HS-CoB → CH_4 + CoB-S-S-CoM (see Figure 7.11). This was required for the coupling of carbon and energy metabolism in methanogens. Although the ion pumping reaction in methanogens is catalyzed by MtrA-H (see **Section 7.2.2**), the methane synthesizing step catalyzed by Mcr ($\Delta G^{0'} = -30$ kJ/mol) provides the thermodynamic pull that drives methanogen metabolism forward. Very specific genetic manipulations and growth conditions are required to force methanogens to grow without producing methane as their main metabolic end product: they can grow as CO-dependent acetogens, as Michael Rother and colleagues have shown. The F_{430}-dependent reaction to synthesize methane (see Figure 7.11), hence the coenzyme F_{430}, was essential for the transition to free living state in methanogens. It was likely the second tetrapyrrole pathway to arise (Figure 7.17).

From UroIII there are three biosynthetic routes to heme, the cofactor of cytochromes. Two are present only in bacteria. The one via siroheme (Figure 7.17) that occurs in archaea and sulfate reducing bacteria is considered to be the most ancient of the heme producing pathways. Siroheme is, like F_{430}, a very specialized tetrapyrrole. It is unique in that when bound in enzymes, with very rare exceptions the fifth coordination site of the central Fe atom is not coordinated by histidine as is typical of hemes, but by a 4Fe4S cluster instead (Figure 7.17). The only known occurrence of siroheme so far is its essential role in the 6-electron transfer reaction that reduces sulfite to hydrogen sulfide in sulfite reductase and the analogous 6-electron transfer reaction that reduces nitrite to ammonia in nitrite reductase.

Sulfite reduction and nitrite reduction have in common that both processes occur as assimilatory and dissimilatory pathways. The assimilatory pathways convert SO_3^{2-} into HS^- and NO_2^- into NH_4^+, respectively, for biosynthesis. The dissimilatory pathways couple SO_3^{2-} and NO_2^- reduction to ion pumping for ATP synthesis. Sulfite was likely present in ancient crust environments because of SO_2 produced by volcanic activity, which readily dissolves in water as H_2SO_3 and HSO_3^-, much like CO_2 dissolves in water as H_2CO_3 and HCO_3^-. The ancestral function of sulfite reduction was probably assimilatory sulfite reduction, accessing S from SO_2 for biosyntheses. Since heme synthesis in archaeal and bacterial sulfate reducers proceeds via siroheme, the path to siroheme was likely the third tetrapyrrole pathway to arise (Figure 7.17). Although siroheme-dependent sulfite reduction typically involves many cytochrome-containing enzymes, as summarized by Rabus and colleagues, there are some archaea, such as *Methanopyrus kandleri*, that possess siroheme (and cobamid and F_{430}) but no cytochromes (heme), compatible with the view that the biosynthetic intermediate siroheme, which still possesses all eight carboxyl groups present in UroIII, was probably the third tetrapyrrole to arise, followed by heme as the fourth (Figure 7.17).

As with amino acids and bases, abiotic porphyrin synthesis has been reported from a variety of abiotic synthesis experiments. That, plus the fact that three biosynthetic routes to heme have evolved, indicates that the synthesis of the product follows a natural tendency of the precursors to react. The starting point of tetrapyrrole biosynthesis, glutamyl-tRNA, also points to an ancient origin of the pathway. Nitrite reduction is probably not an ancient pathway. In early earth crust environments, forms of nitrogen more oxidized than N_2 were unstable and short lived, as nitric oxide (NO) is an extremely strong oxidant (E_0' for the NO/N_2 couple is +1150 mV), stronger than O_2 (+810 mV). Although some nitrite might have been formed at the ocean surface by NO from lightning, it could not stably persist and penetrate through the oceans to the crust (E_0 for the nitrite/nitrate couple is +420 mV). The appearance of nitrite- and nitrate-utilization came later in evolution, after the appearance of O_2 and the expansion of the global nitrogen cycle to its present form.

7.3.7 Using Cytochromes and Quinones to Access FE_3O_4 as an Electron Acceptor

We return to the situation faced by starving fermenters that were literally swimming in non-fermentable substrates surrounded by a suitable electron acceptor (Fe_3O_4), but no means to access the acceptors. Heme is a hydrophobic planar molecule roughly 1 nm (10 Å) in diameter that readily inserts into the hydrophobic interior of protein structures. Proteins that contain heme are called cytochromes. There are several types of cytochromes that are distinguished among other things by the manner in which heme is bound to the protein. The cytochromes of extracellular multiheme cytochromes and nanowires are c-type cytochromes. In c-type cytochromes, heme is covalently bonded via the formation of thioether bonds between the vinyl groups of heme (Figure 7.17) and cysteine residues of the protein. This restrains the heme moieties from diffusion into the environment. Today, nanowires composed of multiheme cytochromes are made both by bacteria and archaea, although the magnetite-reducing bioenergetic reaction of microbes like *Geobacter* and *Shewanella* is so far only known from bacteria. An example of multiheme cytochrome nanowire structures is shown in Figure 7.18.

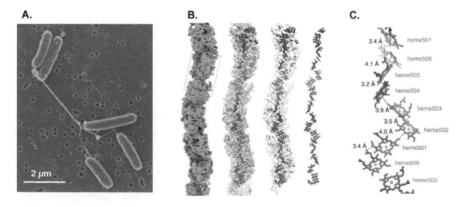

FIGURE 7.18 Electrically conductive extracellular nanowires. (A) Electron micrograph of electrically conductive nanowires from the iron reducing bacterium Shewanella oneidensis. Reproduced with permission from Gorby Y. *et al.* (2006) Electrically conductive bacterial nanowires produced by *Shewanella oneidensis* strain MR-1 and other microorganisms. *Proc. Natl Acad. Sci. USA* 103: 11358–11363. B) Cryo-EM image reconstruction of nanowires composed of the multiheme cytochrome OmcS from Geobacter sulfurreduces. Each OmcS cytochrome protein contains four heme molecules positioned closely enough to one another such that electron transfer from one heme to the next is facile. (C) Closeup of individual heme molecules in OmcS from (B). (B) and (C) drawn with permission from Filman D. J. *et al.* (2019) Cryo-EM reveals the structural basis of long-range electron transport in a cytochrome-based bacterial nanowire, *Commun. Biol.* 2: 219, 1–6.

Nanowires made of multiheme cytochromes are excellent electrical conductors. As in all hemes, the central Fe atom can accept and donate an electron, undergoing Fe^{2+} to Fe^{3+} oxidation changes in the process. In multiheme cytochromes, the electrons from one heme can be passed along to the next, and the next and the next, provided that two conditions are fulfilled. First, the Fe in the donating heme has to be in the reduced state (Fe^{2+}) and the accepting heme has to be in the oxidized state (Fe^{3+}). Second, the heme moieties need to be in close physical proximity to one another because the mechanism of electron transfer from one to the next is tunneling (a quantum chemical process where the electron effectively disappears in the reduced heme and reappears in the oxidized hemes by tunneling through an energy barrier). For tunneling to take place at a rapid rate, which is required for metabolic processes, the heme residues need to be close to one another, within a few angstroms (Å). Tunneling across distances >14 Å is generally too slow to be relevant for biological processes, which is why the prosthetic groups in proteins that transfer electrons individually (typically FeS clusters and heme) are rarely spaced more than 14 Å apart in protein structures. Note the distance between heme molecules in Figure 7.18C. The heme molecules are regularly spaced about 4 Å from one another. Because electrons in reduced heme are delocalized across the macrocyclic ring, the critical value for the efficiency of tunneling is the distance between heme macrocycle edges, not the central Fe atom. For multiheme cytochromes with an edge-to-edge distance between hemes of 3.7 to 4.3 Å, the heme-to-heme electron transfer rate is 10^9 per second. In this way, nanowires made of multiheme cytochromes conduct electrical current.

If synthesized, heme could have paved the way to cytochrome-based extracellular electron transport. How might such a primitive system have operated? Figure 7.19 shows a schematic diagram of D-lactate oxidation with Fe_3O_4 reduction in *Shewanella oneidensis*, which provides a model for considering possible primordial functions of cytochromes and quinones. When *Shewanella oneidensis* grows anaerobically on lactate as the electron donor using Fe_3O_4 as the electron acceptor, the reaction is thought to be sufficiently exergonic for ion pumping and chemiosmotic ATP synthesis. However, *Shewanella oneidensis* grown anaerobically on lactate refrains from ion pumping, instead it obtains its energy from SLP via acetyl-CoA and acetyl phosphate (Figure 7.19).

Lactate is oxidized by a membrane integral D-lactate dehydrogenase that reduces menaquinone (MK), a membrane-soluble hydrophobic two electron carrier (Figure 7.20), which transfers electrons to a membrane integral multiheme cytochrome called CymA. CymA passes the electrons to STC (*small tetraheme cytochrome*), a soluble multiheme cytochrome that resides the periplasmic space between the inner and outer membranes of the bacterium. STC passes the elecrons to heme in the decaheme cytochrome MtrA(10 hemes per monomer), which donates further to MtrC. MtrC is a decaheme cytochrome that resides on the outside surface of the bacterium. It can either reduce Fe_3O_4 directly or transfer electrons to nanowires that contain MtrC, thereby depositing the electrons from lactate onto rocks as the terminal electron acceptor.

The scheme in Figure 7.19 might appear very complicated at first sight, perhaps too complicated to reflect an ancient state of respiration. Yet little in the way of evolutionary invention is required for tetrapyrrole-producing fermenters to reach a state where they could oxidize lactate as in Figure 7.19. The inventions are cytochromes, discussed above (Figure 7.17), and quinones (Figure 7.20).

Quinones are membrane-soluble electron carriers and almost ubiquitous among microbes. Many acetogens and methanogens that grow on H_2 and CO_2 lack both cytochromes and quinones, one of the first reasons why microbiologists suspected that they are ancient. The synthesis of the redox reactive 'head' of quinones starts with chorismate, a ubiquitous intermediate of aromatic amino acid biosynthesis. The lipophilic 'tail' that keeps them dissolved in the membrane stems from isoprenoid synthesis via acetyl CoA and hydroxymethylglutarate (archaea) or via deoxyxylulose-5-phosphate derived from erythrose-4-phosphate and PEP (bacteria). Archaea tend to possess quinone analogs derived from benziothiophene or phenazine that are synthesized via independent biosynthetic pathways, indicating that the synthesis of functional membrane-soluble two-electron redox carriers is a surmountable task in evolutionary terms.

Why might lactate oxidation in *Shewanella* as depicted in Figure 7.19 reflect an ancient, primitive or even ancestral state for respiration? Six reasons speak in favor of its antiquity.

1. The physiology fits the ancient habitat: the electron acceptor Fe_3O_4 (from serpentinization) is present in the hydrothermal system that contained lactate (an accumulated fermentation product).

2. The membrane associated electron transfer chain involves only two components in the membrane, the minimum increment over one component for Rnf, Ech, Mbh, or Mtr, all of which pump. Why would an organism need to evolve components such as in Figure 7.19 to survive if pumping proteins such as Rnf, Ech, or Mbh already exist? Those pumping proteins can be integrated into fermentations, but there are no fermentable substrates left, only unfermentable fermentation end products. Without an electron transfer chain such as in Figure 7.19, growth is not possible for thermodynamic reasons, generating a very strong selection for respiration: not as an improvement of bioenergetic efficiency but as a life-saving solution for cells that otherwise would die.

3. Menaquinone serves solely as an electron carrier, it has no function in pumping. Quinones almost always serve two functions simultaneously: electron transport and pumping (see Figure 7.21). In *Shewanella* growing on lactate and magnetite, menaquinone is only involved in electron transport from inside the cell to the outside, not e^- transport coupled to pumping. One function (electron transport) is simpler than two (electron transport and pumping). The single function almost certainly arose in evolution before the double function did.

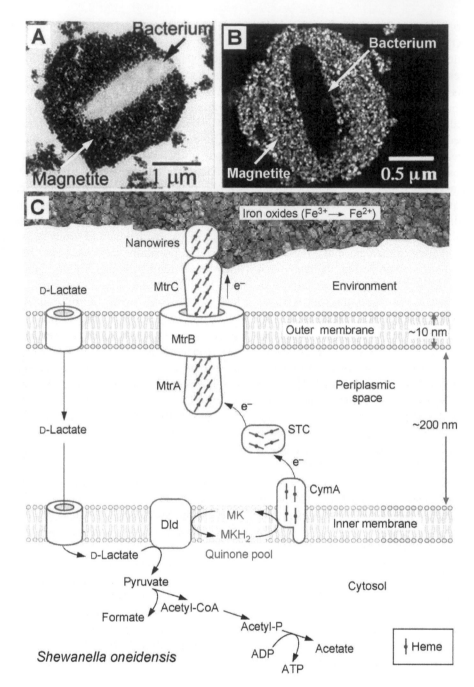

FIGURE 7.19   Using magnetite in ancient cytochrome-dependent electron transfer chains. (A) Electron micrograph of *Shewanella* growing on magnetite as an electron acceptor. Reproduced with permission from: Kim J., Dong H. (2011) Application of electron energy-loss spectroscopy (EELS) and energy-filtered transmission electron microscopy (EFTEM) to the study of mineral transformation associated with microbial Fe-reduction of magnetite. *Clays Clay Minerals*, 59:176–188. (B) *Shewanella* growing on magnetite as an electron acceptor. Reproduced with permission from: Dong H. *et al.* (2000) Mineral transformation associated with the microbial reduction of magnetite, *Chem. Geol.* 169: 299–318. (C) A schematic diagram of Fe_3O_4-dependent lactate oxidation in *Shewanella*. Dld, quinone reducing lactate dehydrogenase; MK, menaquinone; MKH_2, menaquinol; STC small tetraheme cytochrome. *Shewanella* is a Gram negative bacterium that possesses an inner membrane and an outer membrane, the space between the membranes is called the periplasmic space. Main contours of the scheme are synthesized from Beblawy S. *et al.* (2018) Extracellular reduction of solid electron acceptors by *Shewanella oneidensis*, *Mol. Microbiol.* 109: 571–583; Edwards M. J. *et al.* (2020) Role of multiheme cytochromes involved in extracellular anaerobic respiration in bacteria, *Protein Sci.* 29: 830–842; and Bird L. J. *et al.* (2011) Bioenergetic challenges of microbial iron metabolisms, *Trends Microbiol.* 19: 330–340. The figure makes no statement about pumping. ATP synthesis during anaerobic lactate oxidation in *Shewanella* occurs via SLP using the pathway indicated (Hunt K. A. *et al.* (2010) Substrate-level phosphorylation is the primary source of energy conservation during anaerobic respiration of *Shewanella oneidensis* Strain MR-1. *J Bact.* 192: 3345–3351). Dld is related to flavin containing homologs from other bacteria and probably has a flavin cofactor (Pinchuka G. E. *et al.* (2009) Genomic reconstruction of *Shewanella oneidensis* MR-1 metabolism reveals a previously uncharacterized machinery for lactate utilization, *Proc. Natl Acad. Sci. USA* 106: 2874–2879). The composition of *Shewanella* nanowires is still discussed but they contain MtrC, which is part of a larger family of proteins that include OmcS of *Geobacter* (see Figure 7.18). Note that Gram negative bacteria like *Shewanella* possess an outer membrane like hydrophobic layer, the lipopolysaccharide layer, but this is not essential for nanowires to function. Moreover, it is possible that Gram negative cell wall and membrane organization might be ancestral to the bacteria, an issue that is unresolved but immaterial here.

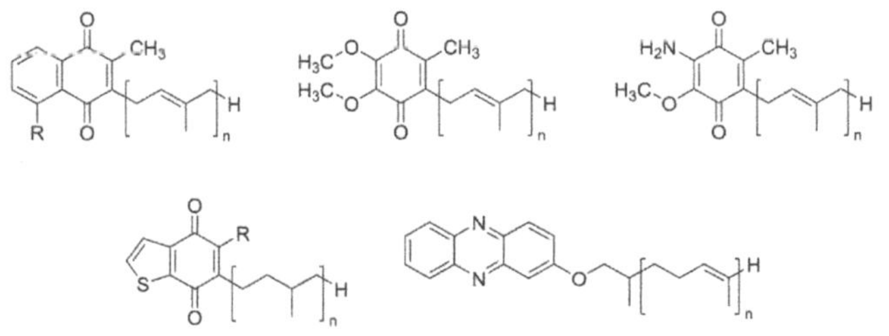

FIGURE 7.20 Quinones and quinone analogues. For mena-quinone (E_0' = −74 mV) and ubiquinione (E_0' = 100 mV) n = 6–8, rhodoquinone (E_0' = −63 mV). The benzothiophene is sulfolobusquinone (E_0' = 100 mV) from the archaeon Sulfolobus. Methanophenazine (E_0' = −165 mV) occurs in methanogens. Redrawn with permission from Berry S. (2002) The chemical basis of membrane bioenergetics, *J. Mol. Evol.* 54: 595–613.

4. There is no specific lock-and-key fit for the terminal acceptor in the active site of an enzyme: physical contact of the electron carrier (reduced heme) with a mineral that can accept the electron is sufficient.
5. The main path and mechanism of electron transfer merely requires physical proximity of heme molecules to within ~4 Å along the electron transport route. It requires neither close positioning of cofactors to within a bond length of ~1 Å of one another as is typical for hydride transfer reactions nor highly specific protein-protein interactions to achieve electron transfer.
6. The measured midpoint potentials for deca-heme cytochromes like MtrC span a wide range from −500 to +100 mV, allowing them to accept electrons from lactate and other unfermentable substrates and donate to a wide range of environmentally available acceptors.

The wide midpoint potential range of multiheme cytochromes is possibly the reason why microbes use heme instead of FeS clusters in extracellular electron transport. Ferredoxins that we have discussed so far have midpoint potentials that are typically too negative (−400 to −500 mV) to support the oxidation of lactate to pyruvate (E_0' = −190 mV). By contrast, the midpoint potentials measured for MtrC purified from *Shewanella* span a broad range from −400 to +100 mV (Table 7.5). This broad range stems from magnetic coupling between low spin heme centers in MtrC. The presence of ten hemes in MtrC allows for a wide range of possibilities for magnetic coupling between hemes. This, in addition to heme-protonation dependent midpoint potential shifts, allows for a wide range of midpoint potentials to be realized within the protein. With pathways for the synthesis of cytochromes and quinones in place, free-living cells could, in principle, access any and all environments where organic substrates, nutrients, and suitable electron acceptors such as magnetite (Fe_3O_4) were available to provide a suitable redox couple for ion pumping and ATP synthesis.

The advent of quinones furthermore enabled a novel mechanism of proton pumping, as shown in Figure 7.21. Figure 7.21A shows the three-dimensional structure of an enzyme named alternative complex III (ACIII) from the bacterium *Rhodothermus marinus*. Figure 7.21B, shows a

TABLE 7.5
Midpoint Potentials [mV] in Lactate Oxidation and Fe(III) Mineral Reduction

Couple	E_o	Reference (See Notes)
Inorganic		
Fe_3O_4/Fe^{2+}	−314	[1,2]
α-$Fe_2O_{3(solid)}$ (haematite)/Fe^{2+}	−287	[1,2]
α-$FeOOH_{solid}$ (goethite)/Fe^{2+}	−274	[1,2]
Ferrihydrite$_{solid}$/Fe^{2+}	−100 to +100	[1,2]
Fe(III) EDTA/Fe(II) EDTA	+96	[1]
$Fe(OH)_3/FeCO_3$	+100 to +200	[3]
Fe(III) citrate/Fe(II) citrate	+385	[1,2]
Fe^{3+}/Fe^{2+} (pH 2)	+770	[1,2]
Organic		
lactate/pyruvate (E_o')	−190	[4]
MK/MKH_2	−67 to −110	[2]
Decaheme cytochrome MtrC	−500 to +100	[5,6]

Note: In magnetite respiration of lactate, electrons flow from lactate/pyruvate (−190 mV) through $MK/MKH2$ (−67 to −110 mV), to MtrC and more positive Fe(III) mineral acceptors; the latter two have very broad midpoint potential ranges. For redox reactions involving iron, the values are for pH 7 unless otherwise indicated. As Straub *et al.* (2001) emphasize, the midpoint potentials of the iron minerals depend upon pH, concentration, temperature and other factors, which are not always uniform across measurements reported in the original literature. References: [1] Straub *et al.* (2001) Iron metabolism in anoxic environments at near neutral pH, *FEMS Microbiol Ecol* 34:181–186; [2] Bird *et al.* (2011) Bioenergetic challenges of microbial iron metabolisms. *Trends Microbiol* 19: 330–340. [3] Heising *et al.* (1999) *Chlorobium ferrooxidans* sp. nov., a phototrophic green sulfur bacterium that oxidizes ferrous iron in coculture with a 'Geospirillum' sp. strain, *Arch Microbiol* 172: 116–124. [4] Weghoff *et al.* (2015) A novel mode of lactate metabolism in strictly anaerobic bacteria. *Environ. Microbiol.* 17: 670–677. See also Thamdrup, B. (2000) Bacterial manganese and iron reduction in aquatic sediments, *Adv. Microb. Ecol.* 16: 41–84. The reaction forming siderite ($FeCO_3$) is relevant because *Shewanella* generates siderite in situ during growth on lactate and magnetite (Kim J., Dong H. (2011) Application of electron energy-loss spectroscopy (EELS) and energy-filtered transmission electron microscopy (EFTEM) to the study of mineral transformation associated with microbial Fe-reduction of magnetite, *Clays Clay Minerals* 59: 176–188). [5] Hartshorne R. S. *et al.* (2007) Characterization of *Shewanella oneidensis* MtrC: A cell-surface decaheme cytochrome involved in respiratory electron transport to extracellular electron acceptors, *J. Biol. Inorg. Chem.* 12: 1083–1094 [6] Choi O., Sang B. I. (2016) Extracellular electron transfer from cathode to microbes: application for biofuel production, *Biotechnol. Biofuels* 9: 11.

schematic representation of the same structure with subunits and redox cofactors indicated. The left-hand side of Figure 7.21B indicates the conformation-induced mechanism of ion pumping, the mechanism that is also used in quinone-*independent* pumping complexes such as Rnf, Mtr, Mbh and Ech (see Figure 7.16): a conformational change in a membrane integral subunit is induced by an exergonic reaction catalyzed by the complex, resulting in the movement of a proton (or Na⁺) across the plasma membrane from the inside to the outside. This generates an ion gradient that

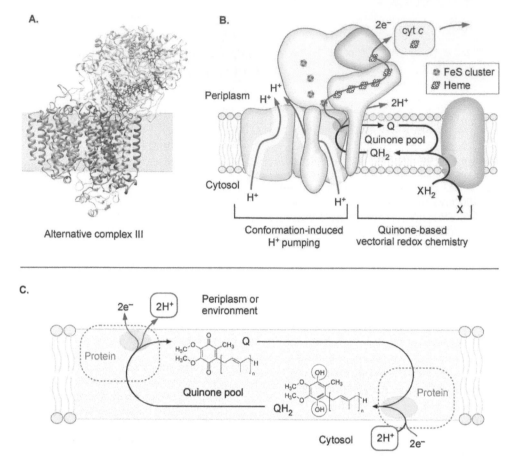

FIGURE 7.21 Quinones enable novel pumping mechanisms. (A) Structure of the alternative complex III (ACIII) from the bacterium *Rhodothermus marinus*. Reproduced with permission from: Calisto F., Pereira M. M. (2021) The ion-translocating NrfD-like subunit of energy-transducing membrane complexes, *Front. Chem.* 9: 663706. (B) Schematic structure of ACIII highlighting the two main mechanisms of ion pumping that cells use. ACIII is taken as an example because it demonstrates both mechanisms in a relatively simple complex. The structure is redrawn from: Sousa J. S. *et al.* (2018) Structural basis for energy transduction by respiratory alternative complex III, *Nature Comms.* 9: 1728. In the case of *Rhodothermus* ACIII, the terminal acceptor that reoxidizes cytochrome c can be molecular oxygen, as in our mitochondria. (C) A schematic representation of quinone-based vectorial chemistry showing the quinol and quinone forms. Similar reduction reactions take place in other quinones and quinone analogs. Proteins are outlined with dotted lines. The donors and acceptors of electrons are highly variable depending on the respiratory chain and the organism. Electron routes indicated in red, proton routes indicated in blue. The electrons in quinol ring are delocalized, obviously, but one pair is indicated in red to emphasize coupled proton and electron transport in quinones.

the ATP synthase can harness. Under physiological conditions, a free energy change of about −20 kJ/mol is needed to pump one proton, a value that is generally regarded as the biological energy quantum, or BEQ. Protons (or Na^+) that are pumped by conformational changes typically traverse the protein complex through a tunnel like opening in the protein structure, usually lined with negatively charged residues that facilitate the movement of the cation from station to station. This is indicated in Figure 7.21B as light shading through the membrane subunits.

On the right side of Figure 7.21B, proton pumping through quinone-based vectorial chemistry is shown. A redox reaction in the cytosol (for example NADH oxidation to NAD^+; generically designated as XH_2 and X in the figure) reduces a quinone in the membrane. The quinol (ubiquinol in the ACIII example) carries two protons and two electrons in the hydrophobic phase of the membrane to the ACIII complex, where it is reoxidized. This is shown in more detail in Figure 7.21C. By physical orientation of the active sites (turquoise ovals) for quinone reduction on the cytosolic side and quinol oxidation reaction on the

periplasmic side of the membrane, protons are translocated (pumped) from the inside of the cell to the outside during quinol oxidation, generating an ion gradient that the ATP synthase can harness. Quinone-based vectorial chemistry does not pump Na^+, but H^+ gradients can be converted into Na^+ gradients with the help of an H^+/Na^+ antiporter. Quinone-based pumping is itself a module of function that is readily incorporated into electron transport chains. Modularity is the underlying theme of membrane bioenergetics, cytochromes and quinones promoted the evolutionary diversification of membrane bioenergetics.

7.3.8 Modularity Promotes Diversity

The modularity in redox-dependent ion pumping proteins shown for Ech and Mbh in Figure 7.16 is just the tip of the iceberg. Modularity leads to many possible combinations of functional components, combinations that quickly become bewilderingly complex, as broadly based comparative work by many researchers, notably Manuela Pereira and colleagues, has shown. This complexity has a simple

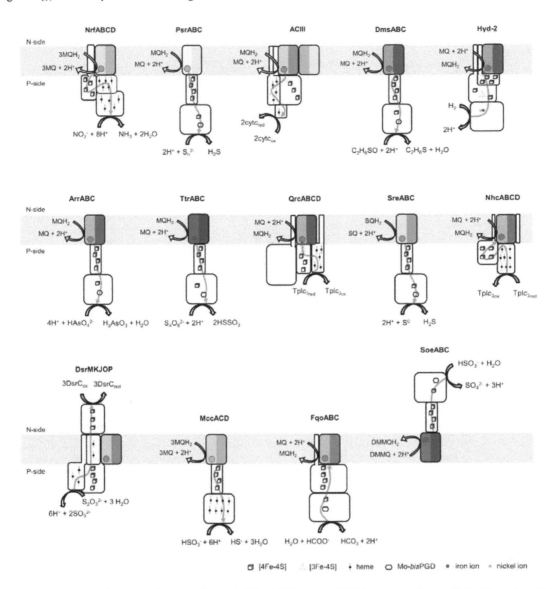

FIGURE 7.22 **Modularity in ion pumping proteins that, like ACIII, use a NrfD type pumping unit.** P-side, extracellular; N-side, cytosol. The movement of ions (H⁺ or Na⁺) in each complex is always from top to bottom across the membrane in the figure. Note the use of symbols for electron carriers at the bottom of the figure. Enzyme name abbreviations (from top left) are as follows with the standard midpoint potential at pH7 for each reaction given in brackets after the name [E_o' in mV]: Nrf, menaquinol:nitrite oxidoreductase [+414]; Psr, menaquinol:polysulfide oxidoreductase [−186]; ACIII, menaquinol:cytochrome c oxidoreductase [+324]; Dms, menaquinol:DMSO oxidoreductase [+234]; Hyd, hydrogen:menaquinone oxidoreductase [+340]; Arr, menaquinol:arsenate oxidoreductase [+134]; Ttr, menaquinol:tetrathionate oxidoreductase [+272]; Qrc, Tplc3:menaquinone oxidoreductase [+340]; Sre, menaquinol:sulfur oxidoreductase [−186]; Nhc, Tplc3:menaquinone oxidoreductase [+340]; Dsr, thiosulfate:DsrC oxidoreductase [+252]; Mcc, menaquinol:hydrogen sulfite oxidoreductase [−42]; Fqo, formate:menaquinone oxidoreductase [+358]; Soe, dimethylmenaquinol:hydrogen oxidoreductase [+507]. Abbreviations: MQ, oxidized menaquinone; MQH2, reduced menaquinone; DMMQ, oxidized demethylmenaquinone; DMMQH2, reduced demethylmenaquinone; cytc, cytochrome c; Tplc3, type I tetraheme cytochrome c3; DsrC, heterodisulfide protein; ox, oxidized; red, reduced. Schematic structures reproduced with permission from: Calisto F., Pereira M. M. (2021) The ion-translocating NrfD-like subunit of energy-transducing membrane complexes, *Front. Chem.* 9: 663706. Midpoint potentials from: Calisto F., Pereira M. M. (2021) Modularity of membrane-bound charge-translocating protein complexes, *Biochem. Soc. Trans.* 49: 2669–2685.

reason: proteins that couple redox reactions to ion pumping form the basis of energy harnessing and ATP synthesis in all life forms today. In the time since the origin of the first free living cells, basic functional modules of ion gradient-generating redox reactions have been invented and recombined in myriad ways to allow energy harnessing from environmentally available redox couples. For example, many membrane-associated complexes use the same membrane-integral structural unit found in ACIII, but in combination with other proteins, as shown in Figure 7.22. The mechanism of quinone-dependent pumping in Figure 7.22 is the

same as in Figure 7.21C. Note however that the orientation in Figure 7.22 (N-side or cytosol, top) differs from that in Figure 7.21C (N-side or cytosol, bottom). Furthermore, in Figure 7.21C quinone oxidation and reduction are shown whereas in Figure 7.22 only the quinone-dependent half reaction catalyzed by the complex is shown.

Figure 7.22 illustrates a segment of known diversity for proteins that use an Nrf (menaquinol:nitrite oxidoreductase) type ion pumping functional module. This provides an impression of how a very diverse spectrum of redox couples can be accessed with recurrent variation on a common

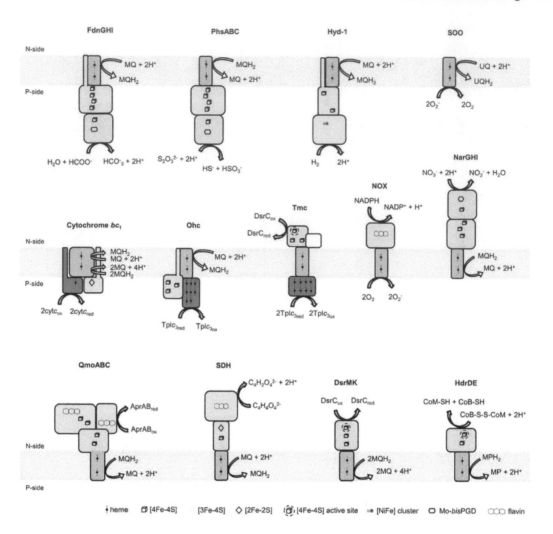

FIGURE 7.23 **Modularity in ion pumping proteins that use a cytochrome b type pumping unit.** P-side, extracellular; N-side, cytosol. The movement of ions (H+ or Na+) in each complex is always from top to bottom across the membrane in the figure. Note the use of symbols for electron carriers at the bottom of the figure. Enzyme name abbreviations (from top left) are as follows with the standard midpoint potential at pH7 for each reaction given in brackets after the name [E_o' in mV]: Fdn, formate:menaquinone oxidoreductase [+358]; Phs, menaquinol:thiosulfate oxidoreductase [−328]; Hyd, hydrogen:menaquinone oxidoreductase [+340]; SOO, superoxide:ubiquinone oxidoreductase [+293]; cytochrome bc1 complex, menaquinol:cytochrome c oxidoreductase [+252]; Ohc, Tplc3:menaquinone oxidoreductase [+340]; Tmc, Tplc3:DsrC oxidoreductase [+264]; NOX, NADPH:oxygen oxidoreductase [+140]; Nar, menaquinol:nitrate oxidoreductase [+507]; Qmo, menaquinol:APS oxidoreductase [+14]; SDH, succinate:menaquinone oxidoreductase [−107]; Dsr, menaquinol:DsrC oxidoreductase [−76]; Hdr, methanophenazine:CoB-S-S-CoM oxidoreductase [+22]. Note that the heterodisulfide reductase shown (HdrDE) is from evolutionarily derived lineages of methanogens that contain cytochromes (heme containing proteins) and quinones. Abbreviations: MQ, oxidized menaquinone; MQH2, reduced menaquinone; UQ, oxidized ubiquinone; UQH2, reduced ubiquinone; MP, oxidized methanophenazine; MPH2, reduced methanophenazine; cytc, cytochrome c; Tplc3, type I tetraheme cytochrome c3; DsrC, heterodisulfide protein; AprAB, adenosine 5'-phosphosulfate reductase; CoM-SH, coenzyme M; CoB-SH, coenzyme B; CoB-S-S-CoM, coenzyme B-coenzyme M heterodisulfide; ox, oxidized; red, reduced. Reproduced with permission from: Calisto F., Pereira M. M. (2021) Modularity of membrane-bound charge-translocating protein complexes, *Biochem. Soc. Trans.* 49: 2669–2685.

theme of modular combinations of functional units to generate a transmembrane ion gradient. Many of the reactions in Figure 7.22 take place at very positive midpoint potentials and were very likely late inventions in Earth history, the figure makes no attempt to order the possible appearance of the structures during evolution. Nonetheless the functional diversification of the underlying units gives an impression of the vast bioenergetic landscape that was possible with the advent of cytochromes and quinones.

The electron donors required to reduce quinones in the bioenergetic membrane can be organic compounds (amino acids, bases, lipids, carbohydrates: organotrophy) or

inorganic compounds (H_2, Fe^{2+}, H_2S: lithotrophy). For the oxidation of organic compounds, the citric acid cycle that generates NADH for the membrane protein NADH:quinone (or menaquinone) oxidoreductase (complex I) is the standard hub to generate NADH for energy metabolism. One of the most widespread and versatile components of ion-pumping, membrane associated redox reactions is the cytochrome *b* pumping unit, which contains two heme molecules in the membrane-bound pumping subunit. It is a component of even a larger diversity of ion pumping complexes that includes the cytochrome *b/c*1 complex, which, in combination with the Rieske iron sulfur protein and a *c*-type cytochrome, functions in mammalian mitochondria and is able to pump up to 4 protons per quinol that is oxidized, utilizing a series of reactions called the Q cycle. The Q cycle entails a mechanism of quinone-based electron bifurcation that shares mechanistic similarity to flavin-based electron bifurcation.

Adding to the structural and functional complexity of ion pumping complexes comes the problem of lateral gene transfer (LGT) during 4 billion years of prokaryote diversification. The evolution of physiology does not follow strict phylogenetic lines or obey broad taxonomic boundaries. As an example, the search for proteins that trace to Luca (Figure 6.8) started with a set of 286,000 protein families distributed across 2000 prokaryotic genomes. Only 11,000 protein families were present in both bacteria and archaea. Of those 11,000, 97% showed evidence of lateral gene transfer across the domain divide separating bacteria and archaea. That is, only 3% of the 11,000, or 355 protein families showed *no evidence* for LGT between bacteria and archaea, and that was for a sample of only 2000 genomes. The more lineages one samples, the more likely it is that new LGT events will be uncovered. For example, cytochromes have been laterally transferred into methanogens in a lineage called Methanosarcinales. There, the cytochrome is utilized in the HdrDE complex (Figure 7.23) to generate an ion gradient during energy metabolism. However, the genes for quinone synthesis were not transferred, such that the Methanosarcinales evolved their own membrane-soluble electron carrier methanophenazine (Figure 7.20).

Lateral gene transfer decouples physiology from phylogeny. For example, archaeal halophiles (Haloarchaea) are derived from methanogens by the acquisition of about 1000 genes from bacteria, including a complete respiratory chain and quinone biosynthesis. That is why we have mostly presented the topic of early evolution here along chemical, mechanistic, thermodynamic and physiological lines, without trying to force a chemical and physiological account of early evolution into agreement with any particular phylogenetic scheme. Phylogenies, especially deep phylogenies, are short lived in the literature because by simply changing one or a few of the many computational parameters that are used to reconstruct phylogenetic trees, one can obtain fundamentally different results with wildly differing biological implications from the same molecular sequence data set. The structure and function of proteins that support physiology are not subject to the latest computational trends among molecular systematists and are therefore more stable as sources of evidence for physiological evolution.

7.3.9 THE END OF THE DARK AGE

Once the first chemolithoautotrophs left the vent where they arose, they rapidly inoculated other vents across the ocean floor in a wave of habitat colonization (Figure 7.14). The same would have occurred for the first fermenters, generating a second wave of colonization. With the ability to perform anaerobic respiration using cytochromes and quinones, a third wave of colonization followed, giving rise to trophic succession in hydrothermal vents: primary production → fermentation → anaerobic respiration. The result was a planet whose crust was entirely inhabited in every habitable

FIGURE 7.24 **Ancient and modern stromatolites.** (A) Reconstruction of a photosynthetic microbial community from 3.4 billion years old deposits in Australia. Scale bar in A 20 cm. (B) A preserved stromatolite from the same location. (C) Living stromatolites from Shark Bay, Australia. (A) and (B) are reproduced with permission from: Allwood A. C. *et al.* (2007) 3.43 billion-year-old stromatolite reef from the Pilbara Craton of Western Australia: Ecosystem-scale insights to early life on Earth, *Precambrian Res.* 158: 198–227.

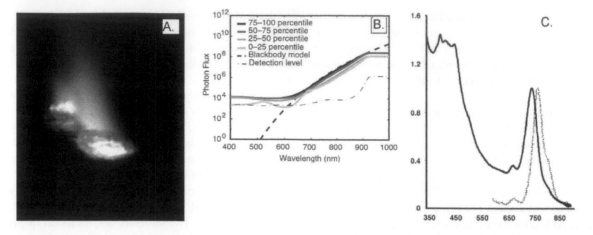

FIGURE 7.25 **Light at hydrothermal vents.** (A) Image of light emission at 870 nm from a 375°C black smoker type hydrothermal vent (vent 'L'), field of vision ca 12 × 15 cm. (B) Emission spectrum for (A), showing correspondence to black body radiation (heavy dotted line). (C) Absorption spectrum (solid line) and fluorescence spectrum (dotted line) for green sulfur bacteria isolated from a hydrothermal vent. Note the absorbance into the infrared spectrum. Vertical axis: arbitrary units. (A) and (B) Reproduced with permission from White S. N. *et al.* (2000) Investigations of ambient light emission at deep sea hydrothermal vents, *J. Geophys. Res.* 107: EPM1–13. (C) Reproduced with permission from Beatty J. T. *et al.* (2005) An obligately photosynthetic bacterial anaerobe from a deep-sea hydrothermal vent, *Proc. Natl Acad. Sci. USA* 102: 9306–9310.

niche, much like today, except there was no photosynthesis. As a consequence, the entire primordial global ecosystem, hence all life, was initially dependent upon H_2-dependent primary production. It is not unlikely that the reverse citric acid cycle evolved during this early colonization period. In contrast to the acetyl-CoA pathway, however, the reverse citric acid cycle requires ATP input, such that cells that use it for CO_2 fixation are dependent upon a separate means of ATP synthesis. In dark environments this would have to be some form of anaerobic respiratory breakdown of organics or redox reactions involving environmentally available donor and acceptors. In either case, ecosystems would be tied to the crust, hence the bottom of the ocean, for lack of means to harness energy sources at the surface. Escape from the crust required the origin of photosynthesis because no electron donor other than H_2 can support primary production over geological time scales. The consequence is that prior to the origin of photosynthesis, all life took place in total darkness, clinging to rocks and a geological source of H_2 for primary production.

Where did photosynthesis arise? One might suggest that when the first aerial land masses appeared, life had access to sunlight and the opportunity for the origin of another important tetrapyrrole, chlorophyll (Figure 7.17), and photosynthetic ATP synthesis. However, sunlight is, and was, very rich in damaging ultraviolet (*uv*) light. Four billion years ago there was no ozone layer like today, making microbial life difficult in the face of sunlight at best. Today, we use uv light to sterilize surfaces in laboratories and hospitals. It is highly mutagenic, and much the same effect could have been expected 4 billion years ago. Nonetheless, we can be quite confident that photosynthesis did arise by at least 3.4 billion years ago because stromatolites ('layer rocks') are found in Western Australia that preserve clear morphological evidence for tidal photosynthetic mats, the general appearance of which is very similar to modern stromatolites, also from Western Australia (Figure 7.24).

One possible site, if not the most likely site, for the origin of photosynthesis is the ocean floor, where light is emitted from hot, black smoker type hydrothermal vents. Euan Nisbet suggested this 30 years ago, based on the observation by Cindy van Dover that certain shrimps isolated from black smokers over 2 km deep possessed unusual light sensing organs. What good, she asked, would light sensing organs be if there was no light? Subsequent studies revealed very low level visible light emitted from the orifice of very hot vents. In 2005, Tom Beatty and colleagues reported the isolation and cultivation of dark green photosynthetic bacteria, green sulfur bacteria, from a deep-sea hydrothermal vent. The bacteria possessed chlorophyll that absorbed wavelengths extending into the far red 750–800 nm.

It is possible that low level light from vents provided the source of photons that enabled use of light as an energy source in an environment that stably supported life, providing opportunity to forge a chlorophyll biosynthetic pathway. In that transition, protoporphyrin IX (Figure 7.17) with zinc in the coordination site might have been a functional intermediate because it absorbs photons in the visible spectrum, generating a very stable triplet ^{3}Zn-protoporphyrin IX. This molecule has an excited state half-life on the order of 10 milliseconds (much longer than required for photochemical reactions) and it has an E_o of −1600 mV, much more than enough to reduce ferredoxin (see Martin *et al.*, 2018). That could have provided the seed of function needed to allow tetrapyrrole-dependent light absorption and photochemistry to replace H_2 and flavin-based electron bifurcation (see **Section 7.2.1**) as the source of reduced ferredoxin for primary production. The new electron source to fill the light-induced electron deficiency in protoporphyrin upon its photoexcitation and electron transfer was likely H_2S, with S^0 being the oxidized end product of photosynthesis, as in the case of green sulfur bacteria that inhabit vents and very low light environments in the depths of the Black sea today. With the backbone of a H_2S-dependent photosynthetic machinery in place, microbes were finally in a position to be able to leave the crust and live from dissolved CO_2 in the ocean using H_2S as an electron donor (H_2S is much more soluble in water than H_2 and can accumulate as the non-volatile sulfide ion, HS^- in the ocean and at the surface). This brought life to the surface and likely fueled the origin of Archaean stromatolites like those in Figure 7.24. The origin of photosynthesis added even more modularity to redox-dependent ion pumping complexes. Oxygenic photosynthesis involving the serial linkage of two photosystems did not evolve for at least another one billion years, roughly 2.4 billion years ago when the first traces of oxygen started appearing in the atmosphere.

* * *

7.4   CHAPTER SUMMARY

Section 7.1 The energy required for the origin of the ribosome and translation was likely obtained via SLP and the energy currency of the primordial translation process at the ribosome was GTP, because translation is GTP dependent in all cells today. At an advanced stage of protomicrobial evolution, the rotor-stator ATP synthase arose and was able to harness geochemical ion gradients, putting a high energy charge on the contents of protocells, thereby promoting further chemical evolution. The substrate specificity of the ATP synthase was likely the cause for the universality of ATP as the energy currency of metabolism, with GTP remaining the energy currency of translation.

Section 7.2 Coupling of ion pumping to exergonic reactions of CO_2 reduction enabled acetogens (Rnf) and methanogens (MtrA-H) to generate their own ion gradients, making them energetically independent of geochemical ion gradients but strictly dependent upon H_2-dependent CO_2 reduction for ATP synthesis (in addition to carbon supply). In order for cells to escape into the wild they needed pathways of carbon and energy metabolism that could emerge from H_2-dependent CO_2 reduction with H_2 being supplied by serpentinization. The first cells to escape into the wild were therefore likely acetogenic bacteria and methanogenic archaea. They were H_2 dependent and required electron bifurcation to use H_2 as a reductant for generating reduced ferredoxin for CO_2 fixation. Upon escape they could have colonized the earth in a very short time, less than 150 times their doubling time in the event of rapid dispersal and spreading. Some vents colonized by chemolithoautotrophs would eventually cease to produce H_2. In such environments, H_2 becomes extremely scarce and H_2 producing fermentations become exergonic. Fermentations require very little in the way of biochemical innovation. Once arisen, they would spread in a second wave to colonize H_2 poor environments or to be in place and be poised when actively serpentinizing systems ceased to produce H_2.

Section 7.3 When fermentations had run their course, what remained were unfermentable substrates in vents and an electron acceptor strong enough to support respiration: solid state magnetite, Fe_3O_4. This situation favored cells that could access Fe_3O_4 as an electron acceptor, requiring the invention of electron transport mechanisms to deposit electrons in the crust. That innovation was possibly the advent of multiheme cytochromes. That required heme synthesis (four enzymatic steps from sirohydrochlorin), which presented less of a challenge than the origin of cobalamin (20 genes), which was already present in fermenting cells. Multiheme cytochromes (conductive nanowires) are very common in modern environments. Together with quinones derived from intermediates of aromatic amino acid biosynthesis, multiheme cytochrome nanowires would have allowed cells to oxidize unfermentable substrates, giving them access to an unoccupied physiological niche and thereby initiating a third wave of global colonization. With cytochrome- and quinone-dependent membrane energetics in place, colonization of many habitats became possible, but primary production remained tied to H_2 producing environments. Cytochrome and quinone-based respiratory chains enabled free-living microbes to access all environmentally available electron donor–acceptor pairs capable of generating ion gradients with ion pumps. The origin of chlorophyll-based photosynthesis, likely occurring with the help of thermal light at hydrothermal vents, allowed cells to access H_2S as a new electron donor in light-dependent reactions. This freed microbial growth from H_2 dependence and permitted colonization of surface waters in the ocean, eventually enabling the colonization of *uv*-protected land and shallow water environments. The advent of oxygenic photosynthesis roughly 2.4 billion years ago marked the end of evolution without oxygen on Earth, a time we can call the Age of Anaerobes. It initiated a long phase of low oxygen stasis called The Pasteurian, during which O_2 did not exceed roughly 1% of its current atmospheric level—the Pasteur point, where fermentations stop and O_2 respirations begin—until the origin of land plants roughly 500 million years ago.

* * *

PROBLEMS FOR CHAPTER 7

1. *Energy harnessing.* What innovations in energy harnessing were necessary for microbes to allow a free living life style?
2. *Energy cost of protein synthesis.* Explain the high ATP cost of protein synthesis.
3. *Cell composition by dry weight.* Summarize the composition of cells by dry weight.
4. *ATP synthase.* Describe the enzymatic mechanism of the ATP synthase.
5. *ATP synthase and ribosome.* Give the approximate number of ATP synthases and ribosomes per cell and specify how many ATP and peptide bonds are produced per second. How much of the ATP produced by ATP synthases is consumed by the ribosomes?
6. *Protein-based ion pumping.* Give arguments why protein-based ion gradient harnessing came before protein-based pumping.
7. *Free energy of a proton gradient.* Calculate the Gibbs free energy of a proton gradient at a hydrophobic layer in a Hadean serpentinizing system (seawater pH 6, hydrothermal vent effluent pH 9, electric membrane potential −300 mV, 80°C; use Equation 7.1).

8. *GTP as energy currency.* Explain why GTP was likely the universal energy currency at the origin of translation.

9. *Triphosphates.* What is the advantage of a triphosphate as universal energy carrier?

10. *High hydrostatic pressure.* Explain why high hydrostatic pressure has little effect on most biochemical processes.

11. *Energy and carbon sources in the absence of H_2.* What were the energy and carbon sources after cutoff of serpentinization in an alkaline hydrothermal vent and after cell escape in the wild? Describe ion gradient harnessing in an alkaline serpentinizing system and amino acid and RNA fermentation.

12. *Fermentation and ion pumping.* Describe ion pumping during fermentation.

13. *Unfermentable substrates.* How do modern microbes extract energy from unfermentable substrates? How could ancient microbes have done that?

14. *Electron acceptor Fe_3O_4.* How do bacteria access solid phase Fe_3O_4 as an electron acceptor? Describe the electron transfer process.

15. *Cytochromes and quinones.* Explain why quinone and cytochrome-based electron transport to external electron acceptors likely arose after fermentations and how it could have given rise to the use of cytochromes in respiratory electron transport chains.

16. *Photosynthesis in hydrothermal vents.* Outline a possible mechanism for the origin of photosynthesis in hot, light emitting black smokers.

17. *Electrons from rocks.* Extract the essential information from the following paper and summarize it critically: Takumi Ishii *et al.* (2015) From chemolithoautotrophs to electrolithoautotrophs: CO_2 fixation by Fe(II)-oxidizing bacteria coupled with direct uptake of electrons from solid electron sources, Frontiers in Microbiology, 6, Article 994, 1–9

Appendix I: Calculation of free reaction enthalpy from reduction potentials

Reduction potentials important for early life are given in Table A1.1. The free reaction enthalpy ΔG of a redox reaction constituting of two half reactions can be calculated from the differences ΔE of their reduction potentials and from the number of transferred electrons n_e according to $\Delta G = -n_e F \Delta E$ and $E = E_0 - (RT/n_e F)2.303\log (a_{Red}/a_{Ox})$. For example, the half reaction $2H^+ + 2e^- \rightarrow H_2$ has a reduction potential of -0.41 V at $a_{Ox}=10^{-7}$ (pH7), $a_{Red} = 1$ atm H_2 and T=298 K (25°C), see Table A1.1, but a redox potential of -0.777 V at 10 atm H_2, T=373 K (100 °C) and pH10 typical for the CO_2 reducing experiments of Preiner et al. (Table 2.1). The reduction potential of the half reaction $CO_2 + 2H^+ + 2e^- \rightarrow HCOOH$ is -0.42 V at pH7, 1 atm CO_2 and T=298K (Table A 1.1) and -0.63 V at 10 atm CO_2, T=373 K (100 °C), pH10 and $a_{HCOOH}=10$mM. Thus $\Delta E= E(CO_2/HCOOH) + (- E(2H^+/H_2)) = + 0.148$ V and $\Delta G= -27.4$ kJ/mol (–20.2/ –13.1 kJ/mol for 1/0.1 atm CO_2). Hence, the reduction of carbon dioxide with hydrogen is exergonic under these conditions. Recall that, by convention, reduction potentials refer to half-reactions written as reductions: oxidant + $ne^- \rightarrow$ reductant, see section 2.4.5.

For obtaining the correct redox equation, the total number of atoms of each kind on each side of the equation and the number of electrons must be balanced. In our cxample $CO_2 + 2H^+ + 2e^- \rightarrow HCOOH$, the number of C-, O- and H-atoms is balanced (1 C, 2H, 2O on each side of the equation). The oxidation state of carbon in neutral CO_2 is +4 (2 O at -2 each=-4) and +2 in neutral HCOOH (-4 from the 2 O, +2 from the 2 H, hence +2 for C), Thus, 2 electrons must be transferred: $n_e=2$.

As another example, consider the reduction of Fe^{3+} with HSO_3^- to Fe^{2+} and SO_4^{2-}. First the number of electrons must be balanced: HSO_3^- (3O=-6,1H=+1, charge -1, S=+4,), SO_4^{2-} (4O=-8, charge -2, S=+6,) such that 2 electrons must be removed to oxidize sulfite to sulfate (*electron balance*: $HSO_3^- + 2 Fe^{3+} \rightarrow SO_4^{2-} + 2 Fe^{2+}$). This equation has 5 positive charges on the left side and 2 positive charges on the right side, hence 3 H^+ must be added on the right side for *charge balance*: $HSO_3^- + 2 Fe^{3+} \rightarrow SO_4^{2-} + 2 Fe^{2+} + 3H^+$. To get the correct *atom balance* we have to add 1 H_2O on the left side: $HSO_3^- + 2 Fe^{3+} + H_2O \rightarrow SO_4^{2-} + 2 Fe^{2+} + 3H^+$. Now the number of electrons, charges and atoms is balanced.

Note: For gas phase reactions, the pressures/activites in the Nernst equation are specified in atmospheres (atm) and the concentrations in mol/l according to electrochemistry convention. The logarithmus can only be calculated from numbers (without units); for the reaction $H_2 \rightarrow 2H^+ + 2e^-$ the experimental pressure is divided by 1 atm H_2 and the H^+ concentration by 1 mol H^+ (pH0) so that there are only numbers left in the logarithmus of the Nernst equation. Remember that the redox potential of the reaction $H_2 \rightarrow 2H^+ + 2e^-$ at 1 atm H_2 and pH0 is zero.

Electrochemical convention: $\Delta G^{0'}$ is used throughout the text for the free enthalpy at standard temperature, pressure and pH7 according to electrochemical convention. The subscript of ΔG is used to design R (free reaction enthalpy) or f (free enthalpy of formation of molecules). For the redox potential $E^{0'}$ or E_0' is used in the literature; we use E_0' throughout the text as in most text books of biochemistry. In case of non-standard conditions we use ΔG and E to designate the free reaction enthalpy and redox potential.

TABLE A1.1
Reduction potentials of selected reactions important for Origins and Early Life

Oxidant	Reductant	E_0' (V)	Ref
Ca^{2+}	Ca	-2.87	1)
Mg^{2+}	Mg	-2.37	1)
Al^{3+}	Al	-1.66	1)
Mn^{2+}	Mn	-1.19	1)
Zn^{2+}	Zn	-0.76	1)
Succinate + CO_2	α- Ketoglutarate	-0.67	6)
Acetate	Acetaldehyde	-0.58	3)
SO_4^{2-}	HSO_3^-	-0.52	2)
Oxalate	Glyoxalate	-0.5	3)
Fe^{2+}	Fe	-0.44	1)
CO_2	Glucose	-0.43	2)
Ferredoxin(ox)	Ferredoxin (red)	-0.43	6)
CO_2	Formate	-0.42	1) 3)
$2H^+$	H_2	-0.41	2) 3) 6)
$S_2O_3^{2-}$	HS– + HSO_3^-	-0.4	2)
a - Ketoglutarate + CO_2	Isocitrate	-0.38	3) 4)

(Continued)

TABLE A1.1 (Continued)

Oxidant	Reductant	E_0' (V)	Ref
CO_2	Methanol	-0.38	2)
Flavodoxin(ox)	Flavodoxin (red)	-0.37	2)
$NADP^+$	NADPH	-0.32	3) 4) 6)
NAD^+	NADH	-0.32	2) 3) 4) 6)
Cytochrome c3 (ox)	Cytochrome c3 (red)	-0.29	2)
Lipoic acid	Dihydrolipoic acid	-0.29	3) 4) 6)
CO_2	Acetate	-0.28	2)
Co^{2+}	Co	-0.28	1)
S	H_2S	-0.28	all 4
Ni^{2+}	Ni	-0.26	1)
CO_2	Methane	-0.24	2)
Acetoacetyl CoA	B-OH-Butyryl CoA	-0.24	3)
FAD	$FADH_2$	-0.22	2) 3) 4) 6)
FMN	$FMNH_2$	-0.22	3)
SO_4^{2-}	HS^-	-0.22	2)
Riboflavin(ox)	Riboflavin(red)	-0.21	3)
Acetaldehyde	Ethanol	-0.20	3) 4) 6)
Glycerolphosphate	Dihydroxyacetone phosphate	-0.19	2)
FMN	FMNH	-0.19	2)
Pyruvate	Lactate	-0.19	2) 3) 4) 6)
Oxaloacetate	Malate	-0.17	3)
Hydroxypyruvate	Glycerate	-0.16	3)
Fe^{3+}-Oxalate	Fe^{2+}-Oxalate	-0.14	3)
CO_2 (gas; pH0)	CO (gas)	-0.12	3)
Flavodoxin(ox)	Flavodoxin(red)	-0.12	2)
HSO_3^-	HS^-	-0.12	2)
Glyoxylate	Glycolate	-0.09	3)
Menaquinone	Menaquinol	-0.08	2)
Adenosine phosphosulfate	$AMP + HSO_3^-$	-0.06	2)
Dehydroascorbate	Ascorbate	-0.06	3)
Rubredoxin(ox)	Rubredoxin(red)	-0.06	2)
Acryloyl-CoA	Propionyl-CoA	-0.02	2)
Crotonyl-CoA	Butyryl-CoA	-0.015	3) 4)
Glycine	$Acetate^- + NH_4^-$	-0.01	2)
Fumarate	Succinate	0.031	2) 3) 6)
Cytochrome b(ox)	Cytochrome b(red)	0.035	2)
Ubiquinone (ox)	Ubiquinone (red)	0.11	2) 6)
$Fe(OH)_3 + HCO_3^-$ (pH 7)	$FeCO_3$	0.2	1) 2) 3)
Cytochrome c (ox)	Cytochrome c (red)	0.25	2) 6)
O_2	H_2O_2	0.30	1) 4)
Cu^{2+}(pH 0)	Cu	0.34	1)
NO_2^-	NO	0.36	1) 2)
Cytochrome f (ox)	Cytochrome f (red)	0.37	3) 4)
Cytochrome a (ox)	Cytochrome a (red)	0.39	2)
NO_3^-	NO_2^-	0.42	3)
NO_3^-	$1/2 N_2$	0.74	2)
Fe^{3+} (pH 2)	Fe^{2+} (pH 2)	0.77	2) 5) 6)
Mn^{4+}	Mn^{2+}	0.80	1) 2)
O_2	$2H_2O$	0.82	1) 3)
ClO^{3-}	Cl^-	1.03	2)
NO	N_2O	1.18	2)
N_2O	N_2	1.36	2)

1) Handbook of Chemistry and Physics 8th Ed, CRC Press (2000), Electrochemical series by P, Vanysek

2) Brock Biology of Microorganisms, 14th Global Ed, Pearson (2015), Redox Tower on p.83 & Table A1.2

3) Handbook of Biochemistry and Molecular Biology 5th Ed., CRC Press (2018), p. 575 – 582

4) Lehninger Principles of Biochemistry 8th Ed., Macmillan Learning (2021), Table 13-7, p.1799ff

5) Table 2.2 from this book

6) Stryer 7th edition p. 529

All oxidations of metal elements are measured at pH 0.

Further Reading

Chapter 1 Geochemistry of the Early Earth

Big Bang Theory

Hawking S. *W. (1998) A Brief History of Time: From the Big Bang to Black Holes*, Bantam Books, New York, Trade Paperback Edition.

Lachieze-Rey M., Gunzig E. (1999) *The Cosmological Background Radiation*, Cambridge University Press, Cambridge.

Stars and Planets

Bethe H. (1939) Energy production in stars, *Phys. Rev.* 55: 434–456.

Fewell *M. P. (1995)* The atomic nuclide with the highest mean binding energy, *Am. J. Phys.* 63: 653–658.

Harrison T. M. *et al.* (2005) Heterogeneous Hadean hafnium: Evidence of continental crust at 4.4 to 4.5 Ga., *Science* 310: 1947–1950.

Kasting J. F. (1993) Earth's early atmosphere, *Science* 259: 920–926.

Lodders K., Fegley B. (1998) *The Planetary Scientist's Companion*, Oxford University Press, New York, NY.

McWilliam A., Rauch M. (2004) *Origin and Evolution of the Elements*, Cambridge University Press, Cambridge.

Stahler S. W., Palla F. (2004) *The Formation of Stars*, Wiley-VCH, Weinheim.

Usoskin I. G. (2013) A history of solar activity over millennia, *Living Rev. Sol. Phys.* 10: 1–94.

Zinnecker H., Yorke H. W. (2007) Toward understanding massive star formation, *Annu. Rev. Astron. Astrophys.* 45: 481–563.

Early and Modern Earth

Arndt N. T., Nisbet E. G. (2012) Processes on the young Earth and the habitats of early life, *Annu. Rev. Earth Planet. Sci.* 40: 521–549.

Baross J. A., Hoffman S. E. (1985) Submarine hydrothermal vents and associated gradient environments as sites for the origin and evolution of life, *Orig. Life* 15: 327–345.

Borg L. E., Carlson R. W. (2023) The evolving chronology of Moon formation. *Annu. Rev. Earth Planet. Sci.* 51: 25–52.

Genda H. *et al.* (2017) The terrestrial late veneer from core disruption of a lunar-sized impactor, *Earth Planet. Sci. Lett.* 480: 25–32.

Harrison T. M. *et al.* (2005) Heterogeneous Hadean hafnium: Evidence of continental crust at 4.4 to 4.5 Ga., *Science* 310: 1947–1950.

Herschy B. *et al.* (2014) An origin-*of-life reactor to simulate alkaline hydrothermal vents, J. Mol. Evol. 79: 213–227.*

Russell M. J., Arndt N. T. (2005) Geodynamic and metabolic cycles in the Hadean, *Biogeosciences* 2: 97–111.

Sossi P. A. *et al.* (2020) Redox state of Earth's magma ocean and its Venus-like early atmosphere, *Sci. Adv.* 6: 1–8.

Wilde S. A. *et al.* (2001) Evidence from detrital zircons for the existence of continental crust and oceans on the Earth 4.4 Gyr ago, *Nature* 409: 175–178.

Zahnle K. *et al.* (2007) Emergence of a habitable planet, *Space Sci. Rev.* 129: 35–78.

Atoms, Molecules, Chemical, and Biochemical Reactions

Atkins P. W. *et al.* (2018) *Atkins' Physical Chemistry*, Oxford University Press, Oxford.

Berg J. M. *et al.* (2012) *Biochemistry*, 7th Edition, W. H. Freeman and Company, New York, NY.

De Meis L., Suzano V. A. (1988) Role of water activity on the rates of acetyl phosphate and ATP hydrolysis, *FEBS Lett.* 232: 73–77.

do Nascimento Vieira A. *et al.* (2020) The ambivalent role of water at the origins of life, *FEBS Lett.* 594: 2717–2733.

Fuchs G. (2011) Alternative pathways of carbon dioxide fixation: Insights into the early evolution of life? *Annu. Rev. Microbiol.* 65: 631–658.

Jarrell K. F. *et al.* (1984) Intracellular potassium concentration and relative acidity of the ribosomal proteins of methanogenic bacteria, *Can. J. Microbiol.* 30: 663–668.

Kim J. D. *et al.* (2013) Anoxic photochemical oxidation of siderite generates molecular hydrogen and iron oxides, *Proc. Natl Acad. Sci. USA* 110: 10073–10077.

Martin W. F. (2020) Older than genes: The acetyl CoA pathway and origins, *Front. Microbiol.* 11: 1–21.

Powner M. W. *et al.* (2009) Synthesis of activated pyrimidine ribonucleotides in prebiotically plausible conditions, *Nature* 459: 239–242.

Preiner M. *et al.* (2020) A hydrogen-dependent geochemical analogue of primordial carbon and energy metabolism, *Nat. Ecol. Evol.* 4: 534–542.

Roeßler M., Müller V. (2001) Osmoadaptation in bacteria and archaea: Common principles and differences, *Environ. Microbiol.* 3: 743–754.

Shock E. L., Boyd E. S. (2015) Principles of geobiochemistry, *Elements* 11: 395–401.

Thermodynamics and Kinetics

Amend J. P., McCollom T. M. (2009) Energetics of biomolecule synthesis on early earth. In *Chemical Evolution II: From the Origins of Life to Modern Society*. Zaikowski, L. *et al.* eds., American Chemical Society, Washington, DC, pp. 63–94.

Bada J. L., Lazcano A. (2002) Some like it hot, but not the first biomolecules, *Science* 296: 1982–1983.

Berg J. M. *et al.* (2012) *Biochemistry*, 7th Edition, W. H. Freeman and Company, New York, NY.

Bracher P. J., Snyder P. W., Bohall B. R., Whitesides G. M. (2011) The relative rates of thiol–thioester exchange and hydrolysis for alkyl and aryl thioalkanoates in water. *Origin Life Evol. Biosph.* 41: 399–412.

Boyd E. S. *et al.* (2020) Bioenergetic constraints on the origin of autotrophic metabolism, *Phil. Trans. Roy. Soc. A.* 378: 20190151.

Chandru K. *et al.* (2016) The abiotic chemistry of thiolated acetate derivatives and the origin of life. *Sci. Rep.* 6: 29883.

Chen B. *et al.* (2017) The Effect of pressure on organic reactions in fluids—a new theoretical perspective, *Angew. Chem. Int. Ed.* 56: 11126–11142.

Standard thermodynamic properties of 2400 chemical substances. www.update.uu.se/~jolkkonen/pdf/CRC_TD.pdf.

Stockbridge R. B. *et al.* (2010) Impact of temperature on the time required for the establishment of primordial biochemistry, and for the evolution of enzymes, *Proc. Natl Acad. Sci. USA* 107: 22102–22105.

Thauer R. K. *et al.* (1977) Energy conservation in chemotrophic anaerobic bacteria, *Bacteriol. Rev.* 41: 100–180.

Wagman D. D. *et al.* (1982) The NBS tables of chemical thermodynamic properties, *J. Phys. Chem. Ref.* 11(Suppl. 2).

Whicher A. *et al.* (2018) Acetyl phosphate as a primordial energy currency at the origin of life, *Orig. Life Evol. Biosph.* 48: 159–179.

Wolfenden R. (2014) Primordial chemistry and enzyme evolution in a hot environment, *Cell. Mol. Life Sci.* 71: 2909–2915.

Wolfenden R., Williams R. (1985) Solvent water and the biological group-transfer potential of phosphoric and carboxylic anhydrides, *J. Am. Chem. Soc.* 107: 4345–4346.

Xavier J. C. *et al.* (2020) Autocatalytic chemical networks at the origin of metabolism, *Proc. R. Soc. B: Biol. Sci.* 287: 20192377.

Transition Metal Catalysts

Böller B. *et al.* (2019) The active sites of a working Fischer-Tropsch catalyst revealed by operando scanning tunneling microscopy, *Nat. Catal.* 2: 1027–1034.

Daumann L. J. (2019) Essential and ubiquitous: The emergence of lanthanide metallobiochemistry, *Angewandte Chemie Int Ed.* 58: 12795–12802.

Ertl G. (1976) Elementary processes at gas/metal interfaces, *Angew. Chem. Int. Ed.* 15: 391–400.

Freund H. J. (2010) Model studies in heterogenous catalysis, *Chem Eur. J.* 16: 9384–9397.

Kotopoulou E. *et al.* (2021) Nanoscale anatomy of iron-silica self-organized membranes: Implications for prebiotic chemistry, *Angew. Chem. Int. Ed.* 60: 1396–1402.

Kubas G. J. (2007) Fundamentals of H_2 binding and reactivity on transition metals underlying hydrogenase function and H_2 production and storage, *Chem. Rev.* 107: 4152–4205.

Schlögl R. (2015) Heterogenous catalysis, *Angew. Chem. Int. Ed.* 54: 3465–3520.

Wintterlin J. *et al.* (1997) Atomic and macroscopic reaction rates of a surface-catalyzed reaction, *Science* 278: 1931–1934.

Zambelli T. *et al.* (1996) Identification of the "active sites' of a surface-catalyzed reaction, *Science* 273: 1688–1690.

Meteorites with Organic Molecules

Allen D. A., Wickramasinghe D. T. (1987) Discovery of organic grains in comet Wilson, *Nature* 329: 615–616.

Green S. (1981) Interstellar chemistry: Exotic molecules in space, *Annu. Rev. Phys. Chem.* 32: 103–138.

Hayatsu, R., Anders E., (2005) Organic compounds in meteorites and their origins. In *Cosmo- and Geochemistry (Part of Topics in Current Chemistry Vol. 99)*, Springer, Berlin. pp. 1–37.

Saladino R. *et al.* (2011) Catalytic effects of Murchison material: Prebiotic synthesis and degradation of RNA precursors, *Orig. Life Evol. Biosph.* 41: 437–451.

Schmitt-Kopplin P. *et al.* (2010) High molecular diversity of extraterrestrial organic matter in Murchison meteorite revealed 40 years after its fall, *Proc. Natl Acad. Sci. USA* 107: 2763–2768.

Sephton M. A. (2002) Organic compounds in carbonaceous meteorites, *Nat. Prod. Rep.* 19: 292–311.

Thomas P. J. *et al.* (1997) *Comets and the Origin and Evolution of Life*, Springer Verlag, Berlin.

Hydrothermal Vents and Serpentinization

Beatty J. T. *et al.* (2005) An obligately photosynthetic bacterial anaerobe from a deep-sea hydrothermal vent, *Proc. Natl Acad. Sci. USA* 102: 9306–9310.

Etiope G. (2017) Abiotic methane in continental serpentinization sites: An overview, *Procedia Earth Planet. Sci.* 17: 9–12.

Horita H., Berndt M. E. (1999) Abiogenic methane formation and isotopic fractionation under hydrothermal conditions, *Science* 285: 1055–1057.

Hsu H. *et al.* (2015) Ongoing hydrothermal activities within Enceladus, *Nature* 519: 207–210.

Kamyshny A. *et al.* (2014) Multiple sulfur isotopes fractionations associated with abiotic sulfur transformations in Yellowstone National Park geothermal springs, *Geochem. Trans.* 15: 1–22.

Kelley D. S. (2005) From the mantle to microbes. The Lost City hydrothermal field, *Oceanography* 18: 32–45.

Kelley D. S. *et al.* (2002) Volcanoes, fluids and life at mid-ocean ridge spreading centers, *Annu. Rev. Earth Planet. Sci.* 30: 385–491.

Koch U. *et al.* (2008) The gas flow at mineral springs and mofettes in the Vogtland/NW Bohemia: An enduring long-term increase, *Geofluids* 8: 274–285.

Kümpel H. J. (1997) *Tides in Water Saturated Rock* (Lecture Notes in Earth Sciences), vol. 66, Springer, New York, pp. 277–291.

Lang S. Q., Brazelton W. J. (2020) Habitability of the marine serpentinite subsurface: A case study of the Lost City hydrothermal field, *Philos. Trans. R. Soc. A* 378: 20180429.

Li J. *et al.* (2020) Recycling and metabolic flexibility dictate life in the lower oceanic crust, *Nature* 579: 250–255.

Martin W. *et al.* (2008) Hydrothermal vents and the origin of life, *Nat. Rev. Microbiol.* 6: 805–814.

Martin W., Russell M. J. (2007) On the origin of biochemistry at an alkaline hydrothermal vent, *Phil. Trans. R. Soc. B* 362: 1887–1925.

McCollom T. M., Bach W. (2009) Thermodynamic constraints on hydrogen generation during serpentinization of ultramafic rocks, *Geochim. Cosmochim. Acta* 73: 856–875.

McCollom T. M., Seewald J. S. (2007) Abiotic synthesis of organic compounds in deep-sea hydrothermal environments, *Chem. Rev.* 107: 382–401.

McCollom T. M., Seewald J. S. (2013) Serpentinites, hydrogen, and life, *Elements* 9: 129–134.

McDermott, J. M. *et al.* (2015) Pathways for abiotic organic synthesis at submarine hydrothermal fields, *Proc. Natl Acad. Sci. USA* 112: 7668–7672.

Preiner M. *et al.* (2018) Serpentinization: Connecting geochemistry, ancient metabolism and industrial hydrogenation, *Life* 8: 1–22.

Rimmer P. R., Shorttle O. (2019) Origin of life's building blocks in carbon- and nitrogen-rich surface hydrothermal vents, *Life* 9: 1–19.

Russell M. J. *et al.* (2010) Serpentinization as a source of energy at the origin of life, *Geobiology* 8: 355–371.

Schwander L. *et al.* (2023) Serpentinization as the source of energy, electrons, organics, catalysts, nutrients and pH gradients for the origin of LUCA and life. *Front. Microbiol.* 14: 1257597.

Seewald J. S. *et al.* (2006) Experimental investigation of single carbon compounds under hydrothermal conditions, *Geochim. Cosmochim. Acta* 70: 446–460.

Shang X. *et al.* (2023) Formation of ammonia through serpentinization in the Hadean Eon, *Sci. Bull.* 68: 1109–1112.

Shock E., Canovas P. (2010) The potential for abiotic organic synthesis and biosynthesis at seafloor hydrothermal systems, *Geofluids* 10: 161–192.

Sleep N. H. (2018) Geological and geochemical constraints on the origin and evolution of life. *Astrobiology* 18: 1–21.

Sleep N. H. *et al.* (2004) H_2-rich fluids from serpentinization: Geochemical and biotic implications, *Proc. Natl Acad. Sci. USA* 101: 12818–12823.

Ueda H. *et al.* (2017) Reactions between olivine and CO_2-rich seawater at 300 C: Implications for H_2 generation and CO_2 sequestration on the early Earth, *Geosci. Front.* 8: 387–396.

Westall F. *et al.* (2018) A hydrothermal-sedimentary context for the origin of life, *Astrobiology* 18: 259–293.

Zhang X. *et al.* (2020) Discovery of supercritical carbon dioxide in a hydrothermal system, *Sci. Bull.* 65: 958–964.

Early Traces of Life

Cappelline E. *et al.* (2019) Early Pleistocene enamel proteome from Dmanisi resolves *Stephanorhinus* phylogeny, *Nature* 574: 103–107.

Dodd M. S. *et al.* (2017) Evidence for early life in Earth's oldest hydrothermal vent precipitates, *Nature* 543: 60–64.

Garcia-Ruiz J. M. *et al.* (2003) Self-assembled silica-carbonate structures and detection of ancient microfossils, *Science* 302: 1194–1197.

Marshall C. P. *et al.* (2011) Haematite pseudomicrofossils present in the 3.5-billion-year-old Apex Chert, *Nat. Geosci.* 4: 240–243.

Nutman A. P. *et al.* (2016) Rapid emergence of life shown by discovery of 3700-million-year-old microbial structures, *Nature* 537: 535–538.

Schopf J. W. *et al.* (2018) SIMS analyses of the oldest known assemblage of microfossils document their taxon-correlated carbon isotope compositions, *Proc. Natl Acad. Sci. USA* 115: 53–58.

Tashiro T. *et al.* (2017) Early trace of life from 3.95 Ga sedimentary rocks in Labrador, Canada, *Nature* 549: 516–518.

Ueno Y. *et al.* (2006) Evidence from fluid inclusions for microbial methanogenesis in the early Archaean era, *Nature* 440: 516–519.

Walter M. R. *et al.* (1980) Stromatolites 3,400–3,500 Myr old from the North Pole area, Western Australia, *Nature* 284: 443–445.

Westall F. *et al.* (2001) Early Archean fossil bacteria and biofilms in hydrothermally-influenced sediments from the Barberton greenstone belt, South Africa, *Precambrian Res.* 106: 93–116.

Phosphates and Sulfides

Gull M. *et al.* (2015) Nucleoside phosphorylation by the mineral schreibersite, *Sci. Rep.* 5: 1–6.

Heinen W., Lauwers A. M. (1996) Organic sulfur compounds resulting from the interaction of iron sulfide, hydrogen sulfide and carbon dioxide in an anaerobic aqueous environment, *Orig. Life Evol. Biosph.* 26: 131–150.

Klein A. H. *et al.* (2007) The intracellular concentration of acetyl phosphate in *Escherichia coli* is sufficient for direct phosphorylation of two-component response regulators, *J. Bacteriol.* 189: 5574–5581.

Mao Z. *et al.* (2023) AMP-dependent phosphite dehydrogenase, a phosphorylating enzyme in dissimilatory phosphite oxidation. *Proc. Natl. Acad. Sci. USA.* 120: e2309743120.

Metcalf W. W., van der Donk W. A. (2009) Biosynthesis of phosphonic and phosphinic acid natural products, *Annu. Rev. Biochem.* 78: 65–94.

Morton S. C. *et al.* (2003) Phosphates, phosphites, and phosphides in environmental samples, *Environ. Sci. Technol.* 37: 1169–1174.

Pasek M. A. (2008) Rethinking early Earth phosphorus geochemistry, *Proc. Natl Acad. Sci. USA* 105: 853–858.

Pasek, M. A. *et al.* (2022) Serpentinization as a route to liberating phosphorus on habitable worlds. *Geochim. Cosmochim. Acta* 336: 332–340.

Romero P. J., de Meis L. (1989) Role of water in the energy of hydrolysis of phosphoanhydride and phosphoester bonds, *J. Biol. Chem.* 264: 7869–7873.

Schink B., Friedrich M. (2000) Phosphite oxidation by sulphate reduction, *Nature* 406: 37.

Wald G. (1962) Life in the second and third periods; or why phosphorus and sulfur for high-energy bonds? In *Horizons in Biochemistry.* Kasha M., Pullman B. eds., Academic Press, New York, NY, pp. 127–142.

Yamagata Y. *et al.* (1991) Volcanic production of polyphosphates and its relevance to prebiotic evolution, *Nature* 352: 516–519.

Other Reading Related to Origins

Eakin R. E. (1963) An approach to the evolution of metabolism, *Proc. Natl. Acad. Sci. USA* 49: 360–366.

Haeckel E. (1902). *Natürliche Schöpfungs-Geschichte. Gemeinverständliche Wissenschaftliche Vorträge über die Entwicklungslehre. Zehnte verbesserte Auflage. Zweiter Theil: Allgemeine Stammesgeschichte*, Georg ReimerVerlag, Berlin.

Haldane J. B. S. (1929). The origin of life, *Rational. Ann.* 148: 3–10.

McMahon S., Ivarsson M. (2019) A new frontier for palaeobiology: Earth's vast deep biosphere, *BioEssays* 41: 1–11.

Mereschkowsky C. (1910). Theorie der zwei Plasmaarten als Grundlage der Symbiogenesis, einer neuen Lehre von der Entstehung der Organismen. *Biol. Centralbl.* 30: 278–288, 289–303, 353–367 [English translation in: Kowallik K. V., Martin W. F. (2021). An annotated English translation of Mereschkowky's 1910 paper on the theory of two plasma lineages. *Biosystems* 199: 104281].

Sabirzyanov A. N. *et al.* (2002) Solubility of water in supercritical carbon dioxide, *High Temp.* 40: 203–206.

Schönheit P. *et al.* (2016) On the origin of heterotrophy, *Trends Microbiol.* 24: 12–25.

Wächtershäuser G. (1992) Groundworks for an evolutionary biochemistry—the Iron Sulfur World, *Prog. Biophys. Mol. Biol.* 58: 85–201.

Wächtershäuser G. (2006) From volcanic origins of chemoautotrophic life to Bacteria, Archaea and Eukarya, *Phil. Trans. R. Soc. B* 361: 1787–1808.

Chapter 2 Origin of Organic Molecules

Dissolved Inorganic Carbon

Diamond L. W., Akinfiev N. N. (2003) Solubility of CO_2 in water from—1.5 to 100°C and from 0.1 to 100 MPa: Evaluation of literature data and thermodynamic modelling, *Fluid Ph. Equilibria* 208: 265–290.

NIST chemistry webbook CO_2 Henry's Law data, Carbon dioxide. nist.gov

Software to calculate DIC (Dissolved Inorganic Carbon) in water. www.aqion.de

Soli A. L. *et al.* (2002) CO_2 system hydration and dehydration kinetics and the equilibrium CO_2/H_2CO_3 ratio in aqueous NaCl solution, *Mar. Chem.* 78: 65–73.

Serpentinization

Baldauf-Sommerbauer G. *et al.* (2016) Sustainable iron production from mineral iron carbonate and hydrogen, *Green Chem.* 18: 6255–6265.

Brazelton W. J. *et al.* (2011) Physiological differentiation within a single-species biofilm fueled by serpentinization, *mBio* 2: e00127–11.

Cody G. D. *et al.* (2000) Primordial carbonylated iron-sulfur compounds and the synthesis of pyruvate, *Science* 289: 1337–1340.

Eickenbusch P. *et al.* (2019) Origin of short-chain organic acids in serpentinite mud volcanoes of the Mariana Convergent Margin, *Front. Microbiol.* 10: 1729, 1–21.

Fuchs G. (2011) Alternative pathways of carbon dioxide fixation: Insights into the early evolution of life? *Annu. Rev. Microbiol.* 65: 631–658.

He C. *et al.* (2010) A mild hydrothermal route to fix carbon dioxide to simple carboxylic acids, *Org. Lett.* 12: 649–651.

He D. *et al.* (2021) Hydrothermal synthesis of long-chain hydrocarbons up to C_{24} with $NaHCO_3$-assisted stabilizing cobalt, *Proc. Natl Acad. Sci. USA* 118: e2115059118.

Horita J., Berndt M. E. (1999) Abiogenic methane formation and isotopic fractionation under hydrothermal conditions, *Science* 285: 1055–1057.

Jagadeesan D. *et al.* (2013) Direct conversion of calcium carbonate to C_1-C_3 hydrocarbons, *RSC Adv.* 3: 7224–7229.

Kelley D. S. *et al.* (2001) An off-axis hydrothermal vent field near the Mid-Atlantic Ridge at 30°N, *Nature* 412: 145–149.

Kelley D. S. *et al.* (2005) A serpentinite-hosted ecosystem: The Lost City hydrothermal field, *Science* 307: 1428–1434.

Klein F. *et al.* (2019) Abiotic methane synthesis and serpentinization in olivine-hosted fluid inclusions, *Proc. Natl Acad. Sci. USA* 116: 17666–17672.

Klein F. *et al.* (2020) Brucite formation and dissolution in oceanic serpentinite, *Geochem. Perspect. Lett.* 16: 1–5.

Krishna Rao J. S. R. (1964) Native nickel-iron alloy, its mode of occurrence, distribution and origin, *Econ. Geol.* 59: 443–448.

Lamadrid H. M. *et al.* (2016) Effect of water activity on rates of serpentinization of olivine, *Nat. Commun.* 8: 1–9.

Lang S. Q. *et al.* (2010) Elevated concentration of formate, acetate and dissolved organic carbon found at the Lost City hydrothermal field, *Geochim. Cosmochim. Acta* 74: 941–952.

Lang S. Q. *et al.* (2018) Deeply-sourced formate fuels sulfate reducers but not methanogens at Lost City hydrothermal field, *Sci. Rep.* 8: 1–10.

Mayhew L. E. *et al.* (2013) Hydrogen generation from low-temperature water-rock reactions, *Nat. Geosci.* 6: 478–484.

McCollom T. M. (2013) Miller-Urey and beyond: What have we learned about prebiotic organic synthesis reactions in the past 60 years? *Annu. Rev. Earth Planet. Sci.* 41: 207–229.

McCollom T. M., Seewald J. S. (2007) Abiotic synthesis of organic compounds in deep-sea hydrothermal environments, *Chem. Rev.* 107: 382–401.

McDermott J. M. *et al.* (2015) Pathways for abiotic organic synthesis at submarine hydrothermal fields, *Proc. Natl Acad. Sci. USA* 112: 7668–7672.

Menez B. *et al.* (2018) Abiotic synthesis of amino acids in the recesses of the oceanic lithosphere, *Nature* 564: 59–65.

Oparin R. *et al.* (2005) Water-carbon dioxide mixtures at high temperatures and pressures: Local order in the water rich phase investigated by vibrational spectroscopy, *J. Chem. Phys.* 123: 224501.

Roldan A. *et al.* (2015) Bio-inspired CO_2 conversion by iron sulfide catalysts under sustainable conditions, *Chem. Commun.* 51: 7501–7504.

Shang X. *et al.* (2023) Formation of ammonia through serpentinization in the Hadean Eon, *Sci. Bull.* 68: 1109–1112.

Suda K. *et al.* (2014) Origin of methane in serpentinite-hosted hydrothermal systems: The CH_4-H_2-H_2O hydrogen isotope systematics of the Hakuba Happo hot spring, *Earth Planet. Sci. Lett.* 386: 112–125.

Tamblyn R., Hermann J. (2023) Geological evidence for high H_2 production from komatiites in the Archaean, *Nat. Geosci.* 16: 1194–1199.

Transition Metal Catalyzed Reduction of Carbon and Nitrogen with Hydrogen

Anderson J. S. *et al.* (2013) Catalytic conversion of nitrogen to ammonia by an iron model complex, *Nature* 501: 84–87.

Bennett B. D. *et al.* (2009) Absolute metabolite concentrations and implied enzyme active site occupancy in *Escherichia coli*, *Nat. Chem. Biol.* 5: 593–599.

Beyazay T. *et al.* (2023) Influence of composition of nickel-iron nanoparticles for abiotic CO_2 conversion to early prebiotic organics. *Angew. Chem.* 62: e202218189.

Brandes J. A. *et al.* (1998) Abiotic nitrogen reduction on the early Earth. *Nature* 395: 365–367.

Dörr M. *et al.* (2003) A possible prebiotic formation of ammonia from dinitrogen on iron sulfide surfaces, *Angew. Chemie Int. Ed.* 42: 1540–1543.

Furdui C., Ragsdale S. W. (2000) The role of pyruvate ferredoxin oxidoreductase in pyruvate synthesis during autotrophic growth by the Wood-Ljungdahl pathway. *J. Biol. Chem.* 275: 28494–28499.

Hoffman B. M. *et al.* (2014) Mechanism of nitrogen fixation by nitrogenase: The next stage, *Chem. Rev.* 114: 4041–4062.

Horita J., Berndt M. E. (1999) Abiogenic methane formation and isotopic fractionation under hydrothermal conditions, *Science* 285: 1055–1057.

Hudson R. *et al.* (2020) CO_2 reduction driven by a pH gradient, *Proc. Natl Acad. Sci. USA* 117: 22873–22879.

Jeoung J.-H. *et al.* (2020) Double-cubane [8Fe9S] clusters: A novel nitrogenase-related cofactor in biology, *Chem. Bio. Chem.* 21: 1710–1716.

Loison A. *et al.* (2010) Elucidation of an iterative process of carbon–carbon bond formation of prebiotic significance, *Astrobiology* 10: 973–988.

Mrnjavac N. *et al.* (2023) The Moon-forming impact and the autotrophic origin of life. *ChemPlusChem*: e202300270.

Niks D., Hille R. (2019) Molybdenum- and tungsten-containing formate dehydrogenases and formylmethanofuran dehydrogenases: Structure, mechanism, and cofactor insertion, *Protein Sci.* 28: 111–122.

Preiner M. *et al.* (2020) A hydrogen-dependent geochemical analogue of primordial carbon and energy metabolism, *Nat. Ecol. Evol.* 4: 534–542.

Smirnov A. *et al.* (2008) Abiotic ammonium formation in the presence of Ni-Fe metals and alloys and its implications for the Hadean nitrogen cycle, *Geochem. Trans.* 9: 1–20.

Spatzal T. *et al.* (2011) Evidence for interstitial carbon in nitrogenase FeMo cofactor, *Science* 334: 940.

Thauer R. K. *et al.* (2010) Hydrogenases from methanogenic archaea, nickel, a novel cofactor, and H_2 storage, *Annu. Rev. Biochem.* 79: 507–536.

Wei J. *et al.* (2017) Directly converting CO_2 into a gasoline fuel, *Nat. Commun.* 8: 1–8.

Electrons, Energy, Redox Potential, Microbial Physiology

Buckel W., Thauer R. K. (2013) Energy conservation via electron bifurcating ferredoxin reduction and proton/Na^+ translocating ferredoxin oxidation, *Biochim. Biophys. Acta, Bioenerg.* 1827: 94–113.

Buckel W., Thauer R. K. (2018) Flavin-based electron bifurcation, a new mechanism of biological energy coupling, *Chem. Rev.* 118: 3862–3886.

Gilbert W. (1986) Origin of life: The RNA world. *Nature* 319: 618.

Kuhn H. *et al.* (2000) *Principles of Physical Chemistry*, John Wiley and Sons, Chichester.

Lengeler J. W. *et al.* (1999) *Biology of the Prokaryotes*. Georg Thieme Verlag, Stuttgart.

Madigan M. T. *et al.* (2012) *Brock Biology of Microorganisms*, 13th Edition, Benjamin Cummings, Boston, MA.

Müller V. *et al.* (2018) Electron bifurcation: A long-hidden energy-coupling mechanism, *Annu. Rev. Microbiol.* 72: 331–353.

Pasek M. A. (2020) Thermodynamics of prebiotic phosphorylation, *Chem. Rev.* 120: 4690–4706.

Pasek M. A. *et al.* (2017) Phosphorylation of the early earth, *Chem. Geol.* 475: 149–170.

Sander R. (2015) Compilation of Henry's law constants (version 4.0) for water as solvent, *Atmos. Chem. Phys.* 15: 4399–4981.

Thauer R. K. *et al.* (1977) Energy conservation in chemotrophic anaerobic bacteria, *Bacteriol. Rev.* 41: 100–180.

Thauer R. K. *et al.* (2008) Methanogenic archaea: Ecologically relevant differences in energy conservation, *Nat. Rev. Microbiol.* 6: 579–591.

Early Physiological Evolution

Brabender M. *et al.* (2024) Ferredoxin reduction by hydrogen with iron functions as an evolutionary precursor of flavin-based electron bifurcation, *Proc. Natl Acad. Sci. USA* 121: e2318969121

Dayhoff M. O., Eck R. V. (1966) Evolution of the structure of ferredoxin based on surviving relics of primitive amino acid sequences, *Science* 152: 363–366.

Fuchs G. (2011) Alternative pathways of carbon dioxide fixation: Insights into the early evolution of life? *Annu. Rev. Microbiol.* 65: 631–658.

Lipmann F. (1965) Projecting backward from the present stage of evolution of biosynthesis. In *The Origin of Prebiological Systems and of their Molecular Matrices*. Fox S. W. ed., Academic Press, New York, NY, pp. 259–280.

Martin W. F. (2012) Hydrogen, metals, bifurcating electrons, and proton gradients: The early evolution of biological energy conservation, *FEBS Lett.* 586: 485–493.

Martin W. F. (2020) Older than genes: The acetyl-CoA pathway and origins. *Front. Microbiol.* 11: 817.

Martin W. F., Russell M. J. (2007) On the origin of biochemistry at an alkaline hydrothermal vent. *Philos. Trans. R. Soc. B Biol. Sci.* 362: 1887–1926.

Martin W. F., Sousa F. L. (2016) Early microbial evolution: The age of anaerobes. *Cold Spring Harbor Persp. Biol.* 8: a018127.

Martin W. F. *et al.* (2018) A physiological perspective on the origin and evolution of photosynthesis. *FEMS Microbiol. Rev.* 42: 205–231.

Schönheit P. *et al.* (2016) On the origin of heterotrophy. *Trends Microbiol.* 24: 12–25.

Weiss M. C. *et al.* (2016) The physiology and habitat of the last universal common ancestor, *Nature Microbiol.* 1: 16116.

Chapter 3 Primordial Reaction Networks and Energy Metabolism

Acetyl-CoA Pathway and Reverse Tricarboxylic Acid Cycle

Amend J. P., McCollom T. M. (2009) Energetics of biomolecule synthesis on early earth. In *Chemical Evolution II: From the Origins of Life to Modern Society*. Zaikowski, L. *et al.* eds., American Chemical Society, Washington, DC, pp. 63–94.

Balch W. E. *et al.* (1979) Methanogens: Reevaluation of a unique biological group, *Microbiol. Rev.* 43: 260–296.

Berg I. A. *et al.* (2010) Autotrophic carbon fixation in archaea, *Nat. Rev.* 8: 447–460.

Fuchs G. (2011) Alternative pathways of carbon dioxide fixation: Insights into the early evolution of life? *Annu. Rev. Microbiol.* 65: 631–658.

Fuchs G., Stupperich E. (1985) Evolution of autotrophic CO_2 fixation. In *Evolution of Prokaryotes* (FEMS Symposium No. 29). Schleifer K., Stackebrandt E. eds., Academic Press, London, pp. 235–251.

Martin W., Russell M. J. (2007) On the origin of biochemistry at an alkaline hydrothermal vent. *Philos. Trans. R. Soc. B: Biol. Sci.* 362: 1887–1926.

McCollom T. M., Amend J. P. (2005) A thermodynamic assessment of energy requirements for biomass synthesis by chemolithoautotrophic micro-organisms in oxic and anoxic environments, *Geobiology* 3: 135–144.

Morra M. J., Freeborn L. L. (1989) Catalysis of amino acid deamination in soils by pyridoxal-5'-phosphate, *Soil Biol. Biochem.* 21: 645–650.

Morton S. C. *et al.* (2003) Phosphates, phosphites, and phosphides in environmental samples, *Environ. Sci. Technol.* 37: 1169–1174.

Rühlemann M. *et al.* (1985) Detection of acetyl coenzyme A as an early CO_2 assimilation intermediate in *Methanobacterium*, *Arch. Microbiol.* 141: 399–406.

Sánchez-Andrea I. *et al.* (2020) The reductive glycine pathway allows autotrophic growth of *Desulfovibrio desulfuricans*, *Nat. Comms.* 11: 5090.

Sousa F. L., Martin W. F. (2014) Biochemical fossils of the ancient transition from geoenergetics to bioenergetics in prokaryotic one carbon compound metabolism, *Biochim. Biophys. Acta* 1837: 964–981.

Thauer R. K. *et al.* (2008) Methanogenic archaea: Ecologically relevant differences in energy conservation, *Nat. Rev. Microbiol.* 6: 579–591.

Wächtershäuser G. (1992) Groundworks for an evolutionary biochemistry: The iron-sulfur world, *Prog. Biophys. Mol. Biol.* 58: 85–201.

Wolfenden R. (2014) Primordial chemistry and enzyme evolution in a hot environment, *Cell. Mol. Life Sci.* 71: 2909–2915.

Zabinski R. F., Toney M. D. (2001) Metal ion inhibition of nonenzymatic pyridoxal phosphate catalyzed decarboxylation and transamination, *J. Am. Chem. Soc.* 123: 193–198.

Experiments Simulating the Acetyl-CoA Pathway Using Transition Metal Catalysis

Cody G. D. *et al.* (2000) Primordial carbonylated iron-sulfur compounds and the synthesis of pyruvate, *Science* 289: 1337–1340

Heinen W., Lauwers A. M. (1996) Organic Sulfur compounds resulting from the interaction of iron sulfide, hydrogen sulfide and carbon dioxide in an anaerobic aqueous environment, *Orig. Life Evol. Biosph.* 26: 131–150.

Huber C., Wächtershäuser G. (1997) Activated acetic acid by carbon fixation on (Fe,Ni)S under primordial conditions, *Science* 276: 245–247.

Huber C., Wächtershäuser G. (2003) Primordial reductive amination revisited, *Tetrahedron Lett.* 44: 1695–1697.

Hudson, R. *et al.* (2020) CO_2 reduction driven by a pH gradient, *Proc. Natl Acad. Sci. USA* 117: 22873–22879

Kitani A. *et al.* (1995) Fe(III)-ion-catalysed non-enzymatic transformation of adenosine diphosphate into adenosine triphosphate part II. Evidence of catalytic nature of Fe ions, *Bioelectrochem. Bioenerget.* 36: 47–51.

Preiner M. *et al.* (2020) A hydrogen-dependent geochemical analogue of primordial carbon and energy metabolism, *Nat. Ecol. Evol.* 4: 534–542.

Shang X. *et al.* (2023) Formation of ammonia through serpentinization in the Hadean Eon, *Sci. Bull.* 68: 1109–1112.

Sousa F. L. *et al.* (2018) Native metals, electron bifurcation, and CO_2 reduction in early biochemical evolution, *Curr. Opin. Microbiol.* 43: 77–83.

Varma S. J. *et al.* (2018) Native iron reduces CO_2 to intermediates and end-products of the acetyl-CoA pathway, *Nat. Ecol. Evol.* 2: 1019–1024.

Weber A. L. (1982) Formation of pyrophosphate on hydroxyapatite with thioesters as condensing agents, *Biosystems* 15: 183–189.

Whicher A. *et al.* (2018) Acetyl phosphate as a primordial energy currency at the origin of life, *Orig. Life Evol. Biosph.* 48: 159–179.

Experiments Simulating Part of the Reverse Tricarboxylic Acid Cycle Using Transition Metal Catalysis

Kaur H. *et al.* (2024) A prebiotic Krebs cycle analogue generates amino acids with H_2 and NH_3 over nickel, *CHEM 10.* http://doi.org/10.1016/j.chempr.2024.02.001

Muchowska K. B. *et al.* (2017) Metals promote sequences of the reverse Krebs cycle, *Nat. Ecol. Evol.* 1: 1716–1721.

Muchowska K. B. *et al.* (2019) Synthesis and breakdown of universal metabolic precursors promoted by iron, *Nature* 569: 104–107.

Muchowska K. B. *et al.* (2020) Nonenzymatic metabolic reactions and life's origins, *Chem. Rev.* 120: 7708–7744.

Organic Catalysts: The Functions of Cofactors

Broderick J. B. *et al.* (2014) Radical S-adenosylmethionine enzymes, *Chem. Rev.* 114: 4229–4317.

Kirschning A. (2021) Coenzymes and their role in the evolution of life. *Angew. Chem. Int. Ed.* 60: 6242–6269.

Martin W., Russell M. J. (2007) On the origin of biochemistry at an alkaline hydrothermal vent. *Philos. Trans. R. Soc. B: Biol. Sci.* 362: 1887–1926.

Metzler D. E., Snell E. E. (1952) Some transamination reactions involving vitamin B_6, *J. Am. Chem. Soc.* 74: 979–983.

Shima S *et al.* (2020) Structural basis of hydrogenotrophic methanogenesis, *Annu. Rev. Microbiol.* 74: 713–733.

Wolfenden R. (2011) Benchmark reaction rates, the stability of biological molecules in water, and the evolution of catalytic power in enzymes, *Annu. Rev. Biochem.* 80: 645–667.

Zabinski R. F., Toney M. D. (2001) Metal ion inhibition of nonenzymatic pyridoxal phosphate catalyzed decarboxylation and transamination, *J. Am. Chem. Soc.* 123: 193–198.

Separation, Accumulation and Autocatalysis in Prebiotic Reaction Networks

Baaske P. *et al.* (2007) Extreme accumulation of nucleotides in simulated hydrothermal pore systems, *Proc. Natl Acad. Sci. USA* 104: 9346–9351.

Duhr S., Braun D. (2006) Why molecules move along a temperature gradient, *Proc. Natl Acad. Sci. USA* 103: 19678–19682.

Hostettmann K. *et al.* (1998) *Preparative Chromatography Techniques Applications in Natural Product Isolation*, Springer, Berlin, Heidelberg.

Incropera F. P., DeWitt D. P. (1990) *Fundamentals of Heat and Mass Transfer*, John Wiley and Sons, Hoboken, NJ.

Krammer H. *et al.* (2012) Thermal, autonomous replicator made from transfer RNA, *Phys. Rev. Lett.* 108: 238104.

Kreysing M. *et al.* (2015) Heat flux across an open pore enables the continuous replication and selection of oligonucleotides towards increasing length, *Nat. Chem.* 7: 203–208.

Mast C. B., Braun D. (2010) Thermal trap for DNA replication, *Phys. Rev. Lett.* 104: 188102.

Mast C. B. *et al.* (2013) Escalation of polymerization in a thermal gradient, *Proc. Natl Acad. Sci. USA* 110: 8030–8035.

Morasch M. *et al.* (2019) Heated gas bubbles enrich, crystallize, dry, phosphorylate and encapsulate prebiotic molecules, *Nat. Chem.* 11: 779–788.

Niether D. *et al.* (2016) Accumulation of formamide in hydrothermal pores to form prebiotic nucleobases, *Proc. Natl Acad. Sci. USA* 113: 4272–4277.

Wang Z. *et al.* (2012) Thermal diffusion of nucleotides, *J. Phys. Chem. B* 116: 7463–7469.

Autocatalytic Reaction Networks

Eakin R. E. (1963) An approach to the evolution of metabolism, *Proc. Natl Acad. Sci. USA* 49: 360–366.

Eigen M. (1971) Self organization of matter and the evolution of biological macromolecules. *Naturwissenschaften* 58: 465–523.

Hordijk W., Steel M. (2004) Detecting autocatalytic, self-sustaining sets in chemical reaction systems, *J. Theor. Biol.* 227: 451–461.

Kauffman S. A. (1986) Autocatalytic sets of proteins, *J. Theor. Biol.* 119: 1–24.

Mossel E., Steel M. (2005) Random biochemical networks and the probability of self-sustaining autocatalysis, *J. Theoret. Biol.* 233: 327–336.

Sousa F. L. *et al.* (2015) Autocatalytic sets in *E. coli* metabolism, *J. Systems. Chem.* 6: 1–21.

Xavier J. C. *et al.* (2020) Autocatalytic chemical networks at the origin of metabolism, *Proc. Roy. Soc. Lond. B.* 287: 20192377.

Scaling up to the Metabolic Pathways of a Cell

Harold F. M. (1986) *The Vital Force: A Study of Bioenergetics*, W. H. Freeman, New York, NY.

Stouthamer A. H. (1973) A theoretical study on the amount of ATP required for synthesis of microbial cell material, *Antonie van Leeuwenhoek* 39: 545–565.

Wimmer J. L. E. *et al.* (2021a) Energy at origins: Favourable thermodynamics of biosynthetic reactions in the last universal common ancestor (LUCA). *Front. Microbiol.* 12: 793664.

Wimmer J. L. E. *et al.* (2021b) The autotrophic core: An ancient network of 404 reactions converts H_2, CO_2, and NH_3 into amino acids, bases, and cofactors, *Microorganisms* 9: 458.

Balance between the Flux of Carbon and Energy in the Model of a Primitive Cell

Daniel S. L. *et al.* (1990) Characterization of the H_2- and CO-dependent chemolithotrophic potentials of the acetogens *Clostridium thermoaceticum* and *Acetogenium kivui*, *J. Bacteriol.* 172: 4464–4471.

Hoehler T. M., Jørgensen B. B. (2013) Microbial life under extreme energy limitation, *Nat. Rev. Microbiol.* 11: 83–94.

Martin W. F. (2020) Older than genes: The acetyl-CoA pathway and origins. *Front. Microbiol.* 11: 817.

McCollom T. M., Amend J. P. (2005) A thermodynamic assessment of energy requirements for biomass synthesis by chemolithoautotrophic micro-organisms in oxic and anoxic environments, *Geobiology* 3: 135–144.

Stouthamer A. H. (1973) A theoretical study on the amount of ATP required for synthesis of microbial cell material, *Antonie van Leeuwenhoek* 39: 545–565.

Stouthamer A. H. (1978) Energy-yielding pathways. In *The Bacteria Vol VI: Bacterial Diversity*. Gunsalus I. C. ed., Academic Press, New York, NY, pp. 389–462.

Szenk M. *et al.* (2017) Why do fast-growing bacteria enter overflow metabolism? Testing the membrane real estate hypothesis, *Cell Syst.* 5: 95–104.

Reconstructing the Physiology of Ancient Lineages from Genomes

Colman D. R. *et al.* (2022) Deep-branching acetogens in serpentinized subsurface fluids of Oman. *Proc. Natl Acad. Sci. USA* 119: e2206845119.

Mei R. *et al.* (2023) The origin and evolution of methanogenesis and Archaea are intertwined. *Proc. Natl Acad. Sci. USA Nexus* 2: pgad023.

Schöne C. *et al.* (2022) Deconstructing *Methanosarcina acetivorans* into an acetogenic archaeon. *Proc. Natl Acad. Sci. USA* 119: e2113853119.

Weiss M. C. *et al.* (2016) The physiology and habitat of the last universal common ancestor, *Nat. Microbiol.* 1: 16116.

Xavier J. C. *et al.* (2021) The metabolic network of the last bacterial common ancestor. *Commun. Biol.* 4: 413.

Chapter 4 Prebiotic Synthesis of Monomers and Polymers

Amino Acid Synthesis

Aubrey A. D. *et al.* (2009) The role of submarine hydrothermal systems in the synthesis of amino acids, *Orig. Life Evol. Biosphere* 39: 91–108.

Barge L. M. *et al.* (2019) Redox and pH gradients drive amino acid synthesis in iron oxyhydroxide mineral systems, *Proc. Natl Acad. Sci. USA* 116: 4828–4833.

Huber C., Wächtershäuser G. (2003) Primordial reductive amination revisited, *Tetrahedron Lett.* 44: 1695–1697.

Huber C. *et al.* (2010) Synthesis of α-amino and α-hydroxy acids under volcanic conditions: Implications for the origin of life, *Tetrahedron Lett.* 51: 1069–1071.

Kaur H. *et al.* (2024) A prebiotic Krebs cycle analogue generates amino acids with H$_2$ and NH$_3$ over nickel, *CHEM 10*. http://doi.org/10.1016/j.chempr.2024.02.001

Kojo S. *et al.* (2004) Racemic D,L-asparagine causes enantiomeric excess of other coexisting racemic D,L-amino acids during recrystallization: A hypothesis accounting for the origin of L-amino acids in the biosphere, *Chem. Commun.* 19: 2146–2147.

Miller S. L. (1953) A production of amino acids under possible primitive earth conditions, *Science* 117: 528–529.

Muchowska K. B. *et al.* (2019) Synthesis and breakdown of universal metabolic precursors promoted by iron, *Nature* 569: 104–107.

Van den Heuvel, R. H. H. *et al.* (2004) Glutamate synthase: A fascinating pathway from L-glutamine to L-glutamate, *Cell. Mol. Life Sci.* 61: 669–681.

Peptide Synthesis

Bandyopadhyay A. K. *et al.* (2019) Stability of buried and networked salt-bridges (BNSB) in thermophilic proteins, *Bioinformation*, 15: 61–67.

Brack A. (1987) Selective emergence and survival of early polypeptides in water, *Orig. Life Evol. Biosp.* 17: 367–369.

Brack A. (1993) Liquid water and the origin of life, *Orig. Life Evol. Biosp.* 23: 3–10.

Chan W. C. (2000) *Fmoc Solid Phase Peptide Synthesis: A Practical Approach*, Oxford University Press, Oxford.

Danger G. *et al.* (2012) Pathways for the formation and evolution of peptides in prebiotic environments, *Chem. Soc. Rev.* 41: 5416–5429.

do Nascimento Vieira A. *et al.* (2020) The ambivalent role of water at the origins of life, *FEBS Lett.* 594: 2717–2733.

Huber C., Wächtershäuser G. (1998) Peptides by activation of amino acids with CO on (Ni, Fe) S surfaces: Implications for the origin of life, *Science* 281: 670–672.

Huber C. *et al.* (2003) A possible primordial peptide cycle, *Science* 301: 938–940.

Jackson C. *et al.* (2022) Adventures on the routes of protein evolution—in memoriam Dan Salah Tawfik (1955–2021), *J. Mol. Biol.* 434: 167462.

Kowallik K. V., Martin W. F. (2021) The origin of symbiogenesis: An annotated English translation of Mereschkowsky's 1910 paper on the theory of two plasma lineages, *Biosystems* 199: 104281.

Leman L. *et al.* (2004) Carbonyl sulfide-mediated prebiotic formation of peptides, *Science* 306: 283–286.

Leman L. J. (2006) Amino acid dependent formation of phosphate anhydrides in water mediated by carbonyl sulfide, *J. Am. Chem. Soc.* 128: 20–21.

Lewinsohn R. *et al.* (1967) Polycondensation of amino acid phosphoanhydrides, III. Polycondensation of alanyl adenylate, *Biochim. Biophys. Acta—Proteins Proteom.* 140: 24–36.

Liu R., Orgel L. E. (1997) Oxidative acylation using thioacids, *Nature* 389: 52–54.

Longo L. M., Blaber M. (2014) Prebiotic protein design supports a halophile origin of foldable proteins, *Front. Microbiol.* 4: 418.

Longo L. M. *et al.* (2013) Simplified protein design biased for prebiotic amino acids yields a foldable, halophilic protein, *Proc. Natl Acad. Sci. USA* 110: 2135–2139.

Martin R. B. (1998) Free energies and equilibria of peptide bond hydrolysis, *Biopolymers* 45: 351–353.

Oren A. (2019) Thermodynamic limits to microbial life at high salt concentrations, *Environ. Microbiol.* 13: 1908–1923.

Radzicka A., Wolfenden R. (1996) Rates of uncatalyzed peptide bond hydrolysis in neutral solution and the transition state affinities of proteases, *J. Am. Chem. Soc.* 118: 6105–6109.

Schreiner E. *et al.* (2011) Peptide synthesis in aqueous environments: The role of extreme conditions and pyrite mineral surfaces on formation and hydrolysis of peptides, *J. Am. Chem. Soc.* 133: 8216–8226.

Weinstock M. T. *et al.* (2014) Synthesis and folding of a mirror-image enzyme reveals ambidextrous chaperone activity. Gregory A. Petsko, Weill Cornell Medical College, *Proc. Natl Acad. Sci. USA* 111: 11679–11684.

Whicher A. *et al.* (2018) Acetyl phosphate as a primordial energy currency at the origin of life, *Orig. Life Evol. Biosph.* 48: 159–179.

Peptides in Autocatalytic Networks

Amend J. P., McCollom T. M. (2009) Energetics of biomolecule synthesis on early earth. In *Chemical Evolution II: From the Origins of Life to Modern Society*. Zaikowski, L. *et al.* eds., American Chemical Society, Washington, DC, pp. 63–94.

Baross J. A., Martin W. F. (2015) The ribofilm as a concept for life's origins, *Cell* 162: 13–15.

Eck R. V., Dayhoff M. O. (1966) Evolution of the structure of ferredoxin based on living relics of primitive amino acid, *Science* 152: 363–366.

Mossel E., Steel M. (2005) Random biochemical networks and the probability of self-sustaining autocatalysis, *J. Theor. Biol.* 233: 327–336.

Say R. F., Fuchs G. (2010) Fructose 1,6-bisphosphate aldolase/phosphatase may be an ancestral gluconeogenic enzyme, *Nature* 464: 1077–1081.

Sugar Synthesis

Haas M. *et al.* (2020) Mineral-mediated carbohydrate synthesis by mechanical forces in a primordial geochemical setting, *Commun. Chem.* 3: 1–6.

Heinz B. *et al.* (1979) Thermal generation of Pteridines and Flavines from Amino Acid mixtures, *Angew. Chem. Intern. Ed.* 18: 478–483.

Keller M. A. *et al.* (2014) Non-enzymatic glycolysis and pentose phosphate pathway-like reactions in a plausible Archean ocean, *Mol. Syst. Biol.* 10: 1–12.

Messner C. B. *et al.* (2017) Nonenzymatic gluconeogenesis-like formation of fructose 1,6-bisphosphate in ice, *Proc. Natl Acad. Sci. USA* 114: 7403–7407.

Muchowska K. B. *et al.* (2020) Nonenzymatic metabolic reactions and life's origins, *Chem. Rev.* 120: 7708–7744.

Weber L. A., Pizzarello S. (2006) The peptide-catalyzed stereospecific synthesis of tetroses: A possible model for prebiotic molecular evolution, *Proc. Natl Acad. Sci. USA* 103: 12713–12717.

Nucleotide Synthesis

Becker S. *et al.* (2019) Unified prebiotically plausible synthesis of pyrimidine and purine RNA ribonucleotides, *Science* 366: 76–82.

Borquez E. *et al.* (2005) An investigation of prebiotic purine synthesis from the hydrolysis of HCN polymers, *Orig. Life Evol. Biosph.* 35: 79–90.

Powner M. W. *et al.* (2009) Synthesis of activated pyrimidine ribonucleotides in prebiotically plausible conditions, *Nature* 459: 239–242.

Powner M. W. *et al.* (2010) Chemoselective multicomponent one-pot assembly of purine precursors in water, *J. Am. Che. Soc.* 132: 16677–16688.

Rimmer P. B., Shorttle O. (2019) Origin of life's building blocks in carbon- and nitrogen-rich surface hydrothermal vents, *Life* 9: 1–19.

Saladino R. *et al.* (2020) Formamide and the origin of life, *Phys. Life Rev.* 9: 84–104.

Sutherland J. D., Whitfield J. N. (1997) Prebiotic chemistry: A bioorganic perspective, *Tetrahedron* 53: 11493–11527.

Xu J. *et al.* (2020) Selective prebiotic formation of RNA pyrimidine and DNA purine nucleosides, *Nature* 62: 60–66.

Yamada Y. *et al.* (1968) Synthesis of adenine and 4,5-dicyanoimidazole from hydrogen cyanide in liquid ammonia, *J. Org. Chem.* 33: 642–647.

Nucleotide Accumulation and RNA Synthesis

Baaske P. *et al.* (2007) Extreme accumulation of nucleotides in simulated hydrothermal pore systems. *Proc. Natl Acad. Sci. USA* 104: 9346–9351.

Burcar B. T. *et al.* (2015) RNA oligomerization in laboratory analogues of alkaline hydrothermal vent systems, *Astrobiology* 15: 509–522.

Cech T. R. (2000) The ribosome is a ribozyme, *Science* 289: 878–879.

Doudna J. A., Cech T. R. (2002) The chemical repertoire of natural ribozymes, *Nature* 418: 222–228.

Ertem G., Ferris J. P. (1997) Template-directed synthesis using the heterogeneous templates produced by montmorillonite catalysis. A possible bridge between the prebiotic and RNA world, *J. Am. Chem. Soc.* 119: 7197–7201.

Ferris J. P. (2005) Mineral catalysis and prebiotic synthesis: Montmorillonite-catalyzed formation of RNA, *Elements* 1: 145–149.

Ferris J. P. *et al.* (2004) Catalysis in prebiotic chemistry: Application to the synthesis of RNA oligomers, *Adv. Space Res.* 33: 100–105.

Leontis N. B., Westhoff E. (2001) Geometric nomenclature and classification of RNA base pairs, *RNA* 7: 499–512.

Lewinsohn R. *et al.* (1967) Polycondensation of amino acid phosphoanhydrides: III. Polycondensation of alanyl adenylate, *Biochim. Biophys. Acta Proteins Proteom.* 140: 24–36.

Limbach P. A. (1994) Summary: The modified nucleosides of RNA, *Nucleic Acids Res.* 22: 2183–2196.

Mast C. B., Braun D. (2010) Thermal trap for DNA replication, *Phys. Rev. Lett.* 104: 188102.

Mast C. B. *et al.* (2013) Escalation of polymerization in a thermal gradient, *Proc. Natl Acad. Sci. USA* 110: 8030–8035.

Morasch M. *et al.* (2019) Heated gas bubbles enrich, crystallize, dry, phosphorylate and encapsulate prebiotic molecules, *Nat. Chem.* 11: 779–788.

Nikolova E. N. *et al.* (2013) A historical account of Hoogsteen base-pairs in duplex DNA, *Biopolymers* 99: 955–968.

Nir E. *et al.* (2002) Properties of isolated DNA bases, base pairs and nucleosides examined by laser spectroscopy, *Eur. Phys. J. D.* 20: 317–329.

Seefeld K. *et al.* (2007) Imino tautomers of gas-phase guanine from mid-infrared laser spectroscopy, *J. Phys. Chem. A* 111: 6217–6221.

In Vitro Selected Ribozymes

Baskerville S., Bartel D. P. (2002) A ribozyme that ligates RNA to protein, *Proc. Natl Acad. Sci. USA* 99: 9154–9159.

Eckland E. H., Bartel D. P. (1996) RNA-catalysed RNA polymerization using nucleoside triphosposphates, *Nature* 382: 373–376.

Horning D. P., Joyce G. F. (2016) Amplification of RNA by an RNA polymerase ribozyme, *Proc. Natl Acad. Sci. USA* 113: 9786–9791.

Khersonsky O., Tawfik D. S. (2010) Enzyme promiscuity: A mechanistic and evolutionary perspective, *Annu. Rev. Biochem.* 79: 471–505.

Müller F. *et al.* (2022) A prebiotically plausible scenario of an RNA–peptide world, *Nature* 605: 279–284.

Sassanfar M., Szostak J. W. (1993) An RNA motif that binds ATP, *Nature* 364: 550–553.

Silverman S. K. (2007) Artificial functional nucleic acids: Aptamers, ribozymes, and deoxyribozymes identified by in vitro selection. In *Functional Nucleic Acids for Sensing and Other Analytical Applications.* Lu Y., Li Y. eds., Springer, New York, NY, pp. 47–108.

Suga H. *et al.* (2011) The RNA origin of transfer RNA aminoacylation and beyond, *Phil. Trans. R. Soc. B* 366: 2959–2964.

Tarasow T. M. *et al.* (1997) RNA-catalysed carbon-carbon bond formation, *Nature* 389: 54–57.

Tsukiji S. *et al.* (2003) An alcohol dehydrogenase ribozyme, *Nat. Struct. Biol.* 10: 713–717.

Turk R. M. *et al.* (2010) Multiple translational products from a five-nucleotide ribozyme, *Proc. Natl Acad. Sci. USA* 107: 4585–4589.

Turk R. M. *et al.* (2011) Catalyzed and spontaneous reactions on ribozyme ribose, *J. Am. Chem. Soc.* 133: 6044–6050.

Wilson D. S., Szostak J. W. (1999) In vitro selection of functional nucleic acids, *Annu. Rev. Biochem.* 68: 611–647.

Wochner A. *et al.* (2011) Ribozyme-catalyzed transcription of an active ribozyme, *Science* 332: 209–212.

Xu J. *et al.* (2014) RNA Aminoacylation mediated by sequential action of two ribozymes and a nonactivated amino acid, *Chem BioChem* 15: 1200–1209.

Zhang B., Cech T. R. (1997) Peptide bond formation by in vitro selected ribozymes, *Nature* 390: 96–100.

Chapter 5 Template-Directed Synthesis of Polymers

General Biochemistry and Molecular Biology

Berg J. M. *et al.* (2015) *Stryer, Biochemistry*, 8th Edition, W. H. Freeman, New York, NY.

Protein Template-Directed Peptide Synthesis in Bacteria

Feinagle E. A. *et al.* (2008) Nonribosomal peptide synthetases involved in the production of medically relevant natural products, *Mol. Pharm.* 5: 191–211.

Kamtekar S. *et al.* (1993) Protein design by binary patterning of polar and nonpolar amino acids, *Science* 262: 1680–1685.

Katchalsky A., Paecht M. (1954) Phosphate anhydrides of amino acids, *J. Am. Chem. Soc.* 76: 6042–6044.

Khersonsky O., Tawfik D. S. (2010) Enzyme promiscuity: A mechanistic and evolutionary perspective, *Annu. Rev. Biochem.* 79: 471–505.

Lee D. H. *et al.* (1996) A self-replicating peptide, *Nature* 382: 525–528.

Lipmann F. (1973) Nonribosomal polypeptide synthesis on poly-enzyme templates, *Acc. Chem. Res.* 6: 361–367.

Mocibob M. *et al.* (2010) Homologs of aminoacyl-tRNA synthetases acylate carrier proteins and provide a link between ribosomal and nonribosomal peptide synthesis, *Proc. Natl Acad. Sci. USA* 107: 14585–14590.

Polycarpo C. *et al.* (2003) Activation of the pyrrolysine suppressor tRNA requires formation of a ternary complex with class I and class II lysyl-tRNA synthetases, *Mol. Cell* 12: 287–294.

Spencer D. F., Gray M. W. (2010) Ribosomal RNA genes in *Euglena gracilis* mitochondrial DNA: Fragmented genes in a seemingly fragmented genome, *Mol. Genet. Genomics* 285: 19–31.

Von Döhren, H. *et al.* (1999) The nonribosomal code, *Chem. Biol.* 6: R273–R279.

Amino Acyl tRNA Synthetases and the Code

Berg P., Ofengand E. J. (1958) An enzymatic mechanism for linking amino acids to RNA, *Proc. Natl Acad. Sci. USA* 44: 78–86.

Caravelli J. *et al.* (1994) The active site of yeast aspartyl-tRNA synthetase: Structural and functional aspects of the amino-acylation reaction, *EMBO J.* 13: 327–337.

Carter Jr. C. W., Wills P. R. (2021) The roots of genetic coding in aminoacyl-tRNA synthetase duality, *Annu. Rev. Biochem.* 90: 349–373.

Carter Jr. W. C., Wolfenden R. (2015) tRNA acceptor stem and anticodon bases form independent codes related to protein folding, *Proc. Natl Acad. Sci. USA* 112: 7489–7494.

Copley S. D. *et al.* (2005) A mechanism for the association of amino acids with their codons and the origin of the genetic code, *Proc. Natl Acad. Sci. USA* 102: 4442–4447.

Davis B. K. (1999). Evolution of the genetic code, *Prog. Biophys. Mol. Biol.* 72: 157–243.

Fujiwara S. *et al.* (1996) Unusual enzyme characteristics of aspartyl-tRNA synthetase from hyperthermophilic archaeon *Pyrococcus* sp. KOD1, *FEBS Lett.* 394: 66–70.

Ibba M., Söll D. (2000) Aminoacyl-tRNA synthesis, *Annu. Rev. Biochem.* 69: 617–650.

Jacubowski H. (2000) Amino acid selectivity in the aminoacylation of coenzyme A and RNA minihelices by aminoacyl-tRNA synthetases, *J. Biol. Chem.* 275: 34845–34848.

Jungck J. R. (1978) The genetic code as a periodic table, *J. Mol. Evol.* 11: 211–224.

Krzyzaniak A. *et al.* (1994) The non-enzymatic specific aminoacylation of transfer RNA at high pressure, *Int. J. Biol. Macromol.* 16: 153–158.

Kumar N., Marx D. (2018) Mechanistic role of nucleobases in self-cleavage catalysis of hairpin ribozyme at ambient versus high-pressure conditions, *Phys. Chem. Chem. Phys.* 20: 20886–20898.

Lee N. *et al.* (2000) Ribozyme-catalyzed tRNA aminoacylation, *Nat. Struct. Biol.* 7: 28–33.

Ramos A., Varani G. (1997) Structure of the acceptor stem of *Escherichia coli* tRNAAla: Role of the G3·U70 base pair in synthetase recognition, *Nucleic Acids Res.* 25: 2083–2090.

Rodin S. *et al.* (1996) The presence of codon-anticodon pairs in the acceptor stem of tRNAs, *Proc. Natl Acad. Sci. USA* 93: 4537–4542.

Root-Bernstein M., Root-Bernstein R. (2015) The ribosome as a missing link in the evolution of life, *J. Theor. Biol.* 367: 130–158.

Schimmel P. *et al.* (1993) An operational RNA code for amino acids and possible relationship to genetic code, *Proc. Natl Acad. Sci. USA* 90: 8763–8768.

Tamura K. (2011) Ribosome evolution: Emergence of peptide synthesis machinery, *J. Biosci.* 36: 921–928.

Wächtershäuser G. (2014) The place of RNA in the origin and early evolution of the genetic machinery, *Life* 4: 1050–1091.

Woese C. R. *et al.* (2000) Aminoacyl-tRNA synthetases, the genetic code and the evolutionary process, *Microbiol. Mol. Biol. Rev.* 64: 202–236.

Wolfenden R. V. *et al.* (1979) Water, protein folding, and the genetic code, *Science* 206: 575–-577.

Possible Origins of (Proto-)RNA Template-Directed Peptide Synthesis

Agmon I. *et al.* (2005) A symmetry at the active site of the ribosome: Structural and functional implications, *Biol. Chem.* 386: 833–844.

Bose T. *et al.* (2021) Origin of life: Chiral short RNA chains capable of non-enzymatic peptide bond formation. *Isr. J. Chem.* 61: 863–872.

Bose T. *et al.* (2022) Origin of life: Protoribosome forms peptide bonds and links RNA and protein dominated worlds, *Nucleic Acids Res.* 50: 1815–1828.

Carter Jr. W. C., Kraut J. (1974) A proposed model for interaction of polypeptides with RNA, *Proc. Natl Acad. Sci. USA* 71: 283–287.

Crick F. H. C. (1966) Codon-anticodon pairing: The wobble hypothesis, *J. Mol. Biol.* 19: 548–555.

Crick F. H. C. (1968) The origin of the genetic code, *J. Mol. Biol.* 38: 367–379.

Francis B. R. (2015) The hypothesis that the genetic code originated in coupled synthesis of proteins and the evolutionary predecessors of nucleic acids in primitive cells, *Life* 5: 467–505.

Frenkel-Pinter M. *et al.* (2020) Mutually stabilizing interactions between proto-peptides and RNA, *Nat. Commun.* 11: 1–14.

Kubyshkin V., Budisa N. (2019) The alanine world model for the development of the amino acid repertoire in protein biosynthesis, *Int. J. Mol. Sci.* 20: 5507.

Kuhn H. (1972) Self-organization of molecular systems and evolution of the genetic apparatus, *Angew. Chem. Int. Ed. Engl.* 11: 798–820.

Kumar N., Marx D. (2018) Mechanistic role of nucleobases in self-cleavage catalysis of hairpin ribozyme at ambient versus high-pressure conditions, *Phys. Chem. Chem. Phys.* 20: 20886–20898.

Orgel L. E. (1968) Evolution of the genetic apparatus, *J. Mol. Biol.* 38: 381–393.

Self-Organization

Baross J. A., Martin W. F. (2015) The ribofilm as a concept for life's origins, *Cell* 162: 13–15.

Eigen M. (1971) Self organization of matter and the evolution of biological macromolecules, *Naturwissenschaften* 58: 465–523.

Eigen M., Schuster P. (1977) A principle of natural self-organization. Part A: Emergence of the hypercycle, *Naturwissenschaften* 64: 541–565.

Kauffman S. A. (1993) *The Origins of Order*, Oxford University Press, Oxford.

Kuhn H., Kuhn C. (2003) Diversified world: Drive to life's origin?! *Angew. Chem. Int. Ed.* 42: 262–266.

Kuhn H., Waser J. (1981) Molecular self-organization and the origin of life, *Angew. Chem. Int. Ed.* 20: 500–520.

Mossel E., Steel M. (2005) Random biochemical networks and the probability of self-sustaining autocatalysis, *J. Theoret. Biol.* 233: 327–336.

RNA Recombination

Altman S. *et al.* (2005) RNase P cleaves transient structures in some riboswitches, *Proc. Natl Acad. Sci. USA* 102: 11284–11289.

Cech T. R. (1990) Self-splicing and enzymatic activity of an intervening sequence RNA from *Tetrahymena* (Nobel Lecture), *Angew. Chem. Int. Ed. Engl.* 29: 759–768.

Gilbert W. (1986) Origin of life: The RNA world, *Nature* 319: 618.

Modified Bases and the Evolution of Translation

Crécy-Lagard V., Jairoch M. (2021) Functions of bacterial tRNA modifications: from ubiquity to diversity, *Trends Microbiol.* 29: 41–53.

Grosjean H. J. *et al.* (1978) On the physical basis for ambiguity in genetic coding interactions, *Proc. Natl Acad. Sci. USA* 75: 610–614.

Lorenz C. *et al.* (2017) tRNA Modifications: Impact on structure and thermal adaptation. *Biomolecules.* 7: 35.

Müller F. *et al.* (2022) A prebiotically plausible scenario of an RNA–peptide world, *Nature* 605: 279–284.

Nainytė M. *et al.* (2020) Amino acid modified RNA bases as building blocks of an early earth RNA-peptide world, *Chem. Eur. J.* 26: 1–6.

Weiss M. C. *et al.* (2016) The physiology and habitat of the last universal common ancestor, *Nat. Microbiol.* 1: 16116.

Ancient Horizontal Gene Transfer and Genome Reduction

Brown J. R. (2003) Ancient horizontal gene transfer, *Nat. Rev. Genet.* 4: 121–132.

Doolittle W. F. (1999) Phylogenetic classification and the universal tree, *Science* 284: 2124–2128.

Koonin E. V., Martin W. (2005) On the origin of genomes and cells within inorganic compartments, *Trends Genet.* 21: 647–654.

Martin W. (1999). Mosaic bacterial chromosomes—a challenge en route to a tree of genomes, *BioEssays* 21: 99–104.

McCutcheon J., von Dohlen C. D. (2011) An interdependent metabolic patchwork in the nested symbiosis of mealybugs, *Curr. Biol.* 21: 1366–1372.

Nagies F. S. P. *et al.* (2020) A spectrum of verticality across genes, *PLoS Genet.* 16: e1009200.

Popa O., Dagan T. (2011) Trends and barriers to lateral gene transfer in prokaryotes. *Curr. Opin. Microbiol.* 14: 615–623.

von Dohlen C. D. *et al.* (2001) Mealybug β-proteobacterial endosymbionts contain γ-proteobacterial symbionts, *Nature* 412: 433–436.

Wagner A. *et al.* (2017) Mechanisms of gene flow in archaea, *Nat. Rev. Microbiol.* 15: 492–501.

Weiss M. C. *et al.* (2018) The last universal common ancestor between ancient Earth chemistry and the onset of genetics, *PLoS Genet.* 14: e1007518.

Chapter 6 Escape and Evolution as Free-Living Microbes

Autotrophic Origins

Corliss J. B. *et al.* (1979) Submarine thermal springs on the Galapagos Rift, *Science* 203: 1073–1083.

Decker K. *et al.* (1970) Energy production in anaerobic organisms, *Angew. Chem. Int. Ed. Engl.* 9: 138–158.

Eakin R. E. (1963) An approach to the evolution of metabolism, *Proc. Natl. Acad. Sci. USA* 49: 360–366.

Eck R. V., Dayhoff M. O. (1966) Evolution of the structure of ferredoxin based on living relics of primitive amino acid, *Science* 152: 363–366.

Martin W., Russell M. J. (2007) On the origin of biochemistry at an alkaline hydrothermal vent, *Philos. Trans. R. Soc. B Biol. Sci.* 362: 1887–1925.

Martin W. F. (2020) Older than genes: The acetyl-CoA pathway and origins. *Front. Microbiol.* 11: 817.

Mereschkowsky C. (1910) Theorie der zwei Plasmaarten als Grundlage der Symbiogenesis, einer neuen Lehre von der Entstehung der Organismen, *Biol. Centralbl.* 30: 278–288, 289–303, 321–347, 353–367 [English translation in: Kowallik K. V., Martin W. F. (2021) The origin of symbiogenesis: An annotated English translation of Mereschkowsky's 1910 paper on the theory of two plasma lineages, *Biosystems* 199: 104281.]

Morowitz H. J. (1993) *Beginnings of Cellular Life: Metabolism Recapitulates Biogenesis*, Yale University Press, New Haven, CT.

Sleep N. H. *et al.* (2004) H_2-rich fluids from serpentinization: Geochemical and biotic implications, *Proc. Natl Acad. Sci. USA* 101: 12818–12823.

Sleep N. H. *et al.* (2011) Serpentinite and the dawn of life. *Philos. Trans. R. Soc. B Biol. Sci.* 366: 2857–2869.

Wächtershäuser G. (1988) Before enzymes and templates: Theory of surface metabolism. *Microbiol. Rev.* 52: 452–484.

Waite J. H. *et al.* (2017) Cassini finds molecular hydrogen in the Enceladus plume: Evidence for hydrothermal processes, *Science* 356: 155–159.

The Ribosome

Agmon I. *et al.* (2005) A symmetry at the active site of the ribosome: Structural and functional implications, *Biol. Chem.* 386: 833–844.

Bose T. *et al.* (2021) Origin of life: Chiral short RNA chains capable of non-enzymatic peptide bond formation. *Isr. J. Chem.* 61: 863–872.

Bose T. *et al.* (2022) Origin of life: Protoribosome forms peptide bonds and links RNA and protein dominated worlds, *Nucleic Acids Res.* 50: 1815–1828.

Fox G. (2010) Origin and evolution of the ribosome, *Cold Spring Harb. Perspect. Biol.* 2: a003483.

Fox G. E. *et al.* (2012) An exit cavity was crucial to the polymerase activity of the early ribosome, *Astrobiology* 12: 57–60.

Illangasekare M. *et al.* (1995) Aminoacyl-RNA synthesis catalyzed by an RNA, *Science* 267: 643–647.

Maier U.-G. *et al.* (2013) Massively convergent evolution for ribosomal protein gene content in plastid and mitochondrial genomes, *Genome Biol. Evol.* 5: 2318–2329.

Petrov A. S. *et al.* (2015) History of the ribosome and the origin of translation, *Proc. Natl Acad. Sci. USA* 112: 15396–15401.

Yarus M. (2011) The meaning of a minuscule ribozyme, *Philos. Trans. R. Soc. B Biol. Sci.* 366: 2902–2909.

DNA

diCenzo G. C., Finan T. M. (2017) The divided bacterial genome: Structure, function, and evolution. *Microbiol. Mol. Biol. Rev.* 81: e00019–17.

Eigen M. (1971) Self organization of matter and the evolution of biological macromolecules, *Naturwissenschaften* 58: 465–523.

Gavette J. V. *et al.* (2016) RNA-DNA Chimeras in the Context of an RNA World Transition to an RNA/DNA World, *Angew. Chem. Int. Ed.* 55: 13204–13209.

Jia H., Gong P. (2019) A structure-function diversity survey of the RNA-dependent RNA polymerases from the positive-strand RNA viruses, *Front. Microbiol.* 10: 1945.

Kidmose R. T. (2010) Structure of the Qβ replicase, an RNA-dependent RNA polymerase consisting of viral and host proteins, *Proc. Natl Acad. Sci. USA* 107: 10884–10889.

Koonin E. V., Martin W. (2005) On the origin of genomes and cells within inorganic compartments. *Trends Genet.* 21: 647–654.

Koonin E. V. *et al.* (2020) The replication machinery of LUCA: Common origin of DNA replication and transcription, *BMC Biol.* 18: 1–8.

Krupovic M. *et al.* (2019) Origin of viruses: primordial replicators recruiting capsids from hosts, *Nat. Rev. Microbiol.* 17: 449–458.

Patzelt D. *et al.* (2016) Gene flow across genus barriers—conjugation of dinoroseobacter shibae's 191-kb killer plasmid into Phaeobacter inhibens and AHL-mediated expression of type IV secretion systems, *Front. Microbiol.* 7: 742.

Santos M. *et al.* (2004) Recombination in primeval genomes: A step forward but still a long leap from maintaining a sizeable genome, *J. Mol. Evol.* 59: 507–519.

Sjöberg B.-M. (2013) Ribonucleotide reductase. In *Encyclopedia of Metalloproteins.* Kretsinger R. H., Uversky V. N., Permyakov E. A. eds., Springer, New York, NY.

Smathers C. M., Robart A. R. (2020) Transitions between the steps of forward and reverse splicing of group IIC introns, *RNA* 26: 664–673.

Stoebe B., Kowallik K. V. (1999) Gene-cluster analysis in chloroplast genomics, *Trends Genet.* 15: 344–347.

Teichert J. S. *et al.* (2019) Direct prebiotic pathway to DNA nucleosides, *Angew. Chem.* 131: 10049–10052.

Wächtershäuser G. (1998) Towards a reconstruction of ancestral genomes by gene cluster alignment, *Syst. Appl. Microbiol.* 21: 473–477.

Wang J. *et al.* (2009) Many nonuniversal archaeal ribosomal proteins are found in conserved gene clusters, *Archaea* 2: 241–251.

Wang Y. (2021) Current view and perspectives in viroid replication, *Curr. Opin. Virol.* 47: 32–37.

Watson J. D., Crick F. H. C. (1953) Molecular structure of nucleic acids, *Nature* 171: 737–738.

Zhang K. *et al.* (2021) Duplex structure of double-stranded RNA provides stability against hydrolysis relative to single-stranded RNA, *Environ. Sci. Tech.* 55: 8045–8053.

Molecular Parasites

Branciamore S. *et al.* (2009) The origin of life: Chemical evolution of a metabolic system in a mineral honeycomb? *J. Mol. Evol.* 69: 458–469.

Bresch G. *et al.* (1980) Hypercycles, parasites and packages, *J. Theor. Biol.* 85: 399–405.

Matsumura S. *et al.* (2016) Transient compartmentalization of RNA replicators prevents extinction due to parasites, *Science* 354: 1293–1296.

Genome Size

Eck R. V., Dayhoff M. O. (1966) Evolution of the structure of ferredoxin based on living relics of primitive amino acid, *Science* 152: 363–366.

Hall D. O. *et al.* (1971) Role of ferredoxins in the origin of life and biological evolution, *Nature* 233: 136–138.

Romero M. L. R. *et al.* (2016) Functional proteins from short peptides: Dayhoff's hypothesis turns 50, *Angew. Chem. Int. Ed.* 55: 15966–15971.

Slesarev A. I. *et al.* (2002) The complete genome of hyperthermophile *Methanopyrus kandleri* AV19 and monophyly of archaeal methanogens. *Proc. Natl Acad. Sci. USA* 99: 4644–4649.

Urzyme AARS

Carter Jr. C. W., Wills P. R. (2021) The roots of genetic coding in aminoacyl-tRNA synthetase duality, *Annu. Rev. Biochem.* 90: 349–373.

Illangasekare M., Yarus M. (1999) Specific, rapid synthesis of Phe-RNA by RNA, *Proc. Natl Acad. Sci. USA* 96: 5470–5475.

Kumar R. K., Yarus M. (2001) RNA-catalyzed amino acid activation, *Biochemistry* 40: 6998–7004.

Li L. *et al.* (2013) Aminoacylating urzymes challenge the RNA world hypothesis, *J. Biol. Chem.* 288: 26856–26863.

Pham Y. *et al.* (2007) A minimal TrpRS catalytic domain supports sense/antisense ancestry of class I and II aminoacyl-tRNA synthetases, *Mol. Cell* 25: 851–862.

Rodin S. N., Ohno S. (1995) Two types of aminoacyl-tRNA synthetase could be originally encoded by complementary strands of the same nucleic acid, *Orig. Life Evol. Biosph.*, 25: 565–589.

Rodin S. N., Rodin A. S. (2006a) Origin of the genetic code: First aminoacyl-tRNA synthetases could replace isofunctional ribozymes when only the second base of codons was established, *DNA Cell Biol.* 25: 365–375.

Rodin S. N., Rodin A. S. (2006b) Partitioning of aminoacyl-tRNA synthetases in two classes could have been encoded in a strand-symmetric RNA World, *DNA Cell Biol.* 25: 617–626.

Weinreb V. *et al.* (2014) Enhanced amino acid selection in fully-evolved tryptophanyl-tRNA synthetase, relative to its Urzyme, requires domain movement sensed by the D1 switch, a remote, dynamic packing motif, *J. Biol. Chem.* 289: 4367–4376.

The First Proteins Were Small: Enzyme Promiscuity and Moonlighting

Boradia V. H. *et al.* (2014) Protein moonlighting in iron metabolism: Glyceraldehyde-3-phosphate dehydrogenase (GAPDH), *Biochem. Soc. Trans.* 42: 1796–1801.

Entelis N. *et al.* (2006) A glycolytic enzyme, enolase, is recruited as a cofactor of tRNA targeting toward mitochondria in Saccharomyces cerevisiae, *Genes Dev.* 20: 1609–1620.

Gibson M. I. *et al.* (2016) A structural phylogeny for understanding 2-oxoacid oxidoreductase function, *Curr. Opin. Struct. Biol.* 41: 54–61.

Huberts D. H. E. W., van der Klei I. J. (2010) Moonlighting proteins: An intriguing mode of multitasking, *Biochim. Biophys. Acta Mol. Cell Res.* 1803: 520–525.

Jeffery C. J. (2020) Enzymes, pseudoenzymes, and moonlighting proteins: Diversity of function in protein superfamilies, *FEBS J.* 287: 4141–4149.

Khersonsky O., Tawfik D. S. (2010) Enzyme promiscuity: A mechanistic and evolutionary perspective, *Annu. Rev. Biochem.* 79: 471–505.

Morita T., *et al.* (2004) Enolase in the RNA degradosome plays a crucial role in the rapid decay of glucose transporter mRNA in the response to phosphosugar stress in Escherichia coli, *Mol. Microbiol.* 54: 1063–1075.

Sirover M. A. (2017) *Glyceraldehyde-3-Phosphate Dehydrogenase (GAPDH). The Quintessential Moonlighting Protein in Normal Cell Function and in Human Disease*, Elsevier, Amsterdam.

Tokuriki N., Tawfik D. S. (2009) Protein dynamism and evolvability, *Science* 324: 203–207.

Amount of Catalysis by an Enzyme

Antoine L. *et al.* (2021) RNA modifications in pathogenic bacteria: Impact on host adaptation and virulence, *Genes* 12: 1125.

Bar-Even A. *et al.* (2011) The moderately efficient enzyme: Evolutionary and physicochemical trends shaping enzyme parameters, *Biochemistry* 50: 4402–4410.

Laurino P., Tawfik D. S. (2017) Spontaneous emergence of S-adenosylmethionine and the evolution of methylation, *Angew. Chem.* 129: 349–351.

Lewis C. A. Jr., Wolfenden R. (2018) Sulfonium ion condensation: The burden borne by SAM synthetase, *Biochemistry* 57: 3549–3551.

Metzler D. E., Snell E. E. (1952) Some transamination reactions involving vitamin B6, *J. Am. Chem. Soc.* 74: 979–983.

Metzler D. E., Snell E. E. (1952b) Deamination of serine: I. catalytic deamination of serine and cysteine by pyridoxal and metal salts, *J. Biol. Chem.* 198: 353–361.

Muchowska K. B. *et al.* (2020) Nonenzymatic metabolic reactions and life's origins, *Chem. Rev.* 120: 7708–7744.

Radzicka A., Wolfenden R. (1995) A proficient enzyme, *Science* 267: 90–93.

Shi L. *et al.* (2015) Chiral pyridxal-catalyzed asymmetric biomimetic transamination of α-Keto acids, *Org. Lett.* 17: 5784–5787

Wimmer J. L. E. *et al.* (2021) Energy at origins: Favourable thermodynamics of biosynthetic reactions in the last universal common ancestor (LUCA), *Front. Microbiol.* 12: 793664.

Wolfenden R. (2011) Benchmark reaction rates, the stability of biological molecules in water, and the evolution of catalytic power in enzymes, *Annu. Rev. Biochem.* 80: 645–667.

Wolfenden R., Snider M. J. (2001) The depth of chemical time and the power of enzymes as catalysts, *Acc. Chem. Res.* 34: 938–945.

Zabinski R. F., Toney M. D. (2001) Metal ion inhibition of nonenzymatic pyridoxal phosphate catalyzed decarboxylation and transamination, *J. Am. Chem. Soc.* 123: 193–198.

Lipids and the Signal Recognition Particle (SRP)

Akopian D. *et al.* (2013) Signal recognition particle: An essential protein-targeting machine, *Annu. Rev. Biochem.* 82: 693–721.

Caforio A. *et al.* (2018) Converting *Escherichia coli* into an archaebacterium with a hybrid heterochiral membrane, *Proc. Natl Acad. Sci. USA* 115: 3704–3709.

He D. *et al.* (2021) Hydrothermal synthesis of long-chain hydrocarbons up to C_{24} with $NaHCO_3$-assisted stabilizing cobalt, *Proc. Natl Acad. Sci. USA* 118: e2115059118.

Kang M. *et al.* (2016) Self-organization of nucleic acids in lipid constructs. *Curr. Opin. Coll. Interface Sci.* 26: 58–65.

Koga Y. *et al.* (1998) Did archaeal and bacterial cells arise independently from noncellular precursors? A hypothesis stating that the advent of membrane phospholipid with enantiomeric glycerophosphate backbones caused the separation of the two lines of descent, *J. Mol. Evol.* 46: 54–63.

Lange B. M. *et al.* (2000) Isoprenoid biosynthesis: The evolution of two ancient and distinct pathways across genomes, *Proc. Natl Acad. Sci. USA* 97: 13172–13177.

Martin W., Russell M. J. (2003) On the origins of cells: A hypothesis for the evolutionary transitions from abiotic geochemistry to chemoautotrophic prokaryotes, and from prokaryotes to nucleated cells, *Philos. Trans. R. Soc. B Biol. Sci.* 358: 59–85.

Scheidler C. *et al.* (2016) Unsaturated C3,5,7,9-monocarboxylic acids by aqueous, one-pot carbon fixation: Possible relevance for the origin of life, *Sci. Rep.* 6: 1–7.

Szostak J. W. (2017) The narrow road to the deep past: in search of the chemistry of the origin of life, *Angew. Chem. Int. Ed.* 56: 2–9.

Tanford C. (1980) *The Hydrophobic Effect: Formation of Micelles and Biological Membranes*, 2nd Edition, Wiley, New York, NY.

Cell Division

Adams D. W., Errington J. (2009) Bacterial cell division: Assembly, maintenance and disassembly of the Z ring, *Nat. Rev. Microbiol.* 7: 642–653.

Albers S. V., Meyer B. H. (2011) The archaeal cell envelope, *Nat. Rev. Microbiol.* 9: 414–426.

Erickson H. P. (1995) FtsZ, a prokaryotic homolog of tubulin, *Cell.* 80: 367–370.

Huecas S. *et al.* (2017) Self-organization of FtsZ Polymers in solution reveals spacer role of the disordered C-terminal tail, *Biophys. J.* 113: 1831–1844.

Koonin E. V., Mulkidjanian A. Y. (2013) Evolution of cell division: From shear mechanics to complex molecular machineries, *Cell* 152: 942–944.

Leaver M. *et al.* (2009) Life without a wall or division machine in *Bacillus subtilis*, *Nature* 457: 849–854.

Lindås A.-C., Bernander R. (2013) The cell cycle of archaea, *Nat. Rev. Microbiol.* 11: 627–638.

Makarova K. S. *et al.* (2010) Evolution of diverse cell division and vesicle formation systems in archaea, *Nat. Rev. Microbiol.* 8: 731–741.

Samson R. Y. *et al.* (2008) A role for the ESCRT system in cell division in archaea, *Science* 322: 1710–1713.

Subedi B. P. *et al.* (2022) Archaeal pseudomurein and bacterial murein cell wall biosynthesis share a common evolutionary ancestry, *FEMS Microb.* 2: xtab012.

Wang X., Lutkenhaus J. (1996) FtsZ ring: The eubacterial division apparatus conserved in archaebacteria, *Mol. Microbiol.* 21: 313–319.

Chapter 7 Harnessing Energy for Escape as Free-Living Cells

Harnessing Geochemical Ion Gradients

Borisowa M. P. *et al.* (1984) H⁺-ATPase in a planar lipid bilayer, *Biol. Membr.* 1: 187–190.

Hirata H. *et al.* (1986) Direct measurement of the electrogenicity of the H⁺-ATPase from thermophilic bacterium PS3 reconstituted in planar phospholipid bilayers, *J. Biol. Chem.* 261: 9839–9843.

Krissansen-Totton J. *et al.* (2018) Constraining the climate and ocean pH of the early Earth with a geological carbon cycle model, *Proc. Natl Acad. Sci. USA* 115: 4105–4110.

Tran Q. H., Unden G. (1998) Changes in the proton potential and the cellular energetics of *Escherichia coli* during growth by aerobic and anaerobic respiration or by fermentation, *Eur. J. Biochem.* 251: 538–543.

A Cell's Energy Budget

Basan M. *et al.* (2015) Overflow metabolism in *Escherichia coli* results from efficient proteome allocation, *Nature* 528: 99–104.

Braun S. *et al.* (2016) Size and carbon content of sub-seafloor microbial cells at Landsort Deep, Baltic Sea, *Front. Microbiol.* 7: 1375.

Buckel W., Thauer R. K. (2013) Energy conservation via electron bifurcation ferredoxin reduction and proton/Na⁺ translocating ferredoxin oxidation, *Biochim. Biophys. Acta, Bioenerg.* 1827: 94–113.

Buckel W., Thauer R. K. (2018) Flavin-based electron bifurcation, ferredoxin, flavodoxin, and anaerobic respiration with protons (Ech) or NAD+ (Rnf) as electron acceptors: A historical review, *Front. Microbiol.* 9: 401.

Harold F. M. (1986) *The Vital Force: A Study of Bioenergetics*, W. H. Freeman, New York, NY.

Hoehler T. M., Jørgensen B. B. (2013) Microbial life under extreme energy limitation, *Nat. Rev. Microbiol.* 11: 83–94.

Russell J. B. (2007) The energy spilling reactions of bacteria and other organisms, *Microb. Physiol.* 13: 1–11.

Russell J. B., Cook G. M. (1995) Energetics of bacterial growth: Balance of anabolic and catabolic reactions, *Microbiol. Rev.* 59: 48–62.

Schaechter M. *et al.* (1958) Dependency on medium and temperature of cell size and chemical composition during balanced grown of *Salmonella typhimurium*, *J. Gen. Microbiol.* 19: 592–606.

Schöne C. *et al.* (2022) Deconstructing *Methanosarcina acetivorans* into an acetogenic archaeon, *Proc. Natl Acad. Sci. USA* 119: e2113853119.

Stouthamer A. H. (1973) A theoretical study on the amount of ATP required for synthesis of microbial cell material, *Antonie van Leeuwenhoek* 39: 545–565.

Szenk M. *et al.* (2017) Why do fast-growing bacteria enter overflow metabolism? Testing the membrane real estate hypothesis, *Cell Syst.* 5: 95–104.

Vadia S., Levin P. A. (2017) Bacterial size: Can't escape the long arm of the law, *Curr. Biol.* 27: R339–R341.

Wolfe A. J. (2005) The acetate switch, *Microbiol. Mol. Biol. Rev.* 69: 12–50.

Ribosome and ATP Synthase: The Main Energy Consumers and Producers

Basu M. K. *et al.* (2009) Domain mobility in proteins: Functional and evolutionary implications, *Brief. Bioinform.* 10: 205–216.

Bennett B. D. *et al.* (2009) Absolute metabolite concentrations and implied enzyme active site occupancy in *Escherichia coli*, *Nat. Chem. Biol.* 5: 593–599.

Harold F. M. (1986) *The Vital Force: A Study of Bioenergetics*, W. H. Freeman, New York, NY.

Meyrat A., von Ballmoos C. (2019) ATP synthesis at physiological nucleotide concentrations, *Sci. Rep.* 9: 1–10.

Milo R., Philips R. (2015a) *Cell Biology by the Numbers*, Taylor and Francis (Chapter 97, Table 3. An average-sized cell has 6800 ribosomes per cell).

Milo R., Phillips R. (2015b) *Cell Biology by the Numbers*, Garland Science.

Senior A. E. *et al.* (2002) The molecular mechanism of ATP synthesis by F_1F_0-ATP synthase, *Biochim. Biophys. Acta Bioenerg.* 1553: 188–211.

Sørensen M. A. *et al.* (1989) Codon usage determines translation rate in *Escherichia coli*, *J. Mol. Biol.* 207: 365–377.

Stouthamer A. H. (1973) A theoretical study on the amount of ATP required for synthesis of microbial cell material, *Antonie van Leeuwenhoek* 39: 545–565.

Tran Q. H., Unden G. (1998) Changes in the proton potential and the cellular energetics of *Escherichia coli* during growth by aerobic and anaerobic respiration or by fermentation, *Eur. J. Biochem.* 251: 538–543.

von Meyenburg K. *et al.* (1984) Physiological and morphological effects of overproduction of membrane-bound ATP synthase in *Escherichia coli* K-12, *EMBO J.* 3: 1791–1797. This is the source for the number **E. coli has about 3200 ATP synthetases per cell.**

Protein-Based Ion Gradient Harnessing

Amend J. P., McCollom T. M. (2009) Energetics of biomolecule synthesis on early earth. In *Chemical Evolution II: From the Origins of Life to Modern Society.* Zaikowski, L. *et al.* eds., American Chemical Society, Washington, DC, pp. 63–94.

Bot C. T., Prodan C. (2010) Quantifying the membrane potential during *E. coli* growth stages, *Biophys. Chem.* 146: 133–137.

Boyd E. S. *et al.* (2020) Bioenergetic constraints on the origin of autotrophic metabolism, *Phil. Trans. Roy. Soc. A* 378: 20190151.

He D. *et al.* (2021) Hydrothermal synthesis of long-chain hydrocarbons up to C_{24} with $NaHCO_3$-assisted stabilizing cobalt, *Proc. Natl Acad. Sci. USA* 118: e2115059118.

Müller V., Hess V. (2017) The minimum biological energy quantum, *Front. Microbiol.* 8: 2019.

Schink B. (1997) Energetics of syntrophic cooperation in methanogenic degradation, *Microbiol. Mol. Biol. Rev.* 61: 262–280.

Silverstein T. P. (2014) An exploration of how the thermodynamic efficiency of bioenergetic membrane systems varies with c-subunit stoichiometry of F_1F_0 ATP synthases, *J. Bioenerg. Biomembr.* 46: 229–241.

Tran Q. H., Unden G. (1998) Changes in the proton potential and the cellular energetics of *Escherichia coli* during growth by aerobic and anaerobic respiration or by fermentation, *Eur. J. Biochem.* 251: 538–543.

Yamamoto M. (2018) Deep-sea hydrothermal fields as natural power plants, *ChemElectroChem* 5: 2162–2166.

Yamamoto M. *et al.* (2013) Generation of electricity and illumination by an environmental fuel cell in deep-sea hydrothermal vents, *Angew. Chem. Int. Ed.* 52: 10758–10761.

Zilberstein D. *et al.* (1984) *Escherichia coli* intracellular pH, membrane potential, and cell growth, *J. Bacteriol.* 158: 246–252.

Aspects of Early Energy Harnessing

Berg J. M. *et al.* (2015) *Stryer, Biochemistry*, 8th Edition, W. H. Freeman, New York, NY.

Brabender M. *et al.* (2024) Ferredoxin reduction by hydrogen with iron functions as an evolutionary precursor of flavin-based electron bifurcation, *Proc. Natl Acad. Sci. USA* 121: e2318969121

Huwiler S. G. et al. (2018) One-megadalton metalloenzyme complex in *Geobacter metallireducens* involved in benzene ring reduction beyond the biological redox window, *Proc. Natl Acad. Sci. USA* 116: 2259–2264.

Martin W. F. (2012) Hydrogen, metals, bifurcating electrons, and proton gradients: The early evolution of biological energy conservation, *FEBS Lett.* 586: 485–493.

Müller V., Hess V. (2017) The minimum biological energy quantum, *Front. Microbiol.* 8: 2019.

Pedersen B. P. *et al.* (2007) Crystal structure of the plasma membrane proton pump, *Nature* 450: 1111–1114.

Schmitt L., Tampé R. (2002) Structure and mechanism of ABC transporters, *Curr. Opin. Struct. Biol.*, 12: 754–760.

Schöcke L., Schink B. (1998) Membrane-bound proton-translocating pyrophosphatase of *Syntrophus gentianae*, a syntrophically benzoate-degrading fermenting bacterium, *Eur. J. Biochem.* 256: 589–594.

Thauer R. K. *et al.* (1977) Energy conservation in chemotrophic anaerobic bacteria, *Bacteriol Rev.* 41: 100–180.

GTP Was the Universal Energy Currency at the Origin of Translation

Basu M. K. *et al.* (2009) Domain mobility in proteins: Functional and evolutionary implications, *Brief. Bioinform.* 10: 205–216.

Bennett B. D. (2009) Absolute metabolite concentrations and implied enzyme active site occupancy in *Escherichia coli*, *Nat. Chem. Biol.* 5: 593–599.

Leipe D. D. *et al.* (2002) Classification and evolution of P-loop GTPases and related ATPases, *J. Mol. Biol.* 317: 41–72.

Mrnjavac N., Martin W. F. (2024) GTP before ATP: The energy currency at the origin of genes. *arXiv* 2403.08744.

Petrychenko V. *et al.* (2021) Structural mechanism of GTPase-powered ribosome-tRNA movement. *Nat. Commun.* 12: 1–9.

Schaupp S. *et al.* (2022) In Vitro biosynthesis of the [Fe]-hydrogenase cofactor verifies the proposed biosynthetic precursors, *Angew. Chem.* e202200994.

Wittinghofer A., Vetter I. R. (2011) Structure-function relationships of the G domain, a canonical switch motif, *Annu. Rev. Biochem.* 80: 943–971.

Wolfe A. J. (2010) Physiologically relevant small phosphodonors link metabolism to signal transduction, *Curr. Opin. Microbiol.* 13: 204–209.

Triphosphates Allow Either Thermodynamic or Kinetic Control, Diphosphates Do Not

Burke C. R., Lupták A. (2018) DNA synthesis from diphosphate substrates by DNA polymerases, *Proc. Natl Acad. Sci. USA* 115: 980–985.

Chen J. *et al.* (1990) Pyrophosphatase is essential for growth of *Escherichia coli*, *J. Bacteriol.* 172: 5686–5689.

Danchin A. *et al.* (1984) Metabolic alterations mediated by 2-ketobutyrate in *Escherichia coli* K12, *Mol. Gen. Genet.* 193: 473–478.

Friedrich M., Schink B. (1993) Hydrogen formation from glycolate driven by reversed electron transport in membrane vesicles of a syntrophic glycolate-oxidizing bacterium, *Eur. J. Biochem.* 217: 233–240.

Gottesman M. E., Mustaev A. (2019). Ribonucleoside-5′-diphosphates (NDPs) support RNA polymerase transcription, suggesting NDPs may have been substrates for primordial nucleic acid biosynthesis, *J. Biol. Chem.* 294: 11785–11792.

Kornberg A. (1962) On the metabolic significance of phosphorolytic and pyrophosphorolytic reactions. In *Horizons in Biochemistry*. Kasha H., Pullman B. eds., Academic Press, New York, NY, pp. 251–264.

Lagunas R. *et al.* (1984) Arsenic mononucleotides. Separation by high-performance liquid chromatography and identification with myokinase and adenylate deaminase, *Biochemistry* 23: 955–960.

Lipmann F. (1965) Projecting backward from the present stage of evolution of biosynthesis. In *The Origin of Prebiological Systems and of their Molecular Matrices*, S. W. Fox ed., Academic Press, New York, NY, pp. 259–280.

Stockbridge R. B., Wolfenden R. (2011) Enhancement in the rate of pyrophosphate hydrolysis by nonenzymatic catalysts and by inorganic pyrophosphatase, *J. Biol. Chem.* 286: 18538–18546.

Westheimer F. H. (1987) Why nature chose phosphates, *Science* 235: 1173–1178.

Wimmer J. L. E. *et al.* (2021) Pyrophosphate and irreversibility in evolution, or why PPi is not an energy currency and why nature chose triphosphates, *Front. Microbiol.* 12: 759359.

Flavin-Based Electron Bifurcation

Brabender M. *et al.* (2024) Ferredoxin reduction by hydrogen with iron functions as an evolutionary precursor of flavin-based electron bifurcation, *Proc. Natl Acad. Sci. USA* 121: e2318969121

Buckel W., Thauer R. K. (2013) Energy conservation via electron bifurcating ferredoxin reduction and proton/Na+ translocating ferredoxin oxidation, *Biochim. Biophys. Acta Bioenerg.* 1827: 94–113.

Buckel W., Thauer R. K. (2018) Flavin-based electron bifurcation, ferredoxin, flavodoxin, and anaerobic respiration with protons (Ech) or NAD+ (Rnf) as electron acceptors: A historical review, *Front. Microbiol.* 9: 401.

Buckel W., Thauer R. K. (2018) Flavin-based electron bifurcation, a new mechanism of biological energy coupling, *Chem. Rev.* 118: 3862–3886.

Müller V. *et al.* (2018). Electron bifurcation: A long-hidden energy-coupling mechanism, *Annu Rev. Microbiol.* 72: 331–353.

Schut G. J. *et al.* (2022) An abundant and diverse new family of electron bifurcating enzymes with a non-canonical catalytic mechanism, *Front. Microbiol.* 13: 946711.

Coupling Pumping to Acetate Synthesis in Bacteria

Kuhns M. *et al.* (2020) The Rnf complex is a Na+ coupled respiratory enzyme in a fermenting bacterium, *Thermotoga maritima*, *Commun. Biol.* 3: 1–10.

Müller V. *et al.* (2018) Electron bifurcation: A long-hidden energy-coupling mechanism, *Annu. Rev. Microbiol.* 72: 331–353.

Schlegel K. *et al.* (2012) Promiscuous archaeal ATP synthase concurrently coupled to Na+ and H+ translocation, *Proc. Natl Acad. Sci. USA* 109: 947–952.

Schuchmann K., Müller V. (2014) Autotrophy at the thermodynamic limit of life: A model for energy conservation in acetogenic bacteria, *Nat. Rev. Microbiol.* 12: 809–821.

Vitt S. *et al.* (2022) Purification and structural characterization of the Na+-translocating ferredoxin:NAD+ reductase (Rnf) complex of Clostridium tetanomorphum. *Nat. Commun.* 13: 6315.

Zhang L., Einsle O. (2022) Architecture of the NADH:ferredoxin oxidoreductase RNF that drives biological nitrogen fixation. *bioRxiv.* https://doi.org/10.1101/2022.07.08.499327

Coupling Pumping to Methane Synthesis in Archaea

Buckel W., Thauer R. K. (2013) Energy conservation via electron bifurcating ferredoxin reduction and proton/Na+ translocating ferredoxin oxidation, *Biochim. Biophys. Acta Bioenerg.* 1827: 94 – 113.

Buckel W., Thauer R. K. (2018) Flavin-based electron bifurcation, ferredoxin, flavodoxin, and anaerobic respiration with protons (Ech) or NAD+ (Rnf) as electron acceptors: A historical review, *Front. Microbiol.* 9: 401.

Gottschalk G., Thauer R. K. (2001) The Na+-translocating methyltransferase complex from methanogenic archaea, *Biochim. Biophys. Acta Bioenerg.* 1505: 28–36.

Schöne C. *et al.* (2022) Deconstructing *Methanosarcina acetivorans* into an acetogenic archaeon, *Proc. Natl Acad. Sci. USA* 119: e2113853119.

Shima S. *et al.* (2020) Structural basis of hydrogenotrophic methanogenesis, *Annu. Rev. Microbiol.* 74: 713–733.

Thauer R. K. *et al.* (2008) Methanogenic archaea: Ecologically relevant differences in energy conservation, *Nat. Rev. Microbiol.* 6: 579–591.

Wongnate T. *et al.* (2016) The radical mechanism of biological methane synthesis by methyl-coenzyme M reductase, *Science* 352: 953–958.

Energy Powers Chemical Reactions of Life

Decker K., *et al.* (1970) Energy production in anaerobic organisms, *Angew. Chem. Int. Ed.* 9: 138–158.

Dimroth P., Schink B. (1998) Energy conservation in the decarboxylation of dicarboxylic acids by fermenting bacteria, *Arch. Microbiol.* 170: 69–77.

Freund F. et al. (2002) Hydrogen in rocks. An energy source of deep microbial communities, *Astrobiology* 2: 83–92.

Morita R. Y. (2000) Is H_2 the universal energy source for long-term survival? *Microb. Ecol.* 38: 307–320.

Morowitz H. J. (1968) *Energy Flow in Biology*, Academic Press, New York, NY.

Schink B. (1997) Energetics of syntrophic cooperation in methanogenic degradation, *Microbiol. Mol. Biol. Rev.* 61: 262–280.

Stams A. J. M., Plugge C. M. (2009) Electron transfer in syntrophic communities of anaerobic bacteria and archaea, *Nat. Rev. Microbiol.* 7: 568–577.

Thauer R. K. et al. (1977) Energy conservation in chemotrophic anaerobic bacteria, *Bacteriol. Rev.* 41: 100–180.

The First Microbes Inhabited Vents in the Crust, Modern Systems Provide Models

Arndt N. T., Nisbet E. G. (2012) Processes on the young earth and the habitats of early life, *Annu. Rev. Earth Planet Sci.* 40: 521–549.

Baross J. A., Martin W. F. (2015) The ribofilm as a concept for life's origins, *Cell* 162: 13–15.

Boyd E. S. et al. (2020) Bioenergetic constraints on the origin of autotrophic metabolism, *Phil. Trans. Roy. Soc. A* 378: 20190151.

Brazelton W. J. et al. (2013) Bacterial communities associated with subsurface geochemical processes in continental serpentinite springs, *Appl. Environ. Microbiol.* 79: 3906–3916.

Colman D. R. et al. (2022) Deep-branching acetogens in serpentinized subsurface fluids of Oman, *Proc. Natl Acad. Sci. USA* 119: e2206845119.

Daniels L. et al. (1987) Bacterial methanogenesis and growth from CO_2 with elemental iron as the sole source of electrons, *Science* 237: 509–511.

Enning D., Garrelfs J. (2014) Corrosion of iron by sulfate-reducing bacteria: New views of an old problem, *Appl. Environ. Microbiol.* 80: 1226–1236.

Fones E. M. et al. (2021) Diversification of methanogens into hyperalkaline serpentinizing environments through adaptations to minimize oxidant limitation, *ISME J.* 15: 1121–1135.

Frouin E. et al. (2018) Diversity of rare and abundant prokaryotic phylotypes in the Prony hydrothermal field and comparison with other serpentinite-hosted ecosystems, *Front. Microbiol.* 9: 102.

Jones W. J. et al. (1989) Comparison of thermophilic methanogens from submarine hydrothermal vents, *Arch. Microbiol.* 151: 314–318.

Kelley D. S. et al. (2003) Volcanoes, fluids, and life at mid-ocean ridge spreading centers, *Annu. Rev. Earth. Planet. Sci.* 30: 385–491.

Lang S. Q., Brazelton W. J. (2020) Habitability of the marine serpentinite subsurface: A case study of the Lost City hydrothermal field, *Philos. Trans. R. Soc. A* 378: 20180429.

McCollom T. M., Seewald J. S. (2013) Serpentinites, hydrogen, and life, *Elements* 9: 129–134.

Merino N. et al. (2020) Single-cell genomics of novel actinobacteria with the Wood–Ljungdahl pathway discovered in a serpentinizing system, *Front. Microbiol.* 11: 1031.

Nealson K. H. et al. (2005) Hydrogen-driven subsurface lithoautotrophic microbial ecosystems (SLiMEs): Do they exist and why should we care? *Trends Microbiol.* 13: 405–410.

Nisbet E. G., Sleep N. H. (2001) The habitat and nature of early life, *Nature* 409: 1083–1091.

Nobu M. K. et al. (2023) Unique H_2-utilizing lithotrophy in serpentinite-hosted systems, *ISME J.* 17: 95–104.

Postec A. et al. (2015) Microbial diversity in a submarine carbonate edifice from the serpentinizing hydrothermal system of the Prony Bay (New Caledonia) over a 6-year period, *Front. Microbiol.* 6: 857.

Schuchmann K., Müller V. (2013) Direct and reversible hydrogenation of CO_2 to formate by a bacterial carbon dioxide reductase, *Science* 342: 1382–1385.

Sleep N. H. et al. (2011) Serpentinite and the dawn of life, *Philos. Trans. R. Soc. B* 366: 2857–2869.

Suzuki S. et al. (2013) Microbial diversity in The Cedars, an ultrabasic, ultrareducing, and low salinity serpentinizing ecosystem, *Proc. Natl Acad. Sci. USA* 110: 15336–15341.

Suzuki S. et al. (2018) Genomic and *in-situ* transcriptomic characterization of the candidate phylum NPL-UPL2 from highly alkaline highly reducing serpentinized groundwater, *Front. Microbiol.* 9: 3141.

Takami H. et al. (2012) A deeply branching thermophilic bacterium with an ancient acetyl-CoA pathway dominates a subsurface ecosystem, *PLoS One* 7: e30559.

Topçuoğlu B. D. et al. (2016) Hydrogen limitation and syntrophic growth among natural assemblages of thermophilic methanogens at deep-sea hydrothermal vents, *Front. Microbiol.* 7: 1240.

Trutschel L. R. et al. (2022) Investigation of microbial metabolisms in an extremely high pH marine-like terrestrial serpentinizing system: Ney Springs, *Sci. Total Environ.* 836: 155492.

Ver Eecke H. C. et al. (2012) Hydrogen-limited growth of hyperthermophilic methanogens at deep-sea hydrothermal vents, *Proc. Natl Acad. Sci. USA* 109: 13674–13679.

Zgonnik V. (2020) The occurrence and geoscience of natural hydrogen: A comprehensive review, *Earth Sci. Rev.* 203: 103140.

Colonizing the Crust

Bar-On Y. M. et al. (2018) The biomass distribution on earth, *Proc. Natl Acad. Sci. USA* 115: 6506–6511.

Bradley J. A. et al. (2019) Survival of the fewest: Microbial dormancy and maintenance in marine sediments through deep time, *Geobiology* 17: 43–59.

Chivian D. et al. (2008) Environmental genomics reveals a single-species ecosystem deep within Earth, *Science* 322: 275–278.

Flemming H. C., Wuertz S. (2019) Bacteria and archaea on Earth and their abundance in biofilms, *Nat. Rev Microbiol.* 17: 247–260.

Heberling C. et al. (2010) Extent of the microbial biosphere in the oceanic crust, *Geochem. Geophys. Geosyst.* 11: Q08003.

Lin L. et al. (2006) Long-term sustainability of a high-energy, low-diversity crustal biome, *Science* 314: 479–482.

Magnabosco C., et al. (2018) The biomass and biodiversity of the continental subsurface, *Nat. Geosci.* 11: 707–717.

Orsi W. D. et al. (2020) Physiological limits to life in anoxic subseafloor sediment, *FEMS Microbiol. Rev.* 44: 219–231.

Parkes R. J. et al. (2014) A review of prokaryotic populations and processes in sub-seafloor sediments, including biosphere:geosphere interactions, *Mar. Geol.* 352: 409–425.

Schrenk M. O., Brazelton W. J., Lang S. Q. (2013) Serpentinization, carbon and deep life, *Rev. Mineral. Geochem.* 75: 575–606.

Sleep N. H. (2018) Geological and geochemical constraints on the origin and evolution of life, *Astrobiology* 18: 1199–1219.

Sleep N. H. et al. (2004) H_2-rich fluids from serpentinization: Geochemical and biotic implications, *Proc. Natl Acad. Sci. USA* 101: 12818–12823.

Whitman W. B. et al. (1998) Prokaryotes: The unseen majority, *Proc. Natl Acad. Sci. USA* 95: 6578–6583.

Yang Y. et al. (2021) The upper temperature limit of life under high hydrostatic pressure in the deep biosphere. *Deep Sea Res. I* 176: 103604.

Ancient Nitrogen and Phosphate

Boyd E. S., Peters J. W. (2013) New insights into the evolutionary history of biological nitrogen fixation, *Front. Microbiol.* 4: 1–12.

Brady M. P. *et al.* (2022) Marine phosphate availability and the chemical origins of life on Earth, *Nat. Commun.* 13: 1–9.

Brandes J. A. *et al.* (1998) Abiotic nitrogen reduction on the early Earth, *Nature* 395: 365–367.

Schink B. *et al.* (2002) *Desulfotignum phosphitoxidans* sp. nov., a new marine sulfate reducer that oxidizes phosphite to phosphate, *Arch. Microbiol.* 177: 381–391.

Shang X. *et al.* (2023) Formation of ammonia through serpentinization in the Hadean Eon, *Sci. Bull.* 68: 1109–1112.

Stüeken E. E. *et al.* (2016) The evolution of Earth's biogeochemical nitrogen cycle, *Earth Sci. Rev.* 160: 220–239.

Fossil Vent Communities

Arndt N. T., Nisbet E. G. (2012) Processes on the young Earth and the habitats of early life, *Annu. Rev. Earth Planet. Sci.* 40: 521–549.

Dodd M. S. *et al.* (2017) Evidence for early life in Earth's oldest hydrothermal vent precipitates, *Nature* 543: 60–64.

Westall F. *et al.* (2015) Archean (3.33 Ga) microbe-sediment systems were diverse and flourished in a hydrothermal context, *Geology* 43: 615–618.

Fermentations Followed Autotrophy

Braun S. *et al.* (2016) Size and carbon content of sub-seafloor microbial cells at Landsort Deep, Baltic Sea, *Front. Microbiol.* 7: 1375.

Buckel W., Thauer R. K. (2013) Energy conservation via electron bifurcating ferredoxin reduction and proton/Na$^+$ translocating ferredoxin oxidation, *Biochim. Biophys. Acta Bioenerg.* 1827: 94–113.

Buckel W., Thauer R. K. (2018) Flavin-based electron bifurcation, ferredoxin, flavodoxin, and anaerobic respiration with protons (Ech) or NAD$^+$ (Rnf) as electron acceptors: A historical review, *Front. Microbiol.* 9: 401.

Chandra T. S., Shethna Y. I. (1977) Oxalate, formate, formamide, and methanol metabolism in *Thiobacillus novellus*, *J. Bacteriol.* 131: 389–398.

Friedrich M., Schink B. (1993) Hydrogen formation from glycolate driven by reversed electron transport in membrane vesicles of a syntrophic glycolate-oxidizing bacterium, *Eur. J. Biochem.* 217: 233–240.

Orell A. *et al.* (2013) Archael biofilms: The great unexplored, *Annu. Rev. Microbiol.* 67: 337–354.

Orsi W. D. *et al.* (2020) Physiological limits to life in anoxic subseafloor sediment, *FEMS Microbiol. Rev.* 44: 219–231.

Sato T., Atomi H. (2011) Novel metabolic pathways in *Archaea*, *Curr. Opin. Microbiol.* 14: 307–314.

Sato T. *et al.* (2007) Archaeal type III RuBisCOs function in a pathway for AMP metabolism, *Science* 315: 1003–1006.

Schönheit P. *et al.* (2016) On the origin of heterotrophy, *Trends Microbiol.* 24: 12–25.

Wolfe A. J. (2005) The acetate switch, *Microbiol. Mol. Biol. Rev.* 69: 12–50.

Wu C.-H. *et al.* (2018) Characterization of membrane-bound sulfane reductase: A missing link in the evolution of modern day respiratory complexes, *J. Biol. Chem.* 293: 16687–16696.

Pumping During Fermentations

Boyd E. S. *et al.* (2014) Hydrogen metabolism and the evolution of biological respiration, *Microbe* 9: 361–367.

Marreiros B. C. *et al.* (2013) A missing link between complex I and group 4 membrane-bound [NiFe] hydrogenases, *Biochim. Biophys. Acta Bioenerg.* 1827: 198–209.

Sapra R. *et al.* (2003) A simple energy-conserving system: Proton reduction coupled to proton translocation, *Proc. Natl Acad. Sci. USA* 100: 7545–7550

Schoelmerich M. C., Müller V. (2019) Energy conservation by a hydrogenase-dependent chemiosmotic mechanism in an ancient metabolic pathway, *Proc. Natl Acad. Sci. USA* 116: 6329–6334.

Schoelmerich M. C., Müller V. (2020) Energy-converting hydrogenases: The link between H$_2$ metabolism and energy conservation, *Cell. Mol. Life Sci.* 77: 1461–1481.

Yu H. *et al.* (2018) Structure of an ancient respiratory system, *Cell* 173: 1636–1649.

Tetrapyrroles, Heme, Cytochromes, and Quinones

Bryant D. A. *et al.* (2020) Biosynthesis of the modified tetrapyrroles—the pigments of life, *J. Biol. Chem.* 295: 6888–6925.

Dailey H. A. *et al.* (2017) Prokaryotic heme biosynthesis: Multiple pathways to a common essential product, *Microbiol. Mol. Biol. Rev.* 81: e00048–16.

Nelson-Sathi S. *et al.* (2012) Acquisition of 1,000 eubacterial genes physiologically transformed a methanogen at the origin of Haloarchaea, *Proc. Natl Acad. Sci. USA* 109: 20537–20542.

Pleyer H. L. *et al.* (2018) A possible prebiotic ancestry of porphyrin-type protein cofactors, *Orig. Life Evol. Biosph.* 48: 347–371.

Rabus R. *et al.* (2015) A post-genomic view of the ecophysiology, catabolism and biotechnological relevance of sulphate-reducing prokaryotes, *Adv. Microb. Physiol.* 66: 55–321.

Simionescu C. I. *et al.* (1978) Porphyrin-like compounds genesis under simulated abiotic conditions, *Orig. Life* 9: 103–114.

Storbeck S. *et al.* (2010) A novel pathway for the biosynthesis of heme in *Archaea*: Genome-based bioinformatic predictions and experimental evidence, *Archaea* 2010: 175050.

Susanti D., Mukhopadhyay B. (2012) An Intertwined evolutionary history of methanogenic archaea and sulfate reduction, *PLoS ONE* 7: e45313.

Svetlitchnaia T. *et al.* (2006) Structural insights into methyltransfer reactions of a corrinoid iron-sulfur protein involved in acetyl-CoA synthesis, *Proc. Natl Acad. Sci. USA* 103: 14331–14336.

Thauer R. K. *et al.* (2008) Methanogenic archaea: Ecologically relevant differences in energy conservation, *Nat. Rev. Microbiol.* 6: 579–591.

Using Cytochromes and Quinones to Access Fe$_3$O$_4$ as an Electron Acceptor

Edwards M. J. *et al.* (2020) Role of multiheme cytochromes involved in extracellular anaerobic respiration in bacteria, *Protein Sci.* 29: 830–842.

Gorby Y. A. *et al.* (2006) Electrically conductive bacterial nanowires produced by *Shewanella oneidensis* strain MR-1 and other microorganisms, *Proc. Natl Acad. Sci. USA* 103: 11358–11363.

Kostka J. E., Nealson K. H. (1995) Dissolution and reduction of magnetite by bacteria, *Environ. Sci. Technol.* 29: 2535–2540.

Lovley D. R., Phillips E. J. (1988) Novel mode of microbial energy metabolism: Organic carbon oxidation coupled to dissimilatory reduction of iron or manganese, *Appl. Environ. Microbiol.* 54: 1472–1480.

Lovley D. R., Walker D. J. F. (2019) *Geobacter* protein nanowires, *Front. Microbiol.* 10: 2078.

Malvankar N. S., Lovley D. R. (2012) Microbial nanowires: A new paradigm for biological electron transfer and bioelectronic, *ChemSusChem* 5: 1039–1046.

Myers C. R., Nealson K. H. (1988) Bacterial manganese reduction and growth with manganese oxide as the sole electron acceptor, *Science* 240: 1319–1321.

Shi L. *et al.* (2016) Extracellular electron transfer mechanisms between microorganisms and minerals, *Nat. Rev. Microbiol.* 14: 651–662.

Stams A. J. M. *et al.* (2006) Exocellular electron transfer in anaerobic microbial communities, *Environm. Microbiol.* 8: 371–382.

Straub K. L. *et al.* (2001) Iron metabolism in anoxic environments at near neutral pH, *FEMS Microbiol. Ecol.* 34: 181–186.

Von Wonderen J. H. *et al.* (2021) Nanosecond heme-to-heme electron transfer rates in a multiheme cytochrome nanowire reported by a spectrally unique His/Met-ligated heme, *Proc. Natl Acad. Sci. USA* 118: 1–9.

Wang F. *et al.* (2019) Structure of microbial nanowires reveals stacked hemes that transport electrons over micrometers, *Cell* 177: 361–369.

Modularity Promotes Diversity of Bioenergetic Systems

Bergdoll L. *et al.* (2016) From low- to high-potential bioenergetic chains: Thermodynamic constraints of Q-cycle function, *Biochim. Biophys. Acta Bioenerg.* 1857: 1569–1579.

Calisto F., Pereira M. M. (2021) Modularity of membrane-bound charge-translocating protein complexes, *Biochem. Soc. Trans.* 49: 2669–2685.

Lengeler J. W. *et al.* eds. (1999) *Biology of the Prokaryotes*, Stuttgart, New York, NY.

Madigan M. T. *et al.* (2019) *Brock, Biology of Microorganisms*, Benjamin Cummings, San Francisco, CA.

Mitchell P. (1975) The protonmotive Q cycle: A general formulation, *FEBS Lett.* 59: 137–139.

Out of the Darkness and into the Light

Allwood A. C. *et al.* (2007) 3.43 billion-year-old stromatolite reef from the Pilbara Craton of Western Australia: Ecosystem-scale insights to early life on Earth, *Precambrian Res.* 158: 198–227.

Beatty J. T. *et al.* (2005) An obligately photosynthetic bacterial anaerobe from a deep-sea hydrothermal vent, *Proc. Natl Acad. Sci. USA* 102: 9306–9310.

Martin W. F. *et al.* (2018) A physiological perspective on the origin and evolution of photosynthesis, *FEMS Microbiol. Rev.* 42: 201–231.

Moalic Y. *et al.* (2021) The piezo-hyperthermophilic Archaeon *Thermococcus piezophilus* regulates its energy efficiency system to cope with large hydrostatic pressure variations, *Front. Microbiol.* 12: 730231.

Nisbet E. G. *et al.* (1995) Origins of photosynthesis, *Nature* 373: 479–480.

Van Dover C. L. *et al.* (1989) A novel eye in eyeless shrimp from hydrothermal vents of the Mid-Atlantic Ridge, *Nature* 337: 458–460.

Van Dover C. L. *et al.* (1996) Light at deep-sea hydrothermal vents, *Geophys. Res. Lett.* 23: 2049–2052.

Solutions to Problems

Chapter 1

1. Solids came from the accretion of solid material from early star explosions and supernovae. Water came from meteorites from the outer part of the solar system and from water inclusions in the Earth's solids.

2. The moon-forming impact transferred rock-bound water, carbon, nitrogen, and sulfur compounds (that came to Earth during accretion) as gases into the atmosphere, mainly water vapor, carbon dioxide, nitrogen, and sulfur gases.

3. 2.435×10^9 years.

4. Electrostatic interactions, hydrogen bonds, dipole and induced dipole interactions, hydrophobic interactions.

5. The hydroxy (–OH), carbonyl (–C=O), carboxyl (–COOH), amine (–NH$_2$), sulfhydryl or thiol (–SH), phosphate (–O–PO$_3^{2-}$) and methyl (–CH$_3$) group.

6. $H_2PO_4^-/H_3PO_4$=72.4, $HPO_4^{2-}/H_2PO_4^- = 6.3 \times 10^{-4}$, $PO_4^{3-}/HPO_4^{2-} = 4.3 \times 10^{-9}$.

7. Water activity is 0.316 Mole fraction of water is 0.316 and salt 0.684.

8. The free reaction enthalpy is −7.4 kJ/mol.

9. The free reaction enthalpy of phosphate transfer is -30.6 kJ/mol and the difference in hydrolysis free enthalpies from Table 1.2 is −31.4 kJ/mol.

10. In physisorption the adsorbate binds only noncovalently to surface atoms whereas in chemisorption the adsorbate binds covalently to surface atoms (formation of a chemical bond).

11. Transition metals generally catalyze well due to incompletely filled d-orbitals for donation and acceptance of electrons from other molecules.

12. 115 ms.

13. Organic material from meteorites gets diluted in the oceans after meteorite impact and is chemically too diverse and reduced to be food or metabolic precursor for chemical evolution.

14. Temperature 90°C and higher, pH 9–11, Lost City's depth below sea level is 750–900 m corresponding to approximately 75–90 bar total pressure, concentrations of H$_2$ up to 15 mM.

Chapter 2

1. $[CO_2]_{aq}$=0.17 mol/l, $[H_2CO_3]$=2.9 $\times$ 10^{-4} mol/l, $[HCO_3]-$ =2.7 $\times$ 10^{-4} mol/l, pH=3.57

2. Reducing agents are Fe^{2+} ions in olivine minerals which react with hot hydrothermal water to hydrogen, iron oxides (magnetite), brucite, and serpentine.

3. Serpentinization consumes water in the pores of serpentinizing rocks so that the concentration of salts and dissolved minerals increases, and thus water activity decreases. Then, new seawater infiltrates due to the concentration difference, and water activity increases again.

4. Transition metals like iron bind H$_2$, dissociate it, and form covalent bonds between the dissociation products and surface atoms of the catalyst (chemisorption, here FeH). The chemically bound H atoms diffuse on the surface and react with other adsorbed species.

5. The acetyl-CoA pathway produces a thioactivated acetyl group and pyruvate from H$_2$ + CO$_2$.

6. $E_0 = -690$ mV.

7. $\Delta G_R^0 = -113.01$ kJ/mol; $E_0' = -0.59$ V ($E_0' = -0.6$ V in Table 2.2)

8. CH$_3$–C=O–COOH + 2H$^+$ + 2e$^-$ → CH$_3$–CH(OH)–COOH, −316.94 − (−352.40) = +35.46 kJ/mol, $E_0' = -0.184$ V ($E_0' = -0.19$ V in Table A1.1)

9. A) In the presence of iron, H$_2$ → 2e$^-$ + 2H$^+$. In alkaline solution, OH$^-$ ions consume the protons produced by the reaction, pulling the equilibrium to the right.
   B) Three reactions, the latter two occur twice per H$_2$ oxidized.
   1. H$_2$ + 2Fe(0) → 2 Fe–H
   2. Fe–H + OH$^-$ → [Fe]$^-$ + H$_2$O
   3. [Fe]$^-$ + ferredoxin → Fe(0) + [ferredoxin]$^-$
   The iron atoms do not leave the solid phase during the reaction, Fe(0) serves as a catalyst.

10. ΔG=11.41 kJ/mol, ΔG=15.40 kJ/mol, ΔG=17.67 kJ/mol; $\Delta G(CO_2)$=10,06 kJ/mol.

11. Ferredoxin (FeS clusters), Flavin, Quinone, Flavoquinone, Nicotinamide, Heme.

12. Large amounts of H$_2$ and NH$_3$/NH$_4^+$ were produced through serpentinization in a high-pressure reactor loaded with natural peridotite powder, H$_2$O, and 5 MPa of N2 at 300°C, corresponding to a total pressure of 20 MPa (200 bar). Addition of carbon dioxide led to somewhat more H$_2$ and ten times more ammonia, which according to the authors is due to the acceleration of serpentinization by CO$_2$. From their data the authors calculate large amounts of ammonia and methane in the atmosphere of the Hadean Eon and take this as support for the experimental conditions of the Miller–Urey experiment. A possible critique would be: The highest concentration of ammonia and H$_2$ will be in peridotite and olivine inclusions at the place of synthesis of these compounds by serpentinization, not diluted in the ocean or atmosphere. Hence the chemical reactions important for the formation of early life will occur there.

13. Metallic sulfide minerals in a black-smoker chimney wall show high electrical conductivity (resistance < 10 Ohm/cm) and the redox potential difference between vent effluent and seawater can generate fuel-type electricity. The anodic electron current generated by the catalytic oxidation of H$_2$S and H$_2$ at the inner chimney wall of the hydrothermal fluid is transmitted over long distances (10 cm

and more!) through the chimney wall to oxidative seawater or to chimney pores with a suitable chemical composition for reduction acting as cathode. Thermo-electricity probably also contributes. The electric current flowing through the conductive minerals of the chimney wall may have induced redox chemistry anywhere in the wall.

A possible critique would be: The authors interpret their findings as a novel concept for the origin of life however there was no or very little O_2 in seawater in the Hadean eon which could be reduced to H_2O as an assumed chemical process at the cathode. Their cathodic current contributes most to the total current but is probably much smaller. A more realistic concept would be the oxidation of very low potential H_2 from alkaline hydrothermal effluent generating the anodic current and reduction of protons in slightly acidic seawater (or reduction of Fe^{3+} in seawater depending on the habitat) generating the cathodic current.

Chapter 3

1. The acetyl-CoA pathway (converts $H_2 + CO_2$ into acetyl-CoA and pyruvate, the rTCA cycle (converts pyruvate into different tricarboxylic acids), Calvin Cycle (converts H_2 carriers + CO_2 into glucose).
2. Carbon dioxide (O=C=O), Formate (O=CH-OH), Formyl (O=CH–), Methenyl (–CH=), Methylene (–CH_2–), Methyl(–CH_3).
3. Dissociation of the reactants to transition metal surface bound H, CO, C, CH, CH_2, CH_3, CH_3C=O, SH^-, PO_3^{2-} and the reaction of the acetyl group to the thioester CH_3COSCH_3. The thioacid CH_3COSH reacts with inorganic phosphate to acetylphosphate CH_3COO-PO_3^{2-}. Acetylphosphate reacts with ADP to ATP.
4. H-CO-COO^- + CH_3CO-COO^- → $[H$-$C^{\delta+}O^{\delta-}$-COO^- + C^-H_2CO-COO^- + $H^+]$ → ^-OOC– CHOH-CH_2-CO-COO^-. Intermediate nucleophilic addition of carbanion C^- to positively charged carbonyl carbon C^+ followed by intramolecular hydrogen rearrangement.
5. Amination of the alpha-ketoacid with activated ammonia to an imine (–C=NH–) and reduction of the imine to the amino acid with electrons from H_2 or from Fe^0.
6. From aspartate, glycine, glutamine, CO_2, and intermediates of the acetyl-Co A pathway.
7. In hot alkaline hydrothermal vent effluent, the redox potential of H_2 is sufficiently negative to synthesize organic compounds, but not in slightly acidic Hadean seawater at 25°C or in modern seawater.
8. Nicotinamide adenine dinucleotide (hydride donor), corrins (for example, methyl transfer), S-adenosyl methionine (for example, radical generation), pyridoxal phosphate (for example, transamination), flavin adenine dinucleotide (for example, transducer of one electron to two electron transfers), thiamine pyrophosphate (for example,

transfer of C2 units), coenzyme A (forms thioesters as the activated forms of organic acids), Biotin (Carboxylation), tetrahydrofolate (C1 carrier in bacteria), tetrahydromethanopterin (C1 carrier in archaea).
9. Bulk flow by momentum, convection, thermodiffusion, concentration diffusion. Balance of gradients (momentum gradient, mass gradient, temperature gradient, concentration gradient).
10. Temperature difference between cold and hot sides of the pores, length-to-width ratios of the pores, initial molecular concentration, magnitude of the Soret coefficient (ratio of thermodiffusion to mass diffusion coefficient)
11. A network is called a reflexively autocatalytic food generated network, when the reaction products increase the rates in the network by feedback sufficiently to make it self-sustaining (given enough input of energy and food molecules).
12. Processes such as cell maintenance, futile cycling, ATP spilling, and ATP uncoupling consume ATP or diminish ATP synthesis efficiency.
13. The reconstruction shows that LUCA lived from the gases H_2, CO_2, N_2, and H_2S, used the Acetyl-CoA pathway to fix CO_2, possessed nitrogenase for N_2 assimilation and subunits of the rotor-stator ATP synthetase for harnessing naturally preexisting ion gradients. LUCA lacked proteins involved in generating ion gradients via chemiosmotic coupling. LUCA used an H^+/Na^+ antiporter because, for example, protonated acetate can easily traverse the membrane and cause a short circuit whereas acetate cannot bind Na^+ strong enough to do that.

Chapter 4

1. Condensation of phosphorylated amino acids by nucleophilic substitution. The amino group must be protected here to exclude its reaction with the activator, say ATP. Condensation of thiolated amino acids by nucleophilic substitution; again, the amino group must be protected. Condensation of COS-activated amino acids by N-carboxyanhydride formation; cyclic intermediate, the amino group must not be protected. Condensation of hydrogencarbonate activated amino acid esters by N-carboxy-anhydride formation; cyclic intermediate, the amino group must not be protected.
2. The separated charges of the NH_3^+ and phosphate group are too far apart in peptides to neutralize each other such that O^- from the phosphate group repels the attacking OH^-.
3. Alternating homochiral leucine-lysine polypeptides fold as random coils in pure water because the positive charges of the lysine amino groups repel each other and hinder the formation of an ordered structure. In salt the charges are sufficiently screened by counterions to allow the formation of thermostable and hydrolysis-stable bilayers of β-sheets.

4. There are several possible ways how homochirality of peptides could have been achieved, one is during peptide synthesis. A tRNA-bound L-amino acid is held in place in the modern peptidyl transferase site by interactions with the terminal D-Rib of the tRNA attached to the growing peptide chain, with a nucleobase and with the C atom of the ester bond of the peptide chain with tRNA, whereas a D-amino acid cannot perform these stabilizing interactions. The amino group of the L-amino acid attacks the C-terminal carbonyl carbon of the tRNA-bound growing peptide chain to extend the chain, whereas the D-configuration leaves the amino group in the wrong place for the peptidyl transfer reaction.

5. Inorganic catalysts (solid phase), inorganic catalysts (aqueous phase), small organic catalysts (organocatalysis), larger organic catalysts (cofactors), and enzymes (soft matter).

6. H_2 as a universal inorganic energy carrier and the catalysts for CO_2 and N_2 fixation to metabolic precursors is constantly produced in serpentinizing olivine micropores. The periodically low water activity there can enhance the polymerization of activated amino acids to thermostable and halophile polypeptides.

7. A variety of primordial polypeptides containing cysteines or side chains picking up environmental sulfides could have bound solved iron ions to form iron-sulfur clusters. The pickup was sustainable if the structures of the peptides were specified by geochemical conditions.

8. Sugar in modern autotrophs is synthesized by gluconeogenesis which starts from pyruvate. The nonenzymatic aldol condensation between G3P and DHAP yields phosphorylated fructose. Pyruvate may react nonenzymatically with glyoxylate at reducing conditions to different forms of the sugar ribose.

9. Biochemistry in cells: Amino acids, ATP, CO_2, NH_3, and formylphosphate-based reactions to purines. Laboratory: Cyanide and formamide-based generation of nucleobases and nucleosides. Cyclization of carbamoylphosphate (from hydrogencarbonate, NH_3, and ATP) with aspartate to orotate as a precursor of uridine- and cytidinetriphosphate similar to the pathway in cells.

10. Short primer RNA strands and mononucleotides can diffuse into swellable clay minerals like montmorillonite, concentrate there, and condense to longer strands in the clay interlayers when the mononucleotides are activated at the 5′-end with a good leaving group like imidazole or pyrophosphate.

11. Constant RNA replication to keep a prebiotic RNA world running requires a constant stream of activated, correctly base pairing, extendable, and hard-to-synthesize monomers and no or little chain termination reactions and is therefore unlikely.

12. Selection from an RNA pool by affinity chromatography and replication of the selected RNA with the polymerase chain reaction.

13. Ribozymes catalyze hydrolysis of cyclic phosphates attached to RNA-ribose at positions 2′ and 3′, RNA self-cleavage, RNA ligation, RNA branching, RNA lariat formation, RNA phosphorylation (with thiol substituted ATP), template-directed RNA polymerization, pyrimidine nucleotide synthesis, peptide bond formation, amino acid adenylation, alcohol oxidation, cofactor synthesis. No catalysis of redox reactions.

Chapter 5

1. NRPS works with a protein complex consisting of several protein modules, one for each amino acid. Each module consists of an amino acid adenylation domain, subsequent thiolation of the activated amino acid with a phosphopantetheinyl cofactor, condensation of the thiolated amino acid to the growing peptide, in specific modules amino acid epimerization, and in the terminal module peptide cyclization (if the peptide is cyclic). Specificity for the added amino acids is achieved by the pattern of the recognition surface of the adenylation and thiolation domains of the respective module. If the synthesized peptide possesses or enhances NRPS activity, NRPS will get autocatalytic.

2. Ribosomes condense amino acids in the order specified by codons of three nucleotides each of a messenger RNA template strand. For this purpose, the ribosome contains three neighboring RNA binding sites A, P, and E, and slides along on mRNA. At the 'Amino acid site' A the anticodon of a transfer-RNA loaded with its specific amino acid by an aminoacyl-tRNA synthetase enzyme binds to the complementary codon of mRNA. At the 'Peptide site' P transpeptidation of an already synthesized peptide chain to the amino acid at site A prolongates the peptide chain. Then, the ribosome moves one codon in the 3′ direction (translocation), so that the free tRNA at site P is now at the 'Exit site' E, where it leaves the ribosome.

3. (a) Polyester template with sidechains for a binary code binds to a complementary nascent strand to which polyester monomers with covalently bound amino acids attach and condense synchronously. (b) RNA template binds amino acids with its nucleobases covalently and specifically. The amino acids condense to a peptide with the amino acid sequence determined by the nucleobase sequence. (c) RNA template binds noncovalently to RNA-amino acid complexes which are formed by Watson-Crick bonding of the nucleotide activator of the amino acid to nucleobases of the RNA transfer hairpin. (d) RNA template binds noncovalently to transfer RNA-amino acid complexes formed by 'urzymes'.

4. Short peptides interacting with only a few bases of the acceptor stem of the RNA hairpin and with amino acids and nucleoside triphosphates already increase the efficiency of aminoacylation

as experiments show. Later, with the increasing size of the RNA hairpins, the anticodon sequence migrated from the acceptor stem to exposed positions of the hairpin at some distance from the aminoacylation site where the anticodon could interact with the codon of an RNA template. A second, non-conserved protein domain extended the proto-AARS to interaction with the anticodon.

5. Transfer RNAs with complementary anticodons are recognized by AARS from complementary genes, that is, their coding sequences trace to the complementary strands of the same double-stranded nucleic acids. Thus, at some early stage of evolution of the code, both single strands of RNA double strands encoded with a part of their sequence tRNAs and with another part of their sequence the corresponding urzymes (proto-AARS).

6. Methylation increases hydrophobicity via the displacement of water molecules by the methyl groups, weakening hydrogen bonds in neighboring base pairs. Methylation acts as a 'lubricant' to form the stressed cloverleaf structure of tRNA and enhance tRNA mobility and rRNA flexibility around the peptidyl transferase site during translation.

7. The 4 nucleotides G, C, A, and T/U combine to $4^3=64$ possible triplet codons, but only 20 different amino acids build proteins. Several codons with different nucleotides in the third position can code for the same amino acid respectively its tRNA. The third nucleobase is not controlled by bases of the 16S-rRNA for correct Watson-Crick pairing with the anticodon of tRNA hence certain deviations are possible (wobble effect). This degeneracy of the genetic code protects somewhat against mutations in the DNA because changes in the third position of a codon can still lead to the same amino acid inserted into the protein (silent mutation).

8. tRNAs have L form, they are all single-stranded, contain many complementary nucleobases for internal base pairing to double helix structures, contain modified bases, the 5′ end is phosphorylated, the carboxyl group of the amino acid esterifies to the 3′ hydroxyl group of the adenosine ribose of the terminal CCA sequence, the anticodon is located in a loop at the 'top' of the L (near to the middle of the tRNA sequence).

9. *Size.* Too large amino acids do not match in the active center of AARS, too small ones are connected to tRNA, but are sorted out by an editing domain: The flexible CCA arm of tRNA can move the amino acid from the activating to the editing site and if the amino acid fits there, it is removed by hydrolysis. *Specific interactions.* Electrostatic interactions and hydrogen bonds of amino acid side chains of AARS bind the respective adenylated amino acid and tRNA specifically. Class I AARSs bind to the minor groove and class II AARS to the major groove of the acceptor stem helix of tRNA. Probably, information from both sides of tRNA is necessary to clearly differentiate between 20 different tRNAs.

10. The reading frame is the DNA sequence of consecutive, non-overlapping codons of three nucleotides between the start and stop codons. Insertion or deletion of a nucleotide in the ORF leads to a completely wrong triplet sequence and thus to a non-functional protein. A silent mutation would be the exchange of the third nucleobase in a triplet with a nucleobase in the degenerate codon which leads to the same amino acid.

11. For irreversible adenylation of the 300 amino acids, 300 molecules ATP are converted to 300 AMP + 300 PPi, and the PPi are hydrolyzed to 2 Pi (corresponding to hydrolysis of ATP to ADP). Thus, 600 molecules of ATP are consumed. 1 molecule GTP is used for binding amino acid-loaded tRNA to the A site of the ribosome and 1 GTP is used as an energy source for the translocation of the ribosome in the direction of the 3′ end of the mRNA. Thus, 598 GTPs are consumed for the 299 peptide bonds and 1 GTP for the initiation of peptide synthesis.

12. The peptide has the sequence (f)Met-Leu-Val-Ala-Gly-Arg (Amino-terminus to carboxy-terminus; N-formylmethionine fMet in prokaryotes, mitochondria, and chloroplasts, Methionine Met in eukaryotes and archaea). Exchange of U by C in position 6 is a silent mutation, the amino acid sequence does not change. The exchange of U by C in position 5 leads to Pro instead of Leu in the peptide.

13. The amino group of aminoacyl-tRNA at site A reacts with the carbonyl group of the ester bond of peptidyl-tRNA in a nucleophilic attack. The intermediate tetrahedral product cleaves the ester bond so that a new peptide bond and free tRNA forms.

14. Recombination: Excision of the inserting RNA strand by self-cleavage of phosphodiester bonds and transesterification to become a freely diffusing lariat, which binds to a second RNA molecule and inserts there via reversal of the above reactions. Recombination does not require external energy input.

15. An estimation of the mutation probability during early RNA translation shows that new mutations occur before a mutation can dominate the pool. Hence for example RNA-peptide aggregation must have improved the accuracy of translation early on.

Chapter 6

1. DNA for stable genetic storage, genomes large enough for a free-living cell, lipids, and cell walls, ATP synthetase, and ion pumping.

2. Black Smokers are on-ridge vents directly above the magma chamber with > 400°C temperature, Lost City consists of off-ridge vents of < 90°C temperature (not from volcanic processes).

3. Several synthesized protoribosomes exhibit A site (attachment of transfer-RNA loaded with its specific amino acid) and P site activity

(transpeptidation of an already synthesized peptide chain to the amino acid at site A to prolong the peptide chain).

4. Cysteine sulfhydryl radicals (R–S·) abstract H and OH radicals from ribose, eliminate water, and add H from R-SH to obtain desoxyribose.

5. Replication of RNAs (+ and – strands) required in translation (RNA-dependent RNA polymerase), synthesis of DNA monomers (ribonucleotide reductase), RNA to DNA information deposition (RNA-dependent DNA polymerase–reverse transcriptase), DNA to RNA transcription (DNA-dependent RNA polymerase), replication of DNA without an RNA intermediate (DNA-dependent DNA polymerase), and gene combinations (DNA ligases, recombinases).

6. The universally conserved SRP (a ribonucleoprotein that targets specific proteins to the prokaryotic plasma membrane) may initially have passed proteins to the hydrophobic layer, restoring contact between the cytosol and the catalytic activity of the solid-state catalyst.

7. The protein FtsZ, a prokaryotic precursor to tubulin, forms a ring along the inside of the cell membrane at the division plane. Upon depolymerization, the filaments shorten and the ring narrows thereby constricting the cell in two.

8. A vesicle membrane becomes deformable at temperatures above phase transition and shear forces from convective flow in narrow cracks and compartments of submerged porous, volcanic rocks are strong enough to pull the vesicles apart.

9. 400 reactions are required for the synthesis of amino acids, bases, and cofactors from H_2, CO_2, and NH_3 which were catalyzed by probably rather short polypeptides demanding say 400 kb (kilobase) to encode them. Another 400 genes encoded by roughly 400 kb are required for lipid and cell wall synthesis, ribosomal proteins, ribosome biogenesis and translation, rRNA and tRNA processing and RNA base modifications, ATP synthetase subunits, RNA and DNA polymerization, membrane proteins, signal recognition particle and diverse cellular functions like proteolysis and folding. A genome size of ~800 kb is a lower and the 1600 kb of a modern methanogen like Methanopyrus is an upper bound for the size of the first microbial genome.

10. Solution: A short self-folded RNA sequence interacts with a short RNA-CCA-(random)peptide analogue to become a stable aggregate. The self-folded RNA sequence replicates and forms a dimer. One entity in the dimer mutates to preferentially bind short RNA-CCA-amino acid monomers (green A site in Figure 5). Now the random peptide and the amino acid monomer are in a suitable stereochemical position to react to an elongated (random) peptide by transesterification. The peptide site is now empty and can bind to another amino acid monomer and elongation continues. Amino acid and peptide sites alternate. (mRNA) template-directed peptide synthesis and correspondingly ribosome motion evolved later. No obvious points for critique.

Chapter 7

1. Development of ATP synthase to extract energy from a geochemical energy gradient and ion pumping with energy from the reaction of H_2 with CO_2 to replace the geochemical ion gradient; generation of reduced ferredoxin from H_2 to become the main reductant.

2. 4 ATP are expended per amino acid polymerized at the ribosome: the PPi-producing step at aminoacyl tRNA synthesis (2 ATP) and the two GTP-consuming steps at translation.

3. A cell is about 50–60% protein (much of that in the ribosomes) and 20% RNA (almost all of that rRNA) by dry weight, the rest are mainly lipids and polysaccharides.

4. The ATP synthase converts the energy of a proton gradient into the physical rotation of its rotor against a stator. The rotor consists of usually 12 protein subunits arranged in a ring c and encompasses an elongated protein working as a motor axis; the stator consists of three α- and three β-subunits of the ATP synthesizing enzyme, the proton channel a fixed in the membrane, and an elongated protein b to connect the proton channel with the enzyme. The rotational force is generated by protons passing through the proton channel into the rotor. During rotation, one proton is released into the cytosol for every proton taken up from the periplasmic space. Each proton released leads to a conformational change that turns the rotor by one c-ring subunit. This rotation causes the motor axis to rotate against the inside of the enzyme, thereby inducing conformational changes in the β-subunits of the enzyme. These conformational changes cause ADP and Pi in the cytosol to bind to one of the β-subunits, to form a covalent bond as ATP, and to release ATP to the cytosol. A full turn of the motor axis requires the traversal of 12 protons through the enzyme, converting 3 ADP+P_i into 3 ATP.

5. ATP-synthases: ~ 3200 per cell, each ~ 100 ATP per second; Ribosomes: ~ 10000 per cell, each ~ 10 peptide bonds per second. ~75% of ATP is consumed by the ribosomes.

6. The ATP synthase is universal (the same rotor-stator principle in bacteria and archaea points to a common ancestor) but the proteins that perform ion-gradient formation (pumping) are not. At hydrothermal vents, there are large geochemical ion gradients that could be harnessed by the ATP synthase of the ancestor of bacteria and archaea before ion pumping developed.

7. −49.2 kJ/mol Gibbs free energy.

8. GTP is the main energy currency of modern ribosome biogenesis and function, both in bacteria and in archaea. The ribosome developed before ATP synthase such that the energy for GTP had to come from substrate-level phosphorylation (probably involving acyl phosphates) and not from ion gradients. This suggests that GTP was the energy currency of translation both in LUCA and at the origin of the ribosome.

9. Adenylation of a reactant with ATP instead of phosphorylation leads to Pyrophosphate (PPi) as product which is an agent of irreversibility due to its fast hydrolysis to 2 Pi catalyzed by inorganic pyrophosphatases. PPases remove a substrate for the back reaction of all PPi forming reactions, making them irreversible (kinetic control).

10. Cell volume decreases by only ~1% at 200 bar hydrostatic water pressure; at 1000 bar (10 km water depth; 4 km sediment depth) the volume decrease is ~ 4%. The increase of the Gibbs energy of gases at high hydrostatic pressure and the lower fluidity of membranes due to lipid compression and changed activity of compressed macromolecules might have some influence on cell metabolism. Experiments show however that microbes isolated from a hydrothermal vent more than 5000 m deep can grow in the laboratory at 1 bar of pressure or at more than 1000 bar of pressure.

11. *Formate as food.* The formate in alkaline serpentinizing systems is thought to stem from abiotic reactions of H_2 with solid-phase carbonates and can serve as a source of carbon and electrons for many acetogens and methanogens instead of H_2 and CO_2.
Ion pumping and ATP synthesis. In alkaline serpentinizing systems protons pumped by alkaliphiles for generating an ion gradient for ATP synthesis can in principle react with hydroxyl ions in the surroundings before reentry into the cell via the ATP synthase. Both the physical proximity of pumping complexes and ATP synthase and proton channeling between them might hinder the pumped protons from diffusion into the surrounding medium.
Amino acid fermentation. In the absence of H_2, amino acid fermentations, which typically produce H_2 as a waste product, become exergonic and can run via simple reversal of amino acid synthesis. Methanogens and acetogens can consume any H_2 produced by fermentations and thus push fermentation. For substrate breakdown, peptidases can be secreted, exit the cell, and digest proteins in the environment; the released amino acids (or small peptides) can be imported and disproportionated. This involves the breakdown of the amino acid down into the corresponding 2-oxoacid via deamination. The 2-oxoacid is oxidatively decarboxylated in a CoA-dependent reaction to a thioester that serves as a substrate for acyl phosphate synthesis and ATP synthesis via SLP.

RNA fermentation. RNA breakdown leads to nucleoside monophosphates (AMP etc.) whose bases are removed via phosphorolysis leading to ribose-1,5-bisphosphate. This is then converted by a simple isomerase into ribulose-1,5-bisphosphate, which breaks down by the addition of CO_2 with ribulose-1,5-bisphosphate carboxylase (Rubisco) into two molecules of 3-phosphoglycerate for fermentative carbon supply and energy metabolism via SLP.

12. The reversible ion pumping hydrogenase Ech can transfer electrons from reduced ferredoxin (electrons obtained from fermentation) to protons to produce H_2 and use that exergonic reaction to generate an ion gradient by pumping H^+ or Na^+ into the hydrogen-producing reaction. Ech can also operate in reverse, harnessing an ion gradient to generate reduced ferredoxin from H_2. Similar to Ech, the protein Mbh in the fermenter *P. furiosus* couples H_2 production to ion gradient formation for ATP synthesis: $Fd^{2-} + 2H^+ \rightarrow Fd + H_2$.

13. Energy is extracted from unfermentable substrates such as lipids and fatty acids (too reduced for disproportionation) by syntrophic interactions and anaerobic respiration. Syntrophic interactions typically involve two microbial partners, one that produces H_2 by oxidation of an organics near to equilibrium and a second partner that scavenges the H_2 as a pulling reaction to improve the thermodynamics of the first. Examples are propionate or butyrate fermenters and methanogens.
In modern anaerobic respiration, electrons are often transferred to elemental sulfur, that can accumulate due to the presence of O_2 which oxidizes reduced sulfur compounds in the environment to S_0. Elemental sulfur is rare at the anaerobic reducing environments of early life but the electron acceptor Fe_3O_4 from serpentinization in hydrothermal vents was probably an abundant alternative.

14. The iron-reducing bacterium *Shewanella oneidensis* uses nanowires from *c* type cytochromes to conduct electrons from anaerobic lactate oxidation to external crystals of the electron acceptor Fe_3O_4. For multiheme cytochrome nanowires with an edge-to-edge distance between hemes of 3.7–4.3 Å, the heme-to-heme electron transfer occurs via electron tunneling and the rate is 10^9 per second. For transfer the iron in the donating heme has to be in the reduced state (Fe^{2+}) and the accepting heme has to be in the oxidized state (Fe^{3+}).

15. In Shewanella, ATP is generated by SLP via acetylphosphate from lactate oxidation, not by chemiosmotic ATP synthesis and ion pumping. Menaquinone is only involved in electron transfer, not in pumping. Both features are ancient. Electron transfer by cytochrome c nanowires merely requires physical proximity of the heme molecules to within ~4 Å, not closer as in protein-protein electron transfer, and thus easier to realize.

The electron acceptor Fe_3O_4 is present in a serpentinizing hydrothermal system and can easily contain lactate as an accumulated fermentation product nearby. Part of the cytochromes transport electrons in Shewanella's inner membrane and periplasmic space and may well have contributed to the development of the electron transport chain inside cells.

16. Isolation and cultivation of green sulfur photosynthetic bacteria was reported from a deep-sea hydrothermal vent. The bacteria possessed chlorophyll that absorbed wavelengths extending into the far red 750–800 nm where Black Smokers glow. Possibly, low-level light from vents provided the source of photons that enabled the use of light as an energy and electron source. How could that initially have developed? Extraction of electrons from chlorophyll (dimers) excited to the singlet state is very difficult due to the short lifetime of singlet-excited chlorophyll; photoexcited triplet states generally have a much longer lifetime. A possible candidate for early photosynthesis is Zn-protoporphyrin IX. It absorbs photons in the visible spectrum, generating a very stable triplet ^{3}Zn-protoporphyrin IX with an excited state half-life on the order of 10 milliseconds and an E_0 of −1600 mV, so that the triplet species can easily reduce ferredoxin. H_2S likely filled the electron hole in the protoporphyrin cation generated by the electron transfer to ferredoxin. Thereby, H_2S became oxidized to elementary sulfur S_0 as the end product of photosynthesis.

17. Solution: The Fe(II)-oxidizing bacterium, *Acidithiobacillus ferrooxidans*, extracts electrons directly from solid electron sources such as conductive minerals and electrodes at a potential of +0.82 V versus Standard Hydrogen Electrode (SHE). The electrons are used for O_2 reduction to H_2O at 0.82 V as a linear sweep of the applied voltage shows (see reduction potentials in the appendix). The down-hill reduction of O_2 to H_2O is coupled by electron bifurcation to an uphill reduction of NAD+ to NADH at −0.32 V (+0.82 +0.32 = 1.14 V uphill boost by use of proton motive force from the exergonic O_2 reduction which pushes electrons up and/or triggers ATP synthesis). Then, NADH is used to start the Calvin cycle. Bacterial growth was observed on the working electrode after the application of voltage and flow of the cathodic current for a few hours; 254 nm irradiation sterilized the electrode and the current dropped to nearly zero. The bacteria are in direct electrical contact with the working electrode by conductive c-type cytochromes (Cyc2) on their outer membrane compartments. The inner electron-transfer chains responsible for the cathodic current generation were identified by photodetachment of CO from CO-poisoned heme groups: an action spectrum of cytochrome *c* and cytochrome *aa*3 was identified by monitoring the wavelength dependence of the cathodic current recovery. Participation of the cytochrome *bc*1 complex in inner electron transfer was shown by suppressing the cathodic current by adding the *bc*1 complex inhibitor Antimycin A.

A possible critique would be: *Acidithiobacillus ferrooxidans* uses O_2 for aspiration and the Calvin cycle for carbon dioxide fixation which are not ancient traits. It can be argued, however, that a similar mechanism with an ancient electron acceptor and direct electron uptake in electrically conductive serpentinizing chimney habitats may have been a route for carbon fixation. See also the conductive mineral magnetite (Fe3O4) as an external electron acceptor in Chapter 7, the reverse process. Overall, this is very much a laboratory experiment and chemical conditions in a vent chimney may have been very different.

Index

9 781032 457673